Multivariable Calculus

(Edition 2015-07-21)

Clark Bray

Duke University

Copyright ©2009, Clark Bray. All rights reserved. Unauthorized duplication is not allowed.

Website: Additional resources and the latest information and updates relevant to this book can be found at:

http://www.math.duke.edu/∼cbray/mv/

Free Video Lectures: The author has recorded video lectures from a multivariable calculus course taught from this book. These lectures are freely available to the public. You can find links to these lectures, and more information about this book, at http://www.math.duke.edu/∼cbray/mv/.

Book Availability: This book can be ordered online from a print-on-demand publisher. You can find a link to where you can place an order, and more information about this book, at http://www.math.duke.edu/∼cbray/mv/.

Image Credits:

Front cover, back cover background: Photograph, Clark Bray, *"Gauss Blur"*

Gauss's divergence theorem has extremely powerful applications in physics and engineering. Of the culminating "boundary theorems" in the final chapter of this book, the divergence theorem draws perhaps the most natural connection between geometric intuition and the physical world.

Back cover, top left inset: NASA and The Hubble Heritage Team (STScI/AURA). Acknowledgment: G.R. Meurer and T.M. Heckman (JHU), C. Leitherer, J. Harris and D. Calzetti (STScI), and M. Sirianni (JHU). Dates: March 1997 and September 2000. Telescope: Hubble Wide Field Planetary Camera 2. Photo of NGC3310 spiral galaxy.

Back cover, bottom left inset: Johan Knapen and Nik Szymanek. Telescope: Jacobus Kapteyn Telescope. B, I, and H-alpha bands. Photo of NGC488 spiral galaxy.

Back cover, right insets: Computer-generated images, from "On Dark Matter, Spiral Galaxies, and the Axioms of General Relativity," H. Bray, April 22, 2010. This paper discusses a reformulation of the axioms of general relativity, and the consequences predicted by these axioms. The images were created by Duke University mathematician and physicist Hubert L. Bray, while supported by NSF grant DMS-0706794.

These images support the possibility that dark matter and spiral galaxies are consequences of the fundamental differential geometry of the universe. Differential geometry, which is a natural sequel to multivariable calculus, can be thought of as being multivariable calculus on curved spaces.

The reformulation in the above paper predicts dark matter as a scalar field satisfying the Klein-Gordon equation, a wave equation. Since wave equations naturally form density waves, these dark matter density waves can then induce density waves in the regular baryonic matter composed of stars, gas, and dust, as shown in the above simulations.

This paper, and the matlab programs used to generate the images, can be found at http://www.math.duke.edu/∼bray/darkmatter/darkmatter.html. Links to lectures on this topic can be found under "Selected Invited Lectures" at http://fds.duke.edu/db/aas/math/faculty/bray.

The simulated image above results from running the Matlab function spiralgalaxy(1, 75000, 1, -0.15, 2000, 1990, 100000000, 8.7e-13, 7500, 5000, 45000000, 50000). The simulated image below results from running the Matlab function spiralgalaxy(0.1, 100000, 1, -0.5, 20000, 19900, 50000000, 8.7e-15, 15000, 5000, 82000000, 20000).

Contents

Preface ix
   0.1 Content ix
   0.2 Permanent Exercise Labels xiii
   0.3 About the Author xiv
   0.4 Acknowledgments xiv

1 Vectors 1
   1.1 Points in $\mathbb{R}^n$ 1
   1.2 Vectors in $\mathbb{R}^n$ 4
      1.2.1 Definition and notation 4
      1.2.2 Some algebra of vectors 7
   1.3 Dot Product 15
      1.3.1 Definition and properties 15
      1.3.2 Components 18
      1.3.3 Work 19
   1.4 Cross Product and Determinant 22
      1.4.1 Right-hand order 22
      1.4.2 Determinant and volume 25
      1.4.3 Cross product 28
      1.4.4 Geometric properties 29
   1.5 Lines and Planes 35
      1.5.1 Lines in $\mathbb{R}^2$ 35
      1.5.2 Planes in $\mathbb{R}^3$ 36
      1.5.3 Lines in $\mathbb{R}^3$ 37
   1.6 Coordinate Systems 43
      1.6.1 Polar coordinates 44
      1.6.2 Cylindrical coordinates 45
      1.6.3 Spherical coordinates 46
      1.6.4 Equations in other coordinate systems 46

2 Multivariable Functions 49
   2.1 Functions 49
      2.1.1 Examples 50
      2.1.2 Pictures 50
      2.1.3 Direct geometric interpretation 51
   2.2 Equations 53
      2.2.1 Solution sets 53
      2.2.2 Cross sections 54

		2.2.3 Transformations of variables	58
		2.2.4 Rotations	64
2.3	Graphs		71
	2.3.1	Single variable functions	71
	2.3.2	Multivariable functions	72
2.4	Level Sets		76
	2.4.1	Construction	76
	2.4.2	Relationship to the graph	77
	2.4.3	Advantages	79
2.5	Parametric Curves and Surfaces		83
	2.5.1	Parametrization	83
	2.5.2	Parametric curves	84
	2.5.3	Examples	87
	2.5.4	Parametric surfaces	89
	2.5.5	Graph parametrizations	91
2.6	Using Functions to Represent Curves and Surfaces		95
	2.6.1	Vertical line test	95
	2.6.2	Examples	96
2.7	Limits and Continuity		103
	2.7.1	Definition	103
	2.7.2	Misinterpretation and counterexample	105
	2.7.3	Alternative interpretation	106
	2.7.4	Continuity	108
	2.7.5	Vector valued functions	109

3 Linear Algebra — 113

3.1	Linear Transformations		113
	3.1.1	Preliminaries	113
	3.1.2	Linear transformations	114
	3.1.3	Standard basis vectors	116
	3.1.4	Examples	117
3.2	Matrices		122
	3.2.1	Matrix notation	122
	3.2.2	Matrix-vector multiplication	123
	3.2.3	Matrix addition and scalar-matrix multiplication	125
	3.2.4	Matrix multiplication	127

4 Derivatives — 133

4.1	Single Variable Derivatives		133
	4.1.1	Interpretations	133
	4.1.2	Examples	134
4.2	Derivatives of Parametric Curves		136
	4.2.1	Definition and interpretation	136
	4.2.2	Examples	137
4.3	Directional Derivatives		142
	4.3.1	Definition	142
	4.3.2	Interpretations	143

CONTENTS

- 4.3.3 Examples . . . 144
- 4.4 Unit Directional Derivatives . . . 149
  - 4.4.1 Derivation . . . 149
  - 4.4.2 Interpretations . . . 149
  - 4.4.3 Examples . . . 151
- 4.5 Partial Derivatives . . . 154
  - 4.5.1 Definition and computation . . . 154
  - 4.5.2 Interpretations . . . 155
  - 4.5.3 Examples . . . 156
  - 4.5.4 Second partials . . . 156
- 4.6 Derivative Transformations . . . 160
  - 4.6.1 Polynomials and directional linearity . . . 160
  - 4.6.2 Differentiability . . . 162
  - 4.6.3 Continuous differentiability . . . 164
  - 4.6.4 A hierarchy of regularity conditions . . . 164
- 4.7 Jacobian Matrices . . . 169
  - 4.7.1 Computations . . . 170
- 4.8 Gradients . . . 175
  - 4.8.1 Definition . . . 175
  - 4.8.2 Geometric interpretations . . . 176
  - 4.8.3 Level sets . . . 177
  - 4.8.4 Examples . . . 179
- 4.9 The Chain Rule . . . 181
  - 4.9.1 Single variable version . . . 181
  - 4.9.2 Multivariable version . . . 182
  - 4.9.3 Intermediate variables . . . 184
  - 4.9.4 Confusion of variables . . . 185
  - 4.9.5 Second derivatives . . . 187
- 4.10 Implicit Differentiation . . . 191
  - 4.10.1 Single variable computations . . . 191
  - 4.10.2 Single variable implicit function theorem . . . 193
  - 4.10.3 Multivariable implicit function theorem . . . 195
- 4.11 Unconstrained Local Extrema . . . 198
  - 4.11.1 Terminology . . . 198
  - 4.11.2 Critical points . . . 200
  - 4.11.3 Second derivative test . . . 204
- 4.12 Constrained Local Extrema and Lagrange Multiplier Theorems . . . 209
  - 4.12.1 Single constraint critical points . . . 209
  - 4.12.2 Two constraint critical points . . . 214
- 4.13 Global Extrema . . . 220
  - 4.13.1 Outline . . . 220
  - 4.13.2 Examples . . . 222

5 Integrals   227
- 5.1 Single Variable Integrals . . . 227
- 5.2 Double Integrals . . . 231
  - 5.2.1 Notation . . . 233

	5.2.2	Definition	234
	5.2.3	Examples	234
5.3	Nested Integrals	239
	5.3.1	Another computation of volume	239
	5.3.2	Switching differentials	245
	5.3.3	Examples other than volume	249
5.4	Non-Rectangular Domains	252
	5.4.1	Examples	252
	5.4.2	Simpler pictures	255
	5.4.3	Choosing the order	259
5.5	Triple Integrals	265
	5.5.1	Definition and interpretation	265
	5.5.2	Nested triple integrals	268
	5.5.3	Non-rectangular domains	268
	5.5.4	Examples	274
5.6	Change of Variables	285
	5.6.1	The substitution rule	285
	5.6.2	Change of variables in $\mathbb{R}^2$	287
	5.6.3	Examples	291
	5.6.4	Change of variables in $\mathbb{R}^3$	294
	5.6.5	Symmetry theorems	297
5.7	Integrals in Coordinate Systems	302
	5.7.1	Coordinate systems as change of variables functions	302
	5.7.2	Integrals "in" coordinate systems	306
5.8	Scalar Line Integrals	320
	5.8.1	Definition	321
	5.8.2	Strategy for computation	322
	5.8.3	Examples	325
5.9	Scalar Surface Integrals	328
	5.9.1	Definition	328
	5.9.2	Strategy for computation	328
	5.9.3	Examples	333
5.10	Vector Fields	338
	5.10.1	Some motivating examples	338
	5.10.2	Definition	341
	5.10.3	More examples	343
5.11	Vector Line Integrals	346
	5.11.1	Motivation and definition	346
	5.11.2	Computation	347
	5.11.3	Examples	350
	5.11.4	The unit tangent vector	353
	5.11.5	Summary	355
5.12	Vector Surface Integrals	359
	5.12.1	Flux	359
	5.12.2	Definition	361
	5.12.3	Computation	362
	5.12.4	Examples	363

         5.12.5 Comparison to line integrals 366
         5.12.6 Flux through a curve 368
  5.13 Coordinate Line Integrals and Surface Integrals 371
         5.13.1 Line integrals 371
         5.13.2 Surface integrals 373

6 Vector Calculus 379
  6.1 Setup and Approach 379
         6.1.1 The general Stokes's theorem, de Rham cohomology, and compromise 379
         6.1.2 Evaluations, domains and integrands 380
         6.1.3 Orientations and boundaries 381
         6.1.4 Diagrams and boundary theorems 386
         6.1.5 The fundamental theorem of calculus 388
  6.2 Fundamental Theorem of Line Integrals 394
         6.2.1 Motivation 394
         6.2.2 Examples 396
         6.2.3 Antigradients 398
  6.3 Green's Theorem 401
         6.3.1 Motivation 401
         6.3.2 Examples 405
         6.3.3 Computing area 407
  6.4 Gauss's Divergence Theorem 410
         6.4.1 Motivation 410
         6.4.2 Flux density 414
         6.4.3 Applications to physics 416
         6.4.4 Gauss's theorem in $\mathbb{R}^2$ 422
  6.5 Stokes's Curl Theorem 427
         6.5.1 Motivation 427
         6.5.2 Examples 431
         6.5.3 Interpretation of curl 433
         6.5.4 Applications to physics 435
  6.6 Path Independence 439
         6.6.1 Gradient fields 439
         6.6.2 Curl 441
  6.7 Surface Independence 444
         6.7.1 Curl fields 444
         6.7.2 Divergence 446
  6.8 Lifetime Theorems 449
         6.8.1 "Life cycle" of an integrand 450
         6.8.2 Lifetime theorems 451
         6.8.3 Examples 452
         6.8.4 Domains 456
  6.9 Computing Line Integrals and Surface Integrals 459
         6.9.1 Diagnosis 459
         6.9.2 Examples 461

Preface

0.1 Content

Multivariable calculus is a beautiful and powerful subject. But there are difficult choices that must be made when creating a multivariable calculus course. All of these choices involve sacrifice in one form or another.

One option is to present a mathematically complete and rigorous course. Such a course would include, among many other things, complete presentations of differentiability, differential forms, and the general Stokes's Theorem of differential forms. And by necessity it would also assume as a prerequisite a reasonably strong understanding of linear algebra. Naturally such a course is a wonderful thing for students that are prepared to handle that level of material. Unfortunately, most students who are not serious math students are simply not prepared to digest the abstractions and rigor involved in such a presentation.

Another option is to make the course better suited to the majority of students in the sciences and engineering by taking steps to avoid the rigor and abstractions that such students find so unpalatable. Unfortunately, in avoiding abstractions, very often such courses also completely bypass important conceptual ideas that are usually expressed with those abstractions. What remains is a course that is boiled down to formulas that students can memorize and work with to a degree, but that fails to communicate some of the most important concepts and connections. Unfortunately, this sacrifice is made all to often.

In this textbook we make a presentation that is a compromise between these two extremes. While we bypass many of the abstractions and most of the rigor, we organize and present ideas in such a way as to leave the student with as many as possible of the important concepts and sophisticated pictures.

Here are some specific points of interest about the approach taken in this text.

1. **The "graph" construction is only one of many important ways to relate a geometric picture to an algebraic item such as a function. Several other types of pictures become important in a multivariable setting, and all must be both explained and distinguished.**

    In most single variable calculus classes, the overwhelmingly most common geometric picture that is drawn to describe a function is the graph of that function. This is a reasonable choice with only single variable calculus in mind, because in such a setting the disadvantages of the graph construction are minimal.

    However in a multivariable setting, the graph quickly becomes not only disadvantaged but sometimes even completely useless, since the graph construction is dimensionally expensive, requiring as many dimensions as the sum of the numbers of independent and dependent variables.

    Several other types of geometric constructions then become critical in order to have any way to represent such functions, and as a consequence notions of calculus are represented in different geometric ways on these different types of pictures. Students must not only become comfortable using each of these types of pictures, but must also understand precisely the process by which each of these types of pictures is constructed in order to be able to understand how and why the several different calculus notions all manifest themselves differently in each of these settings.

    For example, students are often mystified as to why the derivative of a parametric curve function is a tangent vector, while the derivative (gradient) of a function represented by level sets is a perpendicular vector; the answer of course is that these pictures themselves come from different geometric constructions, and the confusion on this point is indicative that this student has not fully digested those differences.

Most textbooks mention that there are these other constructions, but this usually occupies a small part of one section in a chapter. Furthermore, throughout the remainder of the textbook, pictures are drawn without emphasizing each time exactly which of the geometric constructions was used to make this picture, which would allow students to categorize the intuition from that picture appropriately in his or her mind.

This book dedicates extensive discussion at the very beginning of the book to these several geometric constructions, emphasizes the importance of each and provides many examples, and then makes note of these distinctions in each example throughout the remainder of the book.

2. **This book includes a discussion of linear algebra, beyond the limited discussion of the basics in most textbooks, including a presentation of linear transformations, matrices, and their relationships by way of the standard basis vectors.**

   Almost all multivariable calculus textbooks include a short discussion of some basic ideas of linear algebra. This discussion is often limited however to vectors, dot products, cross products, determinants, lines and planes, and perhaps a bit more.

   However, great use can be made of linear algebra in a discussion of multivariable calculus. And while a full presentation of linear algebra requires at least a full course, enough linear algebra to be extremely useful in a multivariable calculus course can be covered very quickly.

   In particular, great use can be made of an understanding of linear transformations and matrices – without discussions of systems of equations, Gaussian elimination, echelon forms, vector subspaces, and other topics that would make the discussion more complete yet more time consuming.

   This book includes a discussion of linear transformations from the definition, provides many examples including discussion of the geometric properties of these functions, and then uses this to motivate and define matrix-vector products and matrix algebra.

   These tools are then put to great use in the presentation of derivatives that follows (see the following item discussing derivatives), giving students a point of view on derivatives that they would not have access to in most textbooks. Furthermore, students using this textbook that do not continue on to other mathematics courses after multivariable calculus will have the advantage of having seen this extra amount of linear algebra, while students using other textbooks will not.

3. **While this book presents many points of view on the derivative of a multivariable function, most fundamentally the presentation is as a relationship between vectors – not as the slope of the tangent line to a cross section of the graph of the function, as is done in most textbooks.**

   This is the first of two main areas in which this textbook aims specifically to fill a gap in the textbook distribution by finding a compromise between (a) presenting important foundational notions, and (b) avoiding intimidating depth, subtlety and rigor.

   Of course a full and complete discussion of multivariable derivatives, done in the more sophisticated textbooks, includes a discussion of linear approximation by linear transformations, tangent planes, the formal definition of differentiability by vanishing relative errors, and the derivation of the Jacobian matrix. These are all beautiful and useful ideas, but as a collection this discussion is too involved and abstract for appropriate use with most science and engineering students.

   Note that the completion of the above discussion leaves surviving students with two important ideas in their minds. First, that the derivative of a multivariable function is a linear transformation that relates a vector in the domain to a vector in the target; second, that we can make a precise quantification of what we mean to say that such a linear transformation provides a "good linear approximation" to a function, which of course is highly wrapped up in the subtleties of the notion of a multivariable limit.

   On the other side of the distribution, the majority of textbooks avoid the subtle notions of differentiability simply by avoiding all discussions of the derivative as being a relationship between vectors. Instead, they present the partial derivative as the most fundamental notion of a derivative in a multivariable setting. This has the advantage that this notion is highly analogous to the notion of derivative that students are accustomed to from single variable calculus, in that it is the slope of a tangent line to the graph of the function; so it makes for an easy first step in discussing derivatives.

However this approach has serious disadvantages, in that it reinforces student dependence on two ideas from single variable calculus that I feel must be broken to make a good transition to multivariable – first, that the derivative is intrinsically based on the geometric construction of the graph, and second that the derivative is intrinsically a single variable notion, in particular a relationship between variables, as opposed to a relationship between vectors. Subsequent points of view on the derivative can be motivated, but students are left with no explanation why velocity vectors, partial derivatives, and gradient vectors have anything to do with each other.

Rather than so avoiding the concepts entirely, let's look again at the two important ideas from the more sophisticated presentation – (a) the derivative is a linear transformation; and (b) multivariable limits can be used to define differentiability precisely by defining what we mean by a "good linear approximation", using vanishing relative error.

The latter is the notion that is the most deep and subtle, that requires the most careful treatment, and also, conveniently, that has the fewest practical applications. The former notion is nowhere near as deep a concept. Still, an understanding of this former concept allows students to have a complete geometric understanding of what a derivative is most fundamentally, and how they should visualize it.

In this book we will approach multivariable derivatives with the goal in mind of emphasizing this idea, that a derivative is a linear transformation, while bypassing the more subtle and intimidating notion of true differentiability.

We do this by establishing immediately that a multivariable derivative is a relationship between vectors. That is, we begin the discussion of derivatives with the directional derivative, defined with derivatives of parametric curves. This can be written down conveniently, and the geometric interpretation underscores the desired relationship between vectors. We can then assert later (without proof) that for most functions, these vectors are related by a linear transformation, and then make use of this fact when developing the chain rule and other derivative topics.

The partial derivative is presented as a special case of the directional derivative. All of the usual interpretations of the partial derivative are given, but importantly it is presented as a computational tool, not as the primary notion of derivative.

4. **Several useful methods of integration (change of variables, integrals in different coordinate systems, line integrals and surface integrals) involve using a new domain to compute an integral defined over a different domain, and keeping track of the ratios of those measures. These can all be similarly motivated by introducing the idea of a "stretching factor" which applies to each using the same conceptual motivation.**

   Most textbooks motivate and derive techniques for computing the above types of integrals in completely different ways, or at best do not underscore the similarities in each to allow students to use their understanding of one to help them reach an understanding of the others. In this book, we provide similar motivations for the computations of the above types of integrals and underscore their similarities in deriving those computational techniques.

   In particular, in each of the above, the given integral can be computed by viewing each piece of the given domain as the image through some appropriate function of a piece of a new, more convenient domain. The relative measures of those pieces of the two domains is computed differently in each of the given cases, but in each case that relationship can be viewed as the amount that the function in question "stretches" length, area or volume.

   In this book, the idea of a stretching factor is first introduced as part of the motivation for the change of variables formula. It is then emphasized that, independent of the specific relationships used in the details of this particular motivation, the idea of a stretching factor is a general idea that can be used elsewhere. It is then emphasized before the derivations of each of other similar methods that this same idea of a stretching factor is being used, and that the only difference is in the details of the computation of the particular stretching factor specific to that particular setting.

   Students can then use the intuition and understanding that they develop from the change of variables to help them understand what is actually going on in the derivations/motivations of the other methods, leading to better conceptual understanding of all of those methods.

5. **This book presents the analogies and relationships between the several special cases of the general Stokes's theorem, and develops ideas and images suggestive of the deeper tools used in the full presentation, while still avoiding the deeper and less intuitively satisfying ideas that make the general Stokes's theorem too abstract and intimidating for appropriate use with science and engineering students.**

A full presentation of the ideas above involves defining chain complexes, differential forms, wedge products, the differential operator, integrals of forms, and then proving the general Stokes's theorem. These are abstract concepts which of course form a beautiful and elegant framework for the ideas. However, the abstraction of these tools makes this approach very difficult for the first-time student of this field, and are far too sophisticated for science and engineering students who do not intend to be mathematics majors.

Instead, these students have a greater need for an intuitive understanding of the several special cases of Stokes's theorem in two and three dimensions (the fundamental theorem of line integrals, Green's theorem, Stokes's curl theorem, and Gauss's divergence theorem), and of the de Rham cohomology theorem (giving relationships between divergence, curl, gradient, and Green's operator).

(We refer to these former theorems in the text as "boundary theorems" since each gives a relationship between a given domain and its boundary; and we refer to these latter theorems as "lifetime theorems", based on a metaphor that is created to simplify the discussion.)

Most multivariable calculus books avoid the intimidating abstractions not only by presenting the boundary theorems as being separate theorems, but also by deriving them in different ways, and even omitting to mention that these theorems have anything to do with each other in the first place. They also very often fail to show that the theorems can be used in conjunction with each other in certain special cases, such as can be done to show some of the lifetime theorems.

In this book, we begin by creating a framework that is suggestive of the structure created by the more sophisticated ideas like differential forms, represented in the form of diagrams (the final forms of which are on page 453) involving different types of integrals and domains. All of the work that is done with the boundary theorems and lifetime theorems is outlined on these diagrams, and is also used to supplement the diagrams. This gives students a visual presentation of the structural similarities of the theorems.

Second, we then motivate each of the boundary theorems by arguments that are highly analogous to an initial motivation of the fundamental theorem of calculus. Each of these theorems then can be seen to be an extension of the fundamental theorem of calculus, and this point of view is supported and emphasized. These analogies are emphasized throughout, so that the student sees these theorems as all being very similar theorems, applying only in different situations.

The similarities between the representations of the theorems on the diagrams and between their derivations makes it easy to claim that these boundary theorems are all in fact special cases of a larger and more sophisticated theorem, too abstract for the text, called the general Stokes's theorem.

This presentation is suggestive of the deeper picture, and shows the analogies and connections between the given theorems, while still not going into any abstract rigor.

With this organization and these diagrams, students have the following advantages.

(a) Students will see that these boundary theorems are all basically the same theorem. This is an important point because it is a deeper truth, which is a fact accessible (presented in this way) to students on this level, and yet it is largely ignored in most of the mass market texts.

(b) Students will have the boundary and lifetime theorems organized with respect to each other. Instead of having these theorems all in their minds in no particular order, students can remember that each theorem has a natural "place" in the diagram.

(c) Students will be able to use their understanding of any one of these boundary theorems to help them in their understanding of any of the other theorems, leading to a better overall understanding of all of them. For example, students often have difficulty understanding Stokes's curl theorem. Within this setting though, such students can start by thinking about, say, Green's theorem, and work through the statement of Stokes's curl theorem by analogy with Green's theorem. Whatever understanding this student has of Green's theorem then will help the student to understand the curl theorem.

(d) Students will be able to use the organized structure of theorems to better understand how to decide which boundary theorem or theorems might be most useful in helping them to solve a particular problem. For example, if a student is asked to evaluate a line integral in $\mathbb{R}^3$, a glance at the diagram gives the student the option of two boundary theorems to consider, those being the fundamental theorem of line integrals and Stokes's curl theorem. Furthermore the diagram presents exactly which issues must be resolved before attempting to apply each theorem – looking at the part of the diagram relating to the fundamental theorem, the student see that there is no question regarding being able to take the boundary of the given curve, but that there is an issue as to whether the vector field is the gradient of some function. The lifetime theorem then indicates how to resolve that issue. Similarly, looking at the part of the diagram relating to Stokes's curl theorem, the student sees that the issue is in determining whether the given curve is or is not the boundary of some surface. Depending on which of these issues resolves positively, the student then has an answer as to which theorem is most likely to allow a convenient solution to the problem.

0.2 Permanent Exercise Labels

In this textbook there are two labels attached to each exercise. In addition to the traditional numerically ordered label, each exercise also has a "permanent exercise label". These permanent labels do not change with new editions; if an instructor uses these labels in his or her syllabus, then used copies of the text will always be consistent with new copies, at least regarding the exercise labels. This allows instructors to continue to use past syllabi without periodic updating, and allows students to maintain a used market for this textbook, despite the appearance of a new edition.

For example, an exercise might be labeled as "**Exercise 24.7.81. (exer-Geom-Flow-57)**". As expected the numerical label indicates that, in this edition of the book, this is the 81st problem in Section 7 of Chapter 24. Of course in a subsequent edition new sections might be inserted, as might new exercises – so in such a subsequent edition the numerical label for this same exercise might instead be "**24.8.85**", making the previous numerical reference useless.

However, the permanent label will remain constant as "(exer-Geom-Flow-57)" in new editions. The "Geom" represents a permanent label for the chapter, and the "Flow" represents a permanent label for the section. Then the "57" indicates that this problem was the 57th problem written, chronologically.

Of course, depending on the locations that new problems are inserted into the exercise sets, the chronological ordering of the permanent labels might not be the same as the numerical ordering on the numerical labels. Instructors are asked to refer students to this explanation at the start of the course to avoid potential confusion on this; and students should make sure to be clear on which labels the instructor has chosen to use.

0.3 About the Author

Clark Bray is an Associate Professor of the Practice in the Department of Mathematics at Duke University. He graduated from Rice University in 1993, cum laude, with majors in Mathematics and Physics, and finished his Ph.D. in Mathematics at Stanford University in 1999, studying algebraic topology with Professor Gunnar Carlsson.

He has taught courses in multivariable calculus at Stanford University and Duke University. At Stanford, he was the primary instructor for the freshman mathematics sequence (including single variable calculus, linear algebra and multivariable calculus) from 2000-2004. As an Associate Professor of the Practice at Duke, the majority of his duties and intellectual energy are dedicated to his teaching, in which he takes great pride and personal satisfaction. In addition to extensive experience teaching multivariable calculus to majors in the sciences and engineering, he was a designer of a new and highly successful course in multivariable calculus for economics majors, which he then taught from 2007 to 2011.

He and his family live in Hillsborough, NC.

0.4 Acknowledgments

My first mathematical influence that I remember was my grandfather, Professor Hubert Evelyn Bray, esteemed member of the faculty at Rice University from 1918 to 1978. I remember as a young child being intrigued by the simple little mathematical curiosities that he showed me. These were my first experiences with creative mathematical thinking, and appreciating the simple elegance that often underlies powerful ideas. I must begin my acknowledgments by thanking him for making this significant impact on me in my formative years.

Also very strong influences on my mathematical development were my parents, by the extensive amount of time during my youth that they spent teaching me mathematics beyond what I was learning in school. For a similar effort I thank my brother, Hubert Bray, who has always been and continues to be a great source of fascinating mathematical ideas and enthusiasm. Of course I must also thank all of my instructors at Rice University and Stanford University.

I am very grateful to the Department of Mathematics at Duke University for its strong support of innovation in teaching. For years I have felt that this book would be a valuable contribution to the teaching of this subject, and it was in this supportive environment that I saw an opportunity to realize my goal of writing it.

Finally, I thank my wonderful wife Holly, in part for her help with editing this text, but mostly for her patience and support, without which this book could not have been completed.

Chapter 1

Vectors

1.1 Points in $\mathbb{R}^n$

Students in this course are already familiar with the xy-plane. This is the collection of points in the plane where the horizontal axis is the x-axis, and the vertical axis is the y-axis. Points are referred to by their coordinates, which are the projections of that point to the corresponding axes. This is represented in Figure 1.1.

Representing a point by its coordinates thus gives a one-to-one correspondence between ordered pairs of numbers (x, y), and geometric points in this plane.

We refer to the collection of all such ordered pairs of numbers as $\mathbb{R}^2$, where the "$\mathbb{R}$" refers to the fact that these numbers are real numbers, and the "2" indicates how many of these numbers are used.

Example 1.1.1. For example, the point P in Figure 1.1 has coordinates $x = 2$ and $y = 1$, as indicated in figure. So we write $P = (x, y)$, referring then to the point P by its corresponding ordered pair in $\mathbb{R}^2$.

Very similarly, we can define $\mathbb{R}^3$ as the set of all ordered triples.

A geometric example that will relate to $\mathbb{R}^3$ is the three-dimensional space defined by the three coordinate axes corresponding to the variables x, y, and z. We call this the xyz-space.

This defines eight octants in the same way that there are four quadrants in the xy-plane. We usually draw the three axes in such a way that the first octant (the one in which all three of the coordinates are positive) is facing toward the viewer (the x-axis is pointing out of the page, and the y-axis is pointing to the right on the page), and the z-axis is pointing upward. This is represented in Figure 1.2.

In analogy with the xy-plane, every point in xyz-space has three coordinates, which again are the projections of that point to the corresponding axes. So again, we have a one-to-one correspondence between lists of numbers and geometric points; in this case the correspondence is between ordered triples of numbers (x, y, z) and points in this space.

Example 1.1.2. For example, the point Q in Figure 1.2 has coordinates $x = 2$, $y = 1$, and $z = 1$, as indicated in figure. So we write $Q = (x, y, z)$, referring then to the point Q by its corresponding ordered triple in $\mathbb{R}^3$.

While we are limited in our ability to visualize more than three dimensions at a time, we can still continue the above pattern and define even higher dimensional spaces. For example, $\mathbb{R}^4$ is the collection of all ordered quadruples of

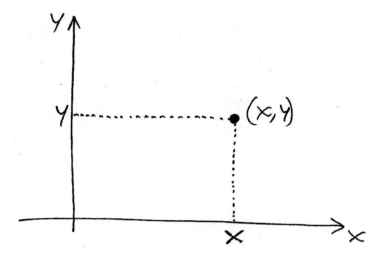

Figure 1.1:

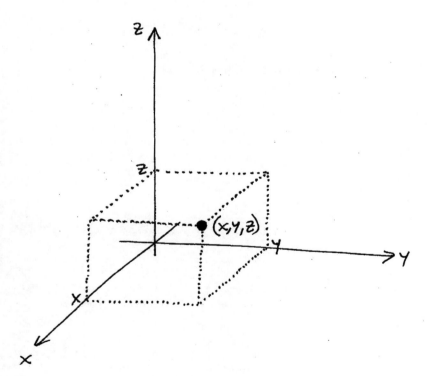

Figure 1.2:

numbers, in which we might represent with (x, y, z, w). We could then say that these are the coordinate[s of] the $xyzw$-space, a four dimensional space that we as three-dimensional beings are not capable of visualizing[.]

In this text, we will not refer extensively to spaces of dimension greater than three. Very often we will write "$\mathbb{R}^n$[" instead] of choosing a specific value for n, so that the statement we are making can refer simultaneously to $\mathbb{R}^2$ and $\mathbb{R}^3$; note [that] in these cases the statement can probably also refer equally well to a higher dimensional space, even though we will [not] make great use of that in this text.

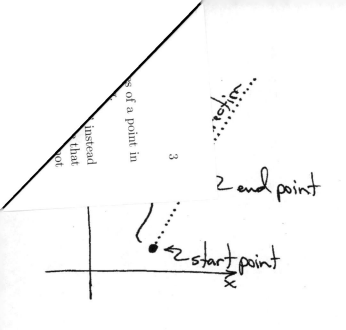
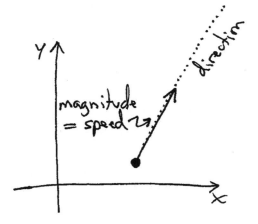

Figure 1.3:

1.2 Vectors in $\mathbb{R}^n$

1.2.1 Definition and notation

In the previous section, we saw how we can use numbers, in the form of coordinates, to describe locations in space.

When talking about positions, very often we need to discuss various types of changes in position. We might want to refer to a discrete movement of a particular distance in a given direction, without having to make reference to the starting point; for example, it makes sense to talk about walking ten meters northwest, even if we do not know where we are currently located. As a different example, we might want to refer to a continuous type of movement, namely, velocity; again, the notion of velocity makes sense independent of knowledge of position.

In each of these cases, the change in position has an associated direction, and an associated magnitude. In the case of a discrete movement, the associated direction is the direction in which that discrete movement was taken; and the associated magnitude is the distance moved in that direction. In the case of velocity, the associated direction is the direction of motion; and the associated magnitude is the speed in that direction. See Figure 1.3.

In addition to types of changes in position, there are many other instances in which we need to refer to something that has an associated direction and an associated magnitude. For example, a force has a direction and a magnitude.

In order to be able to refer to these types of things more conveniently, we create a new mathematical object that is defined as the combination of a direction and a magnitude. We call this a "vector". More specifically, when the direction in question is viewed as a direction in the space $\mathbb{R}^n$, we refer to the vector as a "vector in $\mathbb{R}^n$".

A very natural way to visualize a vector geometrically is with an arrow, because of course an arrow very naturally has a direction (the direction that the arrow is pointing) and a magnitude (the length of the arrow). This will be our standard means of representing vectors in figures throughout this text. In Figure 1.4, we show an arrow representing a vector in $\mathbb{R}^2$ and another arrow representing a vector in $\mathbb{R}^3$.

Very often, when using a variable name to refer to a vector, texts will emphasize that this is a vector by placing a small arrow over the variable name. For example, "$\vec{v}$" is a common example of writing a vector with a variable name. We will use this convention in this text as a visual assistance to the reader, but students should be aware that many texts do not do this, in order to make the notation simpler.

Very critically, note that a vector defined as above does not have an associated location; it only has a direction and a length. So, if there are two arrows pointing in exactly the same direction, and with exactly the same length, these arrows represent the same vector – even if they are drawn in different places on the page. For example, all of the arrows in Figure 1.5 represent the same vector, because they all have the same direction and the same length.

1.2. VECTORS IN $\mathbb{R}^N$

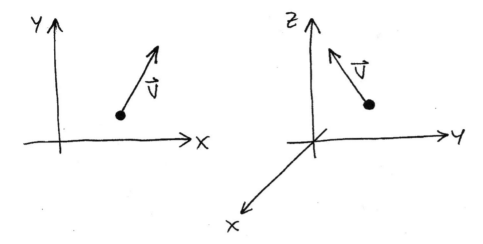

Figure 1.4:

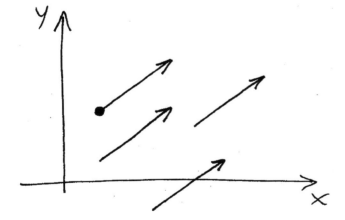

Figure 1.5:

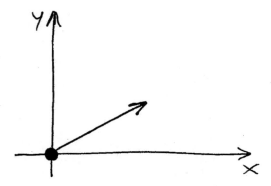

Figure 1.6:

The natural next question then asks how we will represent vectors algebraically. Note that while in some situations it might be reasonably convenient to represent a direction algebraically, very often it simply is not. We would like to have a means of representing vectors algebraically that is more uniformly convenient.

The answer to this question comes directly out of the following observation.

Observation 1.2.1. *There is a natural one-to-one correspondence between vectors in $\mathbb{R}^n$ and points in $\mathbb{R}^n$.*

Before we explain why this relationship exists, let us be sure to note that this does not mean that vectors and points are the same thing. They are definitely different things, but this relationship between them will end up being very useful.

To see why this correspondence exists, we first define the notion of "standard position". We say that a vector is represented by an arrow in standard position if the "tail" of the arrow is located at the origin. See Figure 1.6.

Now, suppose we are given a vector. There is a unique arrow representing this vector in standard position. That arrow will have its tail at the origin of course, and the head of that arrow will be at some other point. There is a unique such point that can be associated to a vector in this way.

Going in the other direction, suppose we are given a point. Again, there is a unique arrow in standard position that we can associate to this point; in this case we take the unique arrow in standard position whose head is located at the given point. This arrow represents a particular vector. There is a unique such vector that can be associated to a point in this way. We will call this the "position vector" for the given point.

The associations in the previous two paragraphs define the one-to-one correspondence in the above observation. See Figure 1.7.

Of course, we already have a very convenient and natural way to represent points algebraically. That is, we represent points with their coordinates. Given the observation above that there is this natural relationship between points and vectors, we can therefore represent a vector with the coordinates of the associated point.

There are some important considerations involved in deciding the precise notation that we will use to represent vectors.

While we must acknowledge that points and vectors are not the same thing, we also note that we can simply always choose to refer to a point by its position vector. So in effect, we can always write down a vector, and the context will make it clear if we are referring to a point by its position vector.

And since every vector can be naturally associated to a unique point, for which we already have the very natural algebraic notation of coordinates, we can just always use that same notation; again, context will make it clear if we are using that notation to describe a vector.

Admittedly, these choices of notation blur the distinction between points and vectors. However, in this course, this will not be a problem; and these notational conventions will allow us more convenience than inconvenience.

For example, consider the point $P \in \mathbb{R}^2$ with $x = 1$ and $y = 3$. This point will have a position vector, which we will call

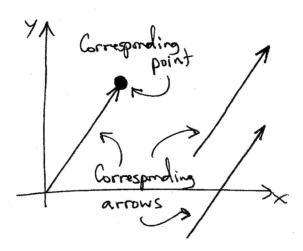

Figure 1.7:

$\vec{p}$. We will often refer to the point by "$\vec{p}$" instead of "P". To define this position vector algebraically, we will write the coordinates of the associated point, as in "$\vec{p} = (1,3)$".

Note that many texts do not blur the distinction between points and vectors, and have a specific notation used to represent vectors algebraically, while still using the coordinates of the associated point. Very commonly, the coordinates of a vector are given by the notation "$\vec{p} = \langle 1, 3 \rangle$". Students should be aware of this notation, but we will not use it in this text.

For reasons that will become clear in Chapter 3 when we begin discussing matrices, it is sometimes convenient to write the coordinates of a vector vertically instead of horizontally. For example, we can write

$$\vec{v} = (v_1, v_2, v_3) = \begin{bmatrix} v_1 \\ v_2 \\ v_3 \end{bmatrix}$$

Unfortunately, this has the simple typesetting problem of taking up several lines of space on the page. We will use this notation often in this text, but also very often we will use the horizontal notation just for efficiency.

We end this subsection with a final reminder about one of the differences between points and vectors. As previously noted, vectors have direction and length, but not location – there are many arrows with the same direction and length (thus representing the same single vector) that are in different locations.

So, it makes sense to talk about a vector as being "parallel" to a line or a plane, but it does not make sense to describe a vector as being "in" a line or a plane. For example, in Figure 1.8, it would be appropriate to say that the vector $\vec{v}$ is "parallel" to the line L; but it would not be correct to say it is "in" the line, because all of the arrows in the figure represent the same vector and most of them are nowhere near L.

By contrast, points have location, but do not have direction or length. A point is just a position in space.

So, we can talk about a point as being "in" a line or a plane. We could also talk about a vector as representing a point "in" a line or a plane, and we might even sloppily say that the vector is "in" the line or plane. See Figure 1.9. But it would not make any sense to refer to a point as being parallel to a line or plane.

Students should be very careful with these points of terminology.

1.2.2 Some algebra of vectors

Recall that we use a vector to represent a direction and a magnitude. When we represent that vector with coordinates, the magnitude of the vector corresponds to the length of the associated arrow. Of course the length of the corresponding

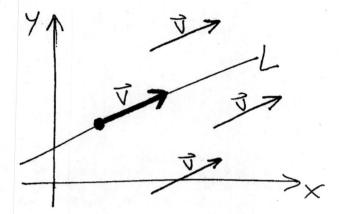

Figure 1.8:

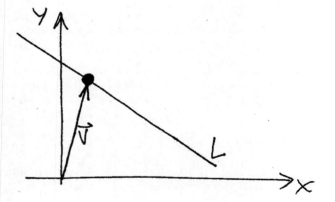

Figure 1.9:

1.2. VECTORS IN $\mathbb{R}^N$

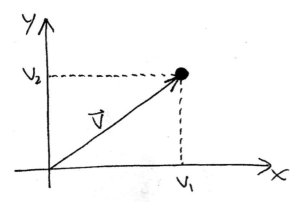

Figure 1.10:

arrow in standard position is the same as the distance to the origin from the corresponding point. And of course, we already have a formula (the Pythagorean theorem) for the distance to the origin from a given point. See Figure 1.10.

We represent the magnitude of a vector $\vec{v}$ with the notation $\|\vec{v}\|$. By the above reasoning, we can compute the magnitude of a vector $\vec{v} = (v_1, v_2)$ in $\mathbb{R}^2$ with the formula

$$\|\vec{v}\| = \sqrt{v_1^2 + v_2^2}$$

It is left to the reader as an exercise in geometry to confirm that the magnitude of a vector $\vec{v} = (v_1, v_2, v_3)$ in $\mathbb{R}^3$ can be computed with the formula

$$\|\vec{v}\| = \sqrt{v_1^2 + v_2^2 + v_3^2}$$

(A similar formula can be used to compute the magnitude of a vector of any dimension.)

If we consider a vector in $\mathbb{R}^1$, $\vec{v} = (v)$, note that we can view the vector $\vec{v}$ simply as a point on the real number line since it has only one coordinate, v. The magnitude of the vector is computed simply as $\|\vec{v}\| = \sqrt{v^2} = |v|$.

Effectively then, we see that the computation of magnitude for vectors corresponds naturally to the computation of the absolute value of a real number. In fact, we could view the absolute value as being simply a special case of the computation of magnitude. Because of this, many texts denote the magnitude of a vector as $|\vec{v}|$ instead of $\|\vec{v}\|$. Students should be aware of this notation. In this text however, we will continue to use the latter, distinct notation.

We will now define the operation of addition as it applies to vectors. Before we do so we must remind ourselves of the original motivation for the definition of vectors, so that we can find a reasonable motivation for what the addition of vectors should represent.

One of the motivating concepts for the definition of the vector was that of the discrete movement in $\mathbb{R}^n$. Given that, we ask ourselves what it should mean to "add" two such discrete movements. A reasonable answer would be that the sum of two discrete movements is the final effective discrete movement caused by performing one of the discrete movements, followed by the other.

We represent this with arrows in Figure 1.11. We start with two vectors, $\vec{v}$ and $\vec{w}$. Arbitrarily we begin at the origin, and first take a discrete movement represented by the vector $\vec{v}$. We accomplish this by placing the vector $\vec{v}$ in standard position, and conclude that this has moved us to the point where the head of the arrow is. From this point we take a discrete movement represented by the vector $\vec{w}$, accomplished by placing an arrow representing $\vec{w}$ at our current position. We conclude that our final location is the point where the head of this last arrow is. Altogether then, the vector $\vec{v} + \vec{w}$ is the vector representing this total displacement from the origin.

If we are given two vectors by their coordinates, there is an easy formula for computing the coordinates of the sum of the two vectors. It is left as an exercise to the reader to confirm that, based on the process described above, given two vectors $\vec{v} = (v_1, v_2)$ and $\vec{w} = (w_1, w_2)$ the sum can be computed as

$$\vec{v} + \vec{w} = (v_1 + w_1, v_2 + w_2)$$

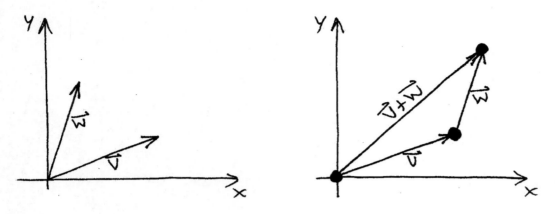

Figure 1.11:

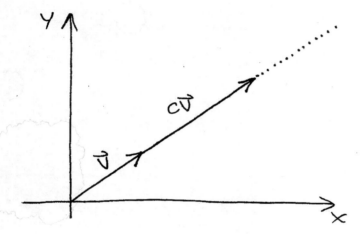

Figure 1.12:

Note, both the geometric construction and the algebraic formula corresponding to the above for vectors in $\mathbb{R}^3$ are naturally analogous.

Having done this, we can also define what it would mean to add a point and a vector. Recall of course that we represent points with their position vectors, so it would be reasonable to consider the position vector instead of the given point, and then add as described above.

Note that this is equivalent to defining the sum of a point and a vector as the ending point achieved by starting at the given point, and taking a discrete movement represented by the given vector.

Either way, the result is that the sum of a point $\vec{p} = (p_1, p_2)$ and a vector $\vec{v} = (v_1, v_2)$ is computed as

$$\vec{p} + \vec{v} = (p_1 + v_1, p_2 + v_2)$$

The fact that the above two formulas are the same is one of the motivations for blurring the notational distinction between points and vectors, discussed in the previous subsection.

We can use a similar line of reasoning to motivate a reasonable definition for the product of a vector and a scalar (that is, a real number).

We begin by considering the product of a vector and a positive scalar. Again recalling that a vector has a direction and a magnitude, it would seem reasonable to define its product with a scalar by multiplying the magnitude by that scalar, and leaving the direction the same. See Figure 1.12.

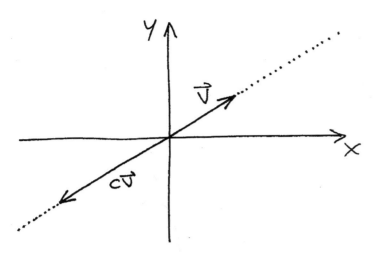

Figure 1.13:

Again, it is not hard to determine from this geometrically stated definition a formula for computing such a product with coordinates. Given a vector $\vec{v} = (v_1, v_2)$ and a scalar c, their product can be computed as

$$c\vec{v} = (cv_1, cv_2)$$

The computation is left as an exercise to the reader.

If we wish to multiply a vector by a negative scalar, of course there is a great algebraic motivation to leave the formula the same as the above. In doing this, though, we should note that the resulting geometric picture is importantly different from the one above. When multiplying a vector by a positive scalar the direction of the vector stays the same; however, if we rely on the same algebraic formula for negative scalars, note that the direction of the resulting product vector is not the same as the original vector – in fact, it is pointing in exactly the opposite direction.

We quickly realize, though, that this is actually very intuitively satisfying. A negative is an opposite of sorts, so having the product point in the opposite direction fits nicely. See Figure 1.13.

The reader can check directly from the above formula that for any vector and any scalar, positive or negative, we have

$$\|c\vec{v}\| = |c|\|\vec{v}\| \tag{1.1}$$

We have now defined what it means to add two vectors, and what it means to multiply a vector by a scalar. What about subtracting two vectors?

In fact, the definitions we have already made completely determine what this must be, because the difference of two vectors can be computed with a combination of the above operations. Specifically, $\vec{v} - \vec{w} = \vec{v} + \left((-1)\vec{w}\right)$.

To see what this means geometrically, one can simply use the geometric constructions above to work this out. The reader is encouraged to do this. But a crafty observation allows us to find a convenient geometric picture more directly. All we have to do is draw our pre-existing geometric picture for the vector sum expression $\vec{w} + \left(\vec{v} - \vec{w}\right) = \vec{v}$, as in Figure 1.14.

Since this figure is valid for all vectors, we can simply reinterpret this figure as a geometric representation of the subtraction of the vectors $\vec{v}$ and $\vec{w}$. Specifically, an arrow representing the vector $\vec{v} - \vec{w}$ is determined by placing both $\vec{v}$ and $\vec{w}$ in standard position, and then drawing the arrow from the head of $\vec{w}$ to the head of $\vec{v}$.

Given all the definitions that we have made in this subsection, and the corresponding computational formulas, students can check the following algebraic properties of vectors.

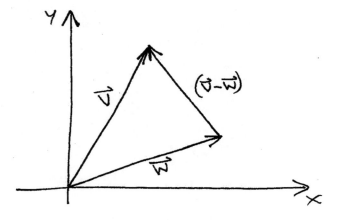

Figure 1.14:

$$\vec{v} + \vec{w} = \vec{w} + \vec{v}$$
$$\vec{u} + \left(\vec{v} + \vec{w}\right) = \left(\vec{u} + \vec{v}\right) + \vec{w}$$
$$c\left(\vec{v} + \vec{w}\right) = c\vec{v} + c\vec{w}$$
$$(c+d)\vec{v} = c\vec{v} + d\vec{v}$$
$$(cd)\vec{v} = c\left(d\vec{v}\right)$$

We have already commented that vectors are very useful for representing many quantities that have both a direction and a magnitude. Sometimes, however, we wish to represent just a direction. There are many vectors that point in a given direction, of course. But there is a unique vector of magnitude 1 pointing in a given direction. We call such a vector a "unit vector", and often use these when we wish to represent a direction without a magnitude.

There are some unit vectors of particular importance, called the "standard basis vectors". These are the vectors whose coordinates are all 0 except for one coordinate, which is 1. In any Euclidean space, the number of standard basis vectors is the same as the dimension of that space.

For example, there are two standard basis vectors in $\mathbb{R}^2$:

$$\vec{e}_1 = (1, 0)$$
$$\vec{e}_2 = (0, 1)$$

(Note that the subscript represents the position of the nonzero coordinate.) Sometimes the notations $\hat{i}$ and $\hat{j}$ are used to represent the unit vectors $\vec{e}_1$ and $\vec{e}_2$, respectively.

Similarly, there are three standard basis vectors in $\mathbb{R}^3$:

$$\vec{e}_1 = (1, 0, 0)$$
$$\vec{e}_2 = (0, 1, 0)$$
$$\vec{e}_3 = (0, 0, 1)$$

Sometimes the notations $\hat{i}$, $\hat{j}$ and $\hat{k}$ are used to represent the unit vectors $\vec{e}_1$, $\vec{e}_2$ and $\vec{e}_3$, respectively.

Students should be careful to observe that we use the same notation to represent standard basis vectors in Euclidean spaces of different dimension; for example in the definitions above, note that $\vec{e}_1$ in $\mathbb{R}^2$ is not the same vector as $\vec{e}_1$ in $\mathbb{R}^3$. The distinction will generally be clear from the context, but students should be aware of this issue.

In Euclidean spaces of higher dimension, the definitions of the standard basis vectors are similar. Note that in the cases of these higher dimensions, it can be much more convenient to use the notation $\vec{e}_i$ for these vectors than to continue to enumerate more and more letters of the alphabet.

1.2. VECTORS IN $\mathbb{R}^N$

Sometimes we will find ourselves in a situation where we have a vector, but we wish to isolate only the direction of the vector. That is, we wish to find a unit vector pointing in the direction of the given vector. There is a very convenient formula for doing this. The unit vector $\vec{u}$ pointing in the direction of the given nonzero vector $\vec{v}$ is

$$\vec{u} = \frac{\vec{v}}{\|\vec{v}\|}$$

We can see immediately that this vector $\vec{u}$ points in the correct direction, because we have obtained it by multiplying the given vector by a positive scalar. And we can confirm that it has unit length by simple computation,

$$\|\vec{u}\| = \left\|\frac{\vec{v}}{\|\vec{v}\|}\right\| = \left|\frac{1}{\|\vec{v}\|}\right| \|\vec{v}\| = \frac{1}{\|\vec{v}\|} \|\vec{v}\| = 1$$

Exercises

Exercise 1.2.1 (**exer-V-V-1**). An arrow representing the vector $(3, 2, 4)$ has its head at the point $(5, 7, 2)$. Where is the tail of this arrow?

Exercise 1.2.2 (**exer-V-V-2**). What vector is represented by the arrow with head at $(4, 3, 2)$ and tail at $(1, 4, 2)$?

Exercise 1.2.3 (**exer-V-V-3**). The plane P is parallel to the xy-plane and intersects the z-axis at the point $(0, 0, 4)$. For each of the following, decide if it is parallel to this plane. (Make sure to make the proper interpretation of the ordered triples based on the phrasing of the question!)
1. $(7, 3, 4)$
2. $(9, 8, 0)$
3. $(-1, 3, 2)$

Exercise 1.2.4 (**exer-V-V-4**). The plane Q is parallel to the xz-plane and intersects the y-axis at the point $(0, 3, 0)$. For each of the following, decide if it is in this plane. (Make sure to make the proper interpretation of the ordered triples based on the phrasing of the question!)
1. $(7, 3, 4)$
2. $(9, 8, 0)$
3. $(-1, 3, 2)$

Exercise 1.2.5 (**exer-V-V-5**). Compute the magnitude of each of the following vectors.
1. $(5, -1)$
2. $(3, 4, 9)$
3. $(-2, -5, -7)$

Exercise 1.2.6 (**exer-V-V-6**). Show that if n and m are integers, then the magnitude of the vector

$$(n^2 - m^2, 2nm)$$

is an integer.

Exercise 1.2.7 (**exer-V-V-7**). Show that if n and m are integers, then the magnitude of the vector

$$(n^2 - m^2, 2nm, \frac{(n^2 + m^2)^2 - 1}{2})$$

is either an integer or half of an integer.

Exercise 1.2.8 (**exer-V-V-8**). Compute each of the vector sums below.

1. $(6,4) + (3,-1) + (2,5)$
2. $(4,3,2) + (-7,8,4)$
3. $(5,6,-3) + (4,3,6) + (1,2,-1)$

Exercise 1.2.9 (exer-V-V-9). Compute each of the scalar-vector products below.
1. $4(5,-2)$
2. $-5(3,-6,2)$
3. $0(4,3,7)$

Exercise 1.2.10 (exer-V-V-10). Use the Pythagorean formula to give an algebraic demonstration of Equation 1.1. (Make sure to explain clearly where the absolute values come from.)

Exercise 1.2.11 (exer-V-V-11). Compute each of the vector differences below.
1. $(6,4) - (3,-1)$
2. $(4,3,2) - (-7,8,4)$
3. $(5,6,-3) - (4,3,6)$

Exercise 1.2.12 (exer-V-V-12). For each of the following vectors, find the unique unit vector corresponding to the direction of that vector.
1. $(3,4)$
2. $(2,3,6)$
3. $(20,30,60)$

Exercise 1.2.13 (exer-V-V-13). Find the unique numbers a, b and c such that $(5,3,6) = a\vec{e}_1 + b\vec{e}_2 + c\vec{e}_3$.

Exercise 1.2.14 (exer-V-V-14). Bob is on a train that is traveling northwest at 60 miles per hour. He aims his BB gun 30 degrees south of east and fires at a tree on the side of the tracks. If the muzzle speed of the BB is 100 miles per hour, how fast is the BB going when it strikes the tree? (Use a calculator to approximate your answer as a decimal.) *(Hint: The "muzzle velocity" of the BB is measured relative to the gun itself; that is, it is the difference between the velocity of the BB and the velocity of the gun.)*

1.3 Dot Product

In the previous section, we motivated and discussed some of the more fundamental algebraic properties of vectors, such as addition and scalar multiplication.

Note that we began our discussion of each of these by first looking for some sort of reasonable motivation, based on what vectors represent in the first place. Having found a way to motivate a natural choice, we could then proceed to finding both a geometric representation for these algebraic operations, and explicit formulas for how to execute these operations given the coordinates of the initial vectors.

Ideally, it would be nice if we could do the same sort of thing with multiplication of vectors. Unfortunately, however, this is not the case.

We will find a useful and meaningful notion of multiplication for vectors (in fact we will find two of them!), but it is difficult to describe initial motivations that both seem reasonable from the beginning and also lead to something that is useful in application.

Instead, for each of these two notions of multiplication, we will begin by simply writing down the algebraic formula. Having done this, we will then show that even though each definition was made without the benefit of motivation, still there are natural geometric interpretations, and useful means of application.

In this section we will discuss the first of these two kinds of vector multiplication, called the "dot product". As promised above, we begin with the algebraic definition, completely without motivation.

1.3.1 Definition and properties

Given two vectors $\vec{v} = (v_1, v_2, v_3)$ in $\mathbb{R}^3$ and $\vec{w} = (w_1, w_2, w_3)$, the dot product $\vec{v} \cdot \vec{w}$ is defined by the expression

$$\vec{v} \cdot \vec{w} = v_1 w_1 + v_2 w_2 + v_3 w_3$$

Similarly, the dot product can be defined for two vectors in any dimension as the sum of the products of corresponding coordinates of the two vectors.

Be sure to note that unlike addition, the dot product of two vectors is not a vector – it is a scalar.

Before we get to the more interesting properties of this dot product, we first enumerate a few algebraic properties that are easily checked from the above definition.

$$\begin{aligned} \vec{v} \cdot \vec{w} &= \vec{w} \cdot \vec{v} \\ \vec{u} \cdot (\vec{v} + \vec{w}) &= \vec{u} \cdot \vec{v} + \vec{u} \cdot \vec{w} \\ c(\vec{v} \cdot \vec{w}) &= (c\vec{v}) \cdot \vec{w} \end{aligned}$$

Note that these algebraic properties are very similar to analogous properties of multiplication of real numbers.

These algebraic properties do not seem very interesting or surprising at this point, because they are so easily checked from the definition. And surely it would seem that there could be any number of other arbitrarily defined "products" that would have these same algebraic properties. While this is certainly true, we will now see that the dot product has many more interesting properties that are much more impressive.

For example, we can also check that

$$\vec{v} \cdot \vec{v} = \|\vec{v}\|^2$$

Of course, this is also very easily checked from the definition. But two points make this property more interesting than the previous. For one thing, the above definition would certainly seem to be the simplest and probably even the only reasonable operation to write down that would have this property.

More importantly though, this formula represents the first connection in this section between the algebraic and the geometric. Specifically, note that the dot product on the left side of the equation is defined algebraically; the right side

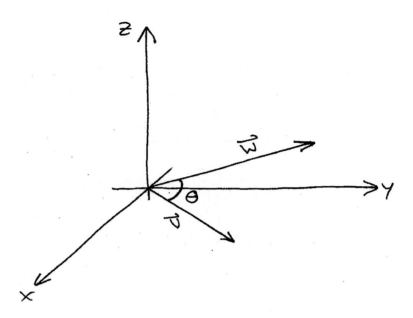

Figure 1.15:

of the equation relates to length, which is an intrinsically geometric notion. This connection to something geometrically relevant gives our algebraic definition of the dot product some utility already.

An even more impressive property of the dot product is the "Cauchy-Schwarz" inequality,

$$|\vec{v} \cdot \vec{w}| \leq \|\vec{v}\|\|\vec{w}\|$$

While one could say that it is less impressive since it is an inequality rather than an equation, note that the dot product in question is between any two vectors, not just a vector with itself.

We will not prove this inequality immediately, even though there are several perfectly reasonable proofs that could be done here. Instead, we will take advantage of the fact that there will be a very easy proof accessible to us once we state and prove the next property that we will discuss very soon. In fact, it could reasonably be said that the Cauchy-Schwarz inequality, to the extent that it pertains to this context, is a special case of that next property we will discuss and prove.

(However, it should be noted that the Cauchy-Schwarz inequality can be expressed in a much more general form, for which we do not have the appropriate tools in this class. In this more general form, the Cauchy-Schwarz inequality can apply to a wide variety of different fields, including functional analysis and statistics. In this more general form it is certainly not a special case of the next property we will discuss.)

Before we move on, let us again make note of the fact that the Cauchy-Schwarz inequality is another connection between the algebraically defined dot product and geometry – again, the left side involves mostly just a dot product, while both of the factors on the right side are lengths, which we again note are geometric concepts. This again lends relevance and natural meaning to our definition of the dot product.

Probably the most powerful property of the dot product that we will make use of in this course is the following amazing fact, which we will state as a theorem.

Theorem 1.3.1. *For any two nonzero vectors $\vec{v}$ and $\vec{w}$,*

$$\vec{v} \cdot \vec{w} = \|\vec{v}\|\|\vec{w}\| \cos \theta$$

where the angle θ is the angle between those two vectors, as in Figure 1.15.

Here we have on the left side exactly the dot product, we have an equation not an inequality, and all three of the factors on the right side again involve geometric notions. This is a remarkably powerful and beautiful equation, and we will make great use of this property in this course.

1.3. DOT PRODUCT

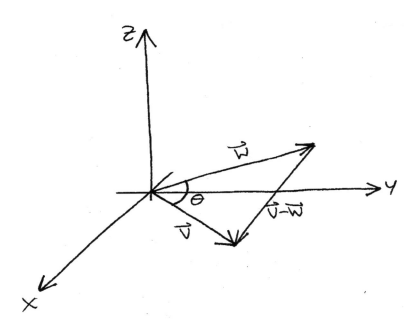

Figure 1.16:

To prove this equation, we will make use of the law of cosines, which students should recall from high school geometry. Specifically, we put both vectors $\vec{v}$ and $\vec{w}$ in standard position, and then complete the triangle by drawing an arrow representing the difference vector $\vec{v} - \vec{w}$, as in Figure 1.16.

The law of cosines, with respect to this triangle and the vertex at the origin, then immediately gives us

$$\|\vec{v} - \vec{w}\|^2 = \|\vec{v}\|^2 + \|\vec{w}\|^2 - 2\|\vec{v}\|\|\vec{w}\|\cos\theta$$

Now let us also consider a different way in which we could represent the left side of the above equation. Since it is the square of the length of a vector we can represent it as a dot product of that vector with itself. That is,

$$\|\vec{v} - \vec{w}\|^2 = (\vec{v} - \vec{w}) \cdot (\vec{v} - \vec{w})$$

Using previously established properties of the dot product, we can simplify the right side to get

$$\begin{aligned}\|\vec{v} - \vec{w}\|^2 &= (\vec{v} - \vec{w}) \cdot (\vec{v} - \vec{w}) \\ \|\vec{v} - \vec{w}\|^2 &= \vec{v} \cdot \vec{v} + \vec{w} \cdot \vec{w} - \vec{v} \cdot \vec{w} - \vec{w} \cdot \vec{v} \\ \|\vec{v} - \vec{w}\|^2 &= \|\vec{v}\|^2 + \|\vec{w}\|^2 - 2\vec{v} \cdot \vec{w}\end{aligned}$$

Now comparing this last equation with the one that we got from the law of cosines, we easily arrive at the desired result.

As previously mentioned, notice that the statement of the Cauchy-Schwarz inequality follows immediately from this result, and the fact that the cosine function is always between -1 and 1. Students should make sure to work out these details for themselves.

There are many different types of applications for which this equation can be very useful. Most obviously, if one knows the necessary geometric information about two vectors, one can use the above formula to compute the dot product, even without knowing the coordinates of the two vectors.

Also, one can use the formula in the other direction. That is, if one knows the coordinates of two vectors, note that it is very easy to compute both the dot product and the lengths of the two vectors; one can then solve for the angle θ with

$$\theta = \arccos\left(\frac{\vec{v} \cdot \vec{w}}{\|\vec{v}\|\|\vec{w}\|}\right)$$

Example 1.3.1. Suppose we wish to compute the angle between the two vectors $\vec{v} = (1, 2, 2)$ and $\vec{w} = (-6, 3, 2)$. We can easily compute their dot product and lengths as

$$\begin{aligned} \vec{v} \cdot \vec{w} &= (1) \cdot (-6) + (2) \cdot (3) + (2) \cdot (2) = 4 \\ \|\vec{v}\| &= \sqrt{1^2 + 2^2 + 2^2} = 3 \\ \|\vec{w}\| &= \sqrt{(-6)^2 + 3^2 + 2^2} = 7 \end{aligned}$$

Plugging this into the above equation, we get

$$\theta = \arccos\left(\frac{4}{3 \cdot 7}\right) = \arccos\left(\frac{4}{21}\right)$$

Finally, we can use this equation to relate the dot product to the geometric notion of two vectors being perpendicular. Of course if two vectors are perpendicular, the angle between them is $\pi/2$, whose cosine is zero, and therefore the dot product of those two vectors is zero. Going in the other direction, if the dot product of two vectors is zero, then either one of the two lengths is zero, or the cosine of the angle is zero, making the angle $\pi/2$ and thus the vectors perpendicular.

We will summarize this below as a corollary to Theorem 1.3.1.

Corollary 1.3.1. *Two nonzero vectors $\vec{v}$ and $\vec{w}$ are perpendicular if and only if their dot product is zero.*

Of course if one of the two vectors is zero, there is no well-defined angle between them anyway, so the question is moot.

Still, it is slightly inconvenient to have to limit the statement to nonzero vectors. We can get around this issue by introducing a new term that includes that possibility. We say that two vectors are "orthogonal" if either they are perpendicular, or at least one of the vectors is zero. The corollary is then restated more concisely as

Corollary 1.3.2. *Two vectors $\vec{v}$ and $\vec{w}$ are orthogonal if and only if their dot product is zero.*

Example 1.3.2. Are the vectors $(2, 3, 1)$ and $(1, -2, 4)$ perpendicular?

Clearly both vectors are nonzero, so this simply boils down to computing the dot product. We easily see that this dot product is zero, so the vectors are indeed perpendicular.

1.3.2 Components

The first application we will talk about is to something called the "component" of a vector in the direction of another vector. In fact it will turn out that not only will we find an application for the dot product – we will also find a geometric interpretation that is reasonably satisfying.

Suppose we are given two vectors $\vec{v}$ and $\vec{w}$. There is a unique way in which the vector $\vec{v}$ can be written as the sum of two vectors $\vec{v}_\|$ and $\vec{v}_\perp$, as in Figure 1.17.

We refer to the length of the vector $\vec{v}_\|$ as "the component of the vector $\vec{v}$ in the direction of the vector $\vec{w}$", denoted as $\text{comp}_{\vec{w}}(\vec{v})$. We might also phrase this as "the $\vec{w}$-component of $\vec{v}$".

We can compute such a component with simple trigonometry applied to Figure 1.17, giving us

$$\text{comp}_{\vec{w}}(\vec{v}) = \|\vec{v}\| \cos\theta$$

1.3. DOT PRODUCT

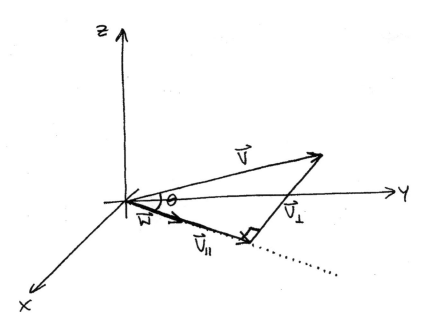

Figure 1.17:

Quickly we notice that this formula resembles the one in Theorem 1.3.1, and so we can rewrite the above equation as

$$\text{comp}_{\vec{w}}(\vec{v}) = \frac{\|\vec{v}\|\|\vec{w}\|\cos\theta}{\|\vec{w}\|} = \frac{\vec{v}\cdot\vec{w}}{\|\vec{w}\|}$$

So we immediately see that dot products can be used to compute components.

With a slight adjustment, we can turn the above equation into a geometric interpretation of the dot product itself. Solving for the dot product, we get

$$\vec{v}\cdot\vec{w} = \|\vec{w}\|\,\text{comp}_{\vec{w}}(\vec{v})$$

Of course the dot product is symmetric, so we can just as well write

$$\vec{v}\cdot\vec{w} = \|\vec{v}\|\,\text{comp}_{\vec{v}}(\vec{w})$$

Taken directly as written, this gives us a geometric interpretation of the dot product. That is, it is the component of one vector in the direction of the other, times the length of that other vector.

Speaking more loosely, one might interpret the above equation to say that the dot product is a way of measuring the combined extent to which two vectors are pointing in the same direction.

Also, note that in the special case where one of the vectors is a unit vector, then the dot product can be concisely interpreted simply as the component of the other vector in that direction.

1.3.3 Work

Work has the same units as energy, and represents the amount of energy required in order to accomplish the given task. Specifically, recall that when a constant force is exerted over a given displacement distance, the work performed is the product of the force and the distance. We write this as

$$W = Fd$$

Here both the force F and the distance d are represented by scalars. Recall also that the scalar F is not actually the entire force, but only the amount of the force in the direction of the displacement.

Of course both force and a displacement are really more naturally represented by vectors, because each has both a magnitude and an associated direction. The force is represented by a vector $\vec{F}$, and the displacement is represented by a vector $\vec{d}$.

We can now rewrite the above equation in terms of the force and displacement vectors. As previously stated the scalar F is the component of the force in the direction of the displacement, so we have

$$F = \text{comp}_{\vec{d}}(\vec{F}) = \frac{\vec{F} \cdot \vec{d}}{\|\vec{d}\|}$$

And of course the distance d is simply the length of the displacement vector, $d = \|\vec{d}\|$. Plugging these in to the original equation for work, we have

$$W = \vec{F} \cdot \vec{d} \tag{1.2}$$

This vector version of the equation for work is a more elegant and convenient formulation, since the dot product automatically accounts for the different directions of the force and the displacement. We will make much use of this formula in Chapter 6.

Example 1.3.3. Suppose that a constant force $\vec{F}$ is exerted over a distance with displacement vector $\vec{d}$. The force vector has a magnitude of 20 N in the direction $30°$ north of East, and the displacement is 3 m east and 1 m south. What is the total amount of work performed?

We can represent the force and displacement vectors by $\vec{F} = (10\sqrt{3}, 10)$ and $\vec{d} = (3, -1)$. Plugging in to equation 1.2, we have

$$W = \vec{F} \cdot \vec{d} = 30\sqrt{3} - 10$$

Exercises

Exercise 1.3.1 (exer-V-DP-1). Compute the following.
1. $(5, 3) \cdot (2, 1)$

2. $(7, -4, 2) \cdot (3, 2, 6)$

3. $(1, 1, 1) \cdot (5, -2, -3)$

Exercise 1.3.2 (exer-V-DP-2). The vector $\vec{x}$ is known to have magnitude equal to 6. What is the dot product of this vector with itself?

Exercise 1.3.3 (exer-V-DP-3). The vector $\vec{v}$ has magnitude 5 and points 14 degrees counterclockwise from the positive part of the x-axis in the xy-plane. The vector $\vec{w}$ has magnitude 3 and points 31 degrees clockwise from the negative part of the y-axis. Compute $\vec{v} \cdot \vec{w}$.

Exercise 1.3.4 (exer-V-DP-4). Find the angle between each of the following pairs of vectors.
1. $(5, 3)$ and $(2, 1)$

2. $(7, -4, 2)$ and $(3, 2, 6)$

3. $(1, 1, 1)$ and $(5, -2, -3)$

Exercise 1.3.5 (exer-V-DP-5). For each of the pairs of vectors below, determine if they are orthogonal.

1.3. DOT PRODUCT

1. $(5, 3)$ and $(3, -5)$
2. $(7, -4, 2)$ and $(3, 2, 6)$
3. $(1, 1, 1)$ and $(5, -2, -3)$

Exercise 1.3.6 (exer-V-DP-6). Show that the angle between two vectors is acute if and only if their dot product is positive, and that the angle between them is obtuse if and only if their dot product is negative.

Exercise 1.3.7 (exer-V-DP-7). Compute the component of the vector $(5, 3, 4)$ in the direction of the vector $(-3, -2, 7)$.

Exercise 1.3.8 (exer-V-DP-8). Show that for any vector $\vec{x} = (x_1, x_2, x_3)$ and any standard basis vector $\vec{e}_i$, the component of the vector in the direction of that standard basis vector is always equal to the corresponding coordinate of the vector.

Exercise 1.3.9 (exer-V-DP-9). The wind is blowing in such a way that resisting it require the exertion of a force of magnitude 6 Newtons in the direction that is 37 degrees west of north. How much work does it require to walk 1000 meters in the direction that is 67 degrees north of east?

Exercise 1.3.10 (exer-V-DP-10). New York City is located at 40 °N. latitude, 74 °W. longitude, and Moscow is located at 55 °N. latitude, 37 °E. longitude. Use the dot product (and a calculator) to answer the following questions:

1. Thinking of the earth as a perfect sphere of radius 3,960 miles, centered at the origin, with the north pole on the positive part of the z-axis and the positive part of the x-axis going through the prime meridian, compute the coordinates of both of these cities in xyz-space.

2. What is the angle between the position vectors for these two cities?

3. The shortest distance along the surface of a sphere from one point to another is the length of an arc on the great circle defined by those two points. Recalling that the length s of an arc of angle α on a circle of radius r is given by $s = r\alpha$, compute the shortest distance along the surface of the earth between New York and Moscow.

1.4 Cross Product and Determinant

The other important kind of multiplication defined for vectors is called the "cross product". The properties of the cross product are completely different from those of the dot product, but still they are very useful and interesting.

It turns out that the cross product is very closely linked to another notion, called the determinant of a matrix. We will talk more about matrices in Chapter 3; for now, all we will say is that a matrix is an array of numbers in a rectangular shape.

There are different ways and orders in which to present the ideas of the cross product, the determinant, their different properties, and the connections between them. Each of these approaches has advantages and disadvantages, but every one of them has the disadvantage of at least one difficult or tedious computation.

In this section we will take an approach that emphasizes geometric intuition. We will present one theorem whose proof would require an undesirable computation; in the interests of focus on the many geometric concepts to be digested in this section, we will simply omit this proof. All of the other facts discussed in this section will be presented with simple, elegant proofs.

1.4.1 Right-hand order

We begin by discussing a subtle but important concept related to ordering. We will begin in $\mathbb{R}^2$, and then move on to $\mathbb{R}^3$.

Suppose we have two vectors in $\mathbb{R}^2$, $\vec{v}$ and $\vec{w}$. Suppose further that these two vectors are not parallel.

There are two different orders that we can write them down – "$\vec{v}, \vec{w}$" and "$\vec{w}, \vec{v}$". Importantly, these orders are different more than just alphabetically. Specifically there is a geometric way of distinguishing these orders.

Suppose we consider the vectors in the order $\vec{v}, \vec{w}$. We begin by taking the first listed vector, $\vec{v}$, and considering the unique line L through the origin determined by this vector. This line L divides the plane into two distinct regions. The second vector, $\vec{w}$, when in standard position, must then point into one of these regions or the other. (Note, we do not need to be concerned about the possibility that it points parallel to L, because we are assuming that these two vectors themselves are not parallel.)

We can distinguish these two possibilities from each other by considering the ways one would most conveniently rotate the first vector to get into the region in question.

If the second vector is in the region that is on the clockwise side of the first vector, we say the vectors are in "clockwise order"; if the second vector is in the region that is on the counterclockwise side of the first vector, we say the vectors are in "counterclockwise order". In Figure 1.18, we show an example where the order $\vec{v}, \vec{w}$ is a counterclockwise order.

It is not hard to convince yourself that if $\vec{v}, \vec{w}$ is a counterclockwise order, then $\vec{w}, \vec{v}$ is a clockwise order, and vice-versa.

This is an easy enough notion to visualize, partly because it is easy to see and draw pictures in two dimensions, and partly because we are all already very familiar with the rotation directions of clockwise and counterclockwise. But what if we want to do something similar in $\mathbb{R}^3$? Is there some way to distinguish orders of vectors in $\mathbb{R}^3$? We will see next that we can, using a different geometric notion.

One of the relationships between clockwise and counterclockwise rotations is that they are "reflections" of each other. That is, if you rotate something clockwise and look at its image in a mirror, the reflection appears to be rotating counterclockwise; and vice versa.

In order to discuss orders in $\mathbb{R}^3$, we will use another pair of geometric notions that have this relationship of reflections.

Suppose we have three vectors in $\mathbb{R}^3$, $\vec{a}, \vec{b}, \vec{c}$, to be thought of in that order; we would like to find a geometric description of this order.

Let us suppose further that there is no single plane to which all three of these vectors are parallel. If this condition

1.4. CROSS PRODUCT AND DETERMINANT

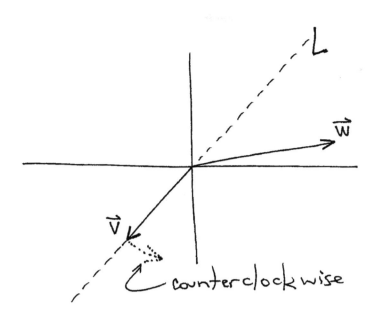

Figure 1.18:

does not hold, we both cannot and need not define the notion of order we will present in this subsection. (Note also, in Chapter 3 we will find another way to phrase this condition.)

The first two of these vectors, $\vec{a}$ and $\vec{b}$, define a unique plane – namely, the plane P passing through the origin that is parallel to both vectors, as in Figure 1.19.

(Note that we need not concern ourselves with the possibility that $\vec{a}$ and $\vec{b}$ might be parallel; if they were, then there would be some plane parallel to all three vectors, contradicting our assumption.)

This plane divides the space into two distinct regions. The third vector $\vec{c}$, when in standard position, must then point into one of these regions or the other.

We distinguish these two possibilities from each other by using the human right hand. Specifically, arrange the right hand so that the index finger is pointing straight forward, parallel to the palm, and the middle finger is pointing perpendicular to the palm. Then, while holding the hand in this position, orient it in such a way that the index finger points in the direction of the first vector $\vec{a}$, and the middle finger points in the direction of the second vector $\vec{b}$. See Figure 1.20.

At this point, note that the thumb is pointing into one of the two regions defined by the plane P.

If the vector $\vec{c}$ is pointing into the same region as the thumb, we say that the vectors $\vec{a}$, $\vec{b}$, $\vec{c}$ are written in "right-hand order". If it is pointing into the other region, then they are not written in right-hand order. (we can refer to this other order as "left-hand order".)

Figure 1.21 shows a case where $\vec{a}$, $\vec{b}$, $\vec{c}$ are written in "right-hand order".

Using one's own right hand, it can easily be checked that the orderings $\vec{a}, \vec{b}, \vec{c}$, and $\vec{b}, \vec{c}, \vec{a}$, and $\vec{c}, \vec{a}, \vec{b}$ are either all in right-hand order, or none of them is in right-hand order. Similarly, the orderings $\vec{a}, \vec{c}, \vec{b}$, and $\vec{c}, \vec{b}, \vec{a}$, and $\vec{b}, \vec{a}, \vec{c}$ are either all in right-hand order, or none of them is in right-hand order. Finally note that if the first three orders are in right-hand order then the other three are in left-hand order, and vice versa.

Note that as promised, we have used two geometric notions that are reflections of each other. That is, the human right hand is a mirror reflection of the human left hand, and vice versa.

Here is an alternative description of right-hand order. With all three vectors in standard position, imagine yourself

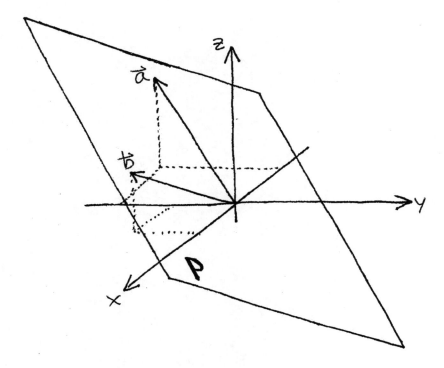

Figure 1.19:

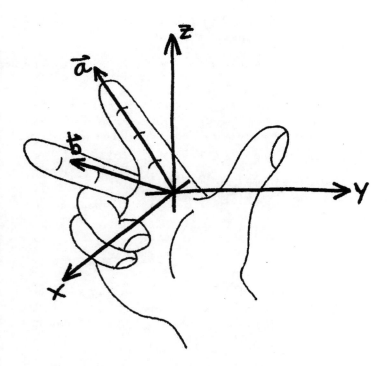

Figure 1.20:

1.4. CROSS PRODUCT AND DETERMINANT

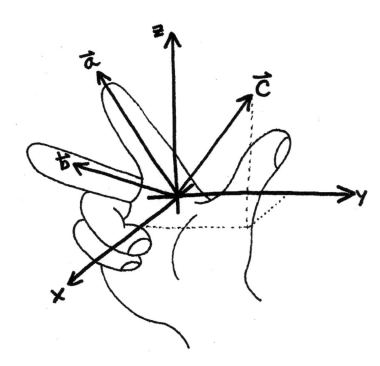

Figure 1.21:

positioned at the head of the third vector and looking at the arrows representing the other two vectors. If the first vector appears that it would most conveniently be rotated counterclockwise to overlap the second vector, then the vectors are in right-hand order (see Figure 1.22). If the first vector appears that it would most conveniently be rotated clockwise to overlap the second vector, then the vectors are in left-hand order.

This notion of right-hand order will be very important when talking about the cross product and the determinant. We will see this in detail very soon.

1.4.2 Determinant and volume

Next, we define the determinant of a 2x2 matrix. Remember, we will have much more to say about matrices in Chapter 3; for now, think of a matrix only as an array of numbers.

For a 2x2 matrix A,

$$A = \begin{pmatrix} a & b \\ c & d \end{pmatrix}$$

we defined the determinant by

$$\det A = ad - bc$$

The determinant of a matrix A is also sometimes written $|A|$.

For a 3x3 matrix, the determinant is computed with the formula

$$\begin{vmatrix} a_1 & a_2 & a_3 \\ b_1 & b_2 & b_3 \\ c_1 & c_2 & c_3 \end{vmatrix} = a_1 \begin{vmatrix} b_2 & b_3 \\ c_2 & c_3 \end{vmatrix} - a_2 \begin{vmatrix} b_1 & b_3 \\ c_1 & c_3 \end{vmatrix} + a_3 \begin{vmatrix} b_1 & b_2 \\ c_1 & c_2 \end{vmatrix} \quad (1.3)$$

(Be sure to note the minus sign in the second term on the right side of the equation.)

This definition can be generalized to a square matrix of any size, but we will not discuss that in this text.

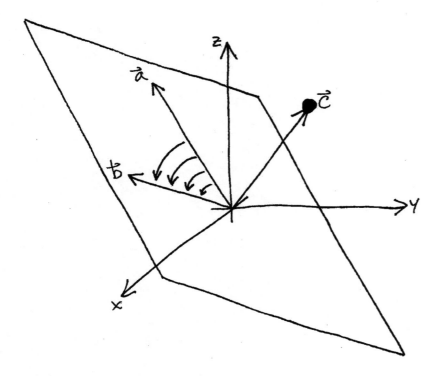

Figure 1.22:

This may seem like an odd and arbitrary definition, but it turns out that the determinant, defined as above, has some remarkable geometric properties. We begin with the following theorems, the latter of which is the most relevant to this section.

Theorem 1.4.1. *Suppose we have two vectors in $\mathbb{R}^2$ in standard position, $\vec{a} = (a_1, a_2)$ and $\vec{b} = (b_1, b_2)$. Form the parallelogram defined by these two vectors, with one vertex at the origin, one vertex at the point defined by the sum of the two vectors, and two of the edges defined by the vectors. (We will use the notation $\|(\vec{a}, \vec{b})$ to represent this parallelogram.)*

Consider the matrix
$$A = \begin{pmatrix} a_1 & a_2 \\ b_1 & b_2 \end{pmatrix}$$
formed by using the coordinates of each vector to make a row of the matrix, specifically in the order that the vectors are listed.

1. *If the vectors are in counterclockwise order, then the determinant of A is positive, and equal to the area of the parallelogram $\|(\vec{a}, \vec{b})$.*

2. *If the vectors are in clockwise order, then the determinant of A is negative, with absolute value equal to the area of the parallelogram $\|(\vec{a}, \vec{b})$.*

3. *If the vectors are in neither order (and thus parallel), then the determinant is zero. Of course in this case the area of $\|(\vec{a}, \vec{b})$ is also zero.*

In all of the cases above, we have that the area of the parallelogram $\|(\vec{a}, \vec{b})$ is equal to $|\det A|$. We denote this by writing

$$|\det(A)| = \text{area}\left(\|(\vec{a}, \vec{b})\right) \tag{1.4}$$

Theorem 1.4.2. *Suppose we have three vectors in $\mathbb{R}^3$ in standard position, $\vec{a} = (a_1, a_2, a_3)$, $\vec{b} = (b_1, b_2, b_3)$, $\vec{c} = (c_1, c_2, c_3)$. Form the parallelepiped defined by these three vectors as shown in Figure 1.23, with one vertex at the origin,*

1.4. CROSS PRODUCT AND DETERMINANT

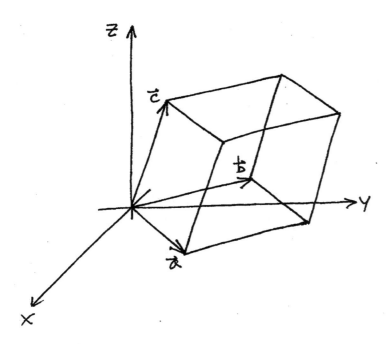

Figure 1.23:

one vertex at the point defined by the sum of the three vectors, and three of the edges defined by the three vectors. (We will use the notation $\|(\vec{a}, \vec{b}, \vec{c})$ to represent this parallelepiped.)

Consider the matrix

$$A = \begin{pmatrix} a_1 & a_2 & a_3 \\ b_1 & b_2 & b_3 \\ c_1 & c_2 & c_3 \end{pmatrix}$$

formed by using the coordinates of each vector to make a row of the matrix, specifically in the order that the vectors are listed.

1. *If the vectors are in right-hand order, then the determinant of A is positive, and equal to the volume of the parallelepiped $\|(\vec{a}, \vec{b}, \vec{c})$.*

2. *If the vectors are in left-hand order, then the determinant of A is negative, with absolute value equal to the volume of the parallelepiped $\|(\vec{a}, \vec{b}, \vec{c})$.*

3. *If the vectors are in neither order (and thus all parallel to the same plane), then the determinant is zero. Of course in this case the volume of $\|(\vec{a}, \vec{b}, \vec{c})$ is also zero.*

In all of the cases above, we have that the volume of the parallelepiped $\|(\vec{a}, \vec{b}, \vec{c})$ is equal to $|\det A|$. We denote this by writing

$$|\det(A)| = vol\Big(\|(\vec{a}, \vec{b}, \vec{c})\Big) \tag{1.5}$$

Since each of the rows of this matrix has elements that come from the coefficients of a given vector, we sometimes indicate this with the shorthand

$$A = \begin{pmatrix} - & \vec{a} & - \\ - & \vec{b} & - \\ - & \vec{c} & - \end{pmatrix} \tag{1.6}$$

This is a very surprising theorem. Most obviously, it is surprising that the algebraically defined determinant is so closely related to the completely geometric notion of volume. Of course this fact also makes the theorem very powerful.

Also, it seems odd that this formula for volume involves only simple arithmetic; surely, one would think, there should be some trig functions involved.

This theorem can be proved without too much difficulty using some techniques beyond the scope of this course. It can also be proved by reordering the topics in this section, providing alternative proofs for these other topics, and then using those to prove this theorem. As mentioned at the beginning of this section, however, this is tedious and is a distraction from the geometric ideas we want the student to digest. We will not present the proof of this theorem in this text.

Example 1.4.1. What is the volume of the parallelepiped determined by the three vectors $(1,3,2)$, $(7,1,4)$, $(0,1,1)$?

We do not know if the vectors are in the right-hand order as they are listed. Still we can use the above theorem – if the determinant ends up being positive, then the vectors must have been in right-hand order, and the volume is that determinant. If the determinant ends up being negative, then the vectors must have been in left-hand order, and the volume is easily recovered simply as the absolute value of that determinant.

So we compute the determinant directly.

$$\begin{vmatrix} 1 & 3 & 2 \\ 7 & 1 & 4 \\ 0 & 1 & 1 \end{vmatrix} = 1 \begin{vmatrix} 1 & 4 \\ 1 & 1 \end{vmatrix} - 3 \begin{vmatrix} 7 & 4 \\ 0 & 1 \end{vmatrix} + 2 \begin{vmatrix} 7 & 1 \\ 0 & 1 \end{vmatrix} = -3 - 21 + 14 = -10$$

We can immediately conclude that the vectors were listed in left-hand order, and that the volume of the parallelepiped is 10.

1.4.3 Cross product

We are now ready to define the cross product of two vectors $\vec{v} = (v_1, v_2, v_3)$ and $\vec{w} = (w_1, w_2, w_3)$ in $\mathbb{R}^3$. We begin simply with the explicit formula,

$$\vec{v} \times \vec{w} = \begin{bmatrix} v_2 w_3 - v_3 w_2 \\ v_3 w_1 - v_1 w_3 \\ v_1 w_2 - v_2 w_1 \end{bmatrix} \tag{1.7}$$

Note that unlike the dot product, the cross product of two vectors is another vector.

The following algebraic properties of the cross product can be checked by direct computation.

$$\begin{aligned} \vec{v} \times \vec{w} &= -\vec{w} \times \vec{v} \\ \vec{u} \cdot (\vec{v} \times \vec{w}) &= \vec{w} \cdot (\vec{u} \times \vec{v}) = \vec{v} \cdot (\vec{w} \times \vec{u}) \\ \vec{v} \times \vec{w} = \vec{0} &\iff \vec{v} \| \vec{w} \quad \text{(or one of the vectors is } \vec{0}) \end{aligned}$$

In the course of the rest of this section, we will see that the cross product has many useful and interesting geometric properties. In fact, even though it is defined above algebraically, it can also be completely characterized by an entirely geometric description. Here we will present one important geometric property, and then make a few necessary algebraic observations. The rest of the geometric properties will be presented in the next subsection.

Theorem 1.4.3. *For any two vectors $\vec{v}$ and $\vec{w}$, the cross product $\vec{v} \times \vec{w}$ is always orthogonal to both $\vec{v}$ and $\vec{w}$.*

We will see in Section 1.5, and on many other occasions throughout this text, that it is often very useful to be able to find a vector perpendicular to two given vectors. This is one of the features that makes the cross product so useful.

This theorem can be proved with a very straightforward computation in light of Corollary 1.3.2 – we need only compute the two dot products

$$\vec{v} \cdot (\vec{v} \times \vec{w}) \quad \text{and} \quad \vec{w} \cdot (\vec{v} \times \vec{w})$$

1.4. CROSS PRODUCT AND DETERMINANT

and confirm that they are both zero. Students are encouraged to perform this computation themselves. Very soon though, we will find a crafty geometric argument that allows us to bypass that algebra.

Before we can move on to the remaining geometric properties of the cross product, we must make an algebraic observation that will connect the cross product to determinants.

We start very simply by noting that each of the three coordinates of the cross product is in the form of a determinant of a 2x2 matrix. With that in mind, we can rewrite the formula for the cross product. For a notational convenience in this case, we will write the cross product in terms of standard basis vectors rather than as an ordered triple of coordinates.

$$\vec{v} \times \vec{w} = \begin{vmatrix} v_2 & v_3 \\ w_2 & w_3 \end{vmatrix} \vec{e}_1 - \begin{vmatrix} v_1 & v_3 \\ w_1 & w_3 \end{vmatrix} \vec{e}_2 + \begin{vmatrix} v_1 & v_2 \\ w_1 & w_2 \end{vmatrix} \vec{e}_3 \tag{1.8}$$

At this point, we see that not only are the individual coordinates of the cross product 2x2 determinants, but also the entire expression for the cross product, written in this form, bears an extremely close resemblance to the expression for a 3x3 determinant.

The problem is that the right-hand side of the equation 1.8 is not entirely made up of mere numbers which we can then assemble into a corresponding matrix; unfortunately, there are standard basis vectors in the place where there are numbers in Equation 1.3.

So, while the cross product is not precisely the determinant of a 3x3 matrix, we will still write it as a "symbolic determinant" of the following "symbolic matrix". This is not a real matrix, as its elements are allowed to be either numbers or vectors, but we can still refer to its symbolic determinant by applying equation 1.3.

$$\vec{v} \times \vec{w} = \begin{vmatrix} \vec{e}_1 & \vec{e}_2 & \vec{e}_3 \\ v_1 & v_2 & v_3 \\ w_1 & w_2 & w_3 \end{vmatrix} \tag{1.9}$$

This equation for the cross product is a very useful memory tool because of its simplicity of structure. It will also end up being useful in computations.

For example, suppose we wish to take a dot product of some vector $\vec{u} = (u_1, u_2, u_3)$ with the cross product of two other vectors, $\vec{v}$ and $\vec{w}$. Of course this is very easily written down directly from Equation 1.8 as

$$\vec{u} \cdot (\vec{v} \times \vec{w}) = \begin{vmatrix} v_2 & v_3 \\ w_2 & w_3 \end{vmatrix} u_1 - \begin{vmatrix} v_1 & v_3 \\ w_1 & w_3 \end{vmatrix} u_2 + \begin{vmatrix} v_1 & v_2 \\ w_1 & w_2 \end{vmatrix} u_3$$

Symbolically, it appears that the three components of the vector $\vec{u}$ have simply taken the places of the three standard basis vectors. Noting that Equation 1.9 comes directly from Equation 1.8, this allows us conveniently to conclude

$$\vec{u} \cdot (\vec{v} \times \vec{w}) = \begin{vmatrix} u_1 & u_2 & u_3 \\ v_1 & v_2 & v_3 \\ w_1 & w_2 & w_3 \end{vmatrix} \tag{1.10}$$

which of course we can rewrite in shorthand as

$$\vec{u} \cdot (\vec{v} \times \vec{w}) = \begin{vmatrix} - & \vec{u} & - \\ - & \vec{v} & - \\ - & \vec{w} & - \end{vmatrix} \tag{1.11}$$

This equation is extremely useful in many circumstances. It is also a very elegant equation, since it is a connection between the dot product, the cross product, and the determinant, which of course has already been related to volume.

1.4.4 Geometric properties

Equation 1.11 gives us a simple geometric argument showing that the cross product is indeed always orthogonal to the two original vectors (we alluded to this after the statement of Theorem 1.4.3). The argument goes as follows.

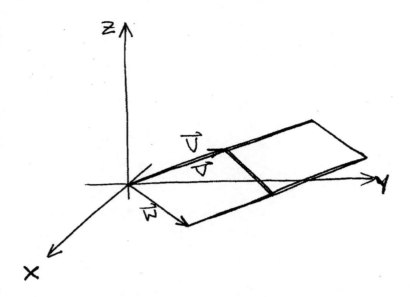

Figure 1.24:

Recall that we must show that both $\vec{v} \cdot (\vec{v} \times \vec{w})$ and $\vec{w} \cdot (\vec{v} \times \vec{w})$ are zero. We consider first $\vec{v} \cdot (\vec{v} \times \vec{w})$. By Equation 1.11 we can rewrite this as

$$\vec{v} \cdot (\vec{v} \times \vec{w}) = \begin{vmatrix} - & \vec{v} & - \\ - & \vec{v} & - \\ - & \vec{w} & - \end{vmatrix} \tag{1.12}$$

According to Theorem 1.4.2, we can relate this to the volume of the parallelepiped $\|(\vec{v}, \vec{v}, \vec{w})$. Of course the three vectors defining this parallelepiped are actually only two, so certainly there is a plane parallel to all three of them. Thus the volume is equal to zero. We can also see this geometrically from the picture in Figure 1.24 of the parallelepiped in question.

So the volume is equal to zero, so the determinant is zero, which of course means that the dot product is zero, which then tells us that the cross product is indeed orthogonal to the vector $\vec{v}$. A similar argument shows that the same is true for $\vec{w}$.

Here is another useful application.

Theorem 1.4.4. *For two vectors $\vec{v}$ and $\vec{w}$ in $\mathbb{R}^3$, we denote the parallelogram defined by those two vectors in standard position by $\|(\vec{v}, \vec{w})$.*

The length of the cross product $\vec{v} \times \vec{w}$ is equal to the area of that parallelogram. We write this as

$$\|\vec{v} \times \vec{w}\| = area\Big(\|(\vec{v}, \vec{w})\Big) \tag{1.13}$$

In most texts, the proof of this theorem is highly algebraic and tedious. Here, because of the approach we are taking to this material, we can present instead a beautiful geometric proof, which also serves as an excellent exercise in working with and visualizing the concepts in this section.

We begin by considering a particular expression; but it is hard to motivate why it is that we begin with this choice. Still, it will turn out that the computation of this expression will yield the desired result. Actually, we will compute this expression in two different ways, and then conclude that the results are equal, leading us to the desired result.

The expression in question is the determinant of a matrix whose rows are defined by the vectors $\vec{v} \times \vec{w}$, $\vec{v}$, and $\vec{w}$.

$$\begin{vmatrix} - & \vec{v} \times \vec{w} & - \\ - & \vec{v} & - \\ - & \vec{w} & - \end{vmatrix} \tag{1.14}$$

1.4. CROSS PRODUCT AND DETERMINANT

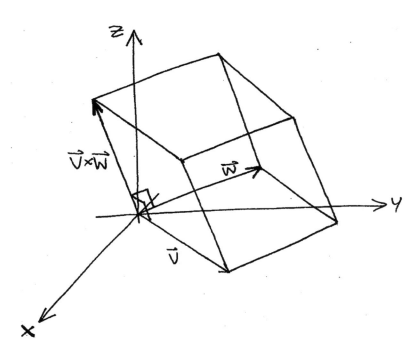

Figure 1.25:

First, we compute this expression by using equation 1.11. We get

$$\begin{vmatrix} - & \vec{v} \times \vec{w} & - \\ - & \vec{v} & - \\ - & \vec{w} & - \end{vmatrix} = (\vec{v} \times \vec{w}) \cdot (\vec{v} \times \vec{w}) = \|\vec{v} \times \vec{w}\|^2 \tag{1.15}$$

Next, we compute the same expression, this time using Theorem 1.4.2. Note that we have already shown above that this determinant must be nonnegative because it is a square, so the theorem then tells us that the determinant is equal to the volume of the parallelepiped $\|(\vec{v} \times \vec{w}, \vec{v}, \vec{w})$.

$$\begin{vmatrix} - & \vec{v} \times \vec{w} & - \\ - & \vec{v} & - \\ - & \vec{w} & - \end{vmatrix} = \text{vol}\Big(\|(\vec{v} \times \vec{w}, \vec{v}, \vec{w})\Big) \tag{1.16}$$

This parallelepiped is represented in Figure 1.25.

We know already that $\vec{v} \times \vec{w}$ is orthogonal to both $\vec{v}$ and $\vec{w}$; so in fact it must also be orthogonal to the entire parallelogram $\|(\vec{v}, \vec{w})$, which of course is one of the faces of the parallelepiped. So we can view this parallelepiped as having as its base the parallelogram $\|(\vec{v}, \vec{w})$, with height represented by the vector $\vec{v} \times \vec{w}$. Simple geometry then tells us that the volume is the area of the base times the height. So we can revise Equation 1.16 to become

$$\begin{vmatrix} - & \vec{v} \times \vec{w} & - \\ - & \vec{v} & - \\ - & \vec{w} & - \end{vmatrix} = \text{vol}\Big(\|(\vec{v} \times \vec{w}, \vec{v}, \vec{w})\Big) = \text{area}\Big(\|(\vec{v}, \vec{w})\Big)\|\vec{v} \times \vec{w}\| \tag{1.17}$$

The results of these two different computations, in equations 1.14 and 1.17, can be combined to yield

$$\|\vec{v} \times \vec{w}\|^2 = \text{area}\Big(\|(\vec{v}, \vec{w})\Big)\|\vec{v} \times \vec{w}\|$$

Canceling one factor of the length of the cross product then yields the desired result. (Unless of course the length is equal to zero, in which case the cancellation is not allowed. However in this case, the cross product itself must be zero which tells us the two vectors must be parallel, and thus that the area is also zero; and thus the result still holds.)

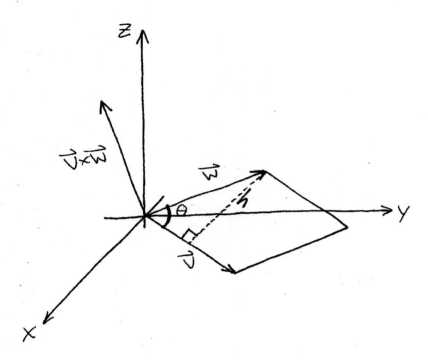

Figure 1.26:

We can write this theorem in a slightly different form by observing that trigonometry gives us another way to compute the area of $\|(\vec{v}, \vec{w})\|$, where as usual we denote the angle between these two vectors by θ, as in Figure 1.26.

We can compute the height h in this figure as $\|\vec{w}\| \sin \theta$ and of course the base is the length of the vector $\vec{v}$. This gives us the following corollary.

Corollary 1.4.1. *For any two vectors $\vec{v}$ and $\vec{w}$ in $\mathbb{R}^3$,*

$$\|\vec{v} \times \vec{w}\| = area\Big(\|(\vec{v}, \vec{w})\|\Big) = \|\vec{v}\|\|\vec{w}\| \sin \theta \tag{1.18}$$

At this point, we have found natural geometric interpretations for both the direction and the length of the cross product. In fact we have a precise geometric description for uniquely determining the length of the cross product; however, even though we have a very good description of the direction of the cross product, we do not yet quite have a geometric description for uniquely determining the direction.

What we know is that the cross product points in a direction that is perpendicular to both original vectors. If these original vectors are not parallel then there is a unique line perpendicular to them both, as indicated in Figure 1.27. But which direction along this line does the cross product point?

The answer to this question comes from the proof of Theorem 1.4.4. Recall (as noted in that proof) that Equation 1.15 shows that the determinant on the left side of that equation is nonnegative. (Of course if this determinant is zero that same equation shows that the cross product itself is zero and our question about direction is moot.)

So the only other possibility is that the determinant is positive. Then Theorem 1.4.2 tells us that the vectors are in right-hand order as listed: $\vec{v} \times \vec{w}$, $\vec{v}$, $\vec{w}$, or equivalently, $\vec{v}$, $\vec{w}$, $\vec{v} \times \vec{w}$.

We restate this as a rule for determining the direction of the cross product. We call this the "right-hand rule". The rule is that with the right hand in the usual position, oriented such that the index finger is pointing in the direction of the first vector $\vec{v}$, and the middle finger is pointing in the direction of the second vector $\vec{w}$, then the cross product $\vec{v} \times \vec{w}$ points in the direction of the thumb.

We can now give a complete description of the cross product in purely geometric terms. In fact this could be taken as an alternative definition for the cross product. The vector $\vec{v} \times \vec{w}$ is the unique vector that:

1.4. CROSS PRODUCT AND DETERMINANT

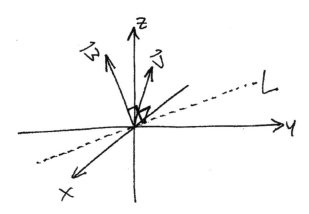

Figure 1.27:

1. is orthogonal to both $\vec{v}$ and $\vec{w}$
2. has length equal to the area of $\|(\vec{v}, \vec{w})\|$
3. satisfies the right-hand rule

Example 1.4.2. Suppose we wish to compute the cross product $\vec{e}_1 \times \vec{e}_2$.

Of course we could compute this from the algebraic definition; instead, here we will draw the correct conclusion directly from the above geometric characterization of the cross product.

The given vectors define the parallelogram $\|(\vec{e}_1, \vec{e}_2)\|$ which is in this case simply the unit square in the xy-plane. The area of this parallelogram is 1, so we know that the length of the cross product vector is 1.

The given vectors are both in the xy-plane, so we know the cross product must point perpendicular to that plane. So it must be parallel to the z-axis. Given that we already know the length is 1, we can already conclude that the cross product must be either $\vec{e}_3$ or $-\vec{e}_3$.

To determine which is the correct answer, we simply apply the right-hand rule. With the hand in the usual position, we point our index finger in the direction of the first vector $\vec{e}_1$, and our middle finger in the direction of the second vector $\vec{e}_2$; the thumb then points in the direction of $\vec{e}_3$, so that is the cross product.

The student can confirm algebraically that this is the correct answer.

Exercises

Exercise 1.4.1 (<u>exer-V-CPD-1</u>). There are six orderings of the three standard basis vectors in $\mathbb{R}^3$. Which of these are in right-hand order, and which are not?

Exercise 1.4.2 (<u>exer-V-CPD-2</u>). Use your right hand to determine by visualization if the vectors $(2, 1, 10)$, $(12, 0, 1)$, $(-1, 11, 0)$ are listed here in right-hand order.

Exercise 1.4.3 (<u>exer-V-CPD-3</u>). Give a geometric explanation of why it must be that if $\vec{a}, \vec{b}, \vec{c}$ are in right-hand order, then $\vec{a}, \vec{b}, -\vec{c}$ must be in left-hand order.

Exercise 1.4.4 (<u>exer-V-CPD-4</u>). Compute the determinants of the following matrices.

1. $\begin{pmatrix} 3 & 4 \\ 5 & -2 \end{pmatrix}$

2. $\begin{pmatrix} 4 & 4 & -1 \\ 7 & -2 & 0 \\ 0 & 12 & 3 \end{pmatrix}$

3. $\begin{pmatrix} 6 & 0 & 0 \\ 0 & 7 & 0 \\ 0 & 0 & 2 \end{pmatrix}$

Exercise 1.4.5 (exer-V-CPD-5). For each of the following lists of vectors, determine (i) the volume of the parallelepiped defined by those vectors, and (ii) if the vectors are listed in right-hand order.
  1. $(4, 4, -1)$, $(7, -2, 0)$, $(0, 12, 3)$
  2. $(4, 7, 0)$, $(4, -2, 12)$, $(-1, 0, 3)$
  3. $(5, 1, 2)$, $(0, 6, -1)$, $(1, -1, 8)$

Exercise 1.4.6 (exer-V-CPD-6). Confirm by direct computation from the definition that $\vec{v} \times \vec{w} = -\vec{w} \times \vec{v}$.

Exercise 1.4.7 (exer-V-CPD-7). Confirm by direct computation from the definition that $\vec{u} \cdot (\vec{v} \times \vec{w}) = \vec{w} \cdot (\vec{u} \times \vec{v}) = \vec{v} \cdot (\vec{w} \times \vec{u})$.

Exercise 1.4.8 (exer-V-CPD-8). Show that if $\vec{v} \| \vec{w}$, then the cross product $\vec{v} \times \vec{w}$ must be zero. (Hint: Write $\vec{w} = k\vec{v}$.)

Exercise 1.4.9 (exer-V-CPD-9). Show that if the cross product $\vec{v} \times \vec{w}$ is zero (with nonzero vectors $\vec{v}$ and $\vec{w}$), then we must have $\vec{v} \| \vec{w}$. (Hint: Each component of the cross product, by being zero, can be interpreted as a statement about the proportions between corresponding coordinates of $\vec{v}$ and $\vec{w}$; assign names to those proportions and then rewrite each of the given vectors as a multiple of the same vector, using those proportions.)

Exercise 1.4.10 (exer-V-CPD-10). Confirm by direct computation from the definitions of dot product and cross product that the cross product $\vec{v} \times \vec{w}$ must be orthogonal to both $\vec{v}$ and $\vec{w}$.

Exercise 1.4.11 (exer-V-CPD-11). Write each of the following cross products as a symbolic determinant and then compute the result.
  1. $(5, 3, 4) \times (-3, -2, 4)$
  2. $(7, 6, 8) \times (2, 4, 5)$
  3. $(1, -2, 4) \times (3, 2, 1)$

Exercise 1.4.12 (exer-V-CPD-12). Compute the areas of the parallelograms defined by each of the following pairs of vectors.
  1. $(5, 3, 4)$ and $(-3, -2, 4)$
  2. $(7, 6, 8)$ and $(2, 4, 5)$
  3. $(1, -2, 4)$ and $(3, 2, 1)$

Exercise 1.4.13 (exer-V-CPD-13). For two vectors $\vec{v}$ and $\vec{w}$, it is known that their dot product is equal to the length of their cross product. What is the angle θ between these two vectors?

Exercise 1.4.14 (exer-V-CPD-14). Find a unit vector $\vec{u}$ that is perpendicular to each of the vectors $\vec{v} = (5, 4, 3)$ and $\vec{w} = (2, 4, 3)$, such that $\vec{u}, \vec{v}, \vec{w}$ are in right-hand order.

Exercise 1.4.15 (exer-V-CPD-16). Given two vectors $\vec{v} = (v_1, v_2)$ and $\vec{w} = (w_1, w_2)$ in $\mathbb{R}^2$, recall that the ordering $\vec{v}, \vec{w}$ is a *counterclockwise ordering* if there is a counterclockwise rotation by an angle less than π that would rotate $\vec{v}$ to point in the same direction as $\vec{w}$. Similarly for a *clockwise ordering*. For example, $\vec{e}_1, \vec{e}_2$ is in counterclockwise order, while $\vec{e}_2, \vec{e}_1$ is in clockwise order.

If $\vec{v}, \vec{w}$ is a counterclockwise ordering, what can you conclude about the ordering in $\mathbb{R}^3$ of the three vectors $\vec{e}_3, (v_1, v_2, 0), (w_1, w_2, 0)$? Give a geometric argument supporting your conclusion. If $\vec{v}, \vec{w}$ is a clockwise ordering, what can you conclude about the ordering in $\mathbb{R}^3$ of the three vectors $\vec{e}_3, (v_1, v_2, 0), (w_1, w_2, 0)$?

Exercise 1.4.16 (exer-V-CPD-17). Use Theorem 1.4.2 and the result of Exercise 1.4.15 to prove the result in Theorem 1.4.1 that shows the relationship between the sign of the determinant of the 2×2 matrix and the orientation of the ordering.

Exercise 1.4.17 (exer-V-CPD-15). Use Theorem 1.4.4 and an argument similar to those in Exercises 1.4.15 and 1.4.16 to prove the result in Theorem 1.4.1 that gives the area of $\|(\vec{v}, \vec{w})\|$.

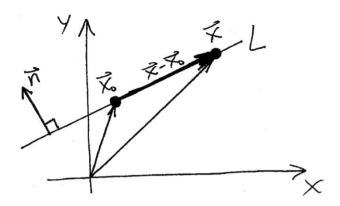

Figure 1.28:

1.5 Lines and Planes

The dot product and the cross product allow us now to have a discussion of lines and planes, and how to represent them algebraically.

1.5.1 Lines in $\mathbb{R}^2$

We begin with lines in $\mathbb{R}^2$. Of course students already know how to represent a line in the xy-plane algebraically – recall that we can write $y = mx + k$, where m is the slope of the line and k is the y-intercept. This is a useful form for the equation of a line, because the constants m and k have these geometric interpretations.

Students have probably also seen the form $ax + by = c$ for the equation of a line; very likely, however, they have probably not seen the geometric interpretation relevant to this form of the equation of a line. We will begin with that here.

Suppose we are given a line L in the xy-plane, but we only know the following two pieces of information about it – the vector $\vec{n} = (a, b)$ is perpendicular to L (we call this a "normal vector"), and the point $\vec{x}_0 = (x_0, y_0)$ is in the line L. From this information, how can we write down an equation representing this line?

That is, we would like to find an equation in the variables x and y such that the equation holds when the point $\vec{x} = (x, y)$ is on L, but does not hold when the point $\vec{x} = (x, y)$ is not on L.

We start with a geometric characterization, namely that the point $\vec{x}$ is on L if and only if the vector $\vec{x} - \vec{x}_0$ is parallel to the line L. This can be seen in Figure 1.28.

Of course the given vector $\vec{n}$ is perpendicular to L, so we can restate this condition by saying that $\vec{x}$ is on L if and only if the vector $\vec{x} - \vec{x}_0$ is perpendicular to the vector $\vec{n}$. So we have

$$\vec{x} \text{ is on } L \iff (\vec{x} - \vec{x}_0) \| L$$
$$\iff (\vec{x} - \vec{x}_0) \perp \vec{n}$$

Conveniently, this is something we can express algebraically with the dot product.

$$\vec{x} \text{ is on } L \iff (\vec{x} - \vec{x}_0) \cdot \vec{n} = 0$$
$$\iff \vec{n} \cdot \vec{x} - \vec{n} \cdot \vec{x}_0 = 0$$
$$\iff \vec{n} \cdot \vec{x} = \vec{n} \cdot \vec{x}_0$$

Writing these out in coordinates and representing the constant $\vec{n} \cdot \vec{x}_0$ by c, this becomes

$$\vec{x} \text{ is on } L \iff ax + by = c \tag{1.19}$$

This derivation then provides us with the desired geometric interpretation. That is, a line with equation $ax + by = c$ is perpendicular to the vector $\vec{n}$ whose coordinates are the coefficients on the left side of the equation, and c is the value of $\vec{n} \cdot \vec{x}$ for all points $\vec{x}$ on the line.

Example 1.5.1. Suppose we wish to find a vector perpendicular to the line with equation $3x - 2y = 5$.

Using the above interpretation, we can immediately say that this line is perpendicular to the vector $\vec{n} = (3, -2)$.

Of course the given equation is equivalent to many other equations, just by multiplying through on both sides by some constant. So the above normal vector is not unique. Of course the normal vector determined from any of these other equivalent of equations would simply be a scalar multiple of the one written down above.

1.5.2 Planes in $\mathbb{R}^3$

By a very similar argument we can derive the equation for a plane in $\mathbb{R}^3$.

Suppose we have a plane P, and all we are given is a normal vector $\vec{n} = (a, b, c)$ perpendicular to the plane, and a point $\vec{x}_0 = (x_0, y_0, z_0)$ in the plane. We would like to find an equation in the variables x, y and z which holds when the point $\vec{x} = (x, y, z)$ is on the plane, and does not hold when the point $\vec{x}$ is not on the plane.

As represented in Figure 1.29, the point $\vec{x}$ is in the plane P if and only if the vector $\vec{x} - \vec{x}_0$ is parallel to the plane P. Again we can state this in terms of a dot product and simplify.

$$\begin{aligned}
\vec{x} \text{ is on } P &\iff (\vec{x} - \vec{x}_0) \| P \\
&\iff (\vec{x} - \vec{x}_0) \perp \vec{n} \\
&\iff (\vec{x} - \vec{x}_0) \cdot \vec{n} = 0 \\
&\iff \vec{n} \cdot \vec{x} - \vec{n} \cdot \vec{x}_0 = 0 \\
&\iff \vec{n} \cdot \vec{x} = \vec{n} \cdot \vec{x}_0
\end{aligned}$$

And again we can write this and coordinates as

$$ax + by + cz = d \tag{1.20}$$

where d represents $\vec{n} \cdot \vec{x}_0$.

Note that we get an equation very similar in form to the previous equation for a line. Also, the coefficients on the left side of the equation have the same interpretation, namely, as the coordinates of a normal vector for the given plane.

While the coordinate form is more familiar, it is very elegant and more directly meaningful to write both the equation for a line and the equation for a plane in the dot product form

$$\vec{n} \cdot \vec{x} = \vec{n} \cdot \vec{x}_0 \tag{1.21}$$

This is the natural starting point for many problems involving lines and planes.

Example 1.5.2. What is the equation of the unique plane passing through the point $(1, 5, 2)$ and perpendicular to the vector $(-2, 1, -4)$?

We can use the above form of the equation for a plane, with $\vec{x}_0 = (1, 5, 2)$ and $\vec{n} = (-2, 1, -4)$. This gives us

$$-2x + y - 4z = (-2, 1, -4) \cdot (1, 5, 2) = -5$$

1.5. LINES AND PLANES

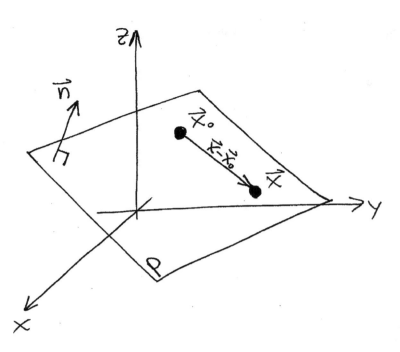

Figure 1.29:

Example 1.5.3. Suppose we wish to find the equation of the unique plane that passes through the three points $\vec{a} = (1, 2, 3)$, $\vec{b} = (2, 8, 5)$, and $\vec{c} = (1, 0, 1)$.

In order to use Equation 1.21, we can use any of the three given points as $\vec{x}_0$. However, we must also find an appropriate normal vector.

In order to find such a normal vector, noticed that without too much trouble we can find two vectors that are parallel to the plane – for example, $\vec{b} - \vec{a}$ and $\vec{c} - \vec{a}$. These vectors are parallel to the plane because each has its head and its tail in the plane.

Since these vectors are both parallel to the plane, their cross product must be perpendicular to the plane, as represented in Figure 1.30. So we can compute an appropriate normal vector by

$$\vec{n} = (\vec{b} - \vec{a}) \times (\vec{c} - \vec{a}) = (1, 6, 2) \times (0, -2, -2) = (-8, 2, -2)$$

We can then use Equation 1.21 to write the equation of the plane as

$$-8x + 2y - 2z = -10$$

1.5.3 Lines in $\mathbb{R}^3$

We must take a different approach when trying to use algebra to represent a line in $\mathbb{R}^3$.

In fact, we will not represent a line in $\mathbb{R}^3$ with an equation. Instead, we will provide a more direct description of the line, by simply enumerating all of the points on that line.

Suppose we have a line L, and we are given a point $\vec{x}_0 = (x_0, y_0, z_0)$ on the line and a vector $\vec{v} = (a, b, c)$ parallel to the line (we call this a "direction vector" for the line).

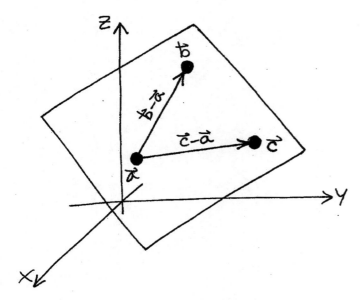

Figure 1.30:

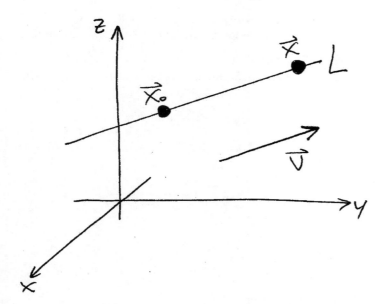

Figure 1.31:

1.5. LINES AND PLANES

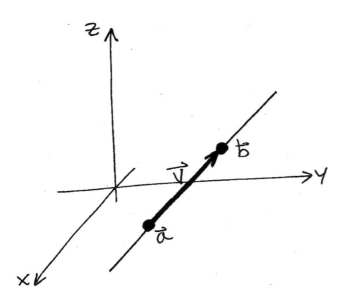

Figure 1.32:

Geometrically, we can see from Figure 1.31 that another point $\vec{x} = (x, y, z)$ is on the line if and only if $\vec{x} - \vec{x}_0$ is parallel to the direction vector $\vec{v}$. We can write this as

$$\begin{aligned} \vec{x} \text{ is on } L &\iff (\vec{x} - \vec{x}_0) \| L \\ &\iff (\vec{x} - \vec{x}_0) \| \vec{v} \\ &\iff (\vec{x} - \vec{x}_0) = t\vec{v} \text{ for some t} \\ &\iff \vec{x} = \vec{x}_0 + t\vec{v} \text{ for some t} \end{aligned}$$

Since every point on the line can be represented in the above form, we can write a line simply as the collection of all such vectors.

$$L = \{\vec{x}_0 + t\vec{v}\} \qquad (1.22)$$

We call this a "parametrization" of the line L. (This is just one example of a parametrization; we will talk much more about parametrizations in general in Section 2.5.) Note, we can use this representation just as well for a line in $\mathbb{R}^2$.

Example 1.5.4. Suppose we wish to find a parametrization of the line that passes through the points $\vec{a} = (4, 3, 1)$ and $\vec{b} = (7, 3, 4)$.

In order to use Equation 1.22, either of the given points could be chosen as $\vec{x}_0$. As is represented in Figure 1.32, we can choose $\vec{v} = \vec{b} - \vec{a}$ as the direction vector.

So we represent the line parametrically by

$$(x, y, z) = (4, 3, 1) + t(3, 0, 3)$$

Note that representing a line parametrically has the advantage that we have a complete enumerated list of all the points on the line. This can be very advantageous, such as when we need to find a point on the line that has a particular property.

Example 1.5.5. Suppose we wish to find a point on the line from Example 1.5.4 that is also on the plane with equation $x + y + 2z = 0$.

Since the parametrization for the line is a list of all the points on the line, and the given equation for the plane is a condition for testing when a point is on that plane, we can just plug in the parametric characterization of the points on the line into the equation for the plane. We get

$$(4 + 3t) + (3 + 0t) + 2(1 + 3t) = 0$$
$$9 + 9t = 0$$
$$t = -1$$

So the point on the line with $t = -1$ is the only point on the line that satisfies the equation for the plane, and thus it is the only point on the line that is also on the plane. Plugging that value of t back into the parametrization, we find that the point in question is $(1, 3, -2)$.

Sometimes however, equations can be more advantageous than a parametrization. We can convert a parametric representation of a line into equations, but we will end up in fact with two equations instead of just one.

If we write out the parametrization in equation 1.22 in coordinates, we get

$$x = x_0 + ta$$
$$y = y_0 + tb$$
$$z = z_0 + tc$$

This says that a point (x, y, z) is on the line if there is a value t for which the above equations produce those coordinates. Of course it must be the same value t in all three of the above equations.

So if we were to solve for t in all three of the above equations, all of the results must be equal. This gives us the following conditions that must be satisfied for the point to be on the line

$$\frac{x - x_0}{a} = \frac{y - y_0}{b} = \frac{z - z_0}{c} \tag{1.23}$$

If for a given point (x, y, z) the above three values are all equal, then we can call that value t and realize the given point from the above parametric equations; so the point is on the line. If they are not all three equal, then there is no single value t that will realize the given point from the above parametric equations; so the point is not on the line.

Of course this is two equations, not just a single equation. These are called the "symmetric equations" for the line.

Example 1.5.6. Is the point $(1, 4, 9)$ on the line represented parametrically by $(x, y, z) = (3, 2, 5) + t(-1, 1, 2)$?

The symmetric equations for this line are

$$\frac{x - 3}{-1} = \frac{y - 2}{1} = \frac{z - 5}{2}$$

Plugging in the given point into these equations, we get all three of the above values equal to 2. So both of the equations above are satisfied, and thus this point is on the line.

The symmetric equations of a line can be interpreted in a very nice geometric way. First, notice that they can be interpreted as three separate equations.

$$\frac{x - x_0}{a} = \frac{y - y_0}{b} \quad \text{and} \quad \frac{x - x_0}{a} = \frac{z - z_0}{c} \quad \text{and} \quad \frac{y - y_0}{b} = \frac{z - z_0}{c} \tag{1.24}$$

Each of these can be thought of as the equation of a plane. Furthermore, because each of these involves only two of the three variables, each of these can be interpreted as the equation of a line in the corresponding coordinate plane. For example, the first equation above is the equation of a line in the xy-plane.

1.5. LINES AND PLANES

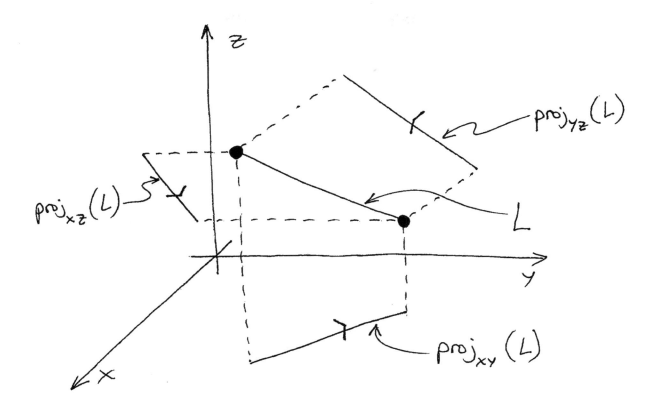

Figure 1.33:

Interpreted like this, each of these equations is in fact the equation of the perpendicular projection of the line L into the corresponding coordinate plane. For example, the equation

$$\frac{x - x_0}{a} = \frac{y - y_0}{b} \tag{1.25}$$

is the equation of the line $\text{proj}_{xy}(L)$ in the xy-plane, as shown in Figure 1.33. We can see this because this equation is the only part of the symmetric equations for L that relates the x and y variables, and the projection to the xy-plane shows exactly the x and y coordinates for points on L. Similarly for the projections into the other two coordinate planes.

And, thought of again as planes in $\mathbb{R}^3$, these three equations are the equations of the three planes through L that are also perpendicular to one of the three coordinate planes. These also are shown in Figure 1.33.

Exercises

Exercise 1.5.1 (exer-V-LP-1). Find two unit vectors that are perpendicular to the line in the xy-plane with equation $3x - 4y = 17$.

Exercise 1.5.2 (exer-V-LP-2). Find an equation (with all integer coefficients) for the line in the xy-plane that passes through the points $(5, 3)$ and $(2, 8)$.

Exercise 1.5.3 (exer-V-LP-3). Find an equation (with all integer coefficients) for the line in the xy-plane that passes through the point $(5, 6)$ and is perpendicular to the vector $(3, 2)$.

Exercise 1.5.4 (exer-V-LP-4). Find two unit vectors that are perpendicular to the plane with equation $6x - 2y + 3z = 17$.

Exercise 1.5.5 (**exer-V-LP-5**). Find an equation (with all integer coefficients) for the plane that passes through the points $(0, 2, 3)$, $(6, 3, 1)$ and $(2, 7, 8)$.

Exercise 1.5.6 (**exer-V-LP-6**). Find an equation (with all integer coefficients) for the plane that passes through the point $(7, 3, 3)$ and is perpendicular to the vector $(2, 6, 4)$.

Exercise 1.5.7 (**exer-V-LP-7**). Find an equation (with all integer coefficients) for the plane that is parallel to the vectors $(4, 2, 6)$ and $(3, -3, 2)$, and passes through the point $(1, 1, 2)$.

Exercise 1.5.8 (**exer-V-LP-8**). Find the parametrization of the line that goes through the points $(5, 3, 6)$ and $(7, 7, 3)$.

Exercise 1.5.9 (**exer-V-LP-9**). Find the parametrization of the line that is parallel to the planes with equations $2x - 4y + z = 10$ and $5x + 6y - z = 1$, and passes through the point $(1, 1, 2)$.

Exercise 1.5.10 (**exer-V-LP-10**). Find the intersection of the plane with equation $3x + 6y - 7z = 12$ and the line represented parametrically by $(8 + t, 9 - t, 3t)$.

Exercise 1.5.11 (**exer-V-LP-11**). Determine if there is an intersection point of the lines represented parametrically by $(3 - 2t, 2 + t, 3 + t)$ and $(3 - s, 1 + s, 8 - 2s)$, and find that point if it exists.

Exercise 1.5.12 (**exer-V-LP-12**). Find an equation (with all integer coefficients) for the plane that is parallel to and equidistant from the lines represented parametrically by $(3 - 3t, 5 - t, -3 + t)$ and $(2 + 4t, 7 + 3t, 12 - t)$. (*Hint: Pick any point on each of these lines – the midpoint between them must be on the desired plane, and you can show this by similar triangles.*)

Exercise 1.5.13 (**exer-V-LP-13**). Find the symmetric equations for the line represented parametrically by $(5 - t, 6 - 3t, 7 + 2t)$.

Exercise 1.5.14 (**exer-V-LP-14**). Find a parametrization of the line with symmetric equations $x + 5 = y - 6 = 3z + 1$.

Exercise 1.5.15 (**exer-V-LP-15**). Find an equation (with all integer coefficients) for the plane that is parallel to and equidistant from the lines represented by the symmetric equations $3x - 7 = 2y + 6 = 6z + 1$ and $x - 5 = 3y + 1 = 3z + 6$.

Exercise 1.5.16 (**exer-V-LP-16**). Suppose that a plane P contains the point $\vec{x}_0$ and has normal vector $\vec{n}$, and let $\vec{x}$ be some point, not necessarily in P. Use a calculation of the $\vec{n}$ component of $\vec{x} - \vec{x}_0$ to derive a formula for the perpendicular distance from $\vec{x}$ to P in terms of $\vec{x}$, $\vec{x}_0$, and $\vec{n}$.

Exercise 1.5.17 (**exer-V-LP-17**). Confirm that your equation from Exercise 1.5.16 correctly computes that every point on the plane P has a perpendicular distance of zero to P.

Exercise 1.5.18 (**exer-V-LP-18**). Use the result of Exercise 1.5.16 to compute the distance from the point $(5, 7, 1)$ to the plane with equation $2x - 3y + 6z = 9$.

Exercise 1.5.19 (**exer-V-LP-19**). Use an argument similar to that in Exercise 1.5.16 to derive a formula for the perpendicular distance from a point $\vec{x}$ to a line L in the xy-plane that contains the point $\vec{x}_0$ and has normal vector $\vec{n}$.

Exercise 1.5.20 (**exer-V-LP-20**). Consider the parametric line L in $\mathbb{R}^3$ represented by $\vec{x}_0 + t\vec{v}$, and a point $\vec{x}$ that is not on L.

1. Let $\vec{p}$ be the point on L that is closest to $\vec{x}$. Show that $\vec{n}_1 = \vec{v} \times (\vec{x} - \vec{x}_0)$ is perpendicular to both $\vec{v}$ and $\vec{x} - \vec{p}$. (*Hint: Write $\vec{x} - \vec{p} = \vec{x} - (\vec{x}_0 + t\vec{v}) = (\vec{x} - \vec{x}_0) + t\vec{v}$.*)

2. Explain how you know that $\vec{x} - \vec{p}$ is perpendicular to both $\vec{v}$ and $\vec{n}_1$, and thus parallel to $\vec{n}_2 = \vec{v} \times \vec{n}_1$.

3. Use a calculation of the $\vec{n}_2$ component of $\vec{x} - \vec{x}_0$ to derive a formula for the perpendicular distance from $\vec{x}$ to L.

Exercise 1.5.21 (**exer-V-LP-21**). Consider the parametric line L in $\mathbb{R}^3$ represented by $\vec{x}_0 + t\vec{v}$. Recall that it has previously been argued that $\vec{x}$ is on the line if and only if $(\vec{x} - \vec{x}_0) \| \vec{v}$. Use this and the results of Exercises 1.4.8 and 1.4.9 to derive a single quadratic equation whose solution set is exactly this line.

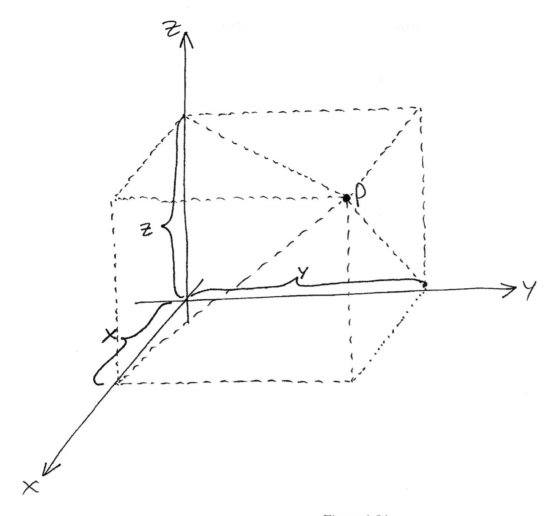

Figure 1.34:

1.6 Coordinate Systems

When we talk about points in $\mathbb{R}^2$ or $\mathbb{R}^3$, we refer to a specific point with a set of coordinates . In $\mathbb{R}^2$ we need two coordinates, and in $\mathbb{R}^3$ we need three coordinates. These coordinates are defined by a set of conventions, defining a "coordinate system".

Most of the time we use the most familiar and often most convenient coordinate system, called the "Cartesian" coordinate system, or sometimes the "rectangular" coordinate system. Under this system, in either $\mathbb{R}^2$ or $\mathbb{R}^3$, the coordinates are defined by the projections of the position vector of the point to specific vectors called the "standard basis vectors". These are unit vectors, all perpendicular to each other; and given these vectors and the definition above, they define the "coordinate axes". See Figure 1.34.

The rectangular coordinate system has some very attractive features that cause us to use it most of the time. For example: every point has a unique set of coordinates; knowing the coordinates gives us a very simple representation of a point as a linear combination of the standard basis vectors; the dot product gives us a simple relationship between rectangular coordinates and angles; ...etc., and there are many more.

But, there are other sets of conventions for defining coordinates, i.e., other coordinate systems. Often, if not most of the time, the rectangular system is the most useful – but in some situations these other systems can be more useful.

We give here a brief summary of three other coordinate systems.

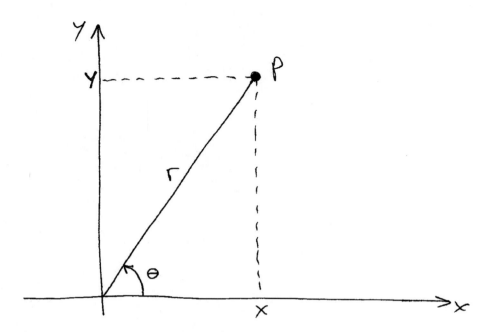

Figure 1.35:

1.6.1 Polar coordinates

The polar coordinate system defines points in $\mathbb{R}^2$ by the following two coordinates. For a point P in $\mathbb{R}^2$:

1. The coordinate "r" is the distance from the origin to P.

2. The coordinate "θ" is the angle, starting from the positive part of the x-axis and moving counterclockwise, to the position vector for P.

This is indicated in Figure 1.35.

Note that because this coordinate system is defined very differently than the rectangular system, it has very different properties. For one thing, there is no "r-axis", nor is there a "θ-axis". Also, points do not have unique sets of coordinates – for example, notice that the angles $\theta = 0$ and $\theta = 2\pi$ represent the same ray from the origin. In fact then one can see that any given point in $\mathbb{R}^2$ has an infinite number of representations in polar coordinates. (Note that we also allow r to be negative, indicating a point a distance $|r|$ from the origin in the direction exactly opposite that indicated by the given angle θ (as if there were an "r-axis"); this gives another infinite collection of representations of any point in polar coordinates.)

Very often, points represented in polar coordinates are written down in a way that is very similar to the way we write down points in rectangular coordinates. Namely, it is common to see "(r, θ)" to represent a point a distance r from the origin in the direction θ.

The problem with this notation is that it has the potential to be confusing. While in some circumstances it may be clear which coordinate system is being used to describe the point in question, in some cases it certainly is not. For example, "(x, y)" would appear to be in rectangular coordinates and "(r, θ)" would appear to be in polar coordinates, but what about "$(1, \pi)$"? The second coordinate is a number which very often refers to an angle, but certainly this is no guarantee; the point $(x, y) = (1, \pi)$ is a point in the plane which might at some point need reference.

To get around this confusion, in any case where there is an ambiguity we will use a subscript to indicate which coordinate system the given ordered pair is using. In particular, we will use the subscript "r" when we are using rectangular coordinates, and the subscript "p" when we are using polar coordinates.

1.6. COORDINATE SYSTEMS

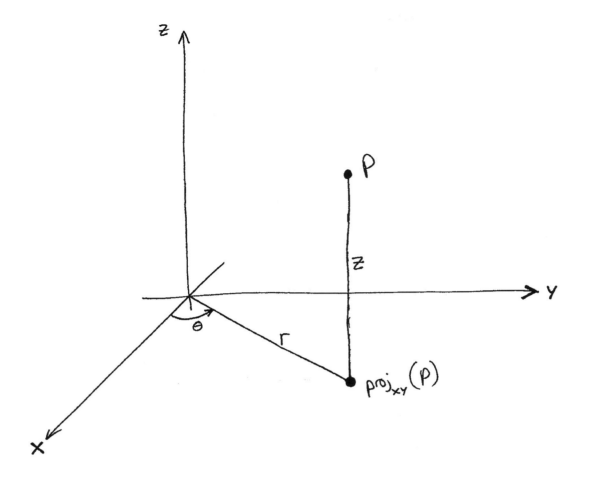

Figure 1.36:

For example, we can now very clearly and conveniently write $P = (x,y)_r = (r,\theta)_p$, indicating that the point P has rectangular coordinates x and y, and polar coordinates r and θ.

There are relationships between the rectangular and polar coordinates. For example, from Figure 1.35, we can see that $x = r\cos\theta$, $y = r\sin\theta$, $r = \sqrt{x^2 + y^2}$, and $\tan\theta = \frac{y}{x}$. So, using the subscripts to denote coordinate systems, we can write

$$P = (x,y)_r = (r\cos\theta, r\sin\theta)_r = (r,\theta)_p \tag{1.26}$$

1.6.2 Cylindrical coordinates

The cylindrical coordinate system defines points in $\mathbb{R}^3$, by using three coordinates. For a point P in $\mathbb{R}^3$:

1. The coordinate "r" is the polar r-coordinate of the projection of P into the xy-plane.

2. The coordinate "θ" is the polar θ-coordinate of the projection of P into the xy-plane.

3. The coordinate "z" is the rectangular z-coordinate of P.

This coordinate system then is sort of like an extension of the polar coordinate system into $\mathbb{R}^3$. See Figure 1.36.

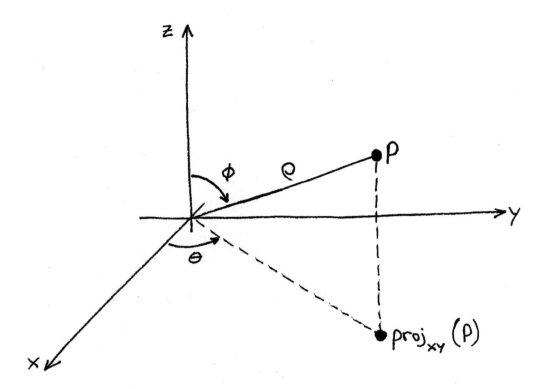

Figure 1.37:

We will represent points in this coordinate system by an ordered triple with a subscript "c" to distinguish it from the rectangular coordinate system. Using this notation, and reasoning from Figure 1.36, we have

$$P = (x, y, z)_r = (r\cos\theta, r\sin\theta, z)_r = (r, \theta, z)_c \tag{1.27}$$

1.6.3 Spherical coordinates

The spherical coordinate system defines points in $\mathbb{R}^3$, by using three coordinates. For a point P in $\mathbb{R}^3$:

1. The coordinate "ρ" is the distance from P to the origin.
2. The coordinate "ϕ" is the angle from the positive part of the z-axis to the position vector for P.
3. The coordinate "θ" is the polar θ-coordinate of the projection of P into the xy-plane.

See Figure 1.37.

Again, from this figure we can see that there are relationships between the spherical coordinates and the rectangular coordinates. First noting that $r = \rho\sin\phi$ and $z = \rho\cos\phi$, and using a subscript of "s" to indicate spherical coordinates, we can write

$$P = (x, y, z)_r = (\rho\sin\phi\cos\theta, \rho\sin\phi\sin\theta, \rho\cos\phi)_r = (\rho, \phi, \theta)_s \tag{1.28}$$

It is also convenient sometimes to note that $\rho^2 = x^2 + y^2 + z^2$.

1.6.4 Equations in other coordinate systems

These different coordinate systems sometimes allow us to write more convenient equations to represent certain types of curves and surfaces.

1.6. COORDINATE SYSTEMS

For example, in polar coordinates, a circle centered at the origin with radius a has the very convenient equation $r = a$, and a line through the origin with angle c from the x-axis has the very convenient equation $\theta = c$. Similarly $r = a$ in cylindrical coordinates defines a cylinder with the z-axis as its axis of symmetry, and $\theta = c$ defines a plane containing the z-axis. In spherical coordinates, $\rho = a$ defines a sphere centered at the origin with radius a.

If a curve or surface in $\mathbb{R}^2$ or $\mathbb{R}^3$ is not one of these listed above, we can still convert the rectangular equation to a different coordinate system simply by using the equations relating those variables.

Example 1.6.1. In polar coordinates, what is the equation for a circle of radius a with center at the point $(a, 0)$?

In rectangular coordinates, of course we know that the equation is $(x-a)^2 + y^2 = a^2$. Since we know that $x = r\cos\theta$ and $y = r\sin\theta$, we can then convert this to polar coordinates simply by plugging in. We get

$$\begin{aligned}
(x-a)^2 + y^2 &= a^2 \\
x^2 - 2ax + a^2 + y^2 &= a^2 \\
x^2 - 2ax + y^2 &= 0 \\
(r\cos\theta)^2 - 2a(r\cos\theta) + (r\sin\theta)^2 &= 0 \\
(r\cos\theta)^2 + (r\sin\theta)^2 &= 2a(r\cos\theta) \\
r &= 2a\cos\theta
\end{aligned}$$

Alternatively we could have noticed that $x^2 + y^2$ was going to come up as part of the calculation of the left side of the initial equation, and used the known relation $x^2 + y^2 = r^2$; the calculation then would have become simply

$$\begin{aligned}
(x-a)^2 + y^2 &= a^2 \\
x^2 - 2ax + a^2 + y^2 &= a^2 \\
r^2 - 2a(r\cos\theta) &= 0 \\
r &= 2a\cos\theta
\end{aligned}$$

Very similarly to Example 1.6.1, it can be shown that a circle of radius a with center at the point $(0, a)$ has polar equation $r = 2a\sin\theta$.

Exercises

Exercise 1.6.1 (exer-V-CS-1). For each of the following points given in polar coordinates, write down the unique rectangular coordinates.
1. $(1, 0)_p$

2. $(1, 2\pi)_p$

3. $(3, \pi)_p$

4. $(-3, \pi/4)_p$

5. $(-2, 7\pi/6)_p$

Exercise 1.6.2 (exer-V-CS-2). For each of the following points given in rectangular coordinates, indicate ALL of the ways to represent the point in polar coordinates.
1. $(1, 0)_r$

2. $(2, 2)_r$

3. $(\sqrt{3}, -1)_r$

4. $(-4\sqrt{2}, 4\sqrt{2})_r$

5. $(5\cos(3\pi/4), 5\sin(3\pi/4))_r$

6. $(5\cos(3\pi/4), 5\sin(-3\pi/4))_r$

Exercise 1.6.3 (exer-V-CS-3). For each of the following points given in cylindrical coordinates, write down the unique rectangular coordinates.

1. $(1, 0, 5)_c$

2. $(1, 2\pi, 4)_c$

3. $(3, \pi, 6)_c$

4. $(-3, \pi/4, 3)_c$

5. $(-2, 7\pi/6, 2)_c$

Exercise 1.6.4 (exer-V-CS-4). For each of the following points given in rectangular coordinates, indicate ALL of the ways to represent the point in cylindrical coordinates.

1. $(1, 0, 5)_r$

2. $(2, 2, 3)_r$

3. $(\sqrt{3}, -1, 2)_r$

4. $(-4\sqrt{2}, 4\sqrt{2}, 3)_r$

5. $(5\cos(3\pi/4), 5\sin(3\pi/4), -1)_r$

6. $(5\cos(3\pi/4), 5\sin(-3\pi/4), -1)_r$

Exercise 1.6.5 (exer-V-CS-5). For each of the following points given in spherical coordinates, write down the unique rectangular coordinates.

1. $(1, 0, 5)_s$

2. $(1, 2\pi, 4)_s$

3. $(3, \pi, 6)_s$

4. $(-3, \pi/4, 3)_s$

5. $(-2, 7\pi/6, 2)_s$

Exercise 1.6.6 (exer-V-CS-6). For each of the following points given in rectangular coordinates, indicate ALL of the ways to represent the point in spherical coordinates.

1. $(0, 0, 5)_r$

2. $(3, 0, 0)_r$

3. $(\sqrt{2}, \sqrt{2}, 2\sqrt{3})_r$

Exercise 1.6.7 (exer-V-CS-7). Derive and simplify the polar equation for the circle of radius a with center at $(0, a)_r$.

Exercise 1.6.8 (exer-V-CS-8). Derive and simplify the cylindrical equation for the paraboloid with rectangular equation $4x^2 + 4y^2 + z = 6$.

Exercise 1.6.9 (exer-V-CS-9). Derive and simplify the cylindrical equation for the sphere of radius a centered at the origin.

Exercise 1.6.10 (exer-V-CS-10). Derive the spherical equation for the sphere of radius a centered at $(0, 0, a)_r$.

Exercise 1.6.11 (exer-V-CS-11). Derive the spherical equation for the sphere of radius a centered at $(0, a, 0)_r$.

Chapter 2

Multivariable Functions

2.1 Functions

One of the fundamental objects of study in mathematics is the "function". A function can be defined formally using set theory; however, for our purposes here we can think of a function as something that takes an input and then gives an output. The input is taken from a set called the "domain", and the output must fall in a set called the "target".

Students at this point in their mathematics educations have seen many functions. Here is a simple example:

$$f(x) = 3x + 5$$

The input to this function "f" can be any real number, so we can say that the domain is the entire real line; the formula for computing the output of this function always yields a real number, so we can say that the target space is also the real line. We denote these choices symbolically by writing

$$f : \mathbb{R}^1 \to \mathbb{R}^1$$

where the first $\mathbb{R}^1$ represents the domain, and the second $\mathbb{R}^1$ represents the target.

Of course we could have made different choices for the domain and target; the only constraint we have is that for every value in the domain the corresponding output value must fall in the target. For example, we could choose the domain to be a subset of the real number line, and the target to be another subset of the real number line, as in

$$f : [0,1] \to [0,10], \quad f(x) = 3x + 5$$

Here the domain is the interval $[0,1]$, and the target is the interval $[0,10]$. Note for every value of x in the domain, the corresponding value $f(x)$ does indeed fall in the target space.

You will also notice that in this case the target space was chosen unnecessarily large. We could have chosen the target space to be any interval that contains all of the output values from the domain. In this case, the complete set of output values of this function from the chosen domain is the interval $[5,8]$. We call this the "image" or "range" of the function f. The target space can be any set containing the image.

Almost every function that students at this point in their mathematics educations have seen is a single variable function; that is, a function in which the input is represented by a single variable, and the output is also represented by a single variable. The function above is an example of a single variable function.

However, there are many functions that are not single variable functions. A function might have several input variables, or several output variables, or even several of each. For example consider the function f defined by the equation

$$f(x, y, z) = (x + y, 3xz - yz)$$

This function has three input variables, and two output variables.

Of course the three input variables could be viewed as the coordinates of a point in three-dimensional space. Similarly, the two output variables could be viewed as the coordinates of a point in two-dimensional space.

So instead of saying that this function takes three input variables and gives two output variables, an alternative point of view would be to say that this function takes as its input a single point in three-dimensional space, and gives as its output a single point in two-dimensional space. We represent that symbolically with

$$f : \mathbb{R}^3 \to \mathbb{R}^2$$

So we can say that the domain of this function is $\mathbb{R}^3$, and the target space for this function is $\mathbb{R}^2$.

2.1.1 Examples

Here we include a few examples of multivariable functions, with different interpretations.

Example 2.1.1. Let $D \subset \mathbb{R}^3$ be a rectangular room, where x, y and z are the coordinates measured in inches from a fixed corner of the room. We can define the function $T : D \to \mathbb{R}^1$ as follows – given a point $\vec{x} \in D$, let $T(\vec{x})$ be the temperature in degrees Kelvin at that point.

Example 2.1.2. Let $R \subset \mathbb{R}^2$ be the region of the uv-plane corresponding to the land surface of the United States, with u and v representing latitude and longitude respectively. We can define the function $h : R \to \mathbb{R}^1$ by letting $h(u,v)$ be the altitude above sea level of the land at those coordinates.

Example 2.1.3. Suppose you take a road trip, starting at time a and ending at time b. We can define the function $\vec{x} : [a,b] \to \mathbb{R}^2$ by letting $\vec{x}(t)$ be the latitude and longitude of your location at time t on your trip.

Example 2.1.4. Again let R be the region of the uv-plane corresponding to the land surface of the United States, with u and v representing latitude and longitude respectively. We can define the function $f : R \to \mathbb{R}^4$ by letting $f(u,v) = (T, p, m, w)$, where T is the temperature, p is the barometric pressure, m is the humidity, and w is the wind speed, all measured at the point $(u,v) \in R$.

Note that the output of this function is a vector, as was the function in Example 2.1.3. In that former example we used an arrow over the name of the function, while in this example we did not.

This is an optional notation, left to the discretion of the user. It has the advantage of emphasizing that the output from the function is to be thought of as a point or a vector. In Example 2.1.3 that was indeed how we wished to view the output of that function. In this example however, we did not need to take that point of view and so we did not use that notation.

2.1.2 Pictures

When we are studying single variable functions, there are pictures that we draw to help us visualize what the function is doing. In the case of single variable functions the overwhelming majority of these pictures are a particular type of picture

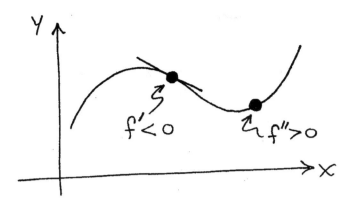

Figure 2.1:

called a "graph". The graph construction is extremely useful for studying single variable functions, allowing us to see the domain, the target, and the image simultaneously, and also drawing a curve that can be naturally associated to the function. Some of the geometric features of this curve can be directly associated to algebraic properties of the function.

For example, the direction in which the curve is sloping is a direct commentary on the value of the derivative of that function; and the direction in which the function is curving is associated with the value of the second derivative of the function. See Figure 2.1.

When one is studying single variable functions these and other useful properties of the graph construction make the graph the most useful geometric picture that one can associate to a single variable function.

While we can and will create an analogous graph construction for multivariable functions in some cases, it will turn out to be the case sometimes that we simply cannot. In those instances, there are other ways of associating geometric pictures to functions and other algebraic things.

It is important to note carefully that these other constructions are NOT graphs; therefore the associations between geometric features and algebraic properties that we are accustomed to in the graph construction are not valid. However, we will be able to make geometric/algebraic associations for these constructions also, appropriate to how they are defined.

We will discuss several of these types of constructions later in this chapter.

2.1.3 Direct geometric interpretation

Before we go on, however, we will present one geometric representation of a multivariable function. This picture can be thought of simply as a direct geometric representation of our point of view that a function takes as its input an element of the domain and returns an element of the target space.

In fact, we simply draw the domain on the left side of the picture, the target space on the right side of the picture, and figuratively represent the function with a curved arrow, representing that it takes a point in the domain and returns a point in the target. We think of points in the domain as being picked up by the function and moved over to points in the target as designated by the function.

See Figure 2.2, showing a function $f : \mathbb{R}^3 \to \mathbb{R}^2$, with the point $\vec{x}$ being taken as an input, and the corresponding output $f(\vec{x})$ in the target.

This picture motivates several pieces of terminology, all meaning that the function f takes the point $\vec{x}$ as an input and returns an output value $f(\vec{x})$. For example, we can say "f takes $\vec{x}$ to $f(\vec{x})$", or "$\vec{x}$ goes to $f(\vec{x})$ by the function f", or "the image of $\vec{x}$ is $f(\vec{x})$".

This geometric representation of a function has several advantages. Most obviously, it is convenient that this is a literal representation of the perspective we are presenting in this text on what a function is most fundamentally. The domain

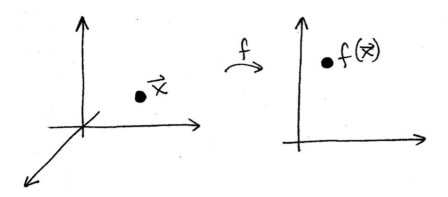

Figure 2.2:

and the target are both represented in this picture, and they are represented as separate spaces, which of course they are. We will also find in Chapter 4 that it will be extremely convenient to see simultaneously both the input and output points themselves, and their immediate surroundings in the domain and target.

A serious disadvantage is that with this picture it is difficult to represent the values of the function for more than a short list of points, because of course each one must be drawn separately.

We will find that every one of the geometric constructions we will use to represent functions has some advantages, and some disadvantages. We study all of them because each is the most useful representation in different circumstances.

Exercises

Exercise 2.1.1 (exer-MVF-F-1). For each of the following, find the largest possible domain for which this would qualify as a function, and then find the corresponding image. *(In this problem we allow ONLY real numbers as inputs and outputs.)*

1. $f(x,y) = \sqrt{4 - x^2 - y^2}$
2. $f(x,y,z) = \cos(x - y^2 + e^z)$
3. $f(x,y,z) = \frac{1}{1-z^2}$

Exercise 2.1.2 (exer-MVF-F-2). Write the formula for the function $S_{12} : \mathbb{R}^3 \to \mathbb{R}^2$ that gives the coordinates of the shadow in the xy-plane projected at noon by a given point in $\mathbb{R}^3$. (Assume that at noon, the sun is directly overhead, in the direction where $\phi = 0$.)

Exercise 2.1.3 (exer-MVF-F-3). Indiana Jones is in the map room at 4pm (assume that at that time, the sun is in the direction where $\phi = \pi/3$ and $\theta = 0$.) He places the base of the staff at a point (x, y) on the map, and the height of the jewel above the map is z. Write the formula for the function $S_4 : \mathbb{R}^3 \to \mathbb{R}^2$ that gives the coordinates in the xy-plane of the focus of the jewel (which is exactly where the shadow would be), indicating the location of the ark.

Exercise 2.1.4 (exer-MVF-F-4). Write the formula for the function $C : \mathbb{R}^2 \to \mathbb{R}^3$ that gives the rectangular coordinates for a point at a given longitude l_1 (measured east from the prime meridian) and latitude l_2 (measured north from the equator). Assume that the earth is a perfect sphere of radius 3,960 miles, centered at the origin, with the north pole on the positive part of the z-axis and the x-axis going through the prime meridian.

Exercise 2.1.5 (exer-MVF-F-5). Write the formula for the function that, for a given point in space, gives the point on the unit sphere closest to that point. Are there any points that cannot be in the domain for this function?

Exercise 2.1.6 (exer-MVF-F-6). Draw the direct geometric interpretation of the statement that the function $f : \mathbb{R}^3 \to \mathbb{R}^2$ takes the point $(1, 2, 1)$ to the point $(-1, 3)$.

2.2 Equations

All too often the words "equation" and "function" are used imprecisely, and sometimes even interchangeably. This is a very dangerous practice, because in fact these are very different things.

An equation is simply a mathematical statement, involving two expressions separated by an equal sign, "=", the interpretation of which is that the two expressions have the same value.

Of course we can use equations as part of our definition of a function, and certainly functions can be parts of the expressions involved in an equation – but we should still carefully note that equations and functions are fundamentally different things.

For example, in the equation

$$f(x, y) = x^2 - y^2$$

the expression on the left-hand side is a shorthand notation representing the output value of the function f corresponding to input variables x and y; the expression on the right-hand side is an explicit description of how to compute this output value in terms of those input variables. This equation serves as part of the definition of a function f.

On the other hand, in the equation

$$x^2 = 2x$$

a statement is being made about a single variable, x. It could be that the expressions in this equation each are to be thought of as separate functions which we are requiring to be equal and for which we wish to find the corresponding value or values of x. Or it could be instead that x is a known value, and the equation is serving simply to state a property that value is known to have.

Another example is the equation

$$(x+1)^2 = x^2 + 2x + 1$$

This equation is true for all values of x. So, rather than being a tool for trying to find x, or a tool for expressing a known fact about x, instead this equation is being used to make a general statement of fact about algebra, for which the variable x is just another tool.

2.2.1 Solution sets

For an equation that involves some set of variables and a collection of chosen values for those variables, that equation will be either true or false. For example, considering the equation

$$x + y = x^2 y + 1$$

choosing the values $x = 1$ and $y = 1$ makes the equation true, while choosing the values $x = 2$ and $y = 3$ makes the equation false. We say that a collection of values is a "solution" to the equation if they make the equation true.

The set of solutions to an equation is called the "solution set" for that equation.

Noting again that a collection of values for a set of variables can be thought of as the coordinates of a point in space of the appropriate dimension, we can alternatively view the solution set for an equation as the collection of points in the appropriate dimension space whose coordinates form a solution to the equation.

In this way we can visualize the solution set as some sort of geometric figure in space.

Example 2.2.1. Suppose we consider the equation $x + y = 1$ for the variables x and y. We know that this is the equation of a line in the xy-plane, which means that every point on this line satisfies the given equation. This line is the solution set for that equation.

Example 2.2.2. Suppose we consider the equation $x^2 + y^2 = 1$ for the variables x and y. We know that this is the equation of a circle in the xy-plane, which means that every point on this circle satisfies the given equation. This circle is the solution set for that equation.

2.2.2 Cross sections

Sometimes, finding the solution set to an equation of three variables can be substantially more difficult than for an equation of two variables. Fortunately we have a few techniques that can help us to do this. These will be the topics of the rest of this section. In each of these techniques, notice that we will be drawing conclusions about the relationships between algebra and geometry.

Suppose we wish to find the solution set S to the equation

$$z = x^2 + y^2$$

This is not an equation that we have dealt with up to this point, so we cannot simply recognize the form as we did for Examples 2.2.1 and 2.2.2.

Specifically, we have three variables in this equation, and all of the equations that we deal with in single variable calculus have only two equations.

Suppose we were to set the variable y equal to some constant c. Then we have two equations,

$$\begin{aligned} z &= x^2 + y^2 \\ y &= c \end{aligned}$$

The set of solutions to these two equations must be a subset of the solution set S. In fact, it is exactly the intersection of S with the solution set to the equation $y = c$, which of course is a vertical plane. We call this intersection then the "cross section of S by the plane $y = c$", or a "cross section of S perpendicular to the y-axis".

The cross section is a useful computational concept here because, algebraically, we can eliminate y from the above pair of equations, resulting in a single equation of two variables describing that cross section. In other words, determining the shape of the cross section, at least in this case, is an easier problem than we started with.

Specifically, eliminating y from the above pair of equations gives us

$$z = x^2 + c^2$$

This is a single equation in the variables x and z, and we can easily see that the values of x and z satisfying this equation, thought of as representing points in the xz-plane, describe a parabola opening in the positive direction on the z-axis, shifted up from the x-axis by the distance c^2.

The cross section itself is the set of points with these same x and z coordinates, and with the fixed value c for y.

So, we conclude that the cross section of S by the plane $y = c$ is a parabola, as described above. This is true for any value of c, and if we choose a few arbitrary values of c for which to take the corresponding y cross sections, we can conclude that the solution set S must look something like Figure 2.3.

By a similar argument, we can conclude the the $x = c$ cross sections of the solution set S must also be parabolas, whose y and z coordinates satisfy the equation

$$z = c^2 + y^2$$

So the same S has cross sections by the planes $x = c$ that are also parabolas, as represented in Figure 2.4

Finally, we can take cross sections perpendicular to the z-axis by setting $z = c$, giving us

$$c = x^2 + y^2$$

2.2. EQUATIONS

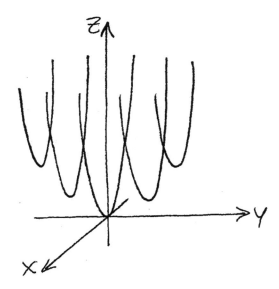

Figure 2.3:

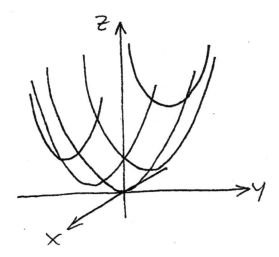

Figure 2.4:

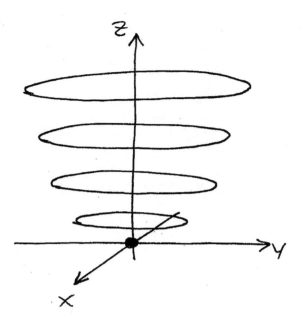

Figure 2.5:

We know that the solution set to this equation is a circle when c is positive, a single point when c is zero, and the empty set when c is negative. So, the horizontal cross sections of the surface S above the xy-plane are circles, as in Figure 2.5.

Combining all of these pictures, we have a very good idea of what this solution set looks like, as represented in Figure 2.6. We call this surface a "paraboloid".

Notice that, as referred to at the beginning of this subsection, we have now made a connection between algebra and geometry. In this case, the algebraic act of setting one variable in an equation equal to a constant is equivalent to the geometric act of taking a cross section perpendicular to the corresponding axis.

We can use cross sections to help us determine the shape of the solution sets to many different equations.

Example 2.2.3. What is the shape of the solution set to the equation $z = x^2 - y^2$?

By setting $y = c$, we have cross sections that are parabolas, but this time the parabolas are shifted down, not up, for larger values of c. Setting $x = c$, we have cross sections that are again parabolas, but these open in the downward direction, and are shifted up for larger values of c.

Setting $z = c$, we have the equation $c = x^2 - y^2$, which we recognize as a hyperbola in the x and y coordinates.

Putting these together, we get the picture in Figure 2.7. Because of the shapes of its cross sections, we call this surface a "hyperbolic paraboloid", along with any surface with equation of the form $z = ax^2 - by^2$ with a and b both of the same sign.

Example 2.2.4. What is the shape of the solution set to the equation $z = 9x^2 + y^2$?

By setting either $y = c$ or $x = c$, we have cross sections that are parabolas opening up and shifted in the same direction. However this time these parabolas do not have the same shapes. So we expect this to look something like a paraboloid, but not exactly.

Setting $z = c$, we have the equation $c = 9x^2 + y^2$, which we recognize as an ellipse in the x and y coordinates.

2.2. EQUATIONS

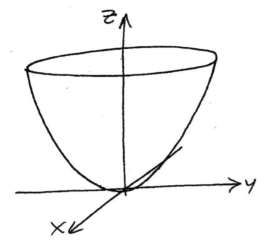

Figure 2.6:

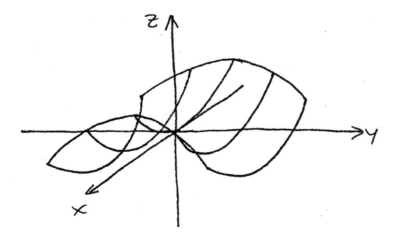

Figure 2.7:

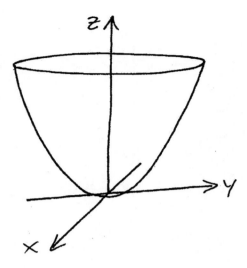

Figure 2.8:

This elliptical horizontal cross section is what allows us most easily to distinguish this figure from the paraboloid with circular horizontal cross section. We represent this surface in Figure 2.8. Because of the shapes of its cross sections, we call this surface an "elliptical paraboloid", along with any surface with equation of the form $z = ax^2 + by^2$ with $a \neq b$ both of the same sign.

In order to be more specific then, sometimes we will use the term "circular paraboloid" for any surface of the form $z = ax^2 + ay^2$.

2.2.3 Transformations of variables

As in the last subsection, we will again be making a connection between algebra and geometry. In this subsection we will talk about how the solution set to an equation changes when we make certain minor changes to the equation itself.

We begin with a quick observation. Notice that with any equation of three variables x, y and z, we can put all of the terms onto the left side of the equation, and then consider that left side to be the formula for some function $F : \mathbb{R}^3 \to \mathbb{R}^1$. So, we can write any such equation in the form

$$F(x, y, z) = 0$$

for some function F. We will also sometimes write this as

$$F(\vec{x}) = 0$$

where as usual $\vec{x} = (x, y, z)$. This will be convenient notation for us in this section.

Suppose we have such an equation given to us, and suppose that we know what the solution set looks like. What happens to the solution set S if we make a minor change to the equation?

For example, what if we go through the equation and replace every "x" with "$x+1$", leaving everything else unchanged? Or what if we replace every "y" with "$-y$", or every "z" with "cz" for some value of c?

The answer to these questions will come from the fact that we can interpret each of these algebraic changes in terms of functions which we can also interpret geometrically. This will allow us to draw a geometric conclusion about the solution set to the new equation.

2.2. EQUATIONS

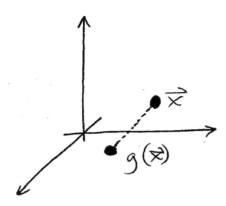

Figure 2.9:

Example 2.2.5. Given an equation

$$F(\vec{x}) = F(x, y, z) = 0 \tag{2.1}$$

with a known solution set S, what can we conclude about the solution set S' to the equation

$$F(x+1, y, z) = 0 \tag{2.2}$$

First, let us observe that we can represent the transformation of this equation by a new function $g : \mathbb{R}^3 \to \mathbb{R}^3$, defined by

$$g(x, y, z) = (x+1, y, z)$$

This function g acts on a point $\vec{x}$ simply by moving it in the positive x-direction by one unit, as represented in Figure 2.9.

Equation 2.2 then becomes

$$F(g(\vec{x})) = 0 \tag{2.3}$$

Note that we can view the left side of this equation either as the function $F \circ g$ evaluated at the point $\vec{x}$, or as the function F evaluated at the point $g(\vec{x})$. This observation about Equation 2.3 allows us to make an immediate statement concerning the solutions to Equation 2.2, which of course is equivalent to Equation 2.3. That is – the point $\vec{x}$ is a solution to Equation 2.2 if and only if $g(\vec{x})$ is a solution to Equation 2.1.

Said in geometric terms, this means that a point $\vec{x}$ is on the set S' if and only if moving it in the positive x-direction by one unit puts it onto the set S. Making the same statement about the entire solution sets – if we take S' and move it in the positive x-direction by one unit, the result is the set S. See Figure 2.10.

Finding S' from S then is just a matter of rephrasing this in the other direction – S' is obtained by moving S in the negative x-direction by one unit. See Figure 2.11.

Example 2.2.6. As a specific case of Example 2.2.5, we know that the equation $x^2 + y^2 + z^2 = 1$ represents the unit sphere centered at the origin. If we replace every "x" in the equation with "$x+1$", the equation becomes

$$(x+1)^2 + y^2 + z^2 = 1 \tag{2.4}$$

and according to the above reasoning the solution set to this equation should be a sphere centered at $(-1, 0, 0)$.

In this case we can confirm this conclusion by noting that the left side of the equation represents the distance between the points (x, y, z) and $(-1, 0, 0)$.

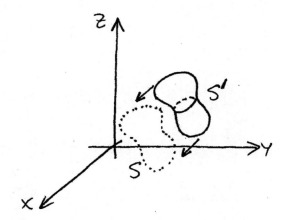

Figure 2.10:

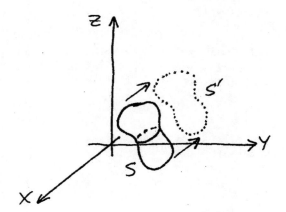

Figure 2.11:

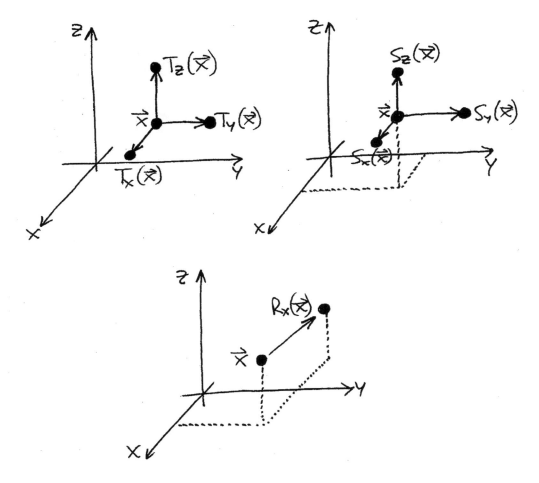

Figure 2.12:

The conclusion of Example 2.2.5 is suggestive of an idea that is valid for a large variety of similar examples. That is, the act of changing an equation by applying a simple transformation to the input variables causes the exact opposite transformation to happen to the solution set.

Here are several such functions, which we represent in Figure 2.12.

We have already seen that the function T_x defined by $T_x(x,y,z) = (x+c, y, z)$ moves points in the positive x-direction by a distance c. Similarly, we can define functions T_y and T_z that move points in the positive y- and z-directions, respectively. We call these "translations".

The function S_x defined by $S_x(x,y,z) = (cx, y, z)$, with $c > 0$, stretches space in the x-direction by a factor c. We will refer to this and the similar functions S_y and S_z simply as "stretches". Note that the combination of stretches in all three directions by the same factor c results in every vector (x, y, z) being multiplied by that factor c. We call this a "scaling transformation" or a "magnification".

The function R_x defined by $R_x(x,y,z) = (-x, y, z)$ simply reflects every point through the yz-plane. We refer to this and the similar functions R_y and R_z as "reflections".

Note that each of the above transformations can be interpreted as acting on the input variables individually.

We can apply arguments similar to those in Example 2.2.5 to conclude what happens to the solution sets when each of the above transformations is applied to the input variables of a given equation. The conclusions are represented in the following chart:

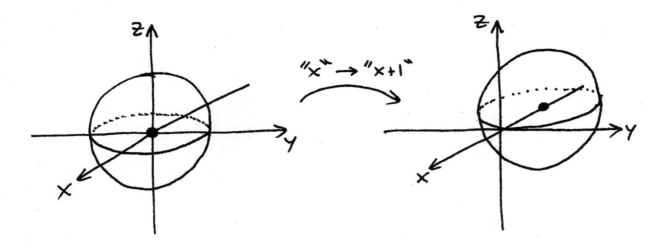

Figure 2.13:

replacing	with	does this to the solution set
"x"	"$(x+c)$"	translates by $-c$ in the x-direction
"y"	"$(y+c)$"	translates by $-c$ in the y-direction
"z"	"$(z+c)$"	translates by $-c$ in the z-direction
"x"	"(cx)"	stretches by $1/c$ in the x-direction
"y"	"(cy)"	stretches by $1/c$ in the y-direction
"z"	"(cz)"	stretches by $1/c$ in the z-direction
"x"	"$(-x)$"	reflects through the yz-plane
"y"	"$(-y)$"	reflects through the xz-plane
"z"	"$(-z)$"	reflects through the xy-plane

These tools allow us to find the solution set to an equation by relating that equation to a similar one whose solution set is already known.

Example 2.2.7. What is the solution set to the equation $(x+1)^2 + y^2 + z^2 = 1$?

(Of course we recognize this as the equation from Example 2.2.6, but here we outline the process where we do not already know the solution set for the equation.)

The first thing we must do is find a similar equation to which we already know the solution set. By "similar", we mean that we must be able to make simple transformations of the variables to turn it into the equation we are interested in. In this case, we note that the equation $x^2 + y^2 + z^2 = 1$ is very similar, and we know that its solution set is the unit sphere centered at the origin.

The next step is simply to apply transformations to the variables in the equation in order to turn it into the desired equation, while simultaneously making the corresponding adjustments to the solution set.

In this case there is only one step required – we replace "x" by "$(x+1)$". The resulting equation is the one we are interested in, and applying the corresponding change to the unit sphere gives us the desired solution set. This is represented in Figure 2.13.

Often the process of turning an equation with known solution set into a given equation requires more than one step. In cases like this, one simply applies one step at a time, being careful to adjust the solution set accordingly in each step.

2.2. EQUATIONS

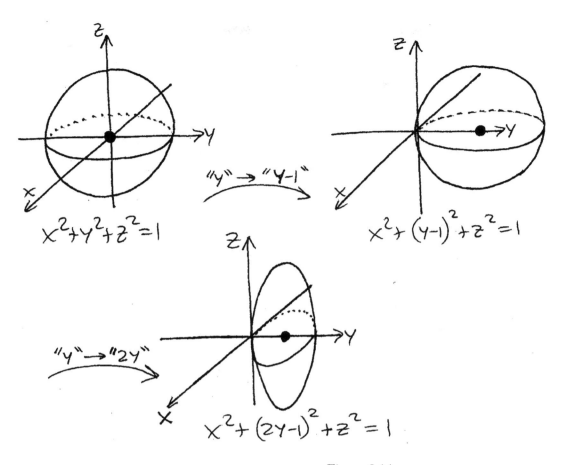

Figure 2.14:

Students must be very careful in situations such as this however, making sure not to apply the steps in the wrong order. This can easily lead to an incorrect conclusion.

Example 2.2.8. What is the solution set to the equation $x^2 + (2y-1)^2 + z^2 = 1$?

Again, we note that this equation is similar in form to $x^2 + y^2 + z^2 = 1$, for which we know the solution set is the unit sphere.

First, let's replace "y" with "$y-1$", giving us $x^2 + (y-1)^2 + z^2 = 1$. The solution set is translated by $+1$ in the y-direction, putting the center of this solution set at $(0,1,0)$, and the two points of intersection with the y-axis are at $(0,0,0)$ and $0,2,0$.

Next, we replace "y" with "$2y$", giving us $x^2 + (2y-1)^2 + z^2 = 1$. The solution set is stretched by a factor of $1/2$ in the y-direction (which we can interpret at "squishing" the sphere). This takes the center to $(0, 1/2, 0)$, and the two intersection points with the y-axis to $(0,0,0)$ and $(0,1,0)$.

These two steps are represented in Figure 2.14.

Note that it is very tempting in Example 2.2.8 to apply the transformations in the opposite order, since the order of operations on the expression $2y - 1$ has the multiplication happening first. Remember though that we are "undoing" the process, so the transformations need to be applied in the opposite order.

If we had done the transformations in the wrong order in Example 2.2.8 – first replacing "y" with "$2y$", and then replacing "y" with "y − 1", then the resulting equation would have been $x^2 + (2(y-1))^2 + z^2 = 1$, which is not the equation we were interested in.

Of course these transformations ideas apply equally well in two dimensions.

Example 2.2.9. **Suppose we consider the ellipse centered at the origin with semi-axes a and b in the directions of the x-axis and y-axis, respectively. If we view this ellipse as the result of stretching the unit circle by a factor of a in the x-direction and by a factor of b in the y-direction, what is the resulting equation for this ellipse?**

Beginning with the equation $x^2 + y^2 = 1$ for the unit circle, we can perform the x-direction stretch by replacing "x" with "$\frac{x}{a}$", and the y-direction stretch by replacing "y" with "$\frac{y}{b}$". The result of these replacements is the equation

$$\frac{x^2}{a^2} + \frac{y^2}{b^2} = 1$$

This is the equation for such an ellipse that students will recognize from their high school algebra courses.

2.2.4 Rotations

Suppose the equation $F(x, y, z) = 0$ we are interested in has the property that the variables x and y appear in the equation only as part of the specific expression "$\sqrt{x^2 + y^2}$". That is, suppose we can write

$$F(x, y, z) = g(\sqrt{x^2 + y^2}, z) = 0 \tag{2.5}$$

for some function g. For example, note that we can rewrite the equation

$$F(x, y, z) = x^2 + y^2 - z = 0 \tag{2.6}$$

as

$$g(\sqrt{x^2 + y^2}, z) = 0$$

where g is the function defined by $g(s, t) = s^2 - t$.

Of course this is only possible in the case of some equations, but when it is, this is another way in which we can reduce the problem of finding the solution set to an equation of three variables to the simpler problem of finding the solution set to an equation of two variables. In this section, we will talk about the geometric interpretation of this relationship.

The important thing to note is that the expression $\sqrt{x^2 + y^2}$ is not arbitrary. Rather, it has a specific geometric interpretation in xyz-space as the distance from the point (x, y, z) to the z-axis. We will call this distance r_z, and rewrite Equation 2.5 as

$$g(r_z, z) = 0 \tag{2.7}$$

This interpretation allows us to draw a strong conclusion about the solution set S to Equation 2.5. That is, if a given point is in S, then any point with the same z-coordinate and the same distance r_z to the z-axis must also be in S. We can make a further geometric interpretation of this, as in the following observation.

Observation 2.2.1. *The solution set S to an equation of the form*

$$F(x, y, z) = g(\sqrt{x^2 + y^2}, z) = 0$$

has rotational symmetry around the z-axis.

This is represented in Figure 2.15.

2.2. EQUATIONS

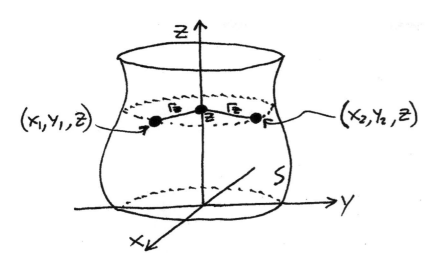

Figure 2.15:

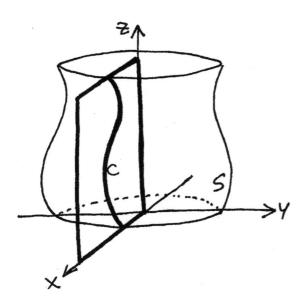

Figure 2.16:

Example 2.2.10. Note that Equation 2.6 is equivalent to the subject of the initial example of Subsection 2.2.2.

Recall in that example we showed that all of the cross sections of the surface perpendicular to the z-axis are circles. As predicted by the above observation, this means that the surface has rotational symmetry around the z-axis. See again Figure 2.6.

Of course, any surface that has rotational symmetry around the z-axis can be thought of as actually being produced by the rotation of a curve around the z-axis. There are many such curves for any given surface, but we can construct a particularly convenient one in this case by looking at the intersection of the surface S with the part of the xz-plane where $x \geq 0$. This is represented in Figure 2.16.

We will call this curve C.

What makes this curve so relevant in this case is that for every point in this half plane, we have $r_z = x$. We can see this

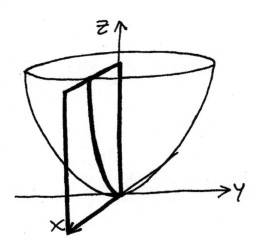

Figure 2.17:

by first noting that $y = 0$ for every point in this half plane, giving us

$$r_z = \sqrt{x^2 + y^2} = \sqrt{x^2} = |x|$$

Then the fact that we are only looking at the half plane with $x \geq 0$ gives us $r_z = x$.

Conveniently then, in this half plane Equation 2.7 for the surface S becomes simply $g(x, z) = 0$. So we can represent the curve C in the xz-plane very conveniently then as the solution to the equation $g(x, z) = 0$ and the inequality $x \geq 0$.

This gives us the following theorem:

Theorem 2.2.1. *The solution set S in xyz-space to an equation of the form*

$$F(x, y, z) = g(\sqrt{x^2 + y^2}, z) = 0 \tag{2.8}$$

is the surface obtained by rotating around the z-axis the curve in the xz-plane defined by the equation $g(x, z) = 0$ and the inequality $x \geq 0$.

We apply this theorem in the following examples.

Example 2.2.11. Suppose we wished to find the solution set to the equation $x^2 + y^2 - z = 0$, but had not already seen this surface in Example 2.2.10 and Subsection 2.2.2.

First, we note that we can write this equation as $g(r_z, z) = 0$ where g is defined by $g(r_z, z) = r_z^2 - z$. Applying Theorem 2.2.1, we can then conclude that the solution set to this equation is the surface obtained by rotating around the z-axis the curve in the xz-plane defined by the equation $x^2 - z = 0$ with $x \geq 0$. This curve is half of a parabola, and rotating it around the z-axis gives us a bowl-like surface that we have already named a "paraboloid". This is represented in Figure 2.17.

Note that this gives us an additional motivation for that name – not only does the paraboloid have parabolic cross sections, but it also is obtained as the rotation of a parabola.

Example 2.2.12. Suppose we wish to find the solution set to the equation $\sqrt{x^2 + y^2} + z^2 - 1 = 0$.

Again applying Theorem 2.2.1, we conclude that this is the surface obtained by rotating around the z-axis the $x \geq 0$ part of the curve in the xz-plane defined by $x + z^2 - 1 = 0$, or $x = 1 - z^2$. Because of the $x \geq 0$ condition this is

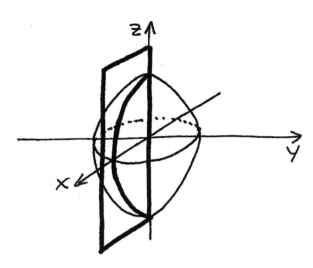

Figure 2.18:

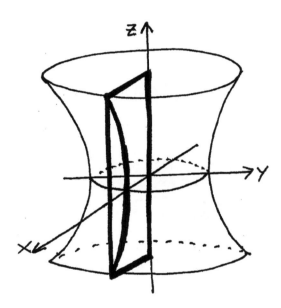

Figure 2.19:

only part of a parabola, and when it is rotated around the z-axis we get a shape sort of like a football, as pictured in Figure 2.18.

Example 2.2.13. Suppose we wish to find the solution set to the equation $x^2 + y^2 - z^2 = 1$.

We recognize again that this can be written in terms of $r_z = \sqrt{x^2 + y^2}$, by writing $g(r_z, z) = r_z^2 - z^2 - 1 = 0$, and so the relevant curve in the xz-plane has equation $x^2 - z^2 - 1 = 0$. This is a hyperbola along the x-axis, and of course we only consider the $x \geq 0$ portion of it. Rotating around the z-axis, the surface is pictured in Figure 2.19.

This surface is the rotation of a hyperbola, and it can also easily be checked that the x- and y- cross sections are all hyperbolas. This motivates the term "hyperboloid" for this surface. More specifically we refer to it as a "hyperboloid of one sheet" since the entire surface is one connected piece; we will see in the next example another type of hyperboloid.

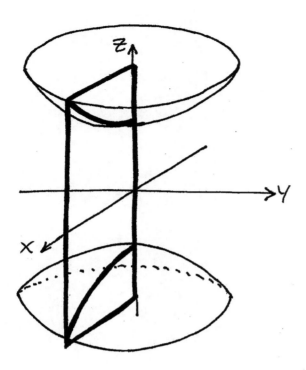

Figure 2.20:

Example 2.2.14. Suppose we wish to find the solution set to the equation $-x^2 - y^2 + z^2 = 1$.

We recognize again that this can be written in terms of $r_z = \sqrt{x^2 + y^2}$, **by writing** $g(r_z, z) = -r_z^2 + z^2 - 1 = 0$, **and so the relevant curve in the** xz-**plane has equation** $-x^2 + z^2 - 1 = 0$. **This is a hyperbola along the** z-**axis, and of course we only consider the** $x \geq 0$ **portion of it. Rotating around the** z-**axis, the surface is pictured in Figure 2.20.**

We call this surface a "hyperboloid of two sheets".

The arguments concluding in Theorem 2.2.1 can be applied in a similar way to surfaces that are rotations around the x- and y- axes. These result in the following very similar theorems.

Theorem 2.2.2. *The solution set S in xyz-space to an equation of the form*

$$F(x, y, z) = g(x, \sqrt{y^2 + z^2}) = 0 \qquad (2.9)$$

is the surface obtained by rotating around the x-axis the curve in the xy-plane defined by the equation $g(x, y) = 0$ and the inequality $y \geq 0$.

Theorem 2.2.3. *The solution set S in xyz-space to an equation of the form*

$$F(x, y, z) = g(y, \sqrt{x^2 + z^2}) = 0 \qquad (2.10)$$

is the surface obtained by rotating around the y-axis the curve in the yz-plane defined by the equation $g(y, z) = 0$ and the inequality $z \geq 0$.

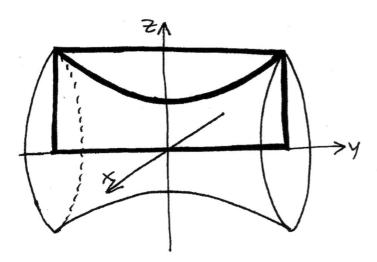

Figure 2.21:

These theorems can be applied to realize a surface as a rotation around the x- and y- axes.

Example 2.2.15. **What shape of surface is described by the equation $x^2 - y^2 + z^2 = 1$?**

Noting that this equation can be written in terms of $r_y = \sqrt{x^2 + z^2}$ as $g(y, r_y) = r_y^2 - y^2 - 1 = 0$ and applying Theorem 2.2.3, we conclude that this surface is obtained by rotating around the y-axis the curve in the yz-plane defined by the equation $z^2 - y^2 - 1 = 0$ and the inequality $z \geq 0$. This curve is part of a hyperbola, and when we rotate around the y-axis we obtain the hyperbola of one sheet as pictured in Figure 2.21.

Note also that the arguments concluding in Theorem 2.2.1 could just as well have been applied to the cross section of the surface S with the yz-plane instead of the xz-plane, giving us another very similar theorem. And likewise for Theorems 2.2.2 and 2.2.3.

In these cases however, the resulting theorem gives us no additional capabilities for finding solution sets, only equivalent alternatives to already stated theorems. So we leave the statement and explanations for these theorems as exercises to the reader.

Exercises

Exercise 2.2.1 (exer-MVF-E-1). Describe all of the horizontal cross sections of the surface that is the solution set to the equation $e^z = 1 + x^2 + y^2$.

Exercise 2.2.2 (exer-MVF-E-2). Describe all of the cross sections perpendicular to each of the three axes, for the surface that is the solution set to the equation $z = xy$.

Exercise 2.2.3 (exer-MVF-E-3). The surface S is described by the equation $x^3 - y^3 + z^2 = 4$. Write the equation for the surface S' that is obtained by each of the following geometric actions:
  1. Shifting S as indicated by the vector $(3, 2, 4)$.

  2. Stretching S by a factor of 5 in the z-direction.

3. Reflecting S through the xy-plane.

Exercise 2.2.4 (exer-MVF-E-4). Beginning with the fact that the unit sphere is the solution set to the equation $x^2 + y^2 + z^2 = 1$, draw the solution set to the equation $(3x + 6)^2 + (y - 2)^2 + 4z^2 = 1$.

Exercise 2.2.5 (exer-MVF-E-5). The surface S is the solution set to the equation $f(x, y, z) = 0$. Describe the sequence of geometric transformations that will turn S into the solution set to the equation $f(3x + 2, y + 2, 4 - z) = 0$.

Exercise 2.2.6 (exer-MVF-E-6). The line L is described by the symmetric equations $x + 1 = 2y - 3 = 3z + 2$.
1. Write down the parametrization of L.

2. Write down the symmetric equations for the line L' obtained by translating the line L as indicated by the vector $(1, 2, 3)$.

3. Write down the parametrization of L'.

4. Based on your conclusions, explain in general how a translation by a vector represents itself in the parametrization of a line. How does this seem different from the representation of a translation in equations?

Exercise 2.2.7 (exer-MVF-E-7). Draw the surface that is the solution set to the equation $x^2 - y^2 + z^2 = 1$.

Exercise 2.2.8 (exer-MVF-E-8). Draw the surface that is the solution set to the equation $x^2 + y^2 - z^2 = 1$.

Exercise 2.2.9 (exer-MVF-E-9). Draw the surface that is the solution set to the equation $\sqrt{y^2 + z^2} = 1 + \sin x$.

Exercise 2.2.10 (exer-MVF-E-10). Draw the surface that is the solution set to the equation $\sqrt{y^2 + z^2} = \sin x$.

Exercise 2.2.11 (exer-MVF-E-11). Draw the surface that is the solution set to the equation $y^2 + z^2 = \sin^2 x$.

Exercise 2.2.12 (exer-MVF-E-12). Draw the surface that is the solution set to the equation $x^2 - 4y^2 + 9z^2 = 1$.

Exercise 2.2.13 (exer-MVF-E-13). Draw the surface that is the solution set to the equation $(x-1)^2 - (2y+2)^2 + (3z+1)^2 = 36$.

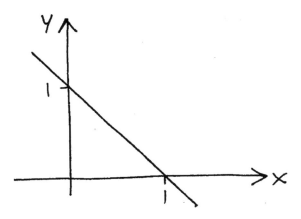

Figure 2.22:

2.3 Graphs

We return now to our study of functions. We have already mentioned that for multivariable functions, there will be several types of pictures that we will draw to help us make geometric associations to the algebraically defined function, in addition to simply drawing the domain and target separately as in Subsection 2.1.3. We begin with the graph construction.

2.3.1 Single variable functions

First let us review the details of the construction process for the graph of a single variable function.

Suppose we have a function $f : \mathbb{R}^1 \to \mathbb{R}^1$ defined with an equation such as $f(x) = x + 1$. Notice that in the definition of this function, the only variable that appears is x. The first step in the graph construction process is to introduce a new variable. It is conventional, but arbitrary, to choose y as this new variable.

In the graph construction, this new variable is set equal to the output value of the function to form the equation

$$y = f(x) \tag{2.11}$$

In this newly formed equation, there are two variables – the input variable, x, and the new variable, y, which we can now think of as the output variable. Taken together, these two variables can be thought of as the coordinates of a point in $\mathbb{R}^2$, which in this case we will refer to as the xy-plane.

Definition 2.3.1. *The "graph" of the function $f : \mathbb{R}^1 \to \mathbb{R}^1$ is defined as the solution set to Equation 2.11 in the xy-plane.*

Note again that even though the original function has only one input variable, the graph is a solution set in two-dimensional space.

Example 2.3.1. Consider the function $f(x) = 1 - x$. Students are already very familiar with what the graph of this function looks like, but let us go through the above procedure in order to clarify the process itself.

We begin by introducing the variable y in forming the equation $y = f(x)$ which we can rewrite as $y = 1 - x$. The graph of the function f is then the solution set to this equation. Of course this equation is equivalent to the equation $x + y = 1$, and we already have seen in Example 2.2.1 that the solution set to that equation is the line below.

So, this line is the graph of the given function.

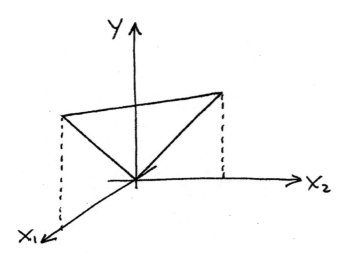

Figure 2.23:

2.3.2 Multivariable functions

At this point, we would like to generalize the construction above so that it can be used for multivariable functions.

Suppose we have a function $f : \mathbb{R}^n \to \mathbb{R}^m$, with n input variables and m output variables. The expression defining the output variables will involve only the n input variables. For convenience of notation, we will choose these input variables to be $x_1, \ldots, x_n$, so the expression for the output of the function is written $f(x_1, \ldots, x_n)$, or equivalently, $f(\vec{x})$.

As in the case of single variable functions, we must now introduce new variables to correspond to each of the output variables. Again for convenience of notation, we will choose these new variables to be $y_1, \ldots, y_m$.

Again, setting these new variables equal to the output variables, we have the following equivalent equations,

$$(y_1, \ldots, y_m) = f(x_1, \ldots, x_n) \quad \text{or} \quad \vec{y} = f(\vec{x}) \tag{2.12}$$

Definition 2.3.2. *The graph of the function $f : \mathbb{R}^n \to \mathbb{R}^m$ is defined to be the solution set to Equation 2.12 in $\mathbb{R}^{n+m}$, the coordinate axis system including both the input variables and the output variables.*

Again, notice that the graph is a set in a space whose dimension is not the same as the number of input variables for the function, or the number of output variables; rather, it is the sum of the number of input and output variables. It is not the domain, nor is it the target. It is a completely new space made up especially for the formation of the graph.

Example 2.3.2. Consider the function $f(x_1, x_2) = x_1 + x_2$. This function has two input variables and one output variable. In order to form the graph, we introduce the new variable y, which we then set equal to the output of the function by writing the equation $y = f(x_1, x_2) = x_1 + x_2$. The graph of this function then is the solution set, in $\mathbb{R}^3$ (the $x_1 x_2 y$-space), of the equation $y = x_1 + x_2$.

Of course for every input pair (x_1, x_2), we get a corresponding output value y and thus a corresponding point. Since we can think of the input pairs as coming from the two-dimensional $x_1 x_2$-plane in $x_1 x_2 y$-space, we get a point on the graph corresponding to every point in the $x_1 x_2$-plane. So we expect the graph to be a surface of some sort.

To be more precise, we recall from Section 1.5 that this solution set is the plane represented in Figure 2.23.

So this plane is the graph of the function.

Note, the domain for the function is $\mathbb{R}^2$, and the target is $\mathbb{R}^1$, but the graph is a set in $\mathbb{R}^3$.

2.3. GRAPHS

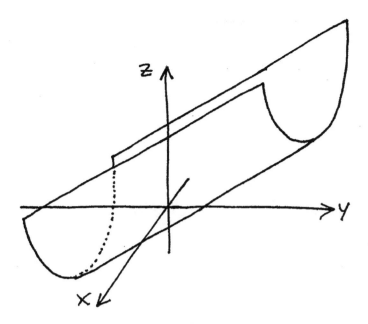

Figure 2.24:

Example 2.3.3. What is the graph of the function $f : \mathbb{R}^2 \to \mathbb{R}^1$ defined by $f(x, y) = x^2 + y$?

The graph of this function is the surface in $\mathbb{R}^3$ that is the solution set to the equation $z = x^2 + y$.

We know from Subsection 2.2.2 that the cross sections of this surface perpendicular to the y-axis are upward parabolas, and that the vertex of the parabola is higher for larger values of y. We also know that the cross sections perpendicular to the x-axis are straight lines with slope 1 in the y-direction.

Combining these cross sections, we conclude that the surface looks like a sloped parabolic trough, as pictured in Figure 2.24.

This surface is sometimes called a "parabolic cylinder".

Example 2.3.4. Consider the function $f(x) = (x - 1, x/2)$. This function has one input variable and two output variables. In order to form the graph, we introduce the two new variables y_1 and y_2, which we then set equal to the outputs of the function by writing $(y_1, y_2) = f(x) = (x - 1, x/2)$, which we can equivalently view as the pair of equations $y_1 = x - 1$ and $y_2 = x/2$. The graph of the function is the set of solutions in the three-dimensional xy_1y_2-space to this pair of equations.

For every value of x of course we get corresponding values of y_1 and y_2, and thus a point in xy_1y_2-space. And we can think of the values of x as corresponding to points on the one-dimensional x-axis in xy_1y_2-space. So we expect this graph to be a one-dimensional curve.

To be precise, recall from Section 1.5 that each of the equations $y_1 = x - 1$ and $y_2 = x/2$ represents a plane in xy_1y_2-space. So, the set of points satisfying both of these equations should be the intersection of two planes, which is indeed a line.

We can even find the parametrization of this line, using the input variable x as our parameter t. This gives us

$$(x, y_1, y_2) = (t, t - 1, t/2) = (0, -1, 0) + t(1, 1, 1/2)$$

The line represented by the above parametrization is the graph of the given function, represented in Figure 2.25.

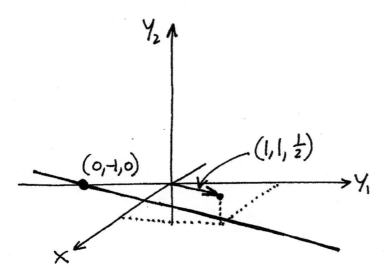

Figure 2.25:

Again, note that the domain is $\mathbb{R}^1$, the target is $\mathbb{R}^2$, but the graph is a set in $\mathbb{R}^3$.

As we have commented before, the graph construction has some wonderful traits. Students have already seen that for single variable functions, there are many associations that can be made between the functions algebraic behavior and certain geometric characteristics of the graph. Also, the graph picture allows us to see the behavior of the entire function all at once, unlike the much greater limitations we had when we drew the domain and target separately.

In a single variable setting, the many advantages of the graph construction make it a natural choice for representing functions in almost every circumstance.

Unfortunately, the graph construction has a tragic flaw that comes to light when we try to use it for multivariable functions. That is, in order to form the graph, we must draw a Euclidean space whose dimension is the sum of the number of input and output variables. It is very easy for this number to be greater than three, thus making the graph impossible to draw.

Example 2.3.5. Suppose we have a function $f : \mathbb{R}^2 \to \mathbb{R}^2$. This function has two input variables and two output variables, making a total of four variables that must be represented in the coordinate axis system in order for us to be able to draw the graph. Of course we cannot draw a four-dimensional space.

This problem is even worse for functions with more than two input variables, or more than two output variables. Such functions are extremely common however and we need to be able to apply the techniques of calculus to them just as much as the lower dimensional functions such as in Examples 2.3.2 and 2.3.4.

Fortunately, as we have commented before, there are other constructions that allow us to form geometric pictures that we can associate to functions.

The graph is still of course a very useful tool in the study of multivariable functions and multivariable calculus. But it is critical that the student realize that the graph by itself is not a sufficient tool for these purposes. Of the many geometric pictures we will see in this text, many of them are not graphs. As we will see in several instances later in this text, their geometric features cannot be interpreted in the same way as graphs.

Throughout this text we will remind the student often whether the figure under consideration is a graph, or one of the other constructions, to make sure that the student is making the correct interpretations. But the student should make

2.3. GRAPHS

sure to actively consider this question regarding every figure, without being reminded. This will be especially important when the student is constructing figures for use in exercises.

Exercises

Exercise 2.3.1 (exer-MVF-G-1). Consider the function $f(x_1, x_2, x_3) = (f_1, f_2, f_3, f_4)$. In what dimension of space does the graph of this function exist? What dimension is that graph within that space?

Exercise 2.3.2 (exer-MVF-G-2). Draw the graph of the function $f : \mathbb{R}^2 \to \mathbb{R}^1$ given by $f(x, y) = 3x + 2y$.

Exercise 2.3.3 (exer-MVF-G-3). Draw the graph of the function $f : \mathbb{R}^2 \to \mathbb{R}^1$ given by $f(x, y) = e^{x^2 + y^2}$.

Exercise 2.3.4 (exer-MVF-G-4). Draw the graph of the function $f : \mathbb{R}^2 \to \mathbb{R}^1$ given by $f(x, y) = e^{-x^2 - y^2}$.

Exercise 2.3.5 (exer-MVF-G-5). Draw the graph of the function $f : \mathbb{R}^2 \to \mathbb{R}^1$ given by $f(x, y) = e^{(x-1)^2 + 4y^2}$.

Exercise 2.3.6 (exer-MVF-G-6). The graph of an unknown function f is a surface in $\mathbb{R}^3$. What must be the dimensions of the domain and the target of this function? Make sure to explain your reasoning.

2.4 Level Sets

Level sets are an alternative construction to graphs when attempting to associate a geometric figure to a given function. Very often, level sets are the primary alternative when the graph is inappropriate for reasons of dimension. But in fact level sets are useful even when there is no dimension issue.

We will also find in Chapter 4 that level sets have a very natural connection to multivariable derivatives.

2.4.1 Construction

This construction can work for any multivariable function $f : \mathbb{R}^n \to \mathbb{R}^m$, but the construction is most useful when the dimension of the domain is greater than the dimension of the target. Here we will describe the construction only for real-valued functions $f : \mathbb{R}^n \to \mathbb{R}^1$, as this is the case in which the construction is most useful.

Definition 2.4.1. *A "level set" of the function $f : \mathbb{R}^n \to \mathbb{R}^1$ is a set of the form*

$$\left\{\vec{x} \in \mathbb{R}^n \,\Big|\, f(\vec{x}) = c\right\}$$

That is, the level set of a function f for a value c is the collection of all points in the domain whose image by f is that value c. Note specifically then, a level set is a subset of the domain of the function f. So, while the graph of a function $f : \mathbb{R}^n \to \mathbb{R}^1$ is a subset of $\mathbb{R}^{n+1}$, a level set is a subset of $\mathbb{R}^n$.

When using level sets, the function in question could represent many different quantities, and in each of those cases every level set represents input points where this quantity takes the same value. This leads to many other choices of terminology specific to those cases. For example if the function represents temperature, the level sets are sometimes called "isotherms" (from "iso" meaning "same"). Similarly when the function represents pressure, the level sets are called "isobars". Suggestive that this makes sense when the function represents any sort of quantity, level sets are also sometimes called "isoquants".

Recall of course that the "image" is the output value that comes from a function for a given input value, and note the above set is the collection of input values that yield a given output value; because of this, we also call a level set a "pre-image", denoted

$$f^{-1}(c) = \left\{\vec{x} \in \mathbb{R}^n \,\Big|\, f(\vec{x}) = c\right\}$$

Any one level set makes a precise indication of the value of the function for every point on that level set, but it says nothing about the value of the function anywhere else. In order to represent as much of a function as possible, usually one draws several level sets of the same function on the same figure. Very often the level sets have similar geometric features, and drawing a few of them can give a good representation of the function.

Example 2.4.1. Consider the function $f(x_1, x_2) = x_1 + x_2$, from Example 2.3.2. We saw in that example that the graph of this function is a plane in $\mathbb{R}^3$; here, we will instead represent it with level sets.

The domain is the $x_1 x_2$-plane, so all of the level sets will be in that plane.

First let's consider the level set $f^{-1}(0)$. By definition this is the set of points satisfying the equation $x_1 + x_2 = 0$, which of course is the line going through the origin with slope -1.

Next we consider the level set $f^{-1}(1)$. By definition this is the set of points satisfying the equation $x_1 + x_2 = 1$, which of course is a line with slope -1, but offset from the level set $f^{-1}(0)$.

In fact we can see that for any value of c, the level set $f^{-1}(c)$ is the set of points satisfying the equation $x_1 + x_2 = c$, all of which are lines with slope -1. We represent this in Figure 2.26.

Note that while we are representing the same function as in Example 2.3.2, this figure resulting from the level set construction is completely different from that resulting from the graph construction.

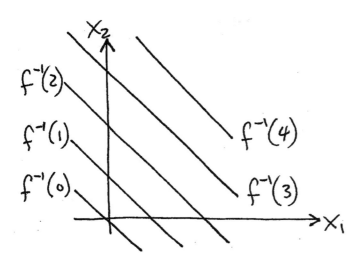

Figure 2.26:

2.4.2 Relationship to the graph

The terminology "level set" comes from the relationship that these pre-images have with the graph of the function. We will illustrate this in the case of a function $f : \mathbb{R}^2 \to \mathbb{R}^1$, with input variables x and y.

Note that the graph construction and the level set construction, while very different, both involve setting the values of the given function equal to something:

- To form the graph, we set $f(x, y)$ equal to a new variable z, and draw the solution set to the resulting equation.
- To form a level set, we set $f(x, y)$ equal to a constant c, and draw the solution set to the resulting equation.

Of course the difference between these constructions is critical and leads to completely different pictures, but they are related. Specifically, note that if we look at the equation $z = f(x, y)$ representing the graph, and then set the output variable z to a constant value c by requiring $z = c$, the resulting equation in x and y is precisely the equation $f(x, y) = c$ for the level set $f^{-1}(c)$.

This observation can be interpreted geometrically. The new equation $z = c$ of course represents a level, horizontal plane in xyz-space. And as we saw in Subsection 2.2.2, requiring that a point on the graph also satisfy this equation is equivalent to taking the cross section of the graph by this level, horizontal plane. See Figure 2.27.

Of course this results in a collection of points in xyz-space, and the level set is in the xy-plane. So we simply ignore the z coordinate, and look at the resulting set in the xy-plane. See Figure 2.28.

To summarize, we can interpret a level set from this point of view instead, as the set of points in the xy-plane with x and y coordinates obtained by taking a level, horizontal cross section of the graph of f. Note the distinction between a level set and a cross section – the cross section is in xyz-space, but the level set is in the xy-plane.

Of course this cross section itself, in xyz-space, is made up of points all with the same z-coordinate, thus making the set a "level" set of points in the xyz-space. This is the motivation for the term "level set". (Remember of course that the cross section is in $\mathbb{R}^3$ and the level set is in $\mathbb{R}^2$.)

Example 2.4.2. Let's look at how the above discussion applies to Example 2.4.1, where we saw that the level sets of this function are a collection of lines of slope -1 in the $x_1 x_2$-plane.

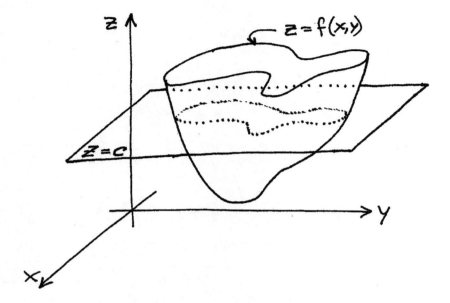

Figure 2.27:

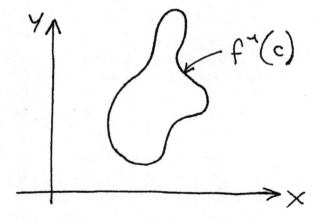

Figure 2.28:

2.4. LEVEL SETS

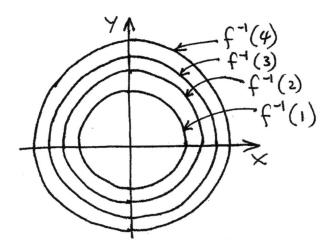

Figure 2.29:

But the function was originally taken from Example 2.3.2 where we saw that the graph is a plane in x_1x_2y-space.

As outlined above, the connection is that if we take that graph in x_1x_2y-space and make horizontal cross sections, in each case we will get a line. Furthermore, if we throw away the z-coordinate and consider each of these lines in the x_1x_2-plane, the slope is always -1.

Specifically, the cross section of the graph plane with the plane $z = c$ always yields exactly the line which is the level set $f^{-1}(c)$.

Example 2.4.3. Consider now the function $f(x, y) = x^2 + y^2$. We know that the graph is the solution set in $\mathbb{R}^3$ to the equation $z = x^2 + y^2$, but what does that surface look like?

We can use level sets to answer that question. Note that we can easily write down all of the level sets for this function – for any value c, the level set $f^{-1}(c)$ is the solution set to the equation

$$x^2 + y^2 = c$$

For positive c these are concentric circles around the origin with radius $\sqrt{c}$, for negative c these are empty sets, and when c is zero this is just the origin. We draw these level sets in Figure 2.29.

We know now though that these level sets are all related to cross sections of the graph. Specifically, the level set $f^{-1}(c)$ is the cross section of the graph at height $z = c$. Since the radius of that cross section is $\sqrt{c}$, we conclude that the height is the square of the distance from the z-axis. This leads us to conclude that the graph is a parabolic bowl, as represented in Figure 2.30.

Note that we arrive here at the same conclusions that we did in Example 2.2.11 and in Subsection 2.2.2.

2.4.3 Advantages

While the graph construction has some highly desirable characteristics, level sets can be highly advantageous in many circumstances, and have some very useful properties.

Most obvious is the dimensional advantage. For instance, in Example 2.4.1 we saw that the level sets of the given function $f : \mathbb{R}^2 \to \mathbb{R}^1$ are curves in $\mathbb{R}^2$, which are easy to draw on a two-dimensional piece of paper.

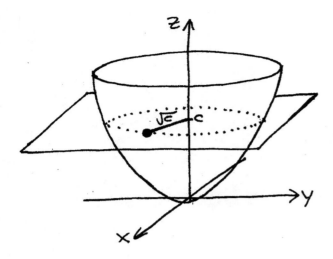

Figure 2.30:

But the graph is a plane in $\mathbb{R}^3$, and it is harder to draw such things. For one thing we cannot draw the figure itself because it is three-dimensional, so we are left only with drawing a projection of the figure. This works reasonably well, but it does require some imagination on the part of the viewer, and so is less than ideal.

Second, we are forced to try to represent a two-dimensional plane with a pencil that is really more naturally suited for drawing one-dimensional curves on the paper. Again, we can get around this using imagination by drawing a seemingly flat rectangle that is part of the plane, and in this case this conveys the desired idea reasonably well. But we must acknowledge that the result is only suggestive of the object in question, and is less than ideal. Certainly also more complex surfaces would be even more difficult to represent in this way.

For functions with three variables the comparison is even stronger. Specifically, note that for a function $f : \mathbb{R}^3 \to \mathbb{R}^1$, the graph is a set in $\mathbb{R}^4$ – and we cannot draw this at all, in any sense.

However, the level sets are surfaces in $\mathbb{R}^3$. Although drawing these surfaces is more difficult and less effective than drawing curves in the plane, it is still something that can be done with reasonable effectiveness.

Example 2.4.4. Let's try to come up with some sort of geometric representation of the function $f(x, y, z) = x^2 + y^2 + z^2$.

As pointed out above, the graph of this function $f : \mathbb{R}^3 \to \mathbb{R}^1$ is four-dimensional, and therefore cannot be drawn in any way.

But level sets give us a reasonable alternative. Note that every level set is a solution set to an equation $x^2 + y^2 + z^2 = c$ for some value of c, and the only non-empty of these sets are concentric spheres around the origin of different radii. So we draw the level sets of this function in Figure 2.31.

Level sets are also more naturally related to equations than functions are, as is pointed out in the following observation.

Observation 2.4.1. *The solution set to any equation can be expressed as a level set of a function.*

We see this very simply by noting that in any equation (let's say involving variables $x_1, \ldots, x_n$), we can simply put all of the terms on the left side of the equation, resulting in an expression on the left side involving those variables. We call this expression f, and then rewrite the equation as

$$f(x_1, \ldots, x_n) = 0$$

Of course the solution set to this equation is precisely the level set $f^{-1}(0)$ of the function f.

2.4. LEVEL SETS

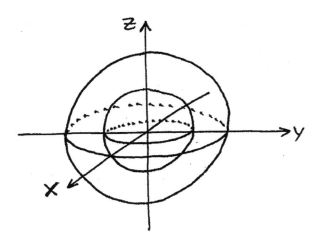

Figure 2.31:

Note that the same cannot be said of graphs – that is, it is certainly not the case the solution set to any equation can be expressed as a graph of a function.

The reason for this is that the equation used to define a graph is in a form where one of the variables is explicitly solved for in terms of the other variables. So, if we begin with an equation for which none of the variables can be solved for, then it simply cannot be put into a form where it can be viewed as a graph.

For example, consider the solution set to the equation $3x^2 + 2y^2 = 12x$, representing an ellipse in the xy-plane. We cannot solve for either variable, so we cannot view this as a graph.

We could also see this geometrically from the vertical line test. No matter which way we order the variables, the ellipse will always fail the vertical line test; therefore it is not the graph of a function.

Example 2.4.5. How can we view the above ellipse as a level set of a function?

As described above, we move everything to the left side of the equation, giving us $3x^2 + 2y^2 - 12x = 0$. Forming the function $f(x, y) = 3x^2 + 2y^2 - 12x$ then, we conclude that the ellipse described by the original equation is the level set $f^{-1}(0)$ of this function f.

As a special case of Observation 2.4.1, we make the following additional observation.

Observation 2.4.2. *For any function $f : \mathbb{R}^n \to \mathbb{R}^1$, the graph of f can be viewed as a level set of some function g.*

Of course this immediately follows from Observation 2.4.1 because every graph is defined from an equation. Note also that this statement cannot be turned around – certainly, there are functions whose level sets cannot be viewed as graphs.

The argument is just a special case of that for the previous observation.

Suppose we have a function $f : \mathbb{R}^n \to \mathbb{R}^1$, defined in terms of the variables $x_1, \ldots, x_n$. The graph of this function is formed by creating new variables y to denote the output of the function, and then looking at the solution set to the equation
$$y = f(x_1, \ldots, x_n)$$
We put everything on the left side of the equation, giving us
$$y - f(x_1, \ldots, x_n) = 0 \tag{2.13}$$
If we then form the new function $g : \mathbb{R}^{n+1} \to \mathbb{R}^1$ defined by
$$g(x_1, \ldots, x_n, y) = y - f(x_1, \ldots, x_n)$$

then certainly the graph of f can be rewritten as the level set $g^{-1}(0)$, because both are equivalent to the expression in Equation 2.13

Exercises

Exercise 2.4.1 (exer-MVF-LS-1). Draw a representative collection of level sets for the function $f : \mathbb{R}^2 \to \mathbb{R}^1$ given by $f(x, y) = 4x^2 + 9y^2$.

Exercise 2.4.2 (exer-MVF-LS-2). Draw a representative collection of level sets for the function $f : \mathbb{R}^2 \to \mathbb{R}^1$ given by $f(x, y) = 3x + 5y$.

Exercise 2.4.3 (exer-MVF-LS-3). Draw a representative collection of level sets for the function $f : \mathbb{R}^2 \to \mathbb{R}^1$ given by $f(x, y) = x + \sin y$.

Exercise 2.4.4 (exer-MVF-LS-4). Draw a representative collection of level sets for the function $f : \mathbb{R}^3 \to \mathbb{R}^1$ given by $f(x, y, z) = x + 2y + 3z$.

Exercise 2.4.5 (exer-MVF-LS-5). Draw a representative collection of level sets for the function $f : \mathbb{R}^3 \to \mathbb{R}^1$ given by $f(x, y, z) = e^{-x^2-y^2} + z$.

Exercise 2.4.6 (exer-MVF-LS-6). Draw a representative collection of level sets for the function $f : \mathbb{R}^3 \to \mathbb{R}^1$ given by $f(x, y, z) = x + y + e^z$.

Exercise 2.4.7 (exer-MVF-LS-7). Find a function g for which one of the level sets is the unit sphere in the xyz-space centered at the point $(3, 2, 5)$.

Exercise 2.4.8 (exer-MVF-LS-8). Find a function h for which one of the level sets is the unit sphere in the xyz-space centered at the point $(3, 2, 5)$ – and which is NOT just a constant plus or minus your answer to Exercise 2.4.7.

Exercise 2.4.9 (exer-MVF-LS-9). Draw the graph and a representative sample of level sets for the function $f : \mathbb{R}^2 \to \mathbb{R}^1$ given by $f(x, y) = x + y^2$. Explain how these level sets relate to the graph.

Exercise 2.4.10 (exer-MVF-LS-10). Find a function h for which one of the level sets is the graph of the function $f : \mathbb{R}^1 \to \mathbb{R}^1$ given by $f(x) = x^2$. What is the domain of h?

Exercise 2.4.11 (exer-MVF-LS-11). Find a function h for which one of the level sets is the graph of the function $f : \mathbb{R}^2 \to \mathbb{R}^1$ given by $f(x, y) = x + y^2$. What is the domain of h?

Exercise 2.4.12 (exer-MVF-LS-12). Find a function h for which one of the level sets is the graph of the function $f : \mathbb{R}^2 \to \mathbb{R}^1$ given by $f(x, y) = xe^{x^2-y^2} - \sin y$. What is the domain of h?

Exercise 2.4.13 (exer-MVF-LS-13). Find a function h for which one of the level sets is the graph of the function $f : \mathbb{R}^4 \to \mathbb{R}^1$ given by $f(x, y, z, w) = x^2 - y^3 + 5z^2 + w^4$. What is the domain of h?

2.5 Parametric Curves and Surfaces

Another alternative to the graph is to represent the function by its "image", which we define below. Like the level set construction, images of functions have the advantage that they require fewer dimensions. We will also see later though that images are still useful even when dimension is not an issue.

Here is the definition.

Definition 2.5.1. *The "image" of the function $f : \mathbb{R}^n \to \mathbb{R}^m$ is the set*
$$\left\{ f(\vec{x}) \in \mathbb{R}^m \,\middle|\, \vec{x} \in \mathbb{R}^n \right\}$$

In simplest terms, this is just the set of all actual output points from the function.

If the function is defined on a domain that is just a subset D of $\mathbb{R}^n$, the definition is similar, again just defining the set of outputs from the function.

Definition 2.5.2. *The "image" of the function $f : (D \subset \mathbb{R}^n) \to \mathbb{R}^m$ is the set*
$$\left\{ f(\vec{x}) \in \mathbb{R}^m \,\middle|\, \vec{x} \in D \right\}$$

We can also call this the image of the domain in the target by the function f.

Recall that the graph of a function is a set in a space defined with both the input and the output variables, while level sets on the other hand are sets contained completely within the domain. Similarly distinct, the image of a function is contained completely within the target space.

There are a few alternative ways to visualize what this construction is doing. We will talk about those throughout this section. First, though, we make a quick and easy observation directly from the definition.

Recall that we can think of a function as picking up a point in the domain, and setting it down at some point in the target; and the image is the set of all of those resulting points in the target. Combining these observations, we can interpret the image as the set in the target space resulting from picking up the entire domain all at once, and placing it into the target space, all at once, in the way prescribed by the function.

From this point of view, we can expect that the dimension of the image sitting inside of the target should probably be the same as the dimension of the domain, because of course it is derived directly from the domain. This is usually the case. So for example we expect the image of a function $f : \mathbb{R}^1 \to \mathbb{R}^n$ to be a one-dimensional curve in $\mathbb{R}^n$.

The image is usually only of interest when the dimension of the domain is smaller than the dimension of the target.

2.5.1 Parametrization

When we studied graphs and level sets we noted that we could use those constructions in either of two opposite ways. On the one hand, given a function, we can form a geometric object that is its graph; instead, given an object that is understood to be a graph, we can try to find the corresponding function.

The image construction that we have just defined allows us to go one of those directions – given a function, we can form the geometric object that is its image. Still we might very naturally wish to turn this around; doing this is called "parametrizing" the object.

For example, a function whose image is a curve is called a parametrization of that curve. A function whose image is a surface is called a parametrization of that surface. We will soon see that given a geometric object, there might be many different parametrizations.

Sometimes we wish to think of one of these geometric objects as being paired with one of its parametrizations. If we have a curve that we wish to think of as paired with a specific parametrization, it is called a "parametric curve". Similarly, a surface paired with a specific parametrization is called a "parametric surface".

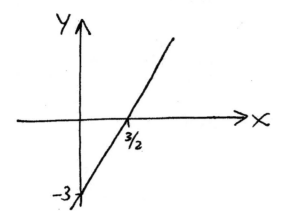

Figure 2.32:

2.5.2 Parametric curves

We will begin with several examples in which the domain is $\mathbb{R}^1$.

Example 2.5.1. Suppose we want to represent the function $f : \mathbb{R}^1 \to \mathbb{R}^2$ defined by $f(t) = (t+1, 2t-1)$.

Of course we can form the graph of this function, as we did for a similar function in Example 2.3.4. As in that example we would find the graph of this function to be a line in $\mathbb{R}^3$.

The image of course is a subset of just the target space $\mathbb{R}^2$, so this figure will be easier to draw. Let's use the xy-plane to be this target space, so the output from the function is $f(t) = (t+1, 2t-1) = (x, y)$.

We need to find the collection of all outputs (x, y) from this function. First, notice that the formula for the function allows us to write down a simple relationship between the x and y coordinates of the output, namely that $y = 2x - 3$. Of course this is the equation of a line, which we can draw very easily, as in Figure 2.32.

So we know that every output point is somewhere on the above line, but it remains to be determined whether every point on the line is an output – or, perhaps just some of them are outputs. In other words, given a point (x_0, y_0) on the given line (in other words, with $y_0 = 2x_0 - 3$), is there actually an input value t for which the function produces (x_0, y_0) as an output?

We can determine that this is indeed the case simply by solving for t in terms of x, as $t = x - 1$. Then given a point (x_0, y_0) on the line, note that the value $t_0 = x_0 - 1$ produces the point

$$f(t_0) = (t_0 + 1, 2t_0 - 1) = \Big((x_0 - 1) + 1, 2(x_0 - 1) - 1\Big) = (x_0, 2x_0 - 3) = (x_0, y_0)$$

as desired.

So, the image of this function is indeed the entire line with equation $y = 2x - 3$. We can also say that this line is parametrized by the function f, and paired together this is a parametric line.

Example 2.5.1 shows that representing a function with its image has an advantage over using a graph, in that the dimension of the picture is lower and thus that the graph is easier to draw. This is even more important for a function $f : \mathbb{R}^1 \to \mathbb{R}^3$, because of course in this case the graph would be an impossible four-dimensional picture, while the image would be in $\mathbb{R}^3$.

The image has a disadvantage though too. Like level sets, the image of a function does not geometrically represent all of

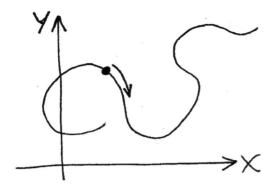

Figure 2.33:

the information about the function. Specifically, even though every output from the function is represented, there is no indication of exactly which input value produced that output point.

We illustrate this problem with the following example.

Example 2.5.2. Consider the function $f : \mathbb{R}^1 \to \mathbb{R}^2$ defined by $f(t) = (t^3 + 1, 2t^3 - 1)$, and again let's find the image of this function. Note the similarity to the function in Example 2.5.1; but of course, these are definitely different functions.

Just as in Example 2.5.1, we notice that every output point satisfies the equation $y = 2x - 3$, so the image is again a subset of that line. Furthermore, every point on that line can again be seen to be included in the image because again we can solve for t in terms of x, this time with $t = \sqrt[3]{x-1}$, and we can confirm that we can use this to find the t_0 that produces any given point (x_0, y_0) on the line.

So in fact, even though this function is clearly different from the one in Example 2.5.1, the image is exactly the same line. Again, this function is a parametrization of the line, and together they are a parametric line.

This is indeed a shortcoming of using the image to represent a function. We can take a different perspective on this issue by making a different interpretation of the image itself.

Notice that in each of the above cases, we used the variable t as the input variable for the given function. This was specifically for the purpose of making the interpretation of the function as representing position (in the target space) as a function of time (the input variable). With this interpretation, we can think of the function as describing the motion of a particle in the target space. See Figure 2.33.

Example 2.5.3. Suppose that we have a snail crawling around on a plane, which we will think of as being the xy-plane. For any given time t, the snail then has a corresponding position (x, y) on the plane. This defines a function $f : \mathbb{R}^1 \to \mathbb{R}^2$, with $f(t) = (x, y)$.

The image of this function f then simply represents the complete set of points where the snail has been throughout its entire crawl. In other words, the image of the function is the set of points defined by the trail of slime that the snail has left behind on the plane, as indicated in Figure 2.34.

This view allows us to take a different perspective on the problem identified by Examples 2.5.1 and 2.5.2.

That is, the problem with the image construction is that while it does represent the complete path that the particle

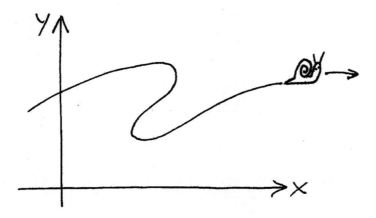

Figure 2.34:

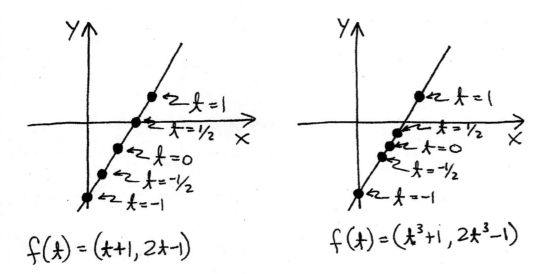

Figure 2.35:

takes, it does not refer in any way to *when* the particle is at any given point on the path, or to the speed it is moving along that path.

In Example 2.5.1, the particle in question is moving along the parametric line at a constant speed, which we see because we can note that both the x- and y-coordinates are moving at constant speeds (we will make this much more precise in Chapter 4). But in Example 2.5.2, even though the particle is tracing out the same path, it is not moving at a constant speed.

This suggests a minor modification that can be added to the image to give a better representation of the parametrization. In order to attempt to denote these differences in speed through the path, we can simply indicate in a few points on the curve at what "time" t the particle is at that point. In Figure 2.35, we do this for the functions in Examples 2.5.1 and 2.5.2.

Note that this distinguishes these two pictures from each other. The first part of the picture suggests that in Example 2.5.1 the particle is moving at a constant speed along the line. In the second part of the picture, we see that the particle defined in Example 2.5.2 seems to be moving fast when t is negative, slows down near $t = 0$, and then speeds up again for positive t. We will be able to confirm these suggestions with a precise analytical discussion in Chapter 4.

2.5. PARAMETRIC CURVES AND SURFACES

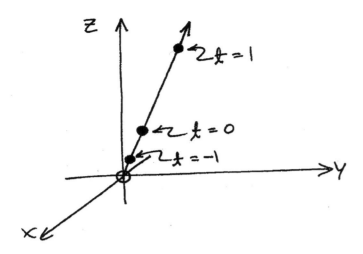

Figure 2.36:

2.5.3 Examples

Here are a few more examples of one-dimensional parametric curves.

Example 2.5.4. Consider the function $f : \mathbb{R}^1 \to \mathbb{R}^3$, defined by $f(t) = (e^t, 2e^t, 3e^t) = (x, y, z)$.

We quickly notice that for every output point, the equations $y = 2x$ and $z = 3x$ are satisfied. Each of these equations represents a plane, and they are not parallel so their intersection must be a line. So we can conclude that the parametric curve must be contained in that line.

Just like in Examples 2.5.1 and 2.5.2 though, we must determine whether the parametric curve includes the entire line, or just part of it. In fact in this case we will see that it includes only part of the line.

As in those previous examples, suppose we have a point on the given line, this time (x_0, y_0, z_0). We need to find a time t_0 such that $f(t_0) = (x_0, y_0, z_0)$. Looking at the x equation and solving for t, we have $t = \ln x$. But of course this has a solution ONLY when x is positive. Similarly for y and z.

So in fact the parametric curve includes ONLY the portion of the given line in the first octant. See Figure 2.36.

Example 2.5.5. Consider the function $f : \mathbb{R}^1 \to \mathbb{R}^2$, defined by $f(t) = (\cos t, \sin t) = (x, y)$.

In this case there is no linear relationship between the coordinates that we can use as a starting point, but the Pythagorean trig identity $\cos^2 t + \sin^2 t = 1$ allows us to conclude that for all of the points (x, y) on this parametric curve, we have $x^2 + y^2 = 1$. So we know that the entire parametric curve is contained on the unit circle.

Note that at $t = 0$, the point is at $(1, 0)$. Between $t = 0$ and $t = \pi/2$, both $\cos t$ and $\sin t$ are positive and at $t = \pi/2$ the point is at $(0, 1)$; so, over that interval of time the point moves counterclockwise across the part of the unit circle in the first quadrant. Similarly we can see that the point continues counterclockwise along the unit circle, returning to the starting point $(1, 0)$ at time $t = 2\pi$.

Of course the function is defined for all values of t, so the function actually represents a particle moving counterclockwise along the unit circle continually. See Figure 2.37.

If we wanted to have a function that represented the unit circle by moving along its length exactly once, we can

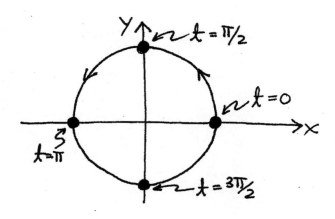

Figure 2.37:

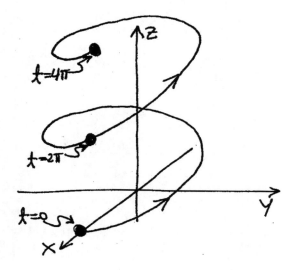

Figure 2.38:

do that simply by restricting the domain of the one above, such as with the function $g : [0, 2\pi) \to \mathbb{R}^2$, defined by $g(t) = (\cos t, \sin t) = (x, y)$.

Example 2.5.6. Suppose we are given the function $f : \mathbb{R}^1 \to \mathbb{R}^3$, defined by $f(t) = (\cos t, \sin t, t) = (x, y, z)$. How can we represent this geometrically?

Of course the graph would be in $\mathbb{R}^4$, so a parametric curve is our only option.

We recognize that, ignoring the z coordinate, the x and y coordinates are identical to those in Example 2.5.5. We could interpret this to say that the projection to the xy-plane of the moving point represented by this function is simply moving around in a circle, as concluded from that example.

On the other hand if we ignore the x and y coordinates, the z-coordinate alone is easy to analyze; the z-coordinate is increasing exactly with time t. So, the height above the xy-plane is increasing exactly with time t.

These two observations taken together tell us exactly what the particle represented by this function is doing. That is, it is moving along a curve called a "helix", as drawn in Figure 2.38.

2.5. PARAMETRIC CURVES AND SURFACES

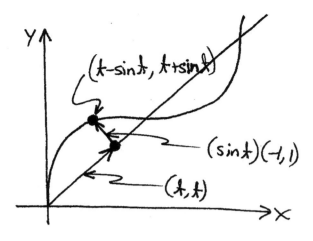

Figure 2.39:

Example 2.5.7. What curve is parametrized by the function $f(t) = (t - \sin t, t + \sin t) = (x, y)$?

Here there are no apparent convenient relationships between the variables x and y. But we can take a different approach using vectors.

First, let's rewrite the given function as $f(t) = (t, t) + (\sin t)(-1, 1)$. Each of these terms is something that we can deal with without much trouble.

For the first term, we observe that the parametric curve $g(t) = (t, t)$ describes a point moving at constant speed along the line passing through the origin with slope **1**.

For the second term, we notice that this is always a scalar multiple of the vector $(-1, 1)$, which points exactly perpendicular to the line above. And the scalar in question is determined by the value of t, in other words, how far $g(t)$ is along the line.

We can then just add these as vectors, and we come up with the curve in Figure 2.39.

2.5.4 Parametric surfaces

Now we will consider a few examples in which the domain is $\mathbb{R}^2$.

Example 2.5.8. Suppose we want to represent the function $f : \mathbb{R}^2 \to \mathbb{R}^3$ defined by $f(u, v) = (u + v, u - v, 3u + v) = (x, y, z)$.

The graph of this function would be a set in $\mathbb{R}^5$, which clearly is not something that will be useful to us. But the image is a subset of $\mathbb{R}^3$, which we can draw acceptably.

We notice that every point (x, y, z) satisfies the equation $z = 2x + y$. So we can immediately conclude that the image of this function must lie entirely in the plane defined by that equation.

Note also that in this case we have the equation written such that we have solved for z. So we could even interpret the plane as being a graph, in this case of the function $g(x, y) = 2x + y$. Of course this function is completely different from the original function f.

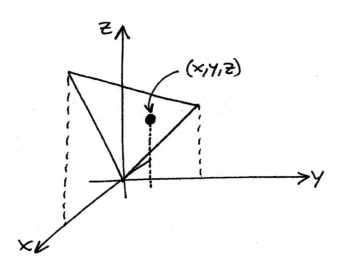

Figure 2.40:

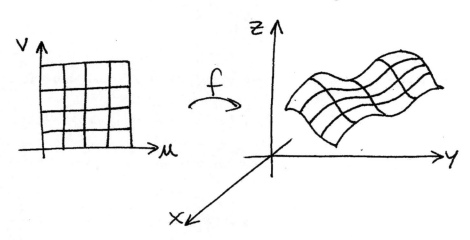

Figure 2.41:

In order to show that the image of f is the entire plane, we just need to find values of u and v that correspond to any given point on the plane. Of course being a graph, every point on the plane is determined by a set of x and y coordinates, so we simply need to show that we can solve for u and v in terms of x and y.

This is not hard to do in this case. We have $u = (x+y)/2$, and $v = (x-y)/2$. So we conclude that the image of the function is the above plane. See Figure 2.40.

We observed at the beginning of this section that we generally expect the dimension of the image of a function to be the same as the dimension of its domain. For a function with a two-dimensional domain then, we expect the image to be a surface. We refer to the image of a function $f : \mathbb{R}^2 \to \mathbb{R}^m$, paired with the function, as a "parametric surface". We also say that the surface is "parametrized" by the function.

We can view the parametric surface as the result of picking up the entire two-dimensional domain, and placing it inside the target space in some way as prescribed by the function. See Figure 2.41.

Example 2.5.9. What is the image of the function

$$f(\phi, \theta) = (\sin\phi\cos\theta, \sin\phi\sin\theta, \cos\phi) = (x, y, z) \tag{2.14}$$

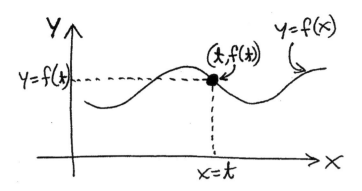

Figure 2.42:

Students can check directly that the above function satisfies the equation $x^2 + y^2 + z^2 = 1$. **So the image of this function is contained entirely on the surface of the unit sphere.**

Of course we can note that this function comes from the definition of spherical coordinates in Section 1.6, with $\rho = 1$. **Choosing the spherical coordinates as the parameters, we see that this function parametrizes the entire unit sphere.**

2.5.5 Graph parametrizations

Here we observe that images have a property similar to that of level sets. That is, we have already shown in Observation 2.4.2 that every graph can be thought of as a level set of a different function. Similarly, here we will show that every graph can be viewed as a image of some different function. That is, every graph can be parametrized by some different function.

We begin by showing this for single variable functions.

Observation 2.5.1. *For any function* $f : \mathbb{R}^1 \to \mathbb{R}^1$, *the graph of* f *can be parametrized by another function* g.

We can see this by simply writing down that new function g, as

$$g(t) = (t, f(t)) = (x, y)$$

It is easily checked that every point (x, y) above satisfies the equation $y = f(x)$, which of course is the equation defining the graph of f.

To see that the parametric curve is indeed the entire graph, we note that it is trivial to solve for t in terms of x. So for any point on the graph there is a corresponding value t that produces that point. See Figure 2.42.

We call this construction the "graph parametrization". A similar observation and proof apply for functions of two variables.

Observation 2.5.2. *For any function* $f : \mathbb{R}^2 \to \mathbb{R}^1$, *the graph of* f *can be parametrized by another function* g.

In this case the function g is

$$g(u, v) = (u, v, f(u, v)) = (x, y, z)$$

Again it is easily checked that every point (x, y, z) above satisfies the equation $z = f(x, y)$, which of course is the equation defining the graph of f.

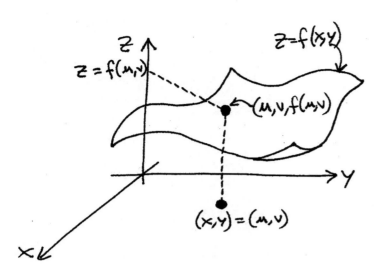

Figure 2.43:

To see that the parametric surface is indeed the entire graph, we note that it is trivial to solve for u and v in terms of x and y. So for any point on the graph there is a corresponding point (u, v) that produces that point. See Figure 2.43.

In each of the above Observations, the idea behind the parametrization of the graph is that every point on the graph is completely defined by the function based on the input variables, and the input variables can simply be reinterpreted as the parameter variables for the new parametrization.

A similar argument shows that the Observations above can be generalized, even to cases where we cannot visualize the graph due to our dimensional restriction.

Observation 2.5.3. *For any function $f : \mathbb{R}^n \to \mathbb{R}^m$, the graph of f can be parametrized by another function g.*

Students are encouraged to show this themselves, working by analogy from the examples above.

Another variation of the above argument allows one to parametrize curves and surfaces that are graphs in other coordinate systems; that is, whenever one can represent the curve or surface by writing one coordinate variable as a function of the other variables in that coordinate system.

Example 2.5.10. Suppose that for a given curve in $\mathbb{R}^2$, we can represent the curve by the equation $r = r(\theta)$. How can we parametrize this curve?

We want to be able to write the position $\vec{x} = (x, y)$ as a function of a single variable. Of course we can write that position in terms of the two polar variables with the standard change of coordinates formula, $(x, y) = (r \cos \theta, r \sin \theta)$; but furthermore since we know r as a function of θ, we can eliminate r and write

$$(x, y) = (r \cos \theta, r \sin \theta) = (r(\theta) \cos \theta, r(\theta) \sin \theta) = \vec{x}(\theta)$$

For example, suppose we consider the unit circle, described in polar coordinates with the equation $r = 1$. Then the parametrization above becomes $\vec{x}(\theta) = (\cos \theta, \sin \theta)$, which we have already seen as a parametrization of the unit circle in Example 2.5.5.

Or suppose we have a more complicated example, such as the curve described in polar coordinates with the equation $r = e^{\sin \theta}$. Then the parametrization above becomes $\vec{x}(\theta) = (e^{\sin \theta} \cos \theta, e^{\sin \theta} \sin \theta)$.

2.5. PARAMETRIC CURVES AND SURFACES

Example 2.5.11. Suppose that for a given surface in $\mathbb{R}^3$, we can represent the surface by the equation $\rho = \rho(\phi, \theta)$. How can we parametrize this surface?

Again we want to be able to write the position $\vec{x} = (x, y, z)$ as a function, this time of two other variables. Again we begin by writing down the change of coordinates formula, and this time note that we can eliminate ρ with the given equation representing the surface, to give

$$\begin{aligned}(x, y, z) &= (\rho \sin\phi \cos\theta, \rho \sin\phi \sin\theta, \rho \cos\phi) \\ &= (\rho(\phi, \theta) \sin\phi \cos\theta, \rho(\phi, \theta) \sin\phi \sin\theta, \rho(\phi, \theta) \cos\phi) \\ &= \vec{x}(\phi, \theta)\end{aligned}$$

We can use this to motivate the parametrization from Example 2.5.9 of the unit sphere, as the unit sphere is represented by the spherical equation $\rho = 1$. (We can also see now that this parametrization covers the entire sphere, because every point on the sphere is easily seen to have corresponding ϕ and θ.)

Or, we can use it to represent something more complicated, such as the surface represented by $\rho = (3 + \sin(4\theta))$. The parametrization above becomes

$$\vec{x}(\phi, \theta) = ((3 + \sin(4\theta)) \sin\phi \cos\theta, (3 + \sin(4\theta)) \sin\phi \sin\theta, (3 + \sin(4\theta)) \cos\phi)$$

Example 2.5.12. Suppose a surface is described by the equation $ye^{xz} - xe^z + z^3 = 0$. How can we parametrize this surface?

While the given algebra is an equation, and it is not given explicitly as a graph, we quickly note that we can indeed solve for y as a function of x and z, giving us $y = e^{-xz}(xe^z - z^3)$. So we can write $\vec{x} = (x, y, z) = (x, e^{-xz}(xe^z - z^3), z)$.

This gives us the position written as a function of two variables, and it would be tempting to have this in its current form represent the parametrization. The problem is that this would have us constructing a function where x and z were both input variables and output variables for the function. In order to get around this problem, we simply rename the input variables as $x = u$ and $z = v$. Then we can happily write our parametrization function as

$$\vec{x}(u, v) = (x, y, z) = (u, e^{-uv}(ue^v - v^3), v)$$

Exercises

Exercise 2.5.1 (exer-MVF-PR-1). Draw the curve that is parametrized by the function given by $\vec{x}(t) = (3t - 4, 2t + 5)$.

Exercise 2.5.2 (exer-MVF-PR-2). Draw the curve that is parametrized by the function given by $\vec{x}(t) = (2t^2 + 1, 3t^2 - 4)$.

Exercise 2.5.3 (exer-MVF-PR-3). Draw the curve that is parametrized by the function given by $\vec{x}(t) = (2t + 1, 3t^2 - 4)$.

Exercise 2.5.4 (exer-MVF-PR-4). Draw the curve that is parametrized by the function given by $\vec{x}(t) = (2\sin t, t, 3\cos t)$.

Exercise 2.5.5 (exer-MVF-PR-5). Draw the curve that is parametrized by the function given by $\vec{x}(t) = (400 \cos t + \cos(12t), 400 \sin t + \sin(12t))$, and describe a simple process that would construct this curve. *(Hint: Think about the moon and the sun.)*

Exercise 2.5.6 (exer-MVF-PR-6). Consider the two functions $f(t) = (t, t^2 - 1)$ and $g(t) = (t^3 - 1, t^6 - 2t^3)$; what is the difference between the two parametric curves represented by these functions?

Exercise 2.5.7 (**exer-MVF-PR-7**). Consider the function $h(t) = (t^2, t^4 - 1)$; what is the difference between the curve parametrized by this function and those in Exercise 2.5.6?

Exercise 2.5.8 (**exer-MVF-PR-8**). What is the relationship between the curves parametrized by $\vec{p}(t) = (t^3, e^t)$ and $\vec{q}(t) = (t^3 + 3, e^t - 4)$?

Exercise 2.5.9 (**exer-MVF-PR-9**). Find a function $\vec{T}$ that parametrizes the curve obtained by translating (by the vector (a, b)) the curve parametrized by $\vec{p}(t) = (p_1(t), p_2(t))$.

Exercise 2.5.10 (**exer-MVF-PR-10**). Find a function $\vec{R}_1$ that parametrizes the curve obtained by reflecting (over the x-axis) the curve parametrized by $\vec{p}(t) = (p_1(t), p_2(t))$. Find another function $\vec{R}_2$ that parametrizes the curve obtained by reflecting (over the y-axis) the curve parametrized by $\vec{p}(t) = (p_1(t), p_2(t))$.

Exercise 2.5.11 (**exer-MVF-PR-11**). Draw the surface that is parametrized by the function given by $\vec{x}(s, t) = (3t - s, 2s + t, 4s)$.

Exercise 2.5.12 (**exer-MVF-PR-12**). Draw the surface that is parametrized by the function given by $\vec{x}(s, t) = (s, t, s^2 + t^2)$.

Exercise 2.5.13 (**exer-MVF-PR-13**). Find a function $\vec{R}_1$ that parametrizes the curve obtained by reflecting (through the xz-plane) the curve parametrized by $\vec{p}(t) = (p_1(t), p_2(t), p_3(t))$.

Exercise 2.5.14 (**exer-MVF-PR-14**). Parametrize the surface described by the equation $z = e^{xy} - y^3$.

Exercise 2.5.15 (**exer-MVF-PR-15**). Parametrize the surface described by the equation $x = y^3 z - yz^3$.

Exercise 2.5.16 (**exer-MVF-PR-16**). Parametrize the surface described by the equation $\rho = \sin \theta \cos(3\phi)$.

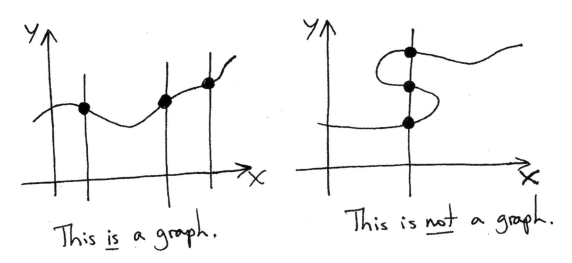

Figure 2.44:

2.6 Using Functions to Represent Curves and Surfaces

We have seen in Sections 2.3, 2.4 and 2.5 that given a function, we have several options for constructions that produce a geometric figure representing that function in some way. Each of these constructions produces a different geometric figure, each with a different interpretation and use.

Here we will consider the opposite question. That is, given a curve or a surface, how can we find a function that represents that geometric figure – using either the graph construction, level set construction, or parametrizing?

This is a question that will come up often in this course. In this section, we present a collection of examples and ideas that are useful when dealing with this question.

2.6.1 Vertical line test

We begin this section with a restatement of an old fact about functions.

Observation 2.6.1 (Vertical Line Test). *A curve in the xy-plane can be viewed as the graph $y = f(x)$ of a function $f : \mathbb{R}^1 \to \mathbb{R}^1$ over some domain in $\mathbb{R}^1$ if and only if there is no vertical line $x = a$ that intersects the curve more than once.*

Of course the idea here is that if there is a vertical line $x = a$ intersecting the curve more than once, say at points (a, y_1) and (a, y_2), then the value $f(a)$ would have to equal both y_1 and y_2 in order for the graph to include both of those points. Since a function can only have a single output value for any given input, this is impossible. See Figure 2.44.

The same is true for functions of several variables. Here we state a version that applies to surfaces in $\mathbb{R}^3$.

Observation 2.6.2 (Vertical Line Test). *A surface in xyz-space can be viewed as the graph $z = f(x, y)$ of a function $f : \mathbb{R}^2 \to \mathbb{R}^1$ over some domain in $\mathbb{R}^2$ if and only if there is no vertical line $(x, y) = (a, b)$ that intersects the surface more than once.*

Again, the idea is that the vertical line tests for situations where an input point, in this case the ordered pair (x, y), might need to give multiple output values, which of course it cannot. See Figure 2.45.

Before we move on, we point out that while we usually think of a graph in xyz-space as representing z as a function of x and y, this is not required. Certainly we could at some point have a function for which x is given as a function of y and z, for example, and the graph would be the set of solutions to the equation $x = f(y, z)$.

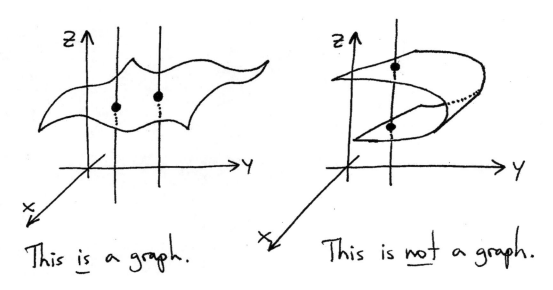

Figure 2.45:

To acknowledge this point, we state another version of the vertical line test

Observation 2.6.3. *A surface in xyz-space can be viewed as the graph $x = f(y, z)$ of a function $f : \mathbb{R}^2 \to \mathbb{R}^1$ over some domain in $\mathbb{R}^2$ if and only if there is no line $(y, z) = (a, b)$ that intersects the surface more than once.*

Of course a similar version can be stated for viewing y as the output variable. We leave this as an exercise for the student.

2.6.2 Examples

Example 2.6.1. How can we represent algebraically the sphere of radius 4 centered at the point $(1, 2, 3)$?

Applying Observation 2.6.2, we can immediately see that we cannot view this surface as the graph of a function of x and y. In fact this surface fails the vertical line test in all three directions, so it cannot be thought of as the graph of a function of any of the three variables.

We can however write this surface as a level set. We observe that the surface is characterized by the fact that every point on this surface is equally distant from a given point; so, the surface is a level set of the function defining distance from that point.

We can write $\sqrt{(x-1)^2 + (y-2)^2 + (z-3)^2} = 4$, or more conveniently in most cases, $(x-1)^2 + (y-2)^2 + (z-3)^2 = 16$. So the surface is the level set $f^{-1}(16)$ of the function

$$f(x, y, z) = (x-1)^2 + (y-2)^2 + (z-3)^2$$

Example 2.6.2. How can we represent algebraically the line in $\mathbb{R}^2$ that passes through the two points $(0, -1)$ and $(1, 1)$, as drawn in Figure 2.46?

This curve passes the vertical line test, so we know that we can represent this curve as the graph of a function. We know that this line has y-intercept equal to -1 and slope equal to 2, so the equation for this line is $y = 2x - 1$.

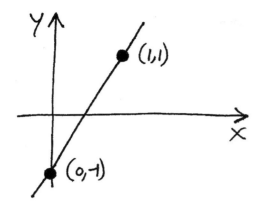

Figure 2.46:

To find a function for which this is the graph, we must put the equation in the form $y = f(x)$, which conveniently it already is. So we conclude that the line is the graph of the function $f(x) = 2x - 1$.

On the other hand, we can also represent this line as a level set. Starting with the equation $y = 2x - 1$, we can put all of the terms onto the left side of the equation to get $y - 2x + 1 = 0$. We can then interpret this as the level set $g^{-1}(0)$ of the function $g(x, y) = y - 2x + 1$.

Finally, we could even parametrize this line. The line passes through the point $(0, -1)$ and is parallel to the vector $(1, 2)$. So, the line is parametrized by the function $h(t) = (0, -1) + t(1, 2)$.

So we have three different functions representing the same line, all by different constructions.

Example 2.6.3. How can we represent algebraically the plane that passes through the points $(1, 0, 0)$, $(0, 1, 0)$ and $(0, 0, 1)$?

We know from the work in Section 1.5 that the equation for this plane is $x + y + z = 1$.

We can write this plane then as the graph $z = f(x, y)$ of the function $f(x, y) = 1 - x - y$.

Or, we can write the plane as the level set $g^{-1}(1)$ of the function $g(x, y, z) = x + y + z$.

Or we can parametrize the plane using the graph parametrization and the function f above, giving us the function $h(u, v) = (u, v, 1 - u - v)$.

Again, this surface is represented by three different functions by way of three different constructions.

Example 2.6.4. How can we represent algebraically the surface S which is the upper half of the unit sphere?

The unit sphere itself has the equation $x^2 + y^2 + z^2 = 1$, and so can be thought of as the level set $g^{-1}(1)$ of the function $g(x, y, z) = x^2 + y^2 + z^2$. But, note that we are not interested specifically in the unit sphere – rather, we want to represent the upper half of the unit sphere. So it will not be convenient to represent this surface as exactly a level set of a function $g : \mathbb{R}^3 \to \mathbb{R}^1$.

Conveniently however, very often we need only have a local representation of a surface, and for this purpose the above function is fine. That is, near any point on the upper half of the unit sphere, the surface can be described as the level set of the function g above. We will see examples of this in Section 4.8.

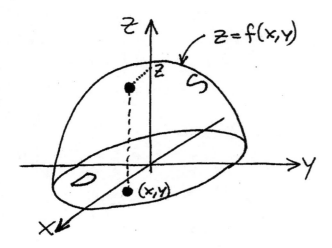

Figure 2.47:

This surface S can be viewed conveniently as a graph though. Solving for z in the above equation for the sphere, we get $z = \pm\sqrt{1 - x^2 - y^2}$ for any (x, y) with $x^2 + y^2 \leq 1$ – which itself cannot be viewed as a function as written because of the "$\pm$". But of course, again, we are not interested in the entire unit sphere, but just the upper part, so we can choose the positive value for z, giving us $z = +\sqrt{1 - x^2 - y^2}$. Therefore we can view S as the graph $z = f(x, y)$ of the function $f : D \to \mathbb{R}^1$, where $D = \{(x, y) | x^2 + y^2 \leq 1\}$ and $f(x, y) = \sqrt{1 - x^2 - y^2}$. See Figure 2.47.

We could parametrize S using the graph parametrization of course. Instead, here we will derive our parametrization from the spherical equation $\rho = 1$, as in Example 2.5.11. This gives us the parametrization

$$\vec{x}(\phi, \theta) = (\sin\phi\cos\theta, \sin\phi\sin\theta, \cos\phi)$$

To use this function to parametrize the surface S, a subset of the unit sphere, we need only restrict the domain of $\vec{x}$ to those values that correspond to the points on S. Note that the points on S are exactly those for which $0 \leq \phi \leq \pi/2$, and furthermore that every point on S has $0 \leq \theta < 2\pi$. So we can parametrize our surface S by the function $\vec{x}_S : D \to \mathbb{R}^3$ where $D = [0, \pi/2] \times [0, 2\pi)$ and $\vec{x}_S(\phi, \theta) = (\sin\phi\cos\theta, \sin\phi\sin\theta, \cos\phi)$.

Example 2.6.5. A bicycle whose wheels have a radius of 16 has a reflector attached to a spoke exactly half of the way from the axle to the tire. As the bicycle rides along the road, the reflector traces out a curve. How can we represent this curve algebraically?

Parametrization will be our most convenient tool for this problem. We set up our picture so that the road is the x-axis, the y-axis is vertical, the bicycle is moving to the right and at the moment when the tire is touching the origin the reflector is directly to the right of the axle. See Figure 2.48.

After the bicycle has moved a bit down the road, the orientation of the wheel has changed, as represented in Figure 2.49.

Let θ represent the angle by which the wheel has rotated, let $\vec{a}$ represent the position of the axle, and let $\vec{x}$ represent the position of the reflector.

Note that the distance the wheel has moved is the same as the length of the arc on the tire where the tire touched the road in that amount of time, which we can write as $d = 16\theta$. This is the x-coordinate then of the axle, which of course is at height 16, so we can write the position vector for the axle as $\vec{a} = (16\theta, 16)$.

We can also conveniently write down the vector $(\vec{x} - \vec{a})$, because we know it has length 8 and it is pointing an angle θ below the horizontal. So we have $(\vec{x} - \vec{a}) = (8\cos\theta, -8\sin\theta)$.

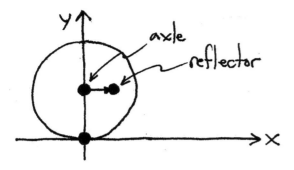

Figure 2.48:

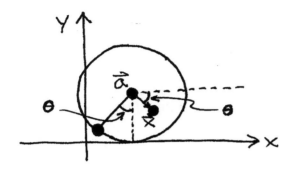

Figure 2.49:

Combining these, we have
$$\vec{x} = \vec{a} + (\vec{x} - \vec{a}) = (16\theta + 8\cos\theta, 16 - 8\sin\theta)$$

This is a parametrization of the curve traced out by the reflector.

Note that in Example 2.6.5, we use an arrow over the name of the function, "$\vec{x}$". The reason for this is that in this case we wish to emphasize the perspective that the value of this function is a vector, and that we will be thinking of it in that way. In fact we thought of the location of the reflector $\vec{x}$ first as being a point, and then subsequently realized that point as a function of θ.

This notation is not required, but in some contexts we will find it to be a useful reminder.

Example 2.6.6. The surface S has the property that every cross section perpendicular to the y-axis at $y = c$ is a circle of radius 3 with center at $(0, c, c)$, as represented in Figure 2.50. How can we represent this surface algebraically?

We know that taking a cross section by the plane $y = c$ is the geometric equivalent of setting y equal to the value c in the original equation. And based on the above description, the result of that is represented by the equation $(x-0)^2 + (z-c)^2 = 9$.

So, we conclude that the equation for this surface is $x^2 + (z-y)^2 = 9$.

Example 2.6.7. The surface S is the hyperboloid of two sheets obtained by rotating around the x-axis the hyperbola in the xz-plane with equation $x^2 - z^2 = 7$. How can we represent this surface algebraically?

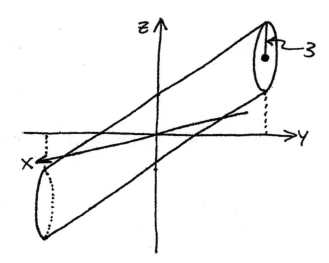

Figure 2.50:

We saw in Subsection 2.2.4 how to take an equation of a certain form and realize that it represents a rotated surface. In this example, we use the same idea in the opposite order.

Note that, because of the symmetries of the hyperbola, the surface obtained by rotating the entire hyperbola around the x-axis is the same as that obtained by rotating only the $z \geq 0$ part.

Given that S is a rotated surface around the x-axis, we know that we can write its equation as $g(x, \sqrt{y^2 + z^2}) = 0$, for some function g. We also know from the above description and the theorems from Subsection 2.2.4 that $g(x, z) = x^2 - z^2 - 7$.

So we can conclude then that the equation for the surface S is $x^2 - (y^2 + z^2) - 7 = 0$, or, $x^2 - y^2 - z^2 = 7$.

Example 2.6.8. The surface S is the ellipsoid with center at $(1, 2, 3)$, radius 4 in the x-direction, radius 5 in the y-direction, and radius 6 in the z-direction. How can we represent it algebraically?

First note that S can be obtained from the unit sphere by first stretching by factors 4, 5, and 6 in the x-, y-, and z-directions respectively, and then translating by the amounts 1, 2, and 3 in the x-, y-, and z- directions respectively.

The unit sphere of course has equation $x^2 + y^2 + z^2 = 1$. Performing the algebraic adjustments corresponding to the stretches (see Subsection 2.2.3), we get

$$\frac{x^2}{4^2} + \frac{y^2}{5^2} + \frac{z^2}{6^2} = 1$$

Then performing the algebraic adjustments corresponding to the translations, we arrive at a final equation for the given ellipsoid,

$$\frac{(x-1)^2}{16} + \frac{(y-2)^2}{25} + \frac{(z-3)^2}{36} = 1 \tag{2.15}$$

On the other hand, we could also use the above observations of the geometric relationship between the unit sphere and S to adjust the parametrization of the sphere from Example 2.6.4.

Of course we have to be careful not to move too quickly here – recall that the rules from Subsection 2.2.3 refer to changes to an equation and the corresponding changes in the solution set to that equation. But a parametric surface is not a solution set to the algebra that represents it, so we cannot use those rules.

The good news is that for a parametric surface the rules are even simpler. The point here is that a function is thought of as a parametrization by the actual points that come out of the function, not points that go into the function. So, if we want to do something to a parametric surface, we simply apply the transformations directly to the parametrization.

We have the parametrization $\vec{h}(\phi, \theta) = (\sin \phi \cos \theta, \sin \phi \sin \theta, \cos \phi)$ for the sphere. Stretching as described above gives us $\vec{h_1}(\phi, \theta) = (4 \sin \phi \cos \theta, 5 \sin \phi \sin \theta, 6 \cos \phi)$, and then translating gives us

$$\vec{h_2}(\phi, \theta) = (4 \sin \phi \cos \theta + 1, 5 \sin \phi \sin \theta + 2, 6 \cos \phi + 3) \tag{2.16}$$

Students can check for themselves that the outputs from the function defined in Equation 2.16 do indeed satisfy Equation 2.15.

Exercises

Exercise 2.6.1 (exer-MVF-UFRCS-1). For each of the following curves and surfaces, determine if it can be viewed in some way as the graph of a function. If it can, identify which ways (which variable is a function of the others) it can be viewed as a graph, and find expressions for the corresponding functions.

1. The part of the unit circle in the upper half plane.
2. The part of the unit circle in the first quadrant.
3. The part of the unit circle in the right half plane.
4. The set of solutions to the equation $3x - 5y = 21$.
5. The set of solutions to the equation $x^2 - y^3 = 0$.
6. The part of the unit sphere with $y \geq 0$ and $z \leq 0$.
7. The part of the unit sphere with $x \leq \frac{1}{2}$.
8. The set of solutions to the equation $x^2 y^3 = z^5$.

Exercise 2.6.2 (exer-MVF-UFRCS-2). For each of the following curves and surfaces, determine if it can be viewed in some way as the graph of a function. If it can, identify which ways (which variable is a function of the others) it can be viewed as a graph.

1. The set of solutions to the equation $e^y + y^3 = \sin x$. (Hint: You can make an argument based on the fact that the left side of this equation is an increasing function of y.)
2. The path of the reflector in Example 2.6.5.

Exercise 2.6.3 (exer-MVF-UFRCS-3). For each of the following curves and surfaces, find a function for which it is a level set.

1. The unit circle.
2. The ellipse centered at the origin with semi-major radius 4 along the x-axis and semi-minor radius 2 along the y-axis.
3. The surface obtained by rotating around the y-axis the curve $x = y^2 + 3$.

4. The surface obtained by rotating around the x-axis the part of the curve $y = \sin x$ with $y \geq 0$.

5. The surface obtained by rotating around the x-axis the entire curve $y = \sin x$. *(Hint: Think about what happens on parts of the curve when y is negative, and how this problem can be removed by squaring.)*

Exercise 2.6.4 (exer-MVF-UFRCS-4). Parametrize each of the following curves and surfaces.

1. The ellipse centered at the origin with semi-major radius 4 along the x-axis and semi-minor radius 2 along the y-axis.

2. The surface obtained by rotating around the y-axis the curve $x = y^2 + 3$.

3. The sphere of radius 2 centered at the point $(3, 2, 5)$.

4. The graph $z = f(x, y)$ of the function $f(x, y) = x^5 - e^y + \sin(xy^2)$.

5. The torus obtained by rotating around the y-axis the circle of radius 3 centered at $(5, 0)$.

Exercise 2.6.5 (exer-MVF-UFRCS-5). Parametrize the path of the reflector on the bicycle in each of the following cases. (In each, assume the radius of the wheel is 16, the wheel is rolling through xyz-space using the xy-plane as the rolling surface, and at the given starting point the reflector is directly beneath the axle of the wheel.)

1. The wheel is moving straight along the x-axis, starting at the origin, and the reflector is a fixed distance of 10 from the axle.

2. The wheel is moving straight along the x-axis, starting at the origin, but the reflector is attached to the wheel by a spring, causing its distance to the axle to change depending on the angle. Specifically, the distance d to the axle is given by $d = 10 - 2\sin\beta$, where β is the angle from the axle between the reflector and the horizontal ($\beta = -\pi/2$ at the start when the reflector is directly beneath the axle).

3. The wheel is moving around the circle of radius 100 centered at the origin, starting at the point $(100, 0)$, and the reflector is a fixed distance of 10 from the axle.

4. The wheel is moving around the circle of radius 100 centered at the origin, starting at the point $(100, 0)$, the reflector is a fixed distance of 10 from the axle, and the wheel is leaning "into the turn" (toward the origin) by $30°$.

2.7. LIMITS AND CONTINUITY

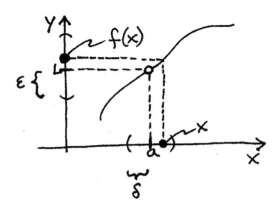

Figure 2.51:

2.7 Limits and Continuity

The notion of a limit is one of the most fundamental and important notions in calculus. In both single variable and multivariable calculus, the limit is required in order to make proper definitions for both the derivative and the integral. In this section we will discuss the definition of the multivariable limit, and the related notion of continuity.

2.7.1 Definition

The definition of a multivariable limit is entirely analogous to that for a single variable limit. In order to extend that analogy, we begin with a reminder of the definition of the limit of a single variable function.

Definition 2.7.1. *Given a function* $f : \mathbb{R}^1 \to \mathbb{R}^1$ *and a point a in the domain, we define the limit by the condition*

$$\lim_{x \to a} f(x) = L \iff \left(\begin{array}{c} \forall \epsilon > 0 \quad \exists \delta > 0 \quad \text{such that} \\ (0 < |x - a| < \delta) \implies (|f(x) - L| < \epsilon) \end{array} \right)$$

Students should have seen this definition in their single variable calculus courses. Just as a reminder, the symbol "$\forall$" is shorthand for the phrase "for all" or "for every", and the symbol "$\exists$" is shorthand for the phrase "there is" or "there exists".

The idea behind this definition is that in order to say that the limit $\lim_{x \to a} f(x)$ exists and equals a value L, we would like to be able to say that the function gets "arbitrarily close" to the value L when x gets "sufficiently close" to a.

Said differently, we would like to be able to say that, no matter how close we want to require the value $f(x)$ to be to the value L (that is, closer than the distance ϵ), we can ensure that this will be the case simply by requiring that x be close to a (that is, closer than the distance δ).

(The additional requirement that $0 < |x - a|$ is simply a way of saying $x \neq a$; that is, we do not want the definition of a limit to depend in any way on the value of the function f at the point a itself.)

Students have probably already seen the geometric representation of this definition and the variables involved, as in Figure 2.51, showing the graph of the function f whose limit is under consideration. Remember though that we are interested in generalizing the notion of a limit to multivariable functions, and for most multivariable functions the graph is hard to draw, if not impossible. So, in Figure 2.52 we make an equivalent representation, showing the domain and the target of the function f as separate spaces.

We can extend this to functions with multivariable domains by making only a simple change.

The input points to such a function are now vector points such as $\vec{x}$ and $\vec{a}$, not just numbers like x and a. But we would like to keep the same interpretation. That is, we would like to be able to say that, no matter how close we want to

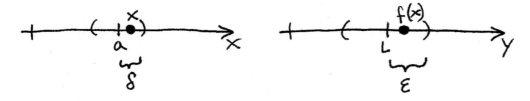

Figure 2.52:

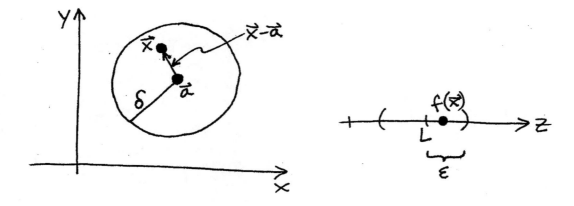

Figure 2.53:

require the value $f(x)$ to be to the value L (that is, closer than the distance ϵ), we can ensure that this will be the case simply by requiring that $\vec{x}$ be close to $\vec{a}$ (that is, closer than the distance δ).

So, we need only to consider how to phrase symbolically that a point $\vec{x}$ is within a given distance δ of a point $\vec{a}$. Of course that distance is just the length of the difference vector $\vec{x} - \vec{a}$.

This leads us to make the following definition for the limit of a function $f : \mathbb{R}^n \to \mathbb{R}^1$.

Definition 2.7.2. *Given a function $f : \mathbb{R}^n \to \mathbb{R}^1$ and a point $\vec{a}$ in the domain, we define the limit by the condition*

$$\lim_{\vec{x} \to \vec{a}} f(\vec{x}) = L \iff \begin{pmatrix} \forall \epsilon > 0 \quad \exists \delta > 0 \quad such\ that \\ (0 < \|\vec{x} - \vec{a}\| < \delta) \implies (|f(\vec{x}) - L| < \epsilon) \end{pmatrix}$$

In Figure 2.53 we make a geometric representation of this definition analogous to Figure 2.52 for single variable limits. Note that the only difference is that in the multivariable case, the set of points within a certain distance of the point $\vec{a}$ is now a solid ball (a disk, when the domain is $\mathbb{R}^2$ as in the figure), as opposed to simply an interval in $\mathbb{R}^1$.

In the case where $\vec{a} = \vec{0} \in \mathbb{R}^2$, there is a very convenient way to rephrase this. Note that in this case the distance $\|\vec{x} - \vec{a}\|$ is given simply by $|r|$, where $x = r\cos\theta$ and $y = r\sin\theta$. Substituting that expression into Definition 2.7.2, we find that we can rephrase the multivariable limit as a single variable limit in the variable r.

$$\lim_{(x,y) \to (0,0)} f(x,y) = \lim_{r \to 0} f(r\cos\theta, r\sin\theta) \tag{2.17}$$

Of course this is only valid when $\vec{a} = \vec{0} \in \mathbb{R}^2$, but this can be very handy in such a circumstance.

Example 2.7.1. **What is the limit**

$$\lim_{(x,y) \to (0,0)} \frac{x^3 y^2}{x^2 + y^2}$$

2.7. LIMITS AND CONTINUITY

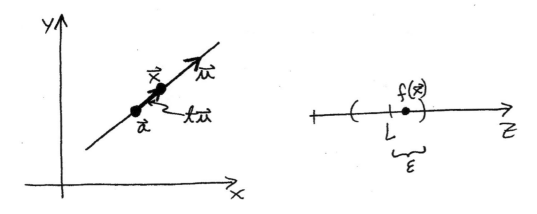

Figure 2.54:

As in Equation 2.17, we can rewrite this as

$$\lim_{r \to 0} \frac{(r\cos\theta)^3(r\sin\theta)^2}{(r\cos\theta)^2 + (r\sin\theta)^2} = \lim_{r \to 0} r^3 \cos^3\theta \sin^2\theta$$

This is now a single variable limit, and we have theorems that allow us to compute those. In this case, we note that $-1 \leq \cos^3\theta \sin^2\theta \leq 1$ and $r^3 \to 0$, so we know that this must be zero.

So, the original multivariable limit is zero.

2.7.2 Misinterpretation and counterexample

A discussion of single variable limits very often includes a presentation of the ideas of left-hand limits and right-hand limits. These can be defined precisely (similarly to the above definition), and are often visualized as letting the input point x get closer and closer to the limit point a, only from the one side or the other.

There is a theorem that relates these one-sided limits to the idea of an actual limit, saying that if these two one-sided limits both exist and equal each other, then the actual limit also exists and equals the same value.

It is extremely tempting to students to try to do the same sort of thing for multivariable limits. Here we will make a formulation of such an attempt.

Since the domain is now more than one dimensional, one could think of approaching the limit point $\vec{a}$ along an infinite number of different lines, rather than just two. We could identify such a line of approach by either one of the two unit vectors pointing along that line from the point $\vec{a}$; and given such a unit vector $\vec{u}$, we can then parametrize that line by $\vec{x} = \vec{a} + t\vec{u}$. Approaching the limit point $\vec{a}$ "along this line" can then be thought of by letting the single parameter t approach zero. This is pictured in Figure 2.54.

Can we say that the multivariable limit $\lim_{\vec{x} \to \vec{a}} f(\vec{x})$ exists if all of the single variable limits of the values of f, approaching along all such lines, all exist and equal the same value? More precisely,

Question: Can we make a definition equivalent to Definition 2.7.2 by saying that

$$\lim_{\vec{x} \to \vec{a}} f(\vec{x}) = L \stackrel{?}{\iff} \left(\forall \text{ unit vectors } \vec{u}, \lim_{t \to 0} f(\vec{a} + t\vec{u}) = L \right)$$

The answer to this question is emphatically NO. This is not an equivalent statement to that in Definition 2.7.2. There are actually many counterexamples showing this statement to be false.

Example 2.7.2. Consider the function $\theta : (\mathbb{R}^2 - \{\vec{0}\}) \to \mathbb{R}^1$ defined by letting $\theta(\vec{x})$ be the angle measured from the positive part of the x-axis to the vector $\vec{x}$, going counterclockwise. In this instance we want to make sure that this function does not take the value 0 though, so we will define $\theta(\vec{x})$ on the positive part of the x-axis itself to be 2π.

We use this function to define the function $f : (\mathbb{R}^2 - \{\vec{0}\}) \to \mathbb{R}^1$ by the formula

$$f(\vec{x}) = \frac{\|\vec{x}\|}{\theta(\vec{x})}$$

We consider now the limit

$$\lim_{\vec{x} \to \vec{0}} f(\vec{x})$$

First, let us consider approaching the origin along straight lines. We note that along any ray extending from the origin, the value of θ is constant. So, for any line extending from the origin, we have

$$\lim_{t \to 0} f(\vec{0} + t\vec{u}) = \lim_{t \to 0} f(t\vec{u})$$
$$= \lim_{t \to 0} \frac{\|t\vec{u}\|}{\theta(t\vec{u})}$$
$$= 0$$

So it is tempting to say that the multivariable limit $\lim_{\vec{x} \to \vec{0}} f(\vec{x})$ is zero.

However, this cannot be the case, as we see by considering the approach to the origin along the spiral curve defined by the equation $\|\vec{x}\| = \theta(\vec{x})$. It is easy to see that at every point on this spiral, the value of f is equal to 1. So there is no way that the limit of this function could equal zero.

The lesson from the above example is that we simply cannot define a multivariable limit in terms of the limits of the function along straight line paths to the given limit point.

2.7.3 Alternative interpretation

Despite the failure of the attempt made in Subsection 2.7.2, it turns out that there is a way to formulate an alternative equivalent to Definition 2.7.2, using single variable limits.

The mistake that was made in the attempt in Subsection 2.7.2 was using only straight line paths to the limit point $\vec{a}$. If instead we use all possible continuous paths, we can make an equivalent formulation of the multivariable limit.

In order to state the theorem, we must first define what we mean by a continuous path.

Definition 2.7.3. *A "continuous path in $\mathbb{R}^n$" is a parametric curve $\vec{x} : \mathbb{R}^1 \to \mathbb{R}^n$ such that all of the component functions $x_i : \mathbb{R}^1 \to \mathbb{R}^1$ are continuous.*

Using this definition, we can state the theorem.

Theorem 2.7.1. *For a function $f : \mathbb{R}^n \to \mathbb{R}^1$, we can reformulate the definition of the limit by saying*

$$\lim_{\vec{x} \to \vec{a}} f(\vec{x}) = L \iff \left(\forall \text{ continuous paths } \vec{x} \text{ with } \vec{x}(0) = \vec{a}, \quad \lim_{t \to 0} f(\vec{x}(t)) = L \right)$$

We will not prove this theorem in this text.

2.7. LIMITS AND CONTINUITY

It should be noted that in most mathematical settings, this formulation is LESS useful than Definition 2.7.2. It should certainly not be thought of as a replacement for that definition.

However, the above formulation has the advantage that it connects the notion of a multivariable limit to the perhaps more familiar notion of a single variable limit. And we will also find that there are many examples of cases where this theorem can be used to show that certain limits do not exist.

This latter use is a matter of considering the cases where the condition in Theorem 2.7.1 fails. We state two such cases.

Theorem 2.7.2. *For a function $f : \mathbb{R}^n \to \mathbb{R}^1$, if there exists a continuous path $\vec{x} : \mathbb{R}^1 \to \mathbb{R}^n$ with $\vec{x}(0) = \vec{a}$, and if $\lim_{t \to 0} f(\vec{x}(t))$ does not exist, then the multivariable limit*

$$\lim_{\vec{x} \to \vec{a}} f(\vec{x})$$

does not exist.

Theorem 2.7.3. *For a function $f : \mathbb{R}^n \to \mathbb{R}^1$, if there exist continuous paths $\vec{x}_1 : \mathbb{R}^1 \to \mathbb{R}^n$, $\vec{x}_2 : \mathbb{R}^1 \to \mathbb{R}^n$ with $\vec{x}_1(0) = \vec{a} = \vec{x}_2(0)$, and if $\lim_{t \to 0} f(\vec{x}_1(t)) \neq \lim_{t \to 0} f(\vec{x}_2(t))$, then the multivariable limit*

$$\lim_{\vec{x} \to \vec{a}} f(\vec{x})$$

does not exist.

Here are some examples of how these theorems can be used.

Example 2.7.3. Compute the limit

$$\lim_{\vec{x} \to \vec{0}} \frac{xy}{x^2 + y^2}$$

In an attempt to use one of the above theorems, in this case we will compute the single variable limits along several different paths to the origin; if any of these single variable limits fails to exist, or if for any two of these paths we get different limits, then the multivariable limit in question cannot exist, by Theorems 2.7.2 and 2.7.3.

First, let's try approaching along the y-axis. We can parametrize this with the observation that $x = 0$, giving us the parametrization $\vec{x}(t) = (0, t)$. Then the limit along this path becomes

$$\lim_{t \to 0} f(\vec{x}(t)) = \lim_{t \to 0} f(0, t) = \lim_{t \to 0} \frac{0}{t^2} = 0$$

We can try more lines, which we could do one at a time, but it is even more convenient to do all of the remaining lines through the origin all at once. This is easily done by noting that every line through the origin (other than the y-axis) can be described by an equation $y = mx$ for some value m, and then can be parametrized by $\vec{x}(t) = (t, mt)$. So for any of these paths, we get

$$\lim_{t \to 0} f(\vec{x}(t)) = \lim_{t \to 0} f(t, mt) = \lim_{t \to 0} \frac{(t)(mt)}{(t)^2 + (mt)^2} = \frac{m}{1 + m^2}$$

This result exists for all values of m, so all of these single variable limits exist. However – we get different values of this expression for different values of m. This means that along different lines, the corresponding single variable limit takes different values. Therefore, by Theorem 2.7.3, the multivariable limit in question does not exist.

Example 2.7.4. Compute the limit

$$\lim_{\vec{x} \to \vec{0}} \frac{x^3 y}{x^7 + y^2}$$

As in Example 2.7.3 we will try different paths to the origin and compute the corresponding single variable limits, until we hopefully find either one that fails to exist or two that take different values.

Again, we do the y-axis first, and easily see that this limit is zero.

Next we try all remaining straight lines again, using the same parametrization as in Example 2.7.3. We compute

$$\lim_{t \to 0} f(\vec{x}(t)) = \lim_{t \to 0} f(t, mt) = \lim_{t \to 0} \frac{mt^4}{t^7 + m^2 t^2} = \lim_{t \to 0} \frac{mt^2}{t^5 + m^2} = 0$$

At this point we have that the limit along all straight lines through the origin is the same value, zero. But we know from the discussion in Subsection 2.7.2 that this is NOT sufficient.

Now we must try some curved paths to the origin. One easy path to try would be along the parabola with equation $y = x^2$. Parametrizing by $\vec{x}(t) = (t, t^2)$, we get

$$\lim_{t \to 0} f(\vec{x}(t)) = \lim_{t \to 0} f(t, t^2) = \lim_{t \to 0} \frac{t^5}{t^7 + t^4} = \lim_{t \to 0} \frac{t}{t^3 + 1} = 0$$

Again we get the same value, and so we must try another curve through the origin. This time let's try the cubic curve $y = x^3$, which we parametrize by $\vec{x}(t) = (t, t^3)$, giving us the limit

$$\lim_{t \to 0} f(\vec{x}(t)) = \lim_{t \to 0} f(t, t^3) = \lim_{t \to 0} \frac{t^6}{t^7 + t^6} = \lim_{t \to 0} \frac{1}{t + 1} = 1$$

Finally we have found a path to the origin along which the single variable limit of the given function is a different value – then applying Theorem 2.7.3, we conclude that the given multivariable limit does not exist.

Students should make certain to understand however that this method of trying different paths can ONLY be used to show that a limit does not exist – because no matter how many paths one uses, one cannot explicitly try ALL such paths; there are just too many of them. If you try lots of paths and keep getting the same single variable limit, you might suspect that the limit exists and equals this value – but this method cannot be used to confirm that your suspicion is correct.

2.7.4 Continuity

As was the case with limits, the notion of continuity for multivariable functions is entirely analogous to continuity for single variable functions.

Definition 2.7.4. *A function $f : \mathbb{R}^n \to \mathbb{R}^1$ is "continuous at the point $\vec{a}$" iff*

$$\lim_{\vec{x} \to \vec{a}} f(\vec{x}) = f(\vec{a})$$

(It is implicit that in order for the two items in the above equation to be equal, they certainly must both exist.)

Continuity is a critical idea that has a wide variety of different uses. One of those uses is in the computation of limits.

Specifically, if we should happen to know that a function is continuous, then Definition 2.7.4 can be rephrased as a formula for computing multivariable limits.

2.7. LIMITS AND CONTINUITY

Theorem 2.7.4. *If a function $f : \mathbb{R}^n \to \mathbb{R}^1$ is continuous at the point $\vec{a}$, then we can compute the limit by*

$$\lim_{\vec{x} \to \vec{a}} f(\vec{x}) = f(\vec{a}).$$

In the absence of other information, this theorem is useless, because of course using this theorem requires the establishment of continuity, and determining continuity from the definition requires the computation of the limit anyway.

The good news here is that an enormous number of these computations have already been done, and we state them here as theorems. We will not prove these theorems in this text.

Theorem 2.7.5. *The following types of functions are all continuous where they are defined: polynomials, trig functions, logarithms, exponentials, rational functions.*

Theorem 2.7.6. *If the functions $f, g : \mathbb{R}^n \to \mathbb{R}^1$ are both continuous at the point $\vec{a}$ and if c is a constant, then the following functions are also continuous at the point $\vec{a}$: $f + g$, fg, cf, and f/g (assuming $g(\vec{a}) \neq 0$).*

These theorems allow us to compute an enormous number of limits simply by computing the value of the function.

Example 2.7.5. Compute the limit

$$\lim_{(x,y,z) \to (2,1,3)} \frac{x^2 + zy^3}{xy + e^z}$$

This denominator is the sum of two continuous functions, so we know from Theorem 2.7.6 that it is continuous. The numerator is a polynomial, so we know from Theorem 2.7.5 that it is continuous. We then have the ratio of two continuous functions, so Theorem 2.7.6 tells us that it is continuous.

So we can compute the limit simply by evaluating the function.

$$\lim_{(x,y,z) \to (2,1,3)} \frac{x^2 + zy^3}{xy + e^z} = \left. \frac{x^2 + zy^3}{xy + e^z} \right|_{(2,1,3)} = \frac{7}{2 + e^3}$$

2.7.5 Vector valued functions

Up to now we have only defined limits and continuity for real valued multivariable functions.

The definition of the limit of a vector valued function $f : \mathbb{R}^n \to \mathbb{R}^m$ can be made explicitly in terms of ϵ's and δ's, but instead here we will make the definition by relating the vector valued functions to its component functions.

Definition 2.7.5. *Given a function $f : \mathbb{R}^n \to \mathbb{R}^m$, written in components as $f(\vec{x}) = (f_1(\vec{x}), \ldots, f_m(\vec{x}))$, and a point $\vec{a}$ in the domain, we define the existence and the value of the limit by the equation*

$$\lim_{\vec{x} \to \vec{a}} f(\vec{x}) = \lim_{\vec{x} \to \vec{a}} \begin{bmatrix} f_1(\vec{x}) \\ \vdots \\ f_m(\vec{x}) \end{bmatrix} = \begin{bmatrix} \lim_{\vec{x} \to \vec{a}} f_1(\vec{x}) \\ \vdots \\ \lim_{\vec{x} \to \vec{a}} f_m(\vec{x}) \end{bmatrix}$$

That is, the limit exists iff each of the component functions has a limit, and the value of the limit is the vector whose components are the limits of those component functions.

We abbreviate this theorem by saying that limits of vector valued functions are computed "componentwise".

Continuity for vector valued functions is defined in the same way as for real valued functions.

Definition 2.7.6. *A function $f : \mathbb{R}^n \to \mathbb{R}^m$ is "continuous at the point $\vec{a}$" iff*
$$\lim_{\vec{x} \to \vec{a}} f(\vec{x}) = f(\vec{a})$$

This could also be viewed as a "componentwise" definition.

These definitions allow us to state another theorem that students can use to determine continuity of a function and thus will help in the computation of limits.

Theorem 2.7.7. *If the function $f : \mathbb{R}^n \to \mathbb{R}^m$ is continuous at the point $\vec{a}$, and if the function $g : \mathbb{R}^m \to \mathbb{R}^k$ is continuous at the point $f(\vec{a})$, then the composition $g \circ f : \mathbb{R}^n \to \mathbb{R}^k$ is continuous at the point $\vec{a}$.*

We show the use of this theorem in the following example.

Example 2.7.6. Compute the limit
$$\lim_{(x,y,z) \to (5,5,2)} e^{xyz}$$

This function is the composition of the polynomial xyz and the exponential function. Both of these are continuous by Theorem 2.7.5, and so by Theorem 2.7.7 we know that the composition is continuous. We can then compute the limit by Theorem 2.7.4, giving us

$$\lim_{(x,y,z) \to (5,5,2)} e^{xyz} = e^{xyz}\Big|_{(5,5,2)} = e^{50}$$

Exercises

Exercise 2.7.1 (exer-MVF-LC-1). In this problem we consider the function $f(x) = x^2$. Suppose we are attempting to use the definition to show that $\lim_{x \to 0} f(x) = 0$, and we are considering the value of $\epsilon = .5$. What is the largest value of δ that will still satisfy the needed implication from the definition?

Exercise 2.7.2 (exer-MVF-LC-2). In this problem we consider the function $f(x) = x^2$. Suppose we are attempting to use the definition to show that $\lim_{x \to 1} f(x) = 1$, and we are considering the value of $\epsilon = .21$. What is the largest value of δ that will still satisfy the needed implication from the definition?

Exercise 2.7.3 (exer-MVF-LC-3). In this problem we consider the function $f(x) = x^2$. Suppose our friend Bob is attempting to use the definition to show that $\lim_{x \to 1} f(x) = 3$ (which of course is not true). How small does ϵ need to be in order for you to be able to be sure that there is no value of δ that will satisfy the needed implication from the definition?

Exercise 2.7.4 (exer-MVF-LC-16). In this problem we consider the function $f(x,y) = 3x^2y - 2e^{xy^2}$. Suppose our friend Bob is attempting to use the definition to show that $\lim_{(x,y) \to (1,0)} f(x,y) = 3$ (which of course is not true). How small does ϵ need to be in order for you to be able to be sure that there is no value of δ that will satisfy the needed implication from the definition?

Exercise 2.7.5 (exer-MVF-LC-4). In this problem we consider the function $f(x,y) = x^2 + y^4$. Suppose we are attempting to use the definition to show that $\lim_{\vec{x} \to \vec{0}} f(\vec{x}) = 0$, and we are considering the value of $\epsilon = .25$. What is the largest value of δ that will still satisfy the needed implication from the definition? Make sure to show that your chosen value does work, and also show that any larger value cannot work. *(Hint: You can make use of the fact that when $|a| < 1$, we have $a^4 < a^2$.)*

Exercise 2.7.6 (exer-MVF-LC-14). Suppose that it is claimed that $\lim_{x \to 1} f(x) = 3$ and it is also claimed that $\lim_{x \to 1} f(x) = 8$. Is it possible that both of these claims could satisfy the definition of a limit? *(Hint: Consider $\epsilon = 2$, and show that there must be a contradiction.)*

2.7. LIMITS AND CONTINUITY

Exercise 2.7.7 (exer-MVF-LC-15). Suppose that it is claimed that $\lim_{\vec{x} \to \vec{a}} f(\vec{x}) = 3$ and it is also claimed that $\lim_{\vec{x} \to \vec{a}} f(\vec{x}) = 8$. Is it possible that both of these claims could satisfy the definition of a limit? *(Hint: Use a technique similar to that from Exercise 2.7.6.)*

For exercises 2.7.8-2.7.15, determine if the given limit exists. If the limit does exist, compute the value of that limit. In each case make sure to justify your claim.

Exercise 2.7.8 (exer-MVF-LC-5).
$$\lim_{\vec{x} \to (3,2)} x^3 e^{x^2 + y^3}$$

Exercise 2.7.9 (exer-MVF-LC-6).
$$\lim_{\vec{x} \to \vec{0}} \frac{x^3 y^2 - xy^4}{x^4 + 2x^2 y^2 + y^4}$$

Exercise 2.7.10 (exer-MVF-LC-7).
$$\lim_{\vec{x} \to \vec{0}} \frac{x^2 y^2 - xy^3}{x^4 + 2x^2 y^2 + y^4}$$

Exercise 2.7.11 (exer-MVF-LC-8).
$$\lim_{\vec{x} \to \vec{0}} \frac{x^2}{y + x^3}$$

Exercise 2.7.12 (exer-MVF-LC-9).
$$\lim_{\vec{x} \to (1,4)} \frac{x^2}{y + x^3}$$

Exercise 2.7.13 (exer-MVF-LC-10).
$$\lim_{\vec{x} \to \vec{0}} \left[\frac{xy + e^{xy}}{\ln(x^2 + y^2)} \right]$$

Exercise 2.7.14 (exer-MVF-LC-11).
$$\lim_{\vec{x} \to \vec{0}} \left[\frac{xy + e^{xy}}{\ln(1 + x^2 + y^2)} \right]$$

Exercise 2.7.15 (exer-MVF-LC-12).
$$\lim_{\vec{x} \to \vec{0}} \ln\left(1 + x^2 + y^2\right)$$

Exercise 2.7.16 (exer-MVF-LC-13). Find a function f such that, in the consideration of the limit $\lim_{\vec{x} \to \vec{0}} f(\vec{x})$, the limit along all straight line paths, all quadratic paths, and all cubic paths give the same limit of zero – but along the quartic path $y = x^4$ the limit is not zero. Make sure to demonstrate that the function you have found satisfies the above requirements.

Chapter 3

Linear Algebra

Before we begin our discussion of the calculus of multivariable functions, there are a few more tools that we need to develop to be able to give the calculus a more appropriate treatment. This is due to a fundamental difference between the derivatives of single variable functions and multivariable functions.

In most single variable calculus courses, the derivative of a single variable function at a given input value is thought of as the slope of the tangent line to the graph of the function, at that given input value. Specifically then, the derivative of such a function is a single real number. Of course real numbers are objects that students are highly familiar with, so no setup is required to be prepared to deal with them.

On the other hand, the derivative of a multivariable function is not quite that simple. While there are several different approaches one can take, and several different variations on the derivative that one can define, we will find in Chapter 4 that the most general and fundamental form of the derivative of a multivariable function is not a number. Instead, it is a different type of mathematical object called a "linear transformation".

Most students at this point in their mathematics educations have not yet seen linear transformations (though many probably have seen a closely related object called a "matrix"). Linear transformations and matrices are some of the fundamental objects of study in a field called "linear algebra", a rich and beautiful field of mathematics with powerful applications in many areas such as engineering, computer science, physics and economics.

A truly solid introduction to linear algebra requires a full semester course. But much less is needed to make great use of the subject as it applies to multivariable calculus. In this chapter, we will introduce some of these fundamental ideas of linear algebra, allowing us to present in Chapter 4 the critical and fundamental perspective that the derivative of a multivariable function is a linear transformation.

3.1 Linear Transformations

3.1.1 Preliminaries

We begin with a reminder about notation. In Subsection 1.2.1 we introduced the notation $\vec{v} = (v_1, v_2, v_3)$ to indicate a vector in $\mathbb{R}^3$ with coordinates v_1, v_2, and v_3. This notation has the advantage that it fits neatly on a typed line along with text. However, as we also commented shortly after, in linear algebra it often turns out to be more natural to write a vector as a column:

$$\vec{v} = \begin{bmatrix} v_1 \\ v_2 \\ v_3 \end{bmatrix}$$

(Admittedly, the reasons for this are based on an arbitrary convention; however, this convention is uniformly accepted in the field, and so we will use both notations regularly in this text.)

As we study linear algebra, we will very often find ourselves taking the sum of scalar multiples of vectors; so, we give this construction a name.

Definition 3.1.1. *A "linear combination" of the vectors $\{\vec{v}_1, \ldots, \vec{v}_k\}$ is a vector that can be written in the form*

$$c_1\vec{v}_1 + \cdots + c_k\vec{v}_k$$

for some real constants $c_1, \ldots, c_k$.

Warning: Vectors have several coordinates, so sometimes we use a subscript to distinguish these coordinates from each other, as we did in the notational comment above. However in other instances we may have an arbitrary collection of vectors, and will use a subscript to distinguish the vectors from each other, as we did in the above definition of linear combinations.

Note that these are completely different uses for subscripts, and by necessity each will be used routinely – so, students must be very careful to notice which role the subscript is playing in any given situation. In order to assist in this, in this course ALL vectors will be written with an arrow over the top, as in "$\vec{v}$" – so for example, we know that $\vec{v}_i$ must represent the ith vector in a list of vectors.

In some texts vectors are represented using boldface instead of the arrow over the top, as in "**v**". Students should be aware however that in other texts, there may be no such notation to distinguish a vector as such, and so context may be the only indication.

3.1.2 Linear transformations

Students in this course have already seen many examples of multivariable functions. A multivariable function $f: \mathbb{R}^n \to \mathbb{R}^m$ is a function that takes in n input variables, and returns m output variables, each of which might depend on all of the inputs. For example, this could be written in an expanded form as

$$f\left(\begin{bmatrix} x_1 \\ \vdots \\ x_n \end{bmatrix}\right) = \begin{bmatrix} f_1\left(\begin{bmatrix} x_1 \\ \vdots \\ x_n \end{bmatrix}\right) \\ \vdots \\ f_m\left(\begin{bmatrix} x_1 \\ \vdots \\ x_n \end{bmatrix}\right) \end{bmatrix}$$

In general, these individual component functions can be any arbitrary real-valued functions.

Here we will discuss a very special category of multivariable functions called linear transformations.

Definition 3.1.2. *A "linear transformation" $T: \mathbb{R}^n \to \mathbb{R}^m$ is a multivariable function such that for any vectors $\vec{x}, \vec{y} \in \mathbb{R}^n$, and any real number c, we have:*

(1) $T(\vec{x} + \vec{y}) = T(\vec{x}) + T(\vec{y})$

(2) $T(c\vec{x}) = cT(\vec{x})$

Without too much trouble one can show that these two conditions are equivalent to the following single condition:

Theorem 3.1.1. *A multivariable function $T: \mathbb{R}^n \to \mathbb{R}^m$ is a linear transformation iff for any vectors $\vec{v}_1, \vec{v}_2 \in \mathbb{R}^n$, and any real numbers c_1, c_2, we have:*

(3) $T(c_1\vec{v}_1 + c_2\vec{v}_2) = c_1T(\vec{v}_1) + c_2T(\vec{v}_2)$

One can also show that this is equivalent to another single condition:

Theorem 3.1.2. *A multivariable function $T: \mathbb{R}^n \to \mathbb{R}^m$ is a linear transformation iff for any vectors $\vec{v}_1, \ldots, \vec{v}_k \in \mathbb{R}^n$, and any real numbers $c_1, \ldots, c_k$, we have:*

3.1. LINEAR TRANSFORMATIONS

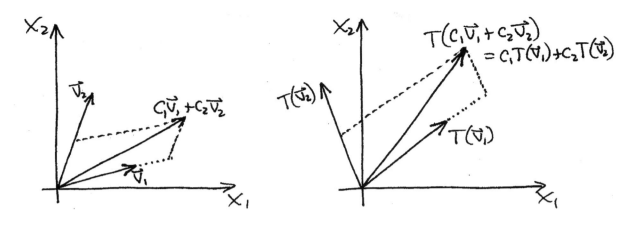

Figure 3.1:

(4) $T(c_1\vec{v}_1 + \cdots + c_k\vec{v}_k) = c_1 T(\vec{v}_1) + \cdots + c_k T(\vec{v}_k)$

We leave these demonstrations to the interested reader.

Thought of in this last way, we can interpret a linear transformation as being a function that "commutes with linear combinations" – that is, you can apply the linear transformation before or after taking a linear combination, and it makes no difference.

To say this in more detail: given a collection of vectors $\{v_i\}$, constants $\{c_i\}$ (to be used as coefficients in a linear combination), and a linear transformation T, one can first take the linear combination of the vectors and then apply the linear transformation (yielding the left side of the equation in Theorem 3.1.2), or first apply the linear transformation to the vectors and then take the linear combination of the images (yielding the right side of the equation in Theorem 3.1.2) – and these computations always give the same result. This is represented in Figure 3.1.

Example 3.1.1. Consider the function $f : \mathbb{R}^2 \to \mathbb{R}^3$ defined by

$$f\left(\begin{bmatrix} x_1 \\ x_2 \end{bmatrix}\right) = \begin{bmatrix} 2x_1 + x_2 \\ x_1 - 3x_2 \\ 5x_2 \end{bmatrix}$$

To see if this is a linear transformation, we check that it satisfies the above condition, with the arbitrary vectors $\vec{x} = \begin{bmatrix} x_1 \\ x_2 \end{bmatrix}$ **and** $\vec{y} = \begin{bmatrix} y_1 \\ y_2 \end{bmatrix}$:

$$\begin{aligned}
f(a\vec{x} + b\vec{y}) &= f\left(a\begin{bmatrix} x_1 \\ x_2 \end{bmatrix} + b\begin{bmatrix} y_1 \\ y_2 \end{bmatrix}\right) \\
&= f\left(\begin{bmatrix} ax_1 + by_1 \\ ax_2 + by_2 \end{bmatrix}\right) \\
&= \begin{bmatrix} 2(ax_1 + by_1) + (ax_2 + by_2) \\ (ax_1 + by_1) - 3(ax_2 + by_2) \\ 5(ax_2 + by_2) \end{bmatrix} \\
&= a\begin{bmatrix} 2x_1 + x_2 \\ x_1 - 3x_2 \\ 5x_2 \end{bmatrix} + b\begin{bmatrix} 2y_1 + y_2 \\ y_1 - 3y_2 \\ 5y_2 \end{bmatrix} \\
&= af(\vec{x}) + bf(\vec{y})
\end{aligned}$$

So we see that this function f is indeed a linear transformation.

Example 3.1.2. Consider the function $g : \mathbb{R}^2 \to \mathbb{R}^2$ defined by

$$g\left(\begin{bmatrix} x_1 \\ x_2 \end{bmatrix}\right) = \begin{bmatrix} 3x_1^2 \\ x_2 \end{bmatrix}$$

It is easy to find a counterexample showing that the needed condition is not satisfied for this function:

$$g\left(1\begin{bmatrix} 1 \\ 0 \end{bmatrix} + 1\begin{bmatrix} 1 \\ 0 \end{bmatrix}\right) = \begin{bmatrix} 12 \\ 0 \end{bmatrix}$$

$$1g\left(\begin{bmatrix} 1 \\ 0 \end{bmatrix}\right) + 1g\left(\begin{bmatrix} 1 \\ 0 \end{bmatrix}\right) = \begin{bmatrix} 6 \\ 0 \end{bmatrix}$$

So, g is not a linear transformation.

There is an immediate theorem that we can state and prove that will be important in our conceptual understanding of linear transformations.

Theorem 3.1.3. *For any linear transformation T, we must have $T(\vec{0}) = \vec{0}$.*

Proof: The proof of this theorem is simply a matter of making a crafty choice for what to plug into the conditions in the definition. Specifically, for any vector $\vec{x}$, choose the multiple 0 and write down the second linearity condition; this gives us

$$T(0\vec{x}) = 0T(\vec{x})$$

which immediately gives us the desired result.

3.1.3 Standard basis vectors

Recall that the standard basis vectors in $\mathbb{R}^n$ are the n vectors whose coordinates are all 0, except for one coordinate which is 1. These vectors are denoted as "$\vec{e}_i$", where i is the index of the sole nonzero coordinate.

Example 3.1.3. The standard basis vectors in $\mathbb{R}^3$ are

$$\vec{e}_1 = \begin{bmatrix} 1 \\ 0 \\ 0 \end{bmatrix} \quad \vec{e}_2 = \begin{bmatrix} 0 \\ 1 \\ 0 \end{bmatrix} \quad \vec{e}_3 = \begin{bmatrix} 0 \\ 0 \\ 1 \end{bmatrix}$$

There is a very useful connection between linear transformations and the standard basis vectors:

Theorem 3.1.4. *If T is a linear transformation and if the images of all of the standard basis vectors $(T(\vec{e}_i))$ are known, then the value of T can be computed for any vector $\vec{x} \in \mathbb{R}^n$.*

Proof: We observe that the vector $\vec{x} = (x_1, \ldots, x_n)$ can be written as a linear combination of the standard basis vectors (using the coordinates of $\vec{x}$ as coefficients), and then apply the linearity condition:

$$\begin{aligned} T(\vec{x}) &= T(x_1\vec{e}_1 + \ldots + x_n\vec{e}_n) \\ &= x_1 T(\vec{e}_1) + \ldots + x_n T(\vec{e}_n) \end{aligned}$$

3.1. LINEAR TRANSFORMATIONS

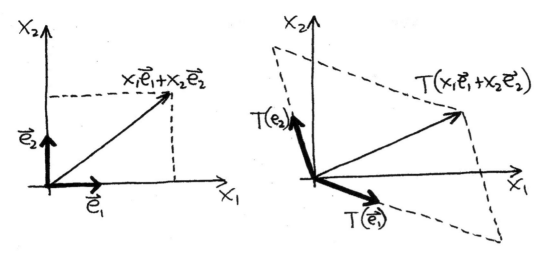

Figure 3.2:

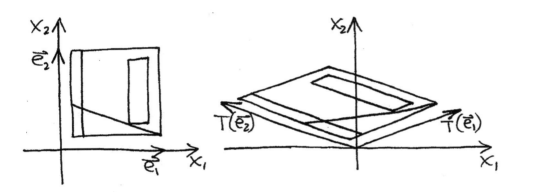

Figure 3.3:

We can interpret this equation as a prescription for how to accomplish the claim of the theorem; to compute the image of a vector $\vec{x}$, simply compute the linear combination of the known images of the standard basis vectors where the coordinates of $\vec{x}$ are used as the coefficients.

This is what we needed to show.

∎

The geometric interpretation of the proof is represented in Figure 3.2.

Theorem 3.1.4 is a powerful result that allows us to draw strong conclusions about linear transformations. In particular, we can use this to help visualize linear transformations.

Linear transformations deform the domain in a uniform, linear way, as is shown in Figure 3.3 depicting a figure in the domain on the left, and on the right the image of that same figure through a linear transformation. Note that even though the image has been stretched and rotated, the relative positions of the objects inside the rectangle are preserved. Note also that, as we concluded in Theorem 3.1.3, every linear transformation must have the property that the image of the origin in the domain is the origin in the target.

3.1.4 Examples

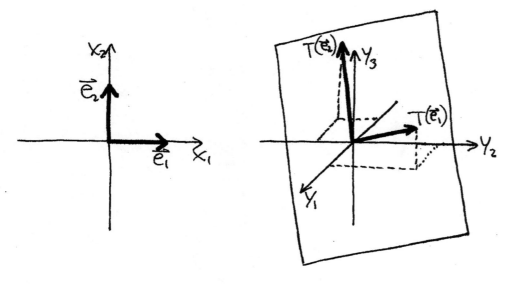

Figure 3.4:

Example 3.1.4. Suppose that we have a linear transformation $T : \mathbb{R}^2 \to \mathbb{R}^3$, and that we know

$$T(\vec{e}_1) = \begin{bmatrix} 1 \\ 2 \\ 1 \end{bmatrix} \quad \text{and} \quad T(\vec{e}_2) = \begin{bmatrix} -1 \\ -1 \\ 3 \end{bmatrix}$$

What is the image of this linear transformation?

We know that every vector in the domain is a linear combination of the standard basis vectors $\vec{e}_1$ and $\vec{e}_2$; therefore, we know that every image point is a linear combination of the images (in the target) of those standard basis vectors – which are given (See Figure 3.4). So, the complete image is just the collection of all linear combinations of those image vectors, which is just the plane that is defined by those two vectors. (Sidenote – this is also called the "span" of those image vectors.)

Example 3.1.5. In Subsection 2.2.3 we saw several functions that we used to help us understand solution sets to equations. Some of those are linear transformations and some of them are not.

Translations, for example, are not linear transformations. This is easy to see because the image of the origin in the domain is not the origin in the target.

Both stretches and reflections, however, are linear transformations. We leave it to the reader to check that these functions satisfy the linearity conditions.

Example 3.1.6. What can we say about the function $R : \mathbb{R}^2 \to \mathbb{R}^2$ defined by

$$R\left(\begin{bmatrix} x \\ y \end{bmatrix}\right) = \begin{bmatrix} -y \\ x \end{bmatrix}$$

We can easily check that this is a linear transformation by confirming that the condition in Theorem 3.1.1 is satisfied; we leave this as an exercise to the reader.

3.1. LINEAR TRANSFORMATIONS

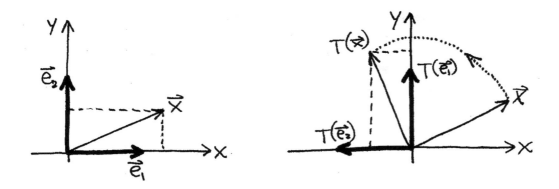

Figure 3.5:

Knowing that this is a linear transformation, we know that it is some sort of a uniform deformation of the plane, and that it is completely defined by the images of the two standard basis vectors. We compute then

$$R(\vec{e}_1) = \begin{bmatrix} 0 \\ 1 \end{bmatrix} \quad \text{and} \quad R(\vec{e}_2) = \begin{bmatrix} -1 \\ 0 \end{bmatrix}$$

As we see in Figure 3.5, each of these images can be thought of as the result of rotating the given standard basis vector counterclockwise by an angle $\pi/2$. Certainly such a rotation is a uniform deformation of the plane, and we know that linear transformations are uniquely determined by the images of the standard basis vectors, so we can conclude that the function R rotates every input vector by the same angle.

(This can be easily checked by noting that for any input vector $\vec{x}$, the given formula for R is easily confirmed to satisfy $\vec{x} \cdot R(\vec{x}) = 0$.)

Example 3.1.7. How can we find a formula for a general rotation R_θ in $\mathbb{R}^2$, rotating counterclockwise by an angle θ?

We observed in Example 3.1.6 that a rotation is a linear transformation. So, in light of Theorem 3.1.4, we need only compute the images of the standard basis vectors.

Simple trigonometry as in Figure 3.6 tells us that

$$R_\theta(\vec{e}_1) = \begin{bmatrix} \cos\theta \\ \sin\theta \end{bmatrix} \quad \text{and} \quad R_\theta(\vec{e}_2) = \begin{bmatrix} -\sin\theta \\ \cos\theta \end{bmatrix}$$

Then we can compute $R_\theta(\vec{x})$ by

$$\begin{aligned} R_\theta(\vec{x}) &= x R_\theta(\vec{e}_1) + y R_\theta(\vec{e}_2) \\ &= x \begin{bmatrix} \cos\theta \\ \sin\theta \end{bmatrix} + y \begin{bmatrix} -\sin\theta \\ \cos\theta \end{bmatrix} \\ &= \begin{bmatrix} x\cos\theta - y\sin\theta \\ x\sin\theta + y\cos\theta \end{bmatrix} \end{aligned}$$

We can use this to rotate any general vector by an angle θ.

(Again, we can make a quick check that this is correct. Students can confirm for themselves that for any vector $\vec{x}$, we have $\vec{x} \cdot R_\theta(\vec{x}) = \|\vec{x}\|^2 \cos\theta$, as we would expect.)

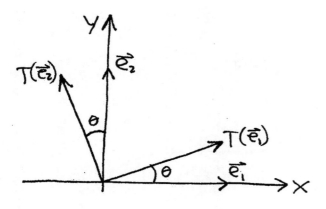

Figure 3.6:

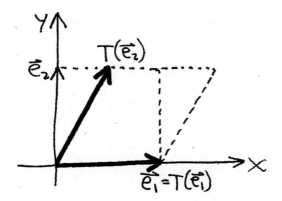

Figure 3.7:

Example 3.1.8. What can we say about the function $S : \mathbb{R}^2 \to \mathbb{R}^2$ defined by

$$S\left(\begin{bmatrix} x \\ y \end{bmatrix}\right) = \begin{bmatrix} x + cy \\ y \end{bmatrix}$$

Again it is simple computation to confirm that this function is a linear transformation. Knowing this, we need only compute the images of the standard basis vectors, which we represent in **Figure 3.7**.

$$S(\vec{e}_1) = \begin{bmatrix} 1 \\ 0 \end{bmatrix} \quad \text{and} \quad S(\vec{e}_2) = \begin{bmatrix} c \\ 1 \end{bmatrix}$$

We can see then that this linear transformation leaves the y-coordinate unchanged, but slides points in the x-direction by an amount that is proportional to that y-coordinate. We call this a "shearing transformation".

Exercises

3.1. LINEAR TRANSFORMATIONS

Exercise 3.1.1 (exer-LA-LT-1). Find a way to write the vector $\vec{x} = (3, 5, 2)$ as a linear combination of the vectors $\vec{e}_1 = (1, 0, 0)$, $\vec{e}_2 = (0, 1, 0)$, $\vec{e}_3 = (0, 0, 1)$.

Exercise 3.1.2 (exer-LA-LT-2). Find a way to write the vector $\vec{x} = (3, 5, 2)$ as a linear combination of the vectors $\vec{u} = (1, 1, 0)$, $\vec{v} = (0, 1, 1)$, $\vec{w} = (1, 0, 1)$.

Exercise 3.1.3 (exer-LA-LT-3).
1. Find a way to write the vector $\vec{x} = (3, 5, 2)$ as a linear combination of the vectors $\vec{u} = (1, 1, 0)$, $\vec{v} = (0, 1, 1)$

2. Find a way to write the vector $\vec{0} = (0, 0, 0)$ as a linear combination of the vectors $\vec{u} = (1, 1, 0)$, $\vec{v} = (0, 1, 1)$, $\vec{p} = (-1, 0, 1)$.

3. Noting that $\vec{x} = \vec{x} + k\vec{0}$, use the above information to find an infinite number of different ways to write $\vec{x}$ as a linear combination of $\vec{u}, \vec{v}, \vec{p}$.

Exercise 3.1.4 (exer-LA-LT-4). Find a way to write the vector $\vec{y} = (2, 3, 4)$ as a linear combination of the vectors $\vec{u} = (1, 1, 0)$, $\vec{v} = (0, 1, 1)$, $\vec{p} = (-1, 0, 1)$, or show that it is not possible to do this.

Exercise 3.1.5 (exer-LA-LT-5). Use Definition 3.1.2 to determine if the following functions are linear transformations. If the function is not a linear transformation, find explicit vectors that fail the necessary conditions.

1. $f(x, y) = (3x - 2y, 2x + 3y)$
2. $g(x, y, z) = (2x - z, x + z, z - y)$
3. $h(x, y, z) = (2x - z, x + z, z - y + 1)$
4. $m(x, y) = (x^2, y^2)$
5. $n(x, y) = (e^x, 3y)$

Exercise 3.1.6 (exer-LA-LT-6). Confirm that the function R from Example 3.1.6 is a linear transformation.

Exercise 3.1.7 (exer-LA-LT-7). Find the vectors that are obtained by rotating the vectors below by the indicated angles.

1. $(1, 0)$, counterclockwise by $\pi/2$.
2. $(3, 6)$, counterclockwise by $\pi/6$.
3. $(3, 6)$, clockwise by $\pi/6$.
4. $(4, 1)$, clockwise by $7\pi/4$.

Exercise 3.1.8 (exer-LA-LT-8). The linear transformation $L : \mathbb{R}^2 \to \mathbb{R}^3$ sends $\vec{e}_1$ to $(3, 2, 5)$ and sends $\vec{e}_2$ to $(-1, -2, 4)$. What is $L(-3, 2)$?

Exercise 3.1.9 (exer-LA-LT-9). Find an explicit formula for the function L in Exercise 3.1.8.

Exercise 3.1.10 (exer-LA-LT-10). The linear transformation L acts on points in $\mathbb{R}^3$ by the following process – first, the point is rotated around the z-axis by $\pi/2$, clockwise as seen from above; then the result is reflected through the yz-plane; finally, this result is rotated around the x-axis by $\pi/2$, counterclockwise as seen from the positive part of the x-axis.

Find the formula for this function L. (Hint: Use Theorem 3.1.4.)

Exercise 3.1.11 (exer-LA-LT-11). Prove Theorem 3.1.1.

Exercise 3.1.12 (exer-LA-LT-12). Prove Theorem 3.1.2.

Exercise 3.1.13 (exer-LA-LT-13). Show that the scaling and reflection functions defined in Section 2.2 on page 61 are linear transformations, while the translation functions are not.

3.2 Matrices

3.2.1 Matrix notation

As we saw in Example 3.1.1, one can represent a linear transformation by giving an explicit formula for evaluating it. But as we just saw with Theorem 3.1.4, we can also prescribe a linear transformation uniquely by simply listing the images of the standard basis vectors – and then in light of the theorem it is understood that we can evaluate the linear transformation on an arbitrary vector by taking appropriate linear combinations.

For example, let's consider the linear transformation from that first example, which we will rename here as T in order to emphasize that it is a linear transformation:

$$T\left(\begin{bmatrix} x_1 \\ x_2 \end{bmatrix}\right) = \begin{bmatrix} 2x_1 + x_2 \\ x_1 - 3x_2 \\ 5x_2 \end{bmatrix}$$

This formula identifies the linear transformation uniquely, but we could do the same simply by listing

$$T(\vec{e}_1) = \begin{bmatrix} 2 \\ 1 \\ 0 \end{bmatrix} \quad \text{and} \quad T(\vec{e}_2) = \begin{bmatrix} 1 \\ -3 \\ 5 \end{bmatrix}$$

with the understanding that the value of T on an arbitrary vector $\vec{x}$ is computed with

$$T(\vec{x}) = x_1 T(\vec{e}_1) + x_2 T(\vec{e}_2)$$

from Theorem 3.1.4.

If we make a point of always listing the image vectors in the natural order $T(\vec{e}_1), \ldots, T(\vec{e}_n)$, then we don't even need to identify them individually since it is understood that the first column vector is $T(\vec{e}_1)$, the second is $T(\vec{e}_2)$, ...etc. So, applying this to the above linear transformation T, we could abbreviate our representation as simply

$$A = \begin{pmatrix} 2 & 1 \\ 1 & -3 \\ 0 & 5 \end{pmatrix}$$

This is called a "matrix". (It is a very common and useful representation of a linear transformation. It should be pointed out here also that matrices have many other uses and interpretations, many of which are equivalent to this one; those other uses and points of view will be discussed in any text on linear algebra.)

Definition 3.2.1. *For any linear transformation $T : \mathbb{R}^n \to \mathbb{R}^m$, the "matrix representing T" is the matrix*

$$A = \begin{pmatrix} | & & | \\ T(\vec{e}_1) & \cdots & T(\vec{e}_n) \\ | & & | \end{pmatrix}$$

whose columns are the images by T of the standard basis vectors of $\mathbb{R}^n$.

The reader will note that we gave the matrix the name A, even though it is a representation of the function that has already been named T. The reason for this is that even though we have motivated matrices simply as a shorthand for linear transformations, we will be developing algebraic properties of matrices so that we can think of them as separate objects from the transformations that they represent.

Note that in general there is an immediate relationship between the dimensions of the matrix, and the dimensions of the domain and target of the linear transformation. Suppose we consider the linear transformation $T : \mathbb{R}^n \to \mathbb{R}^m$. There are n standard basis vectors in the domain and thus there are n images of standard basis vectors in the target; so the matrix

3.2. MATRICES

will have n columns. Each of these columns, being an image and thus a vector in the target, will have m components; since each column has m components, this means that the matrix will have m rows.

This matrix with m rows and n columns is referred to as an "$m \times n$" matrix. Notice that as they are written on the page, the m and the n appear here in the opposite order than they appear in the description of the function, $T : \mathbb{R}^n \to \mathbb{R}^m$.

Example 3.2.1. How can we represent as a matrix the linear transformation defined by

$$T\left(\begin{bmatrix} x \\ y \\ z \end{bmatrix}\right) = \begin{bmatrix} 2x - z \\ x + 3y \end{bmatrix}$$

Working straight from Definition 3.2.1, we compute the images of the standard basis vectors as

$$T(\vec{e}_1) = \begin{bmatrix} 2 \\ 1 \end{bmatrix} \quad \text{and} \quad T(\vec{e}_2) = \begin{bmatrix} 0 \\ 3 \end{bmatrix} \quad \text{and} \quad T(\vec{e}_3) = \begin{bmatrix} -1 \\ 0 \end{bmatrix}$$

and assemble them into the matrix

$$\begin{pmatrix} 2 & 0 & -1 \\ 1 & 3 & 0 \end{pmatrix}$$

Example 3.2.2. What linear transformation is represented by the matrix

$$\begin{pmatrix} 9 & 2 & 1 \\ 3 & 4 & 5 \\ 5 & 7 & 8 \end{pmatrix}$$

Again going straight from Definition 3.2.1, we know that we must have

$$T(\vec{e}_1) = \begin{bmatrix} 9 \\ 3 \\ 5 \end{bmatrix} \quad \text{and} \quad T(\vec{e}_2) = \begin{bmatrix} 2 \\ 4 \\ 7 \end{bmatrix} \quad \text{and} \quad T(\vec{e}_3) = \begin{bmatrix} 1 \\ 5 \\ 8 \end{bmatrix}$$

Then using the linearity condition we have

$$\begin{aligned} T(\vec{x}) &= xT(\vec{e}_1) + yT(\vec{e}_2) + zT(\vec{e}_3) \\ &= x \begin{bmatrix} 9 \\ 3 \\ 5 \end{bmatrix} + y \begin{bmatrix} 2 \\ 4 \\ 7 \end{bmatrix} + z \begin{bmatrix} 1 \\ 5 \\ 8 \end{bmatrix} \\ &= \begin{bmatrix} 9x + 2y + z \\ 3x + 4y + 5z \\ 5x + 7y + 8z \end{bmatrix} \end{aligned}$$

3.2.2 Matrix-vector multiplication

As of now a matrix is just a shorthand for a linear transformation, in that its columns represent the images of the standard basis vectors. We have not yet defined any algebraic properties of matrices.

Of course we can make up any definitions we want – but we want to make sure that the algebraic properties we define are natural in some sense that will be useful to us.

For example – we know that we can apply a linear transformation to a vector, and that the result of this is another vector. Since a matrix represents a linear transformation, this suggests that we should define some sort of way that a matrix can be applied to a vector, yielding another vector – such that we get the same thing that we would have had we applied the linear transformation. Since notationally we write a matrix with numbers, we arbitrarily choose to refer to this as "multiplication" of the matrix and the vector.

Specifically then, if A is the matrix representing the linear transformation T, we want to define "matrix-vector multiplication" in such a way that we get

$$T(\vec{x}) = A\vec{x}$$

Before we write this down as a definition, let's denote this motivation as:

Interpretation 0: The matrix-vector product $A\vec{x}$ is the result of the application of the linear transformation T (corresponding to A) to the vector $\vec{x}$.

If we let $\vec{a}_i = T(\vec{e}_i)$ so that the vectors $\vec{a}_i$ represent the columns of A, giving us

$$A = \begin{pmatrix} | & & | \\ \vec{a}_1 & \cdots & \vec{a}_n \\ | & & | \end{pmatrix} \quad \text{and} \quad \vec{x} = \begin{bmatrix} x_1 \\ \vdots \\ x_n \end{bmatrix}$$

then the above equation can be rewritten as

$$\begin{aligned} T(\vec{x}) &= A\vec{x} \\ x_1 T(\vec{e}_1) + \cdots + x_n T(\vec{e}_n) &= A\vec{x} \\ x_1 \vec{a}_1 + \cdots + x_n \vec{a}_n &= A\vec{x} \end{aligned}$$

We use this as our definition of matrix-vector multiplication.

Definition 3.2.2. *Given a matrix A with m rows and n columns, and a vector $\vec{x} \in \mathbb{R}^n$, we define the matrix-vector product $A\vec{x}$ as*

$$A\vec{x} = \begin{pmatrix} | & & | \\ \vec{a}_1 & \cdots & \vec{a}_n \\ | & & | \end{pmatrix} \begin{bmatrix} x_1 \\ \vdots \\ x_n \end{bmatrix} = x_1 \vec{a}_1 + \cdots + x_n \vec{a}_n$$

The verbal interpretation is:

Interpretation 1: The matrix-vector product $A\vec{x}$ is a linear combination of the columns of A, where the components of $\vec{x}$ are used as the coefficients.

Writing out all of the individual terms of each column vector, and expanding this out and regrouping, it can be shown that we can restate this definition instead in terms of the rows of A (which we denote with capital A's to distinguish them from the columns); we leave this algebra as an exercise to the interested reader. The result is the following equivalent definition:

Equivalent Definition: Given a matrix A with m rows and n columns, and a vector $\vec{x} \in \mathbb{R}^n$, we define the matrix-vector product $A\vec{x}$ as

$$A\vec{x} = \begin{pmatrix} \text{—} & \vec{A}_1 & \text{—} \\ & \vdots & \\ \text{—} & \vec{A}_m & \text{—} \end{pmatrix} \begin{bmatrix} | \\ \vec{x} \\ | \end{bmatrix} = \begin{bmatrix} \vec{A}_1 \cdot \vec{x} \\ \vdots \\ \vec{A}_m \cdot \vec{x} \end{bmatrix}$$

This gives us another interpretation:

Interpretion 2: The matrix-vector product $A\vec{x}$ has as its components the dot products of the rows of A with the vector $\vec{x}$.

3.2. MATRICES

Each of these interpretations is important, and will be used extensively in this course.

Example 3.2.3. Suppose we wish to compute the matrix-vector product

$$\begin{pmatrix} 9 & 2 & 1 \\ 3 & 4 & 5 \\ 5 & 7 & 8 \end{pmatrix} \begin{bmatrix} 1 \\ 2 \\ 3 \end{bmatrix}$$

We will compute this straight from Definition 3.2.2, as

$$= 1 \begin{bmatrix} 9 \\ 3 \\ 5 \end{bmatrix} + 2 \begin{bmatrix} 2 \\ 4 \\ 7 \end{bmatrix} + 3 \begin{bmatrix} 1 \\ 5 \\ 8 \end{bmatrix} = \begin{bmatrix} 16 \\ 26 \\ 43 \end{bmatrix}$$

Example 3.2.4. Suppose we wish to compute the same matrix-vector product as in Example 3.2.3,

$$\begin{pmatrix} 9 & 2 & 1 \\ 3 & 4 & 5 \\ 5 & 7 & 8 \end{pmatrix} \begin{bmatrix} 1 \\ 2 \\ 3 \end{bmatrix}$$

but this time by using the Equivalent Definition above. We get

$$\begin{bmatrix} \begin{bmatrix} 9 & 2 & 1 \end{bmatrix} \cdot \begin{bmatrix} 1 \\ 2 \\ 3 \end{bmatrix} \\ \begin{bmatrix} 3 & 4 & 5 \end{bmatrix} \cdot \begin{bmatrix} 1 \\ 2 \\ 3 \end{bmatrix} \\ \begin{bmatrix} 5 & 7 & 8 \end{bmatrix} \cdot \begin{bmatrix} 1 \\ 2 \\ 3 \end{bmatrix} \end{bmatrix} = \begin{bmatrix} 16 \\ 26 \\ 43 \end{bmatrix}$$

3.2.3 Matrix addition and scalar-matrix multiplication

In the last section we developed an algebraic operation (which we decided to call "matrix-vector multiplication") that corresponds to the application of a linear transformation to a vector. We are now going to do something very similar to motivate and define two more algebraic operations involving matrices – matrix addition and scalar-matrix multiplication.

First, we need to define some operations on linear transformations themselves. If we have linear transformations T and S, with matrices A and B respectively, then we can (a) add those transformations, and (b) multiply one of them by a scalar c. In particular, we define the results of those operations by requiring the same results for their actions on vectors. Namely,

Definition 3.2.3. *Given linear transformations T and S with the same domains and targets, the sum $T + S$ and the scalar product cT are functions defined by the formulas*

$$\begin{aligned} (T+S)(\vec{x}) &= T(\vec{x}) + S(\vec{x}) \\ (cT)(\vec{x}) &= cT(\vec{x}) \end{aligned}$$

These new functions are also linear transformations (this is left as an exercise to the interested reader).

Very naturally we would want to define our new matrix operations $(A+B)$ and (cA) so that they correspond to these new linear transformations. Again, we use this to motivate our first interpretation of what these operations should be:

Interpretation 0: The sum of two matrices is the matrix corresponding to the sum of the two corresponding linear transformations; the product of a scalar and a matrix is the matrix corresponding to the product of that scalar and the corresponding linear transformation.

Algebraically, this is

$$\begin{aligned}(T+S)(\vec{x}) &= (A+B)\vec{x} \\ (cT)(\vec{x}) &= (cA)\vec{x}\end{aligned}$$

and we can use this setup to compute the individual columns of $(A+B)$ and (cA):

$$\begin{aligned}j\text{th column of } (A+B) &= (A+B)\vec{e}_j = (T+S)(\vec{e}_j) = T(\vec{e}_j) + S(\vec{e}_j) = A\vec{e}_j + B\vec{e}_j \\ &= j\text{th column of } A + j\text{th column of } B\end{aligned}$$

$$\begin{aligned}j\text{th column of } (cA) &= (cA)\vec{e}_j = (cT)(\vec{e}_j) = cT(\vec{e}_j) = c(A\vec{e}_j) \\ &= c(j\text{th column of } A)\end{aligned}$$

Since addition and scalar multiplication of vectors is performed component-wise, this gives us:

Interpretation 1: The sum $(A+B)$ of two matrices A and B is the matrix whose elements are the sums of the corresponding elements of A and B; the scalar-matrix product (cA) is the matrix whose elements are the products of the scalar c with the corresponding elements of A.

We write these out explicitly as our definitions.

Definition 3.2.4. *If we write*

$$A = \begin{pmatrix} a_{11} & \cdots & a_{1n} \\ \vdots & & \vdots \\ a_{m1} & \cdots & a_{mn} \end{pmatrix} \qquad B = \begin{pmatrix} b_{11} & \cdots & b_{1n} \\ \vdots & & \vdots \\ b_{m1} & \cdots & b_{mn} \end{pmatrix}$$

then

$$A+B = \begin{pmatrix} a_{11}+b_{11} & \cdots & a_{1n}+b_{1n} \\ \vdots & & \vdots \\ a_{m1}+b_{m1} & \cdots & a_{mn}+b_{mn} \end{pmatrix} \qquad cA = \begin{pmatrix} ca_{11} & \cdots & ca_{1n} \\ \vdots & & \vdots \\ ca_{m1} & \cdots & ca_{mn} \end{pmatrix}$$

Example 3.2.5. The sum of the matrices

$$\begin{pmatrix} 3 & -4 & 2 \\ 5 & 3 & -6 \end{pmatrix} \quad \text{and} \quad \begin{pmatrix} 1 & 3 & -6 \\ 7 & 2 & 2 \end{pmatrix}$$

is computed straight from Definition 3.2.4 to be

$$\begin{pmatrix} 4 & -1 & -4 \\ 12 & 5 & -4 \end{pmatrix}$$

and similarly the scalar-matrix product

$$5 \begin{pmatrix} 3 & -4 & 2 \\ 5 & 3 & -6 \end{pmatrix}$$

is computed straight from Definition 3.2.4 to be

$$\begin{pmatrix} 15 & -20 & 10 \\ 25 & 15 & -30 \end{pmatrix}$$

3.2.4 Matrix multiplication

In the last two sections we developed algebraic properties of matrices by analogy with the application of a linear transformation to a vector. We are now going to do something very similar to motivate and define an algebraic property of matrices that naturally corresponds to the composition of linear transformations.

If we have a linear transformation $T : \mathbb{R}^n \to \mathbb{R}^m$ represented by a matrix A, and another linear transformation $S : \mathbb{R}^m \to \mathbb{R}^k$ represented by a matrix B, then we can compose them to form a new linear transformation. The output from the transformation T is used as the input to the transformation S, so we represent this composition by the diagram

$$\mathbb{R}^n \xrightarrow{T} \mathbb{R}^m \xrightarrow{S} \mathbb{R}^k$$

The composition then is the function that takes a vector all the way through this diagram, so we have

$$S \circ T : \mathbb{R}^n \to \mathbb{R}^k$$

This composition transformation can thus be applied to a vector in $\mathbb{R}^n$, and yields a vector in $\mathbb{R}^k$. Notice that because arrows go to the right, while functions are written on the left of the vector they are applied to, the functions are written on the page in the opposite order in the two notations above. Students should be especially careful always to be aware of this potential confusion!

By definition of composition of functions, we have

$$(S \circ T)(\vec{x}) = S(T(\vec{x}))$$

If the composition $S \circ T$ is represented by the matrix C, then we can rewrite this in matrix notation as

$$C\vec{x} = B(A\vec{x})$$

Notationally, this is highly suggestive that we should define the matrix C to be the "matrix-matrix product" of the matrices A and B. This gives us

$$(S \circ T)(\vec{x}) = S(T(\vec{x})) \iff (BA)\vec{x} = B(A\vec{x})$$

So as we did with matrix-vector products, before we write down our definition, let's use our motivation above to write down an interpretation of what this matrix multiplication should be:

Interpretation 0: The matrix product BA is the matrix that represents the composition of the two linear transformations S and T (represented by B and A, respectively).

There are several ways to represent this algebraically. For each matrix A and B, we will need to refer both to the rows and the columns; here we will use capital letters to denote the rows, and lower case letters to denote the columns:

$$A = \begin{pmatrix} \text{---} & \vec{A}_1 & \text{---} \\ & \vdots & \\ \text{---} & \vec{A}_m & \text{---} \end{pmatrix} = \begin{pmatrix} | & & | \\ \vec{a}_1 & \cdots & \vec{a}_n \\ | & & | \end{pmatrix}$$

$$B = \begin{pmatrix} \text{---} & \vec{B}_1 & \text{---} \\ & \vdots & \\ \text{---} & \vec{B}_k & \text{---} \end{pmatrix} = \begin{pmatrix} | & & | \\ \vec{b}_1 & \cdots & \vec{b}_m \\ | & & | \end{pmatrix}$$

Let's compute the jth column of the product BA. Using previous ideas, we compute as follows:

$$\begin{aligned} j\text{th column of } BA &= (S \circ T)(\vec{e}_j) \\ &= S(T(\vec{e}_j)) \\ &= B(A\vec{e}_j) \\ &= B\vec{a}_j \end{aligned}$$

Using this result and our previous interpretations of matrix-vector multiplication, we can make the following interpretations (respectively) of matrix-matrix multiplication:

Interpretion 1: The columns of the matrix product BA are linear combinations of the columns of B, where the components of the corresponding column of A are used as coefficients.

Interpretion 2: An element of the matrix product BA is a dot product of the corresponding row of B with the corresponding column of A.

A third interpretation of matrix multiplication can be shown by explicit expanding and regrouping from the above; again, those details are left to the interested reader:

Interpretion 3: The rows of the matrix product BA are linear combinations of the rows of A, where the components of the corresponding row of B are used as coefficients.

For notational convenience, we use interpretation 2 to write down the definition:

Definition 3.2.5. *The matrix product*

$$BA = \begin{pmatrix} - & \vec{B}_1 & - \\ & \vdots & \\ - & \vec{B}_k & - \end{pmatrix} \begin{pmatrix} | & & | \\ \vec{a}_1 & \cdots & \vec{a}_n \\ | & & | \end{pmatrix} = \begin{pmatrix} \vec{B}_1 \cdot \vec{a}_1 & \cdots & \vec{B}_1 \cdot \vec{a}_n \\ \vdots & & \vdots \\ \vec{B}_k \cdot \vec{a}_1 & \cdots & \vec{B}_k \cdot \vec{a}_n \end{pmatrix}$$

is the matrix whose element in the ith row and jth column is

$$a_{ij} = \vec{B}_i \cdot \vec{a}_j$$

Example 3.2.6. Compute the matrix product

$$\begin{pmatrix} 1 & 3 \\ 2 & 4 \end{pmatrix} \begin{pmatrix} 5 & 7 \\ 6 & 8 \end{pmatrix}$$

We will perform this computation from the points of view of each of the interpretations above.

Interp. 1: We compute the columns individually; the first column is

$$5 \begin{bmatrix} 1 \\ 2 \end{bmatrix} + 6 \begin{bmatrix} 3 \\ 4 \end{bmatrix} = \begin{bmatrix} 23 \\ 34 \end{bmatrix}$$

and the second column is

$$7 \begin{bmatrix} 1 \\ 2 \end{bmatrix} + 8 \begin{bmatrix} 3 \\ 4 \end{bmatrix} = \begin{bmatrix} 31 \\ 46 \end{bmatrix}$$

So the product is

$$\begin{pmatrix} 1 & 3 \\ 2 & 4 \end{pmatrix} \begin{pmatrix} 5 & 7 \\ 6 & 8 \end{pmatrix} = \begin{pmatrix} 23 & 31 \\ 34 & 46 \end{pmatrix}$$

Interp. 2: We compute the individual elements of the matrix by dot products, as

$$\begin{pmatrix} 1 & 3 \\ 2 & 4 \end{pmatrix} \begin{pmatrix} 5 & 7 \\ 6 & 8 \end{pmatrix} = \begin{pmatrix} [1 \ 3] \cdot \begin{bmatrix} 5 \\ 6 \end{bmatrix} & [1 \ 3] \cdot \begin{bmatrix} 7 \\ 8 \end{bmatrix} \\ [2 \ 4] \cdot \begin{bmatrix} 5 \\ 6 \end{bmatrix} & [2 \ 4] \cdot \begin{bmatrix} 7 \\ 8 \end{bmatrix} \end{pmatrix} = \begin{pmatrix} 23 & 31 \\ 34 & 46 \end{pmatrix}$$

Interp. 3: We compute the rows individually; the first row is

$$1 \begin{bmatrix} 5 & 7 \end{bmatrix} + 3 \begin{bmatrix} 6 & 8 \end{bmatrix} = \begin{bmatrix} 23 & 31 \end{bmatrix}$$

3.2. MATRICES

and the second row is

$$2 \begin{bmatrix} 5 & 7 \end{bmatrix} + 4 \begin{bmatrix} 6 & 8 \end{bmatrix} = \begin{bmatrix} 34 & 46 \end{bmatrix}$$

So the product is

$$\begin{pmatrix} 1 & 3 \\ 2 & 4 \end{pmatrix} \begin{pmatrix} 5 & 7 \\ 6 & 8 \end{pmatrix} = \begin{pmatrix} 23 & 31 \\ 34 & 46 \end{pmatrix}$$

Combining several of the ideas in this chapter, we can also make the following observation which we state as a theorem.

Theorem 3.2.1. *A function T is a linear transformation iff each component function T_i can be expressed with a dot product, as in*

$$T_i(\vec{v}) = \vec{r}_i \cdot \vec{v} \tag{3.1}$$

for some vector $\vec{r}_i$.

The idea behind the proof of this theorem is that every linear transformation is written as a matrix product, and every matrix product has its components computed as dot product. So, the vectors $\vec{r}_i$ in the theorem are in fact the row vectors of the matrix corresponding to the linear transformation T. We leave the details of this proof to the interested reader.

We have now developed all of the algebraic properties we need for matrices. At this point, anything that we can do with linear transformations we can also do correspondingly with matrices. We can thus think of (for the purposes of this course) linear transformations and matrices as being the same thing.

Exercises

Exercise 3.2.1 (exer-LA-M-1). For each of the linear transformations described below, write down the matrix representing that linear transformation.

1. L(x,y) = (3x-2y, 2x+6y, x-y)

2. T(x,y,z) = (z-3x+y, -5y-6z)

3. the linear transformation $f : \mathbb{R}^2 \to \mathbb{R}^2$ for which $f(\vec{e}_1) = (7, 2)$ and $f(\vec{e}_2) = (1, 0)$.

Exercise 3.2.2 (exer-LA-M-2). We consider here the matrix A and the vector $\vec{v}$ below. For each of the following questions show the work in your computation, and confirm that all three methods yield the same answer.

$$A = \begin{pmatrix} 2 & 3 & 1 \\ 6 & -1 & 3 \\ 1 & 0 & -2 \end{pmatrix} \quad \text{and} \quad \vec{v} = \begin{bmatrix} 2 \\ 4 \\ 1 \end{bmatrix}$$

1. Compute the product $A\vec{v}$ directly from Interpretation 0 of matrix-vector multiplication.

2. Compute the product $A\vec{v}$ directly from Interpretation 1 of matrix-vector multiplication.

3. Compute the product $A\vec{v}$ directly from Interpretation 2 of matrix-vector multiplication.

Exercise 3.2.3 (exer-LA-M-3). We consider the matrices A and B and the vector $\vec{v}$ below.

$$A = \begin{pmatrix} -2 & 1 & 0 \\ 3 & -2 & -4 \\ -1 & 2 & -1 \end{pmatrix} \quad B = \begin{pmatrix} 5 & -1 & 2 \\ -2 & 5 & 4 \\ 5 & 2 & 0 \end{pmatrix} \quad \text{and} \quad \vec{v} = \begin{bmatrix} 1 \\ 3 \\ 7 \end{bmatrix}$$

1. Confirm by direct computation that $(A+B)\vec{v} = A\vec{v} + B\vec{v}$.
2. Confirm by direct computation that $(5A)\vec{v} = 5(A\vec{v})$.
3. Confirm by direct computation that $(2B)\vec{v} = 2(B\vec{v})$.

Exercise 3.2.4 (**exer-LA-M-4**). We consider again the matrices A and B and the vector $\vec{v}$ from Exercise 3.2.3.

1. Compute the product matrix AB directly from Interpretation 0 of matrix multiplication (find the linear transformations represented by A and B, compose them, and then find the matrix for that composition).
2. Compute the product matrix AB directly from Interpretation 1 of matrix multiplication.
3. Compute the product matrix AB directly from Interpretation 2 of matrix multiplication.
4. Compute the product matrix AB directly from Interpretation 3 of matrix multiplication.
5. Confirm by direct computation that $(AB)\vec{v} = A(B\vec{v})$.

Exercise 3.2.5 (**exer-LA-M-5**). Repeat the five computations in Exercise 3.2.4 for the product BA.

Exercise 3.2.6 (**exer-LA-M-6**). We consider the matrix A below.

$$A = \begin{pmatrix} a_{11} & a_{12} & a_{13} \\ a_{21} & a_{22} & a_{23} \\ a_{31} & a_{32} & a_{33} \end{pmatrix}$$

1. What is the product of A and the column vector $\vec{e}_1 = (1, 0, 0)$?
2. What is the product of A and the column vector $\vec{e}_2 = (0, 1, 0)$?
3. What is the product of A and the matrix C whose first column is $\vec{e}_1$ and whose second column is $\vec{e}_2$?
4. What is the product of A and the matrix D whose first column is $\vec{e}_2$ and whose second column is $\vec{e}_1$?
5. What is the product of A and the matrix E whose first column is $\vec{e}_3$, whose second column is $\vec{e}_1$, and whose third column is $\vec{e}_2$?
6. There is a unique matrix I_R for which $AI_R = A$; what is that matrix?

Exercise 3.2.7 (**exer-LA-M-7**). We consider again the matrix A from Exercise 3.2.6.

1. What is the product of the row vector $(1\,0\,0)$ and the matrix A? (Note that in this product we list the row vector on the left and the matrix A on the right.)
2. What is the product of the row vector $(0\,1\,0)$ and the matrix A?
3. What is the product FA, where F is the matrix whose first row is $(1\,0\,0)$ and whose second row is $(0\,1\,0)$?
4. What is the product GA, where G is the matrix whose first row is $(0\,1\,0)$ and whose second row is $(1\,0\,0)$?
5. What is the product HA, where H is the matrix whose first row is $(0\,0\,1)$, whose second row is $(1\,0\,0)$, and whose third row is $(0\,1\,0)$?
6. There is a unique matrix I_L for which $I_L A = A$; what is that matrix?

Exercise 3.2.8 (**exer-LA-M-8**). The "n by n identity matrix" is defined as the matrix I_n such that, for every other n by n matrix A, we must have $AI_n = I_n A = A$. Describe how to write down this matrix explicitly, and explain how you know that, as you have described it, it must have the given property.

Exercise 3.2.9 (**exer-LA-M-9**). Compute the product

$$\begin{pmatrix} 5 & 7 \\ 2 & 3 \end{pmatrix} \begin{pmatrix} 3 & -7 \\ -2 & 5 \end{pmatrix}$$

3.2. MATRICES

Exercise 3.2.10 (exer-LA-M-10). Compute the product

$$\begin{pmatrix} a & b \\ c & d \end{pmatrix} \begin{pmatrix} d & -b \\ -c & a \end{pmatrix}$$

Exercise 3.2.11 (exer-LA-M-11). A square matrix B is called a "right inverse matrix" for the square matrix A if the product AB is the identity matrix of the same dimensions. Write out an explicit formula for the right inverse of the matrix

$$A = \begin{pmatrix} a & b \\ c & d \end{pmatrix}$$

(assuming that $ad - bc \neq 0$) and confirm that your matrix has the required property.

Exercise 3.2.12 (exer-LA-M-12). A square matrix C is called a "left inverse matrix" for the square matrix A if the product CA is the identity matrix of the same dimensions. Write out an explicit formula for the left inverse of the matrix

$$A = \begin{pmatrix} a & b \\ c & d \end{pmatrix}$$

(assuming that $ad - bc \neq 0$) and confirm that your matrix has the required property.

What do you notice about this left inverse and the right inverse found in Exercise 3.2.11?

Chapter 4

Derivatives

4.1 Single Variable Derivatives

4.1.1 Interpretations

We are accustomed to several interpretations of the notion of the derivative in single variable calculus. Most notably, students often think of the derivative as the slope of the tangent line to the graph of the function. But there are other interpretations that can be made of single variable derivatives, and these others will be more useful as we try to generalize in a multivariable setting. We begin this chapter by discussing two of these other interpretations.

First, we recall that a derivative can be used to compute linear approximations. This can be expressed by

$$f(x + \Delta x) \approx f(x) + \left(\frac{df}{dx}\right) \Delta x$$

Moving the term $f(x)$ over to the left side of the equation, this gives us another interpretation of the derivative $\frac{df}{dx}$ – that is, for small values of Δx, we have

$$\Delta f \approx \left(\frac{df}{dx}\right) \Delta x \qquad (4.1)$$

Here is a second interpretation we can make. Suppose we think of the value x as moving along the x-axis over time, so that x is a function of t. Then if f is a function of x, the chain rule tells us that

$$\frac{df}{dt} = \left(\frac{df}{dx}\right) \frac{dx}{dt} \qquad (4.2)$$

Each of equations 4.1 and 4.2 can be phrased in words, giving us the following two interpretations – we can view the derivative of a single variable function f as:

1. the thing you multiply by a discrete change in x to approximate the discrete change in f; or

2. the thing you multiply by the rate of change of x to compute the rate of change of f

These two interpretations each relate a "change" in the input of the function to the corresponding "change" in the output of the function. In the first interpretation, "change" refers to discrete jumps – the input value changes by some amount Δx, and the output changes by some amount Δf. In the second interpretation, "change" refers to a continuous rate of change – the input value has a rate of change $\frac{dx}{dt}$, and the output has a rate of change $\frac{df}{dt}$.

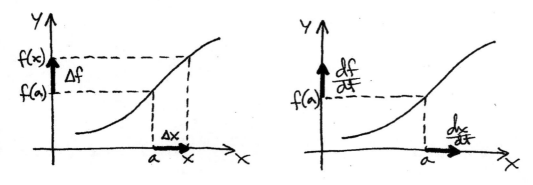

Figure 4.1:

4.1.2 Examples

Example 4.1.1. Suppose we know that the function f has derivative $\frac{df}{dx}(4) = 3$. Approximately how much does the function change between $x = 4$ and $x = 4.01$?

Using the first of the above interpretations, we approximate the change Δf by

$$\Delta f \approx \left(\frac{df}{dx}\right)\Delta x = (3)(.01) = .03$$

Of course this is only an estimate, but it is the best estimate we can make given only the above information.

Example 4.1.2. Again suppose we know that the function f has derivative $\frac{df}{dx}(4) = 3$. If the value of x is 4 at time $t = 0$ and is increasing at a rate of 6 with respect to t, how fast is the value of $f(x)$ increasing at that moment?

Using the second of the above interpretations, we can compute $\frac{df}{dt}$ by

$$\frac{df}{dt} = \left(\frac{df}{dx}\right)\frac{dx}{dt} = (3)(6) = 18$$

In each of the above interpretations, these changes could instead be thought of as vectors. We could visualize the value Δx as an arrow from our initial input value to our new input value, and Δf as an arrow from our initial output value to our new output value. Similarly, if we think of the point x as moving along the x-axis, then $\frac{dx}{dt}$ is an arrow representing that velocity, and likewise $\frac{df}{dt}$ is an arrow representing the velocity of the output value. See Figure 4.1

We can summarize both of the interpretations above by saying that the derivative is the answer to the following question: Given a point a in the domain, and a vector $\vec{v}$ representing a change from that point (either discrete or continuous), what is the corresponding change in the function?

In this chapter we will be using the above two interpretations of the single variable derivative to motivate analogous ideas for multivariable derivatives. We will find that there are several different ways of doing this – for different kinds of functions, and interpreted with different of the geometric constructions from Chapter 2. But in each of these cases, make sure to note that we will define the derivative of a function as something that relates input changes to output changes of that function.

4.1. SINGLE VARIABLE DERIVATIVES

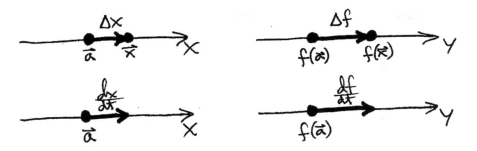

Figure 4.2:

In the multivariable setting it is more difficult to draw graphs, since we quickly run out of dimensions that we can draw. So, to allow us to more easily transition to analogous multivariable pictures, in Figure 4.2 we represent the same single variable picture as above, but with the domain and the target drawn separately.

Exercises

Exercise 4.1.1 (exer-Ders-SV-1). Suppose the production (P) from a factory is determined as a function of labor (L) by $P(L) = L - \ln L$. At a given moment the labor in the factory is $L = 4$.

1. If labor goes up by .01, what is the linear approximation of how much the production will change?
2. If labor instead goes up by .02, what is the linear approximation of how much the production will change?
3. If labor instead goes down by .03, what is the linear approximation of how much the production will change?

Exercise 4.1.2 (exer-Ders-SV-2). The temperature (T) at a position x along a highway is given by $T(x) = 85 + 15\sin(2\pi x/48)$. Here we consider a car that is at the position where $x = 16$.

1. If the car is moving with velocity 2, how fast is the temperature changing?
2. If the car is moving with velocity 14, how fast is the temperature changing?
3. If the car is moving with velocity -5, how fast is the temperature changing?

4.2 Derivatives of Parametric Curves

4.2.1 Definition and interpretation

Suppose we have a function $\vec{r} : \mathbb{R}^1 \to \mathbb{R}^n$ representing a point moving along a given path (parametrized by $\vec{r}$).

$$\vec{r}(t) = \begin{bmatrix} r_1(t) \\ \vdots \\ r_n(t) \end{bmatrix}$$

How should we define the derivative of a function like this? And, how does that derivative relate to the motion of the point in question?

Regarding the first question, we actually have a nice piece of good luck –that is, because there is only one independent variable, we can write down the same definition for the derivative of a parametric curve that we did for a single variable function.

Definition 4.2.1. *The derivative of a function $\vec{r} : \mathbb{R}^1 \to \mathbb{R}^n$ is*

$$\frac{d\vec{r}}{dt} = \lim_{h \to 0} \frac{\vec{r}(t+h) - \vec{r}(t)}{h}$$

(Of course the interpretations we make of this expression will not be the same as those for a single variable function, since of course we are no longer looking at a single variable function. Furthermore, we are interested in an interpretation of this derivative regarding the image of this function, not the graph.)

This expression involves several operations on vectors – subtraction, scalar multiplication, and limits – and conveniently, all of these are computed "componentwise"; namely, we can compute the operation simply by doing the same to all of the components individually. This allows us to perform the following computation:

$$\begin{aligned}
\frac{d\vec{r}}{dt} &= \lim_{h \to 0} \frac{\vec{r}(t+h) - \vec{r}(t)}{h} \\
&= \lim_{h \to 0} \frac{\begin{bmatrix} r_1(t+h) \\ \vdots \\ r_n(t+h) \end{bmatrix} - \begin{bmatrix} r_1(t) \\ \vdots \\ r_n(t) \end{bmatrix}}{h} \\
&= \lim_{h \to 0} \frac{\begin{bmatrix} r_1(t+h) - r_1(t) \\ \vdots \\ r_n(t+h) - r_n(t) \end{bmatrix}}{h} \\
&= \lim_{h \to 0} \begin{bmatrix} \frac{r_1(t+h) - r_1(t)}{h} \\ \vdots \\ \frac{r_n(t+h) - r_n(t)}{h} \end{bmatrix} \\
&= \begin{bmatrix} \lim_{h \to 0} \frac{r_1(t+h) - r_1(t)}{h} \\ \vdots \\ \lim_{h \to 0} \frac{r_n(t+h) - r_n(t)}{h} \end{bmatrix} \\
&= \begin{bmatrix} \frac{dr_1}{dt} \\ \vdots \\ \frac{dr_n}{dt} \end{bmatrix}
\end{aligned}$$

So, we conclude that the derivative of a function parametrizing a curve has components that are just the derivatives of the individual component functions. Or we sometimes shorten this by saying simply that the derivative of a parametric

4.2. DERIVATIVES OF PARAMETRIC CURVES

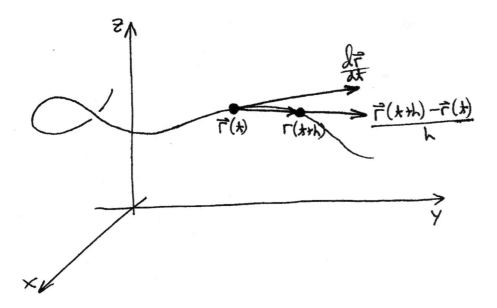

Figure 4.3:

curve is "computed componentwise".

$$\frac{d\vec{r}}{dt} = \begin{bmatrix} \frac{dr_1}{dt} \\ \vdots \\ \frac{dr_n}{dt} \end{bmatrix}$$

How does this relate to the motion of the point in question? In Figure 4.3 we draw a picture representing the expression in Definition 4.2.1.

The numerator is the difference of two vectors that are very close to each other on the curve; so, the difference is very close to being tangent to the curve. In fact, in the limit, it is a tangent vector.

As for the length, note that we can write

$$\left\|\frac{d\vec{r}}{dt}\right\| = \lim_{h \to 0} \frac{\|\vec{r}(t+h) - \vec{r}(t)\|}{|h|}$$

In this form we see that the length of the derivative vector is just the limit of the ratio of distance and time; that is, the length of the derivative vector is the instantaneous speed of the point.

So the derivative vector is a vector that is tangent to the curve, with length equal to the speed of motion. Therefore the derivative of the parametrization is in fact exactly the velocity vector of the moving point. We represent this in Figure 4.4.

4.2.2 Examples

Example 4.2.1. Say a bird is flying with its position as a function of time given by

$$\vec{r}(t) = \begin{bmatrix} \cos t \\ \sin t \\ t \end{bmatrix}$$

What is the bird's velocity as a function of time?

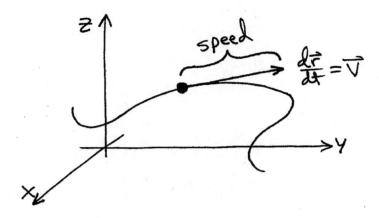

Figure 4.4:

To compute the velocity, we need only compute the derivative vector; and we can compute the derivative vector by differentiating $\vec{r}(t)$ componentwise. So, we get

$$\textbf{velocity} = \frac{d}{dt}\vec{r}(t) = \begin{bmatrix} \frac{d}{dt}(\cos t) \\ \frac{d}{dt}(\sin t) \\ \frac{d}{dt}(t) \end{bmatrix} = \begin{bmatrix} -\sin t \\ \cos t \\ 1 \end{bmatrix}$$

Example 4.2.2. Suppose we know that a moving particle has initial position given by $\vec{x}(0) = (3,2,5)$, and that its velocity function is given by

$$\vec{v}(t) = \begin{bmatrix} \sin t \\ \cos(2t) \\ \cos(3t) \end{bmatrix}$$

What is the position of this particle as a function of time?

We know that the given velocity function is the derivative of the desired position function; so, we can simply antidifferentiate the velocity function to find the position function. As derivatives are computed componentwise, so are antiderivatives.

$$\begin{aligned}
\vec{x}(t) &= \int \vec{v}(t)\, dt \\
&= \int \begin{bmatrix} \sin t \\ \cos(2t) \\ \cos(3t) \end{bmatrix} dt \\
&= \begin{bmatrix} \int \sin t\, dt \\ \int \cos(2t)\, dt \\ \int \cos(3t)\, dt \end{bmatrix} \\
&= \begin{bmatrix} -\cos t + c_1 \\ \frac{1}{2}\sin(2t) + c_2 \\ \frac{1}{3}\sin(3t) + c_3 \end{bmatrix}
\end{aligned}$$

4.2. DERIVATIVES OF PARAMETRIC CURVES

We can evaluate the constants by using our initial condition $\vec{x}(0) = (3, 2, 5)$, which tells us

$$\vec{x}(0) = \begin{bmatrix} -\cos 0 + c_1 \\ \frac{1}{2}\sin(0) + c_2 \\ \frac{1}{3}\sin(0) + c_3 \end{bmatrix}$$

$$\begin{bmatrix} 3 \\ 2 \\ 5 \end{bmatrix} = \begin{bmatrix} -1 + c_1 \\ c_2 \\ c_3 \end{bmatrix}$$

$$\begin{bmatrix} 4 \\ 2 \\ 5 \end{bmatrix} = \begin{bmatrix} c_1 \\ c_2 \\ c_3 \end{bmatrix}$$

So we then have

$$\vec{x}(t) = \begin{bmatrix} -\cos t + 4 \\ \frac{1}{2}\sin(2t) + 2 \\ \frac{1}{3}\sin(3t) + 5 \end{bmatrix} \qquad (4.3)$$

Example 4.2.3. Suppose we are given the acceleration function $\vec{a}(t)$, along with initial conditions for the velocity and position functions. Can we recover the entire position function? The answer is yes, and the process is similar to that in Example 4.2.2.

Since $\vec{a} = \vec{v}'$, the first step is to antidifferentiate the acceleration vector to obtain the velocity vector. Of course there will be integration constants, and those can be evaluated using the given initial velocity.

Next, we antidifferentiate the velocity vector to obtain the position vector. Again there will be integration constants, and those can be evaluated using the given initial position.

Note that there are some similarities between the interpretation of the derivative of a parametric curve and the usual interpretation of the derivative of a single variable function – but there are several critical differences. Of course for one thing, here the derivative is a vector, whereas we usually think of a single variable derivative as just a number. Also, here we are looking at the image of the function in order to get the interpretation of the derivative as a velocity vector, whereas we usually think of single variable derivatives in the context of graphs.

This notion of the derivative of a parametric curve will turn out to be very useful as we proceed to discussing the derivatives of a general multivariable function.

Exercises

Exercise 4.2.1 (exer-Ders-PC-1). Compute the derivatives of the following parametric curves.

1. $\vec{x}(t) = (t^2, \cos(e^t))$
2. $\vec{r}(t) = (\cos t, \sin t, t)$
3. $\vec{p}(t) = (\tan(3t - 5), 4e^{t^2 - t}, \arcsin(\frac{1}{2}\sin t))$

Exercise 4.2.2 (exer-Ders-PC-2). Compute the speed of each of the parametric curves in Exercise 4.2.1 at the point where $t = 0$.

Exercise 4.2.3 (**exer-Ders-PC-3**). Find the unit vector representing the direction of motion of each of the parametric curves in Exercise 4.2.1 at the point where $t = 0$.

Exercise 4.2.4 (**exer-Ders-PC-15**). Use the fact that parametric derivatives are computed componentwise to show that
$$\left(\vec{x}(t) + \vec{y}(t)\right)' = \vec{x}' + \vec{y}' \quad \text{and} \quad \left(k\vec{x}(t)\right)' = k\vec{x}'$$

Exercise 4.2.5 (**exer-Ders-PC-16**). Use the single variable product rule and the fact that parametric derivatives are computed componentwise to show that
$$\left(f(t)\vec{x}(t)\right)' = f'\vec{x} + f\vec{x}'$$

Exercise 4.2.6 (**exer-Ders-PC-17**). Show that
$$\left(\vec{x}(t) \cdot \vec{y}(t)\right)' = \vec{x}' \cdot \vec{y} + \vec{x} \cdot \vec{y}' \quad \text{and} \quad \left(\vec{x}(t) \times \vec{y}(t)\right)' = \vec{x}' \times \vec{y} + \vec{x} \times \vec{y}'$$

Exercises 4.2.7 through 4.2.12 develop the idea of curvature for parametric curves in $\mathbb{R}^2$. In each of these exercises you may assume the results of the previous exercises.

Exercise 4.2.7 (**exer-Ders-PC-4**). For a parametric curve $\vec{x}(t)$ in the xy-plane, we define the angle ϕ to be the angle of the velocity vector $\vec{v} = \vec{x}'$, measured counterclockwise from the positive part of the x-axis, so that $dy/dx = \tan\phi$. Find a formula for ϕ' in terms of the derivatives of $x(t)$ and $y(t)$. (Hint: Note that $dy/dx = y'/x'$.)

Exercise 4.2.8 (**exer-Ders-PC-5**). We define the *curvature*, κ, at a point on the curve to be $|d\phi/ds|$, where s represents the length along the curve itself. Show that the curvature can be written as
$$\kappa = \frac{|x'y'' - x''y'|}{((x')^2 + (y')^2)^{3/2}}$$
(Hint: Use Exercise 4.2.7 and note that $d\phi/ds = \phi'/s'$, and that $s' = v = \|\vec{v}\|$.)

Exercise 4.2.9 (**exer-Ders-PC-18**). Use the graph parametrization to find a formula for the curvature of the graph $y = f(x)$ of the function $f : \mathbb{R}^1 \to \mathbb{R}^1$, in terms of the derivatives of that function f.

Exercise 4.2.10 (**exer-Ders-PC-6**). We define the unit tangent vector $\vec{T}$ to be $\vec{v}/\|\vec{v}\|$. Show that $\vec{T} = (\cos\phi, \sin\phi)$.

Exercise 4.2.11 (**exer-Ders-PC-7**). Show that $d\vec{T}/d\phi$ is always a unit vector perpendicular to $\vec{T}$.

Exercise 4.2.12 (**exer-Ders-PC-8**). Show that $\|d\vec{T}/ds\| = \kappa$. (Hint: Use the chain rule to relate this question to Exercises 4.2.8 and 4.2.11.)

Exercises 4.2.13 through 4.2.18 develop the idea of curvature for parametric curves in $\mathbb{R}^3$. In each of these exercises you may assume the results of the previous exercises.

Exercise 4.2.13 (**exer-Ders-PC-9**). Motivated by the result of Exercise 4.2.12, for curves in $\mathbb{R}^3$ we define the *curvature*, κ, to be $\|d\vec{T}/ds\|$. We then define $\vec{N}$ (the *principal unit normal vector*) to be the unit vector in the direction of $d\vec{T}/ds$, giving us $d\vec{T}/ds = \kappa\vec{N}$.

Use the fact that $\vec{T} \cdot \vec{T}$ is a constant to show that $d\vec{T}/ds$ is orthogonal to $\vec{T}$ itself.

(This justifies calling $\vec{N}$ a "normal" vector, meaning "perpendicular".)

Exercise 4.2.14 (**exer-Ders-PC-10**). The acceleration vector $\vec{a}$ is a linear combination of the unit tangent vector and the principal unit normal vector, which we write as $\vec{a} = a_T \vec{T} + a_N \vec{N}$. We refer to the coefficients a_T and a_N as the *tangential component of acceleration* and *normal component of acceleration*, respectively.

Differentiate both sides of the equation $\vec{v} = v\vec{T}$ with respect to t to show that $a_T = dv/dt$ and $a_N = \kappa v^2$. (Hint: Use the product rule, and then rewrite $d\vec{T}/dt$ as $(d\vec{T}/ds)(ds/dt)$.)

4.2. DERIVATIVES OF PARAMETRIC CURVES

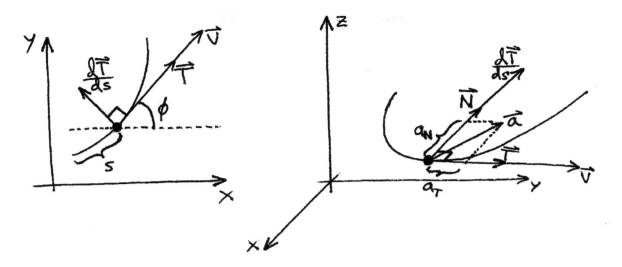

Figure 4.5: Diagram for use with Exercises 4.2.7 through 4.2.18

(The tangential component of acceleration, as simply the rate of change of the speed, can be thought of as the rate of change of the *length* of the velocity vector; the normal component of acceleration, sometimes referred to as "centripetal" acceleration, relates to the rate of change of the *direction* of the velocity vector.)

Exercise 4.2.15 (exer-Ders-PC-11). Show that $\vec{T} \cdot \vec{T} = 1$, $\vec{T} \cdot \vec{N} = 0$, $\vec{T} \times \vec{T} = \vec{0}$, and $\|\vec{T} \times \vec{N}\| = 1$.

Exercise 4.2.16 (exer-Ders-PC-12). Use the equations $\vec{v} = v\vec{T}$ and $\vec{a} = a_T \vec{T} + a_N \vec{N}$ and the results of Exercise 4.2.15 to show that $\vec{v} \cdot \vec{a} = v a_T$ and $\|\vec{v} \times \vec{a}\| = v a_N$.

Exercise 4.2.17 (exer-Ders-PC-13). Use the results of Exercises 4.2.14 and 4.2.16 to find formulas for a_T, a_N, and κ in terms of $\vec{v}$ and $\vec{a}$.

Exercise 4.2.18 (exer-Ders-PC-14). Compute the curvature of the path parametrized by $\vec{x}(t) = (t^2 + 2t, t - 4, e^t - t^3)$ at the point where $t = 0$.

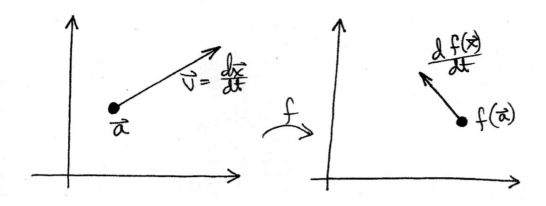

Figure 4.6:

4.3 Directional Derivatives

4.3.1 Definition

In this section we consider now a general multivariable function $f : \mathbb{R}^n \to \mathbb{R}^m$. In light of the interpretations of the single variable derivative that we made in Section 4.1, we can try to do the same thing for a multivariable function $f : \mathbb{R}^n \to \mathbb{R}^m$ – given a point $\vec{a}$ in the domain, and a vector $\vec{v}$ representing a change from that point, what is the corresponding change in the function?

Let's phrase this question more precisely, arbitrarily thinking in terms of continuous changes, i.e. velocities, instead of discrete changes. Suppose we have a fixed point $\vec{a}$ in the domain, and a given vector $\vec{v}$. Suppose that $\vec{x}$ is a moving point that is momentarily at the point $\vec{a}$ and is moving with velocity $\frac{d\vec{x}}{dt} = \vec{v}$. What is the velocity of the image point – that is, what is the velocity of the output $f(\vec{x})$ in the target? More algebraically phrased, what is $\frac{df(\vec{x})}{dt}$? We lay out this setup in Figure 4.6.

In order to get started on this, let's actually write down a parametric curve that describes the point $\vec{x}$ that moves in the domain like we described above. In other words, let's write down a parametric curve $\vec{x}(t)$ that is at the point $\vec{a}$ (say, when $t = 0$), and that has velocity vector given by $\vec{v}$.

There are many possible such curves of course, but the easiest one to write down is

$$\vec{x}(t) = \vec{a} + t\vec{v}$$

From our previous discussion of parametric curves, it is easily checked that $\vec{x}(0) = \vec{a}$ and $\frac{d\vec{x}}{dt}(0) = \vec{v}$.

As previously stated we are interested in the motion of the image of this point in the target – that is, of the point $f(\vec{x})$. As $\vec{x}$ follows this path in the domain, it's image follows some path given parametrically by $f(\vec{x}(t))$ in the target. So, the velocity of $f(\vec{x})$ is just the derivative of that parametrization. See Figure 4.7.

We can now write down exactly the velocity of the output value $f(\vec{x})$ when $\vec{x} = \vec{a}$, that is, when $t = 0$:

$$\frac{df(\vec{x})}{dt}(0) = \frac{d}{dt}\bigg|_{t=0} f(\vec{a} + t\vec{v})$$

This is what we will call the "directional derivative".

Definition 4.3.1. *The directional derivative of f at the point $\vec{a}$ with velocity $\vec{v}$ is*

$$D_{\vec{v}}f(\vec{a}) = \frac{d}{dt}\bigg|_{t=0} f(\vec{a} + t\vec{v})$$

4.3. DIRECTIONAL DERIVATIVES

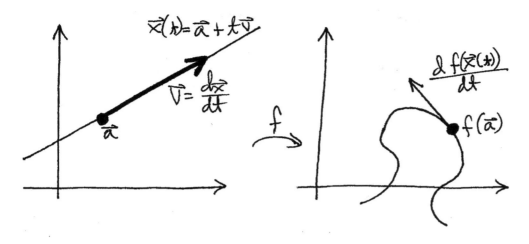

Figure 4.7:

Be sure to note that the way we have this defined, computing the directional derivative requires not only a function f and an input point $\vec{a}$, but also a vector $\vec{v}$ representing the velocity (or change) in the domain.

Students should be aware that the term "directional derivative" is used in a different way in some other textbooks. Specifically, in some textbooks the term is reserved for use only with unit vectors $\vec{v}$. While it is true that additional interpretations can be made in this case, it certainly remains that the directional derivative is an extremely useful construction in the more general case as well. In this textbook we will use the term in the more general sense, as defined above. In Section 4.4 we will consider the interesting special case where $\vec{v}$ is a unit vector.

(Of course with single variable derivatives, the analogue of $\vec{v}$ (representing speed or change in the domain) is needed to be able to use the derivative to compute anything, but the way we defined single variable derivatives allows us still to view the derivative as being a separate, independent object. We will see soon that something very similar can be done for the directional derivative.)

4.3.2 Interpretations

We have the following two interpretations of the directional derivative, which the reader will note are extremely similar to the interpretations of single variable derivatives made in Section 4.1:

(1) The directional derivative $D_{\vec{v}}f(\vec{a})$ is the rate of change $\frac{df}{dt}$ of the output of the function, given that the rate of change $\frac{d\vec{x}}{dt}$ of the input $\vec{x}$ of the function is the vector $\vec{v}$.

$$\frac{df}{dt} = D_{\vec{v}}f(\vec{a}) \quad \text{assuming that} \quad \frac{d\vec{x}}{dt} = \vec{v}$$

(2) The directional derivative $D_{\vec{v}}f(\vec{a})$ is an approximation to the discrete change Δf of the output of the function, given that the discrete change $\Delta \vec{x}$ of the input $\vec{x}$ of the function is the vector $\vec{v}$.

$$\Delta f \approx D_{\vec{v}}f(\vec{a}) \quad \text{assuming} \quad \Delta \vec{x} = \vec{v}$$

Note that the directional derivative makes sense in several settings:

(1) If $f : \mathbb{R}^n \to \mathbb{R}^1$, then $D_{\vec{v}}f(\vec{a})$ is a scalar, representing the velocity of the output as it moves through the target $\mathbb{R}^1$ (as the input moves through the domain with velocity $\vec{v}$). See Figure 4.8.

Of course in this case f has just a one-dimensional output, so the directional derivative is just the rate of change of that single output variable.

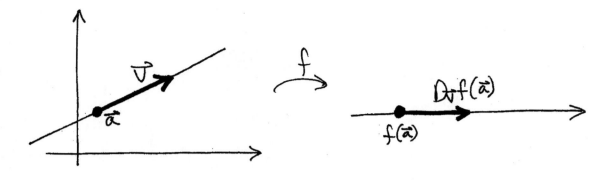

Figure 4.8:

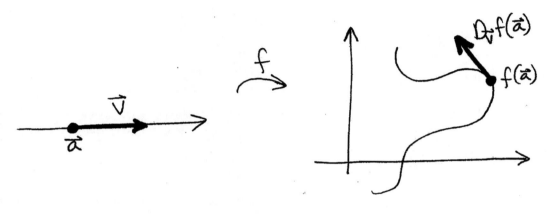

Figure 4.9:

(2) If $f : \mathbb{R}^1 \to \mathbb{R}^m$, then $D_v f(a)$ is a vector, representing the velocity of the output vector as it moves through the target $\mathbb{R}^m$ (as the input moves with velocity v, which of course in this case is just a scalar). See Figure 4.9.

In this case the outputs of f form a parametric curve in $\mathbb{R}^m$, parametrized by the function f. The interested reader can check that in this context the directional derivative (with $v = 1$) is just the velocity vector of this parametric curve, at the point in question.

(3) If $f : \mathbb{R}^n \to \mathbb{R}^m$, then $D_{\vec{v}} f(\vec{a})$ is a vector, representing the velocity of the output vector as it moves through the target $\mathbb{R}^m$ (as the input moves through the domain with velocity $\vec{v}$). See Figure 4.10.

As in case (2) the output values of f (as the input moves through the domain with velocity $\vec{v}$ at the point $\vec{a}$) form a parametric curve in $\mathbb{R}^m$, and the directional derivative is just the velocity vector of this parametric curve.

Note that in each of cases (2) and (3), the directional derivative is a vector; this is to be expected, because the output of f consists of vectors, and the directional derivative refers to changes in the output of f.

4.3.3 Examples

Example 4.3.1. Say we have the function $f : \mathbb{R}^3 \to \mathbb{R}^1$ **given by**

$$f\left(\begin{bmatrix} x \\ y \\ z \end{bmatrix}\right) = x^2 + yz$$

4.3. DIRECTIONAL DERIVATIVES

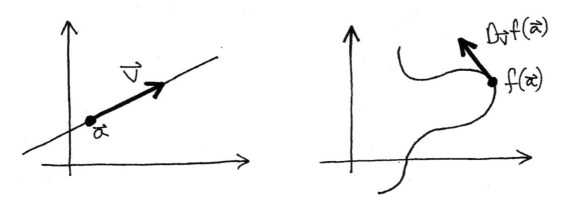

Figure 4.10:

Suppose that $\vec{x}$ is at the point $\vec{a} = \begin{bmatrix} 1 \\ 2 \\ 0 \end{bmatrix}$, moving with velocity $\vec{v} = \frac{d\vec{x}}{dt} = \begin{bmatrix} 2 \\ 1 \\ 3 \end{bmatrix}$. What is the rate of change of $f(\vec{x})$?

The rate of change of f is the directional derivative, $D_{\vec{v}}f(\vec{a})$, which we can compute directly from the definition:

$$
\begin{aligned}
\frac{df}{dt} = D_{\vec{v}} f(\vec{a}) &= \frac{d}{dt}\bigg|_{t=0} f(\vec{a} + t\vec{v}) \\
&= \frac{d}{dt}\bigg|_{t=0} f\left(\begin{bmatrix} 1 \\ 2 \\ 0 \end{bmatrix} + t \begin{bmatrix} 2 \\ 1 \\ 3 \end{bmatrix} \right) \\
&= \frac{d}{dt}\bigg|_{t=0} f\left(\begin{bmatrix} 1 + 2t \\ 2 + t \\ 3t \end{bmatrix} \right) \\
&= \frac{d}{dt}\bigg|_{t=0} (1 + 2t)^2 + (2 + t)(3t) \\
&= \frac{d}{dt}\bigg|_{t=0} 7t^2 + 10t + 1 \\
&= (14t + 10)\bigg|_{t=0} \\
&= 10
\end{aligned}
$$

Example 4.3.2. Say we have the function $f : \mathbb{R}^3 \to \mathbb{R}^2$ given by

$$f\left(\begin{bmatrix} x \\ y \\ z \end{bmatrix}\right) = \begin{bmatrix} xy \\ yz \end{bmatrix}$$

Suppose that $\vec{x}$ is at the point $\vec{a} = \begin{bmatrix} 1 \\ 2 \\ 0 \end{bmatrix}$, moving with velocity $\vec{v} = \frac{d\vec{x}}{dt} = \begin{bmatrix} 2 \\ 1 \\ 3 \end{bmatrix}$. What is the velocity of $f(\vec{x})$?

Again, the velocity of f is the directional derivative, $D_{\vec{v}}f(\vec{a})$, which we can compute directly from the definition:

$$\begin{aligned}
\frac{df}{dt} = D_{\vec{v}}f(\vec{a}) &= \frac{d}{dt}\bigg|_{t=0} f(\vec{a}+t\vec{v}) \\
&= \frac{d}{dt}\bigg|_{t=0} f\left(\begin{bmatrix}1\\2\\0\end{bmatrix} + t \begin{bmatrix}2\\1\\3\end{bmatrix}\right) \\
&= \frac{d}{dt}\bigg|_{t=0} f\left(\begin{bmatrix}1+2t\\2+t\\3t\end{bmatrix}\right) \\
&= \frac{d}{dt}\bigg|_{t=0} \begin{bmatrix}(1+2t)(2+t)\\(2+t)(3t)\end{bmatrix} \\
&= \frac{d}{dt}\bigg|_{t=0} \begin{bmatrix}2t^2+5t+2\\3t^2+6t\end{bmatrix} \\
&= \begin{bmatrix}4t+5\\6t+6\end{bmatrix}\bigg|_{t=0} \\
&= \begin{bmatrix}5\\6\end{bmatrix}
\end{aligned}$$

Note that in this case the velocity is a vector, because the function is vector-valued.

Example 4.3.3. Because of the position of the sun at a certain moment, a point at location (x, y, z) in space casts a shadow onto the xy-plane at the point $(x - 3z, y + 2z, 0)$. What is the velocity of the shadow of the bird of Example 4.2.1 at the time $t = \pi$?

The location of the shadow can be thought of as the value of the function

$$f(x, y, z) = (x - 3z, y + 2z, 0)$$

and then the velocity of the shadow can be computed simply as the directional derivative of this function, at the bird's position and with the bird's velocity.

Using the results of that previous example, we know that the velocity of the bird at time $t = \pi$ is given by

$$\vec{v} = \begin{bmatrix}-\sin\pi\\ \cos\pi\\ 1\end{bmatrix} = \begin{bmatrix}0\\-1\\1\end{bmatrix}$$

and its position at that time is

$$\vec{a} = \begin{bmatrix}\cos\pi\\ \sin\pi\\ \pi\end{bmatrix} = \begin{bmatrix}-1\\0\\\pi\end{bmatrix}$$

4.3. DIRECTIONAL DERIVATIVES

The directional derivative can then be computed as

$$\begin{aligned}
D_{\vec{v}}f(\vec{a}) &= \frac{d}{ds}\bigg|_{s=0} f(\vec{a}+s\vec{v}) \\
&= \frac{d}{ds}\bigg|_{s=0} f\left(\begin{bmatrix}-1\\0\\\pi\end{bmatrix} + s\begin{bmatrix}0\\-1\\1\end{bmatrix}\right) \\
&= \frac{d}{ds}\bigg|_{s=0} f\left(\begin{bmatrix}-1\\-s\\\pi+s\end{bmatrix}\right) \\
&= \frac{d}{ds}\bigg|_{s=0} \begin{bmatrix}-1-3(\pi+s)\\-s+2(\pi+s)\\0\end{bmatrix} \\
&= \begin{bmatrix}-3\\1\\0\end{bmatrix}
\end{aligned}$$

(Note we use "s" here for the directional derivative to avoid confusion with the "t" that had already been chosen to parametrize the path of the bird.)

As was the case with parametric curves, we can relate the directional derivative of a vector-valued function f to the directional derivatives of its component functions; as with parametric curves, we note that all of the operations involved in the computation of the directional derivative are componentwise operations, and so the directional derivative can be computed componentwise. In particular, if we have

$$f(\vec{x}) = \begin{bmatrix} f_1(\vec{x}) \\ \vdots \\ f_n(\vec{x}) \end{bmatrix}$$

then a computation much like the one immediately following Definition 4.2.1 for the derivative of a parametric curve shows that the directional derivative can be rewritten as

$$D_{\vec{v}}f(\vec{a}) = \begin{bmatrix} D_{\vec{v}}f_1(\vec{a}) \\ \vdots \\ D_{\vec{v}}f_n(\vec{a}) \end{bmatrix}$$

showing that the directional derivative too can be computed componentwise.

Exercises

Exercise 4.3.1 (exer-Ders-DD1-1). In each of the items below, find a parametric curve (in terms of the input variable t) for which the position and velocity at $t=0$ are as indicated.

1. $\vec{x}(0) = \vec{a} = (3,2)$, and $\vec{x}'(0) = \vec{v} = (1,4)$
2. $\vec{x}(0) = \vec{a} = (3,2)$, and $\vec{x}'(0) = \vec{v} = (5,12)$

3. $\vec{x}(0) = \vec{a} = (0,5)$, and $\vec{x}'(0) = \vec{v} = (-1,-3)$

Exercise 4.3.2 (exer-Ders-DD1-2). Suppose we consider the function $P(x,y) = (x^2 y, x - 3y)$. If we consider the curves found in Exercise 4.3.1 to be moving through the domain of this function, find the formulas for the values of this function P in terms of t for each of those curves.

Exercise 4.3.3 (exer-Ders-DD1-3). Evaluate $\frac{dP}{dt}\big|_{t=0} = D_{\vec{v}} P(\vec{a})$ for each of the functions determined in Exercise 4.3.2.

Exercise 4.3.4 (exer-Ders-DD1-4). Find the directional derivative $D_{\vec{v}} f(\vec{a})$ for the function $f(x,y,z) = x^2 - 2y^2 + 4z$ at the point $\vec{a} = (0,1,3)$ with velocity $\vec{v} = (2,3,2)$.

Exercise 4.3.5 (exer-Ders-DD1-5). Find the directional derivative $D_{\vec{v}} f(\vec{a})$ for the function $f(x,y,z) = (e^{x-2y-z^2}, \sin(z-y-3x), 3x+4y)$ at the point $\vec{a} = (6,2,3)$ with velocity $\vec{v} = (0,-2,1)$.

Exercise 4.3.6 (exer-Ders-DD1-6). Compute *individually* the directional derivatives of each of the components of the function f from Exercise 4.3.5, and confirm that these agree with those determined in that Exercise.

Exercise 4.3.7 (exer-Ders-DD1-7). The position on the ground in the xy-plane that is hit by a laser in orbit is given by $(x,y) = (3t + \tan(\phi), -2t + \tan(\theta))$, where t, ϕ, and θ are input variables controlled by the operator of the laser. What is the velocity of the hit point if the input variables are at values $(5, \pi/4, \pi/3)$ and changing with velocity $(1, 3, -2)$?

Exercise 4.3.8 (exer-Ders-DD1-8). What is the speed of the hit point in Exercise 4.3.7?

4.4 Unit Directional Derivatives

In this section we present a construction that is a special case of the directional derivative. In this special case, we will be able to make an interpretation that has some advantages and some disadvantages.

4.4.1 Derivation

We will begin with the important observation that the directional derivative is "scalar multiplicative". That is, when the velocity in the domain is multiplied by a scalar, then the value of the directional derivative (the velocity in the target) is multiplied by that same scalar.

$$D_{k\vec{v}}f(\vec{a}) = kD_{\vec{v}}f(\vec{a}) \tag{4.4}$$

This is not hard to show by a simple application of the chain rule, using the new variable $s = tk$:

$$\begin{aligned} D_{k\vec{v}}f(\vec{a}) &= \frac{d}{dt}\bigg|_{t=0} f(\vec{a} + t(k\vec{v})) \\ &= \frac{d}{dt}\bigg|_{t=0} f(\vec{a} + s\vec{v}) \\ &= \left(\frac{d}{ds}\bigg|_{s=0} f(\vec{a} + s\vec{v})\right)\left(\frac{ds}{dt}\right) \\ &= (D_{\vec{v}}f(\vec{a}))(k) \end{aligned}$$

Suppose now that we rewrite the velocity vector $\vec{v}$ in the domain as the product of its scalar speed v and its unit vector direction $\vec{u}$, as $\vec{v} = v\vec{u}$. We can then use the result above to conclude

$$D_{\vec{v}}f(\vec{a}) = D_{\vec{u}}f(\vec{a})(v) \tag{4.5}$$

Remembering that this directional derivative represents the rate of change of the value of the function, and that $v = \|\frac{d\vec{x}}{dt}\|$, we can rewrite this as

$$\frac{df}{dt} = (D_{\vec{u}}f(\vec{a}))\left(\left\|\frac{d\vec{x}}{dt}\right\|\right) \tag{4.6}$$

We call $D_{\vec{u}}f(\vec{a})$ the "unit directional derivative", because as written the velocity vector $\vec{u}$ is a unit vector. The above equation will allow us to make an interpretation of the meaning of this unit directional derivative.

4.4.2 Interpretations

Equation 4.6 gives us an alternative point of view on computing the rate of change of f. Thinking back to one of our interpretations of single variable derivatives, recall that we viewed the derivative as a factor – namely, the derivative was the factor that you multiply by the rate of change of the input to get the rate of change of the output.

The equation above is of a similar form – and so we can use it to make similar interpretations to those that we had for single variable derivatives. We represent this geometrically in Figure 4.11.

(1) The unit directional derivative $D_{\vec{u}}f(\vec{a})$ is the factor that you multiply by the speed $\|\frac{d\vec{x}}{dt}\|$ of the input point $\vec{x}$ to find the velocity $\frac{df}{dt}$ of the output (where $\vec{u}$ is the (unit vector) direction of motion of the input $\vec{x}$).

$$\frac{df}{dt} = \left(D_{\vec{u}}f(\vec{a})\right)\left\|\frac{d\vec{x}}{dt}\right\| \quad \text{assuming that } \frac{d\vec{x}}{dt} \text{ is in the } \vec{u} \text{ direction.}$$

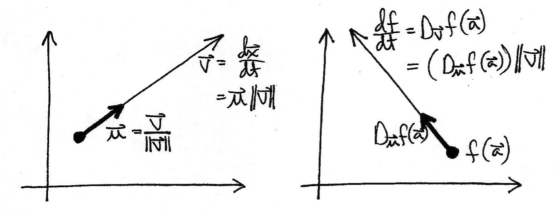

Figure 4.11:

(2) The unit directional derivative $D_{\vec{u}}f(\vec{a})$ is the factor that you multiply by the magnitude $\|\Delta\vec{x}\|$ of a discrete change $\Delta\vec{x}$ of the input point $\vec{x}$ to approximate the discrete change Δf of the output (where $\vec{u}$ is the (unit vector) direction of $\Delta\vec{x}$).

$$\Delta f \approx \left(D_{\vec{u}}f(\vec{a})\right)\|\Delta\vec{x}\| \quad \text{assuming } \Delta\vec{x} \text{ is in the } \vec{u} \text{ direction.}$$

Note that the computation above shows that in order to be able to make this interpretation, we must have $\vec{u}$ be a unit vector.

Of course as a special case of the directional derivative, it is still a directional derivative and so those interpretations still hold as well.

When applying the directional derivative to a problem that involves the input to a function moving through the domain, we now have a choice – we can apply the directional derivative directly with the velocity $\vec{v}$, or we can determine the direction vector $\vec{u}$ for the motion and then use the unit directional derivative and multiply by the speed v. The only difference between using the directional derivative and the unit directional derivative is a scalar factor, that factor being the speed of motion in the domain.

$$D_{\vec{v}}f(\vec{a}) = v D_{\vec{u}}f(\vec{a})$$

The choice of which to use is largely a question of convenience based on how a question is phrased. If the velocity of the input is either given or most easily written down as a vector, then the plain directional derivative is likely the most convenient tool. On the other hand if the velocity of the input is either given or most easily written down with a unit direction vector and a speed, then one might prefer to use the unit directional derivative.

Recall that the unit directional derivative does have the previously mentioned similarity in interpretation to single variable derivatives. So, for real valued functions, we can make one more interpretation of unit directional derivatives, motivated by our view of a single variable derivative as the slope of a tangent line to a graph. We represent this interpretation geometrically in Figure 4.12.

Consider the graph of the function $f : \mathbb{R}^n \to \mathbb{R}^1$, and then look at the cross section of that graph by the plane extending vertically from the line in the domain defined by the point $\vec{a}$ and the direction $\vec{u}$. That cross section will be a curve, and that curve has a tangent line. Direct computation show that the slope of that tangent line is the same as the unit directional derivative. This is left as an exercise for the reader to verify.

(3) For $f : \mathbb{R}^n \to \mathbb{R}^1$, the unit directional derivative $D_{\vec{u}}f(\vec{a})$ is the slope of the tangent line to the curve that is the cross section of the graph of f by the plane extending vertically from the line in the domain defined by the point $\vec{a}$ and the direction $\vec{u}$.

4.4. UNIT DIRECTIONAL DERIVATIVES

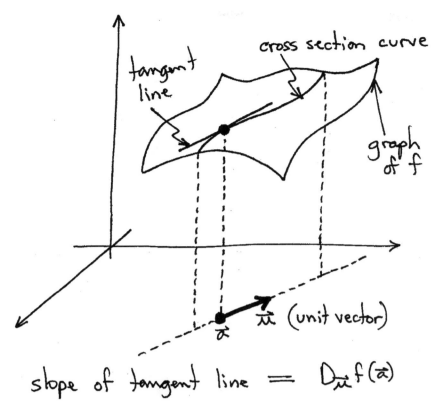

Figure 4.12:

4.4.3 Examples

Example 4.4.1. Bob's position in a forest is given by the coordinates x and y, measured in miles, where the x axis points due east and the y axis points due north. The concentration of mosquitos is given by the function f defined by $f(x, y) = e^{-x^2 - y^2}$. If Bob is at the point $(2, 1)$ and moving due west at a speed of 3.5 miles per hour, how fast is the concentration of mosquitos changing?

Of course we could represent Bob's motion with a velocity vector and then compute the corresponding directional derivative. But in this case the unit direction itself is given explicitly and is convenient to write down as simply $\vec{u} = (-1, 0)$. So we compute the desired derivative by

$$\begin{aligned}
\frac{df}{dt} &= (v)\left(D_{\vec{u}} f(\vec{a})\right) \\
&= (3.5) \left.\frac{d}{dt}\right|_{t=0} f(2-t, 1) \\
&= (3.5) \left.\frac{d}{dt}\right|_{t=0} e^{-t^2 + 4t - 5} \\
&= (3.5)(-2t + 4) e^{-t^2 + 4t - 5}\Big|_{t=0} \\
&= 14 e^{-5}
\end{aligned}$$

Note, in this case, this is algebraically more convenient than using the plain directional derivative, because the bulk of these calculations avoid dealing with fractions.

Example 4.4.2. An ant is at the point $(2,0)$ on the graph of the function f given by $f(x,y) = x^2 - 2y^2$. If he starts walking in the direction of the vector $(1,1)$, what will be the initial slope of his path?

The question asks for the slope of a graph in a given direction, so we know that the unit directional derivative is our answer. However, in order to be able to compute the unit directional derivative, we must have a unit vector $\vec{u}$ representing the direction in question. So we must divide $\vec{v} = (1,1)$ by its own length to get

$$\vec{u} = \frac{\vec{v}}{\|\vec{v}\|} = \frac{(1,1)}{\sqrt{2}}$$

We could write down this unit vector in terms of components and compute directly, but in this case the vector $\vec{v}$ is more convenient to write down than $\vec{u}$. So, even though we are actually interested in the unit directional derivative, we compute it in terms of the plain directional derivative.

$$\begin{aligned}
D_{\vec{u}}f(\vec{a}) &= \frac{1}{\sqrt{2}} D_{\vec{v}}f(\vec{a}) \\
&= \frac{1}{\sqrt{2}} \frac{d}{dt}\bigg|_{t=0} f(2+t, t) \\
&= \frac{1}{\sqrt{2}} \frac{d}{dt}\bigg|_{t=0} (4 + 4t - t^2) \\
&= \frac{1}{\sqrt{2}} (4 - 2t)\bigg|_{t=0} \\
&= \frac{4}{\sqrt{2}}
\end{aligned}$$

Example 4.4.3. Suppose we wish to compute the value of the directional derivative $D_{\vec{v}}f(\vec{a})$, where $\vec{v} = (2,7)$. Of course this could be computed directly from the definition.

Another option would be to use the unit directional derivative. This method would involve computing the length of $\vec{v}$, dividing $\vec{v}$ by that length, computing the unit directional derivative, and then multiplying the result by that same length.

But note that in this case, the unit vector is more inconvenient to write down than the original vector $\vec{v}$ itself. So, not only would this option involve several additional steps, but it would make the rest of the calculation more inconvenient as well.

In this case, it would be pointless to use the unit directional derivative; the desired directional derivative would be substantially easier to compute directly.

Exercises

Exercise 4.4.1 (exer-Ders-DD2-1). Use the unit directional derivative to compute the answer to Exercise 4.3.7.

Exercise 4.4.2 (exer-Ders-DD2-2). What is the slope of the tangent line to the graph $z = f(x,y)$ of the function f defined by $f(x,y) = x^3 - xy$, at the point on the graph above $(2,1)$, in the vertical plane that contains the line in the xy-plane that passes through $(2,1)$ and is parallel to the vector $(-4,3)$.

Exercise 4.4.3 (exer-Ders-DD2-3). Recompute the answer found in Example 4.3.1, this time using the unit directional derivative instead of the directional derivative.

Exercise 4.4.4 (exer-Ders-DD2-4). Recompute the answer found in Example 4.3.2, this time using the unit directional derivative instead of the directional derivative.

Exercise 4.4.5 (exer-Ders-DD2-5). Recompute the answer found in Example 4.3.3, this time using the unit directional derivative instead of the directional derivative.

Exercise 4.4.6 (exer-Ders-DD2-6). The temperature (in degrees Fahrenheit) at a location (x, y) on the hood of a car sitting in the sun is given by $T(x, y) = 175 + 25 \sin(2\pi x) - 10 \cos(4\pi y + \frac{\pi}{2})$. At the point $(1, 3)$, use the unit directional derivative to determine how fast the temperature changes, per unit distance traveled, when moving in the direction of the vector $(1, 0)$.

Exercise 4.4.7 (exer-Ders-DD2-7). The depth of the ocean from a point (x, y) on the surface (x, y, and depth are measured in miles) is given by $2 - \sin(x^2 + 3x - 2y)$. At the point $(1, 3)$, use the unit directional derivative to determine the rate of change of depth per unit distance traveled in the direction of the vector $(12, 5)$.

Exercise 4.4.8 (exer-Ders-DD2-8). Recompute the answer to Exercise 4.4.7 by using the plain directional derivative with the vector $(12, 5)$ and then at the end dividing by the speed.

In which of your two solutions is the algebra the most convenient, in your opinion, and why?

4.5 Partial Derivatives

4.5.1 Definition and computation

The partial derivative is a special case of the unit directional derivative. In addition to requiring that $\|\vec{u}\| = \|\frac{d\vec{x}}{dt}\| = 1$, we further require that the unit vector $\vec{u}$ be a standard basis vector, $\vec{e}_i$.

Of course this doesn't affect our interpretations above at all. Thought of as a directional derivative, this can still be interpreted as the rate of change $\frac{df}{dt}$ of the output of the function in the case where the input $\vec{x}$ of the function is moving with velocity $\frac{d\vec{x}}{dt} = \vec{e}_i$. Thought of as a unit directional derivative, this can still be interpreted as the factor that you multiply by the speed $\|\frac{d\vec{x}}{dt}\|$ of the input $\vec{x}$ to get the velocity $\frac{df}{dt}$ of the output, when the input is moving in the $\vec{e}_i$ direction.

But given this restriction that $\vec{x}$ is moving in the direction of a standard basis vector, we get a very convenient computational advantage. The point here is that because of this restriction, there is only one of the input variables that is changing at all, because standard basis vectors have all coordinates equal to zero except for one. So we can write the following.

$$\begin{aligned}
D_{\vec{e}_i} f(\vec{a}) = \frac{d}{ds}\bigg|_{s=0} f(\vec{a} + s\vec{e}_i) &= \lim_{h \to 0} \frac{f(\vec{a} + h\vec{e}_i) - f(\vec{a})}{h} \\
&= \lim_{h \to 0} \frac{f\left(\begin{bmatrix} a_1 \\ \vdots \\ a_i \\ \vdots \\ a_n \end{bmatrix} + h \begin{bmatrix} 0 \\ \vdots \\ 1 \\ \vdots \\ 0 \end{bmatrix}\right) - f\left(\begin{bmatrix} a_1 \\ \vdots \\ a_i \\ \vdots \\ a_n \end{bmatrix}\right)}{h} \\
&= \lim_{h \to 0} \frac{f\left(\begin{bmatrix} a_1 \\ \vdots \\ a_i + h \\ \vdots \\ a_n \end{bmatrix}\right) - f\left(\begin{bmatrix} a_1 \\ \vdots \\ a_i \\ \vdots \\ a_n \end{bmatrix}\right)}{h}
\end{aligned}$$

Written in this form, we can take the following point of view on what we are doing in this final expression – first we fix the values of all of the coordinates except for the ith variable, thus viewing the result as a single variable function of the single variable x_i; then, we take the single variable derivative of that function with respect to its single variable, x_i.

Taking this view allows us to compute these partial derivatives very easily. For example, using the above point of view, let's compute the partial derivative of $f(x,y) = x^2 y^3 + y^2$, in the $\vec{e}_1$ direction. First, we view this function as being a function only of x – so for this computation, we will view y as being a constant. Then we take the single variable derivative of this with respect to its single variable, x.

$$\frac{d}{dx}(x^2 y^3 + y^2) = 2xy^3 + 0 = 2xy^3$$

Note that the y^3 in the first factor is treated as a constant, and so there is no use of the product rule; note also that the y^2 is treated as a constant, and so its derivative is zero.

This point of view makes the computation of partial derivatives very convenient, since we are in fact just taking a plain single variable derivative – and thus we can use all of the many convenient rules for differentiation.

Because this partial derivative can be thought of as a single variable derivative, with respect to only one of the input variables, we give it a notation similar to the Leibniz notation for single variable derivatives. Of course we have to

4.5. PARTIAL DERIVATIVES

acknowledge that it is not precisely a single variable derivative, because in fact there really are other variables present, even if we aren't treating them as variables. So, the notation is altered slightly to acknowledge this point. Instead of using the lower case "d", we use the similar but different symbol "∂".

Definition 4.5.1. *The partial derivative of f at the point $\vec{a}$ with respect to the input variable x_i is*

$$\frac{\partial f}{\partial x_i}(\vec{a}) = D_{\vec{e}_i} f(\vec{a}) = \frac{d}{ds}\bigg|_{s=0} f(\vec{a} + s\vec{e}_i) = \lim_{h \to 0} \frac{f(\vec{a} + h\vec{e}_i) - f(\vec{a})}{h}$$

In some instances, it can be convenient to use an alternative notation for the partial derivative. For example, instead of writing $\frac{\partial f}{\partial x}$, it is instead sometimes more convenient to write "f_x". And in instances where the output value of the function is associated with a variable, such as if we write $z = f(x, y)$, we might also write this partial derivative as "$\frac{\partial z}{\partial x}$" or "$z_x$".

If the function f has multiple output components, then the partial derivatives of f can be computed componentwise. Students can note this as a consequence of the similar observation made about directional derivatives since a partial derivative is a special case of a directional derivative. Specifically then, if

$$f(\vec{x}) = \begin{bmatrix} f_1(\vec{x}) \\ \vdots \\ f_n(\vec{x}) \end{bmatrix}$$

then we can write a partial derivative of f as

$$\frac{\partial}{\partial x_i} f(\vec{a}) = \begin{bmatrix} \frac{\partial}{\partial x_i} f_1(\vec{a}) \\ \vdots \\ \frac{\partial}{\partial x_i} f_n(\vec{a}) \end{bmatrix}$$

4.5.2 Interpretations

Since the partial derivative is a directional derivative, we still have all of those previous interpretations:

(1) The partial derivative $\frac{\partial f}{\partial x_i}$ is the factor that you multiply by the speed $\|\frac{d\vec{x}}{dt}\|$ of the input point $\vec{x}$ to find the rate of change $\frac{df}{dt}$ of the output (when the input is moving in the $\vec{e}_i$ direction).

$$\frac{df}{dt} = \left(\frac{\partial f}{\partial x_i}\right) \left\|\frac{dx}{dt}\right\| \quad \text{assuming that } \frac{d\vec{x}}{dt} \text{ is in the } \vec{e}_i \text{ direction.}$$

(2) The partial derivative $\frac{\partial f}{\partial x_i}$ is the factor that you multiply by the magnitude $\|\Delta \vec{x}\|$ of a discrete change $\Delta \vec{x}$ of the input point $\vec{x}$ to approximate the discrete change Δf of the output (when Δx is in the $\vec{e}_i$ direction).

$$\Delta f \approx \left(\frac{\partial f}{\partial x_i}\right) \|\Delta x\| \quad \text{assuming the change is only in the } i^{th} \text{ variable.}$$

(3) The partial derivative $\frac{\partial f}{\partial x_i}$ is the rate of change $\frac{df}{dt}$ of the ouput of the function, given that the rate of change $\frac{d\vec{x}}{dt}$ of the input $\vec{x}$ of the function is the unit vector $\vec{e}_i$.

(4) The partial derivative $\frac{\partial f}{\partial x_i}$ is an approximation to the discrete change Δf of the output of the function, given that the discrete change $\Delta \vec{x}$ of the input $\vec{x}$ of the function is the unit vector $\vec{e}_i$.

(5) For $f : \mathbb{R}^n \to \mathbb{R}^1$, the partial derivative $\frac{\partial f}{\partial x_i}$ is the slope of the tangent line to the curve that is the cross section of the graph of f by the plane extending vertically from the line in the domain passing through the point $\vec{a}$ in the direction of the x_i-axis.

4.5.3 Examples

Example 4.5.1. What is the slope of the tangent line at the point $(2, 1)$ to the cross section in the x-direction of the graph of the function f defined by $f(x, y) = x^2 y^3 + e^{\sin y^2}$?

Using the interpretation above of the partial derivative, we know that the answer to this question is simply $\frac{\partial f}{\partial x}(2, 1)$, which we can compute directly by

$$\frac{\partial f}{\partial x}(2, 1) = 2xy^3 \Big|_{(2,1)} = 4$$

Example 4.5.2. A bird is flying through the air such that at a particular time she is at the point $\vec{a} = (3, 4, 2)$ and moving with velocity $\vec{v} = (0, 5, 0)$. The humidity is given by the function f defined by $f(x, y, z) = \sin^2(x + 2y + 3z)$. How fast is the humidity of the air around the bird changing?

We can write the answer to this question immediately as $D_{\vec{v}} f(\vec{a})$. Noting that the velocity vector $\vec{v}$ can be written as $\vec{v} = 5\vec{e}_2$, we can translate this expression using a unit directional derivative and then a partial derivative to get

$$\begin{aligned} D_{\vec{v}} f(\vec{a}) = 5 D_{\vec{e}_2} f(\vec{a}) &= 5 \frac{\partial f}{\partial y}(\vec{a}) \\ &= (5)\left(2\sin(x + 2y + 3z)\cos(x + 2y + 3z)(2) \Big|_{(3,4,2)} \right) \\ &= 10 \sin(17) \cos(17) \end{aligned}$$

4.5.4 Second partials

Given a function $f : \mathbb{R}^n \to \mathbb{R}^1$, note that a partial derivative of that function is also a function $\frac{\partial f}{\partial x_i} : \mathbb{R}^n \to \mathbb{R}^1$. So, in fact, we can take a partial derivative of that partial derivative function. The result is called a "second partial derivative" of f.

For example, suppose we consider the function $f(x, y) = x^2 y^3$. The partial derivatives of this function are

$$\frac{\partial f}{\partial x} = 2xy^3 \quad \text{and} \quad \frac{\partial f}{\partial y} = 3x^2 y^2 \tag{4.7}$$

Taking the partials of these then gives us

$$\frac{\partial}{\partial x}\left(\frac{\partial f}{\partial x}\right) = 2y^3 \qquad \frac{\partial}{\partial x}\left(\frac{\partial f}{\partial y}\right) = 6xy^2$$

$$\frac{\partial}{\partial y}\left(\frac{\partial f}{\partial x}\right) = 6xy^2 \qquad \frac{\partial}{\partial y}\left(\frac{\partial f}{\partial y}\right) = 6x^2 y$$

Notationally, sometimes these second partials are written as single fractions, such as

$$\frac{\partial}{\partial y}\left(\frac{\partial f}{\partial x}\right) = \frac{\partial^2 f}{\partial y \partial x} \quad \text{and} \quad \frac{\partial}{\partial y}\left(\frac{\partial f}{\partial y}\right) = \frac{\partial^2 f}{\partial y^2} \tag{4.8}$$

In the case of real valued functions f, we can use previous interpretations of partial derivatives to make a geometric interpretation of these second partials. For example, let's consider the second partial $\frac{\partial^2 f}{\partial y^2} = \frac{\partial}{\partial y}\left(\frac{\partial f}{\partial y}\right)$.

4.5. PARTIAL DERIVATIVES

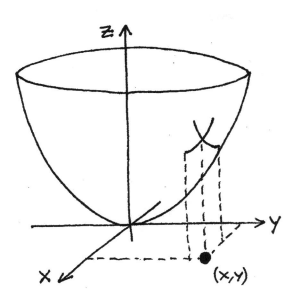

Figure 4.13:

Applying interpretation (5), we have that $\frac{\partial f}{\partial y}$ represents the slope of the tangent line to the graph, in the y direction. Applying interpretation (3) to this, we interpret that $\frac{\partial^2 f}{\partial y^2} = \frac{\partial}{\partial y}\left(\frac{\partial f}{\partial y}\right)$ represents the rate that the slope of this tangent line changes as we move in the y direction.

Said differently then, we get a result that is highly familiar from single variable calculus. That is, the second partial $\frac{\partial^2 f}{\partial y^2} = \frac{\partial}{\partial y}\left(\frac{\partial f}{\partial y}\right)$ is related to the concavity of the y cross section of the graph.

This should not be surprising of course, because as both of the partial derivatives are with respect to y, the other variable x is treated as a constant for the entire computation. So only the y cross section of the graph is relevant. Furthermore that cross section is thought of as representing a function of a single variable y for each partial derivative, so the familiar notion of concavity is again the interpretation.

A similar argument can be made to interpret $\frac{\partial^2 f}{\partial x^2} = \frac{\partial}{\partial x}\left(\frac{\partial f}{\partial x}\right)$.

Example 4.5.3. Suppose we consider the second partials $\frac{\partial^2 f}{\partial x^2}$ and $\frac{\partial^2 f}{\partial y^2}$ of the function $f(x,y) = x^2 + y^2$.

Direct computation gives us $\frac{\partial^2 f}{\partial x^2} = 2$ and $\frac{\partial^2 f}{\partial y^2} = 2$. These are both positive, so we interpret that the graph of this function f has concave up cross sections in both the x direction and the y direction, at every point on the graph. This is represented in Figure 4.13.

Of course this is consistent with our previous knowledge that the graph of this function is an upward paraboloid.

Things get a bit more interesting though when we consider "mixed" partials; that is, second partials where the two partial derivatives are taken with respect to different variables. For example, let's now try to interpret $\frac{\partial^2 f}{\partial y \partial x} = \frac{\partial}{\partial y}\left(\frac{\partial f}{\partial x}\right)$.

Again we begin by noting that $\frac{\partial f}{\partial x}$ represents the slope of the tangent line to the graph, in the x direction. Applying interpretation (3) to this, we interpret that $\frac{\partial^2 f}{\partial y \partial x} = \frac{\partial}{\partial y}\left(\frac{\partial f}{\partial x}\right)$ represents the rate that the slope of this tangent line changes as we move in the y direction.

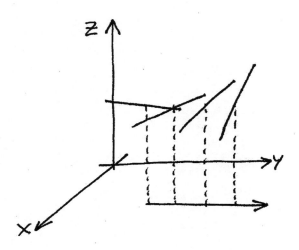

Figure 4.14:

Critically, here we are thus considering how the slope in the x direction changes as we move in the y direction. Neither of the variables then is held constant for the entire computation – so we cannot simply consider a single cross section of the graph, and we will not be able to relate our final interpretation to anything from single variable calculus.

We can still take another geometric step though. That is, let's consider what it might look like if the slope in the x direction changes as we move in the y direction. As is shown in Figure 4.14, the idea is that this indicates a sort of a "twist" of the surface.

Similarly we would interpret that $\frac{\partial^2 f}{\partial x \partial y} = \frac{\partial}{\partial x}\left(\frac{\partial f}{\partial y}\right)$ represents the rate that the slope of the tangent line in the y direction changes as we move in the x direction, again representing a twist.

Thinking about it geometrically, one notices that the sort of twist of the graph corresponding to a value of $\frac{\partial^2 f}{\partial x \partial y}$ would in fact also create a twist corresponding to some value of $\frac{\partial^2 f}{\partial y \partial x}$. Not only is this good geometric intuition, but in fact it corresponds to a surprising and powerful result about these mixed partials.

Theorem 4.5.1. *If the second partials $\frac{\partial^2 f}{\partial x_i \partial x_j}$ of a function f are all continuous, then the order of the variables in a second partial does not matter. That is, for all i and j, we have*

$$\frac{\partial^2 f}{\partial x_i \partial x_j} = \frac{\partial^2 f}{\partial x_j \partial x_i} \tag{4.9}$$

Consider for example the case of the function $f(x,y) = x^2 y^3$ discussed at the beginning of this subsection. Note that all of the second partials are polynomials and therefore are continuous functions. So, as promised by Theorem 4.5.1, the two mixed partials are in fact equal.

Exercises

Exercise 4.5.1 (exer-Ders-PD-1). For each of the functions below, compute all of the first partial derivatives.

1. $f(x,y) = x^2 y^3 - x^4 y$
2. $g(x,y,z) = z e^{x^2 - y^2} + \sin xyz$

4.5. PARTIAL DERIVATIVES

3. $h(u, v, w) = v^u - u^{vw}$

Exercise 4.5.2 (exer-Ders-PD-2). For each of the functions in Exercise 4.5.1, compute all of the second partial derivatives.

Exercise 4.5.3 (exer-Ders-PD-3). Use logarithmic differentiation to compute the partial derivatives of $f(s,t) = (s^2 t^2)^{st^2}$.

Exercise 4.5.4 (exer-Ders-PD-4). Use the interpretation of partial derivative as a directional derivative to compute the partial derivative with respect to x of the function $g(x, y) = x^2 y^3 - y^{23}$ at the point $(3, 4)$. Confirm that your answer is correct by computing the same partial derivative, interpreting it as a single variable derivative.

Exercise 4.5.5 (exer-Ders-PD-5). What is the slope of the tangent line to the cross section of the graph of the function $f(x,y) = \sin 2\pi x - 3\pi y^2$ by the plane perpendicular to the x-axis that passes through the point $(-1, 3)$ in the xy-plane?

Exercise 4.5.6 (exer-Ders-PD-6). Confirm by direct computation that all corresponding mixed second partials of the function $h(x, y, z) = xy^3 - e^{yz} + z\sin xy$ are indeed equal, as required by Theorem 4.5.1.

Exercise 4.5.7 (exer-Ders-PD-7). In Section 5.10 we will define and discuss an object called a *vector field*, which is represented by a function $\vec{F} : \mathbb{R}^n \to \mathbb{R}^n$. Then in Chapter 6 we will discuss many of the remarkable theorems that relate to these objects.

Involved in these discussions will be an expression called the *divergence* of the vector field. For a vector field $\vec{F}$ with component functions P, Q, and R, we define the divergence by

$$\text{div}\vec{F} = \frac{\partial P}{\partial x} + \frac{\partial Q}{\partial y} + \frac{\partial R}{\partial z}$$

Compute the divergence of each of the following vector fields.

1. $\vec{F} = (x^2 y, e^{yz}, \sin x + y)$
2. $\vec{G} = \left(\frac{-x}{(x^2+y^2+z^2)^{3/2}}, \frac{-y}{(x^2+y^2+z^2)^{3/2}}, \frac{-z}{(x^2+y^2+z^2)^{3/2}}\right)$, $(\vec{x} \neq \vec{0})$
3. $\vec{H} = \left(\frac{y^2 e^z}{\ln(1+y^2+z^4)}, \frac{x^3 z - z^4}{e^x}, \frac{xye^y - x^2 \sin y}{4x}\right)$

Exercise 4.5.8 (exer-Ders-PD-8). Also involved in the discussions described in Exercise 4.5.7 is an expression called the *curl* of a vector field. The curl of a vector field $\vec{F} = (P, Q, R)$ is defined by

$$\text{curl}\vec{F} = \left(\frac{\partial R}{\partial y} - \frac{\partial Q}{\partial z}, \frac{\partial P}{\partial z} - \frac{\partial R}{\partial x}, \frac{\partial Q}{\partial x} - \frac{\partial P}{\partial y}\right)$$

Compute the curl of each of the following vector fields.

1. $\vec{F} = (xy - z^3, 3x^2 + 4yz, -2xyz)$
2. $\vec{G} = (-y, x, 0)$
3. $\vec{H} = \left(\frac{-x}{(x^2+y^2+z^2)^{3/2}}, \frac{-y}{(x^2+y^2+z^2)^{3/2}}, \frac{-z}{(x^2+y^2+z^2)^{3/2}}\right)$

Exercise 4.5.9 (exer-Ders-PD-9). The *Laplacian operator*, represented with the symbol Δ, is defined by $\Delta u = \frac{\partial^2 u}{\partial x^2} + \frac{\partial^2 u}{\partial y^2} + \frac{\partial^2 u}{\partial z^2}$. Show that for the function $u(x, y, z) = 1/\sqrt{x^2 + y^2 + z^2}$, we have $\Delta u = 0$ everywhere except at the origin.

Exercise 4.5.10 (exer-Ders-PD-10). The *three dimensional heat operator* is defined by $(\partial_t - \Delta)u = \frac{\partial u}{\partial t} - \Delta u$. In a uniform closed system heat disperses by a function $u = u(x, y, z, t)$ in such a way that the heat equation $(\partial_t - \Delta)u = 0$ is satisfied for all values of $t > 0$.

Show that the *Gauss-Weierstrass heat kernel* defined by $K(x, y, z, t) = t^{-3/2} e^{-(x^2+y^2+z^2)/4t}$ satisfies the heat equation. (Note that for very small values of t this kernel represents a system where heat is heavily concentrated near the origin; and as t increases, this kernel represents that heat flowing and spreading out, becoming less concentrated.)

Exercise 4.5.11 (exer-Ders-PD-11). A function $u = u(x, t)$ represents a wave in $\mathbb{R}^1$ if the *one dimensional wave equation*, defined by $\frac{\partial^2 u}{\partial t^2} - c^2 \frac{\partial^2 u}{\partial x^2} = 0$, is satisfied for some value c.

Show that for any smooth function f of one variable, the function $W(x, t) = f(x - ct)$ satisfies the wave equation.

Using the ideas of translation from Section 2.2, determine the velocity of this wave.

4.6 Derivative Transformations

From one point of view, we might say that we have been successful at this point in defining derivatives for multivariable functions, since we have a constructive definition for the derivatives of such functions in any vector direction.

$$D_{\vec{v}} f(\vec{a}) = \frac{d}{dt}\bigg|_{t=0} f(\vec{a} + t\vec{v}) \tag{4.10}$$

This construction allows us to relate changes in the input of a function to the corresponding changes in the output of the function. The special case of unit directional derivatives allows us to relate this to the graph of such a function (as the slope of the corresponding cross section), and the special case of partial derivatives can be computed with great ease.

However, as it turns out, this is not the entire story. There is a very surprising and powerful relationship that we have not yet discovered, between values of the directional derivative for different vectors.

4.6.1 Polynomials and directional linearity

We will introduce this relationship by a particularly convenient example, namely, that of a polynomial function $p: \mathbb{R}^n \to \mathbb{R}^1$ and the computation of the directional derivative for a general vector $\vec{v}$. Before we do this for a general polynomial, we choose a specific one from which to work by example.

Example 4.6.1. **Consider the polynomial function** $p: \mathbb{R}^2 \to \mathbb{R}^1$ **given by** $p(x_1, x_2) = x_1 x_2 + x_2^2$, **and the point** $\vec{a} = (1, 2) \in \mathbb{R}^2$. **Let's find a general formula for the directional derivative** $D_{\vec{v}} f(\vec{a})$ **in terms of the components of** $\vec{v} = (v_1, v_2)$.

We parametrize the motion in the domain by $\vec{x} = \vec{a} + t\vec{v} = (1 + tv_1, 2 + tv_2) = (x_1, x_2)$. **Then we can write** $p(\vec{x}(t))$ **as**

$$p(\vec{x}(t)) = (1 + tv_1)(2 + tv_2) + (2 + tv_2)^2 = 6 + (2v_1 + 5v_2)t + (v_1 v_2 + v_2^2)t^2$$

We can then compute the directional derivative directly.

$$\begin{aligned} D_{\vec{v}} p(\vec{a}) &= \frac{d}{dt}\bigg|_{t=0} p(\vec{x}(t)) \\ &= \frac{d}{dt}\bigg|_{t=0} 6 + (2v_1 + 5v_2)t + (v_1 v_2 + v_2^2)t^2 \\ &= (2v_1 + 5v_2) + 2(v_1 v_2 + v_2^2)t \bigg|_{t=0} \\ &= (2v_1 + 5v_2) \end{aligned}$$

This computation is valid for any velocity vector $\vec{v}$. **Notice that the result can be viewed as a dot product of** $\vec{v}$ **with the vector** $(2, 5)$.

We can do this same computation for an arbitrary polynomial, and we will find that we come to a very similar answer.

We can write a general formula for a multivariable polynomial (of arbitrary degree k) in the variables $x_1, \ldots, x_n$ as a sum of monomials, each of which has a coefficient, and a total exponent less than or equal to k. We denote this by

$$p(\vec{x}) = \sum_{k_1 + \ldots + k_n \leq k} c_{k_1 \ldots k_n} x_1^{k_1} \cdots x_n^{k_n} \tag{4.11}$$

(This notation we need to write down a general polynomial is a bit cumbersome, but as we will soon see it will not make much difference in the computation.)

4.6. DERIVATIVE TRANSFORMATIONS

With a given starting point $\vec{a}$ and velocity vector $\vec{v}$, we can write down the components x_i of $\vec{x}(t) = \vec{a} + t\vec{v}$ as $x_i = a_i + tv_i$. With these expressions for the components of $\vec{x}$, we can then compute $p(\vec{x}(t))$ directly from Equation 4.11.

While the result of that computation would be hard to write down precisely, one thing that can be seen easily is that it is a gigantic polynomial of the variable t; we can then write this as

$$p(\vec{x}(t)) = b_0 + b_1 t + b_2 t^2 + \cdots + b_k t^k \tag{4.12}$$

where the coefficients b_j are some combination of the $c_{k_1 \cdots k_n}$, a_i, and v_i.

Most of these coefficients b_j are very difficult to write down simply because of the complexity of the polynomial p and the sheer number of terms involved. But, a clever observation will allow us to write down at least the form of the coefficient b_1.

Note that because the inputs to the polynomial p are $x_i = a_i + tv_i$, the variable t always appears paired with one of the coefficients v_i. So, any term in $p(\vec{x}(t))$ that has a single power of t must also have as a factor a single power of a single coefficient v_i. This tells us that the coefficient b_1 must be a linear combination of the coefficients $v_1, \ldots, v_n$, which we write as

$$b_1 = d_1 v_1 + \cdots + d_n v_n \tag{4.13}$$

(In specific examples, these coefficients d_i can be computed by direct computation.)

This observation will be particularly handy as we proceed on to compute the directional derivative

$$D_{\vec{v}} p(\vec{a}) = \frac{d}{dt}\bigg|_{t=0} p(\vec{x}(t)) = \frac{d}{dt}\bigg|_{t=0} p(\vec{a} + t\vec{v}) \tag{4.14}$$

Using equations 4.12 and 4.13, we get

$$\begin{aligned} D_{\vec{v}} p(\vec{a}) &= \frac{d}{dt}\bigg|_{t=0} b_0 + b_1 t + b_2 t^2 + \cdots + b_k t^k \\ &= b_1 + 2b_2 t^1 + \cdots + k b_k t^{k-1} \bigg|_{t=0} \\ &= b_1 \\ &= d_1 v_1 + \cdots + d_n v_n \end{aligned}$$

Note that this is consistent with the result of Example 4.6.1.

Of course, we still do not know what these coefficients d_i are in general; in fact we will soon see that there is a remarkably convenient and important formula for computing them. But the important observation to make at this point from the above computation is the form of the result, not the specifics. That is, it appears to be a dot product of the vector $\vec{v}$ with some other vector, which we might call $\vec{d}$.

We state this result as a theorem.

Theorem 4.6.1. *For any multivariable polynomial function p, the directional derivative at $\vec{a}$ with velocity vector $\vec{v}$ can be computed as a dot product of $\vec{v}$ with some other vector that depends only on the polynomial p and the point $\vec{a}$.*

$$D_{\vec{v}} p(\vec{a}) = \vec{d} \cdot \vec{v} \tag{4.15}$$

Of course in light of Theorem 3.2.1, this means that the directional derivative of the polynomial p at a point $\vec{a}$ is a linear transformation on the velocity vector $\vec{v}$. This is a useful notion, so we make the following definition.

Definition 4.6.1. *A function $f : \mathbb{R}^n \to \mathbb{R}^m$ is "directional linear" at the point $\vec{a}$ in the domain iff there exists a linear transformation that can be used to compute all of the directional derivatives at that point. That is, there exists a linear transformation T such that for all $\vec{v}$ we have*

$$D_{\vec{v}} f(\vec{a}) = T(\vec{v})$$

We can then restate Theorem 4.6.1 very simply as

Theorem 4.6.2. *All polynomials are directional linear at every point in the domain.*

4.6.2 Differentiability

It turns out that this notion of directional linearity is similar to another even more useful notion, called "differentiability". We will see momentarily that both of these notions involve approximating changes in a function with a linear transformation – but, they are slightly different.

We will begin with the definition, discuss an interpretation, and make the connection to directional linearity.

Definition 4.6.2. *A function $f : \mathbb{R}^n \to \mathbb{R}^m$ is "differentiable at the point $\vec{a}$" if there exists a linear transformation T for which*

$$\lim_{\vec{x} \to \vec{a}} \frac{\left\| \left(f(\vec{x}) - f(\vec{a}) \right) - T(\vec{x} - \vec{a}) \right\|}{\|\vec{x} - \vec{a}\|} = 0$$

Note, since T approximates the change in the value of the function between the points $\vec{x}$ and $\vec{a}$, the numerator is simply the size of the error in that approximation.

Of course for any continuous function f, both $(f(\vec{x}) - f(\vec{a}))$ and $T(\vec{x} - \vec{a})$ will approach zero, and thus so will that error, no matter what linear transformation is chosen. So, merely requiring the numerator to approach zero would have accomplished nothing.

Given this, we can reasonably hope for more than just the size of the error approaching zero. Instead, let's consider the "relative error", which is the ratio of the size of the error (the numerator above) and the size of the displacement (the denominator above). For a "good" approximation, perhaps this relative error should also approach zero.

This is one motivation for the above definition of differentiability.

It can be shown that with this stronger requirement that the relative error approach zero, there is at most one linear transformation T that can satisfy the above definition for a given function f and point $\vec{a}$. Many functions are differentiable, as we will soon see. For example, all polynomials are differentiable.

It can also be shown that if they both exist, the linear transformations in Definitions 4.6.1 and 4.6.2 must be the same. This is the important connection between directional linearity and differentiability.

The difference between these conditions has to do with the multivariable limit. It turns out that the notion of directional linearity is equivalent to the requirement that the expression above has limits equal to zero along all straight lines through $\vec{a}$. Of course we remember from Section 2.7 that this is NOT equivalent to saying that the limit itself exists and equals zero.

In fact, there are examples of directional linear functions which are not differentiable. But, as with the similar result for limits, all differentiable functions are directional linear. This fact is worth stating as a theorem.

Theorem 4.6.3. *All differentiable functions are directional linear.*

We will not prove this theorem in this text.

It turns out that this notion of differentiability is the precise condition that is necessary for the statement of many theorems.

If a function f is differentiable at the point $\vec{a}$, then on the one hand we can view the linear transformation T from Definition 4.6.2 simply as a tool used to define the property of differentiability. On the other hand, we could instead view this definition as (conditionally) asserting the existence of a new object, T, which can be used to compute the values of directional derivatives.

Taking this point of view, we make the following definition.

Definition 4.6.3. *For a function f that is differentiable at the point $\vec{a}$, we call the linear transformation T satisfying the definition "the derivative transformation of f at $\vec{a}$". It is denoted as*

$$D_{f,\vec{a}}$$

4.6. DERIVATIVE TRANSFORMATIONS

For a differentiable function then, this definition and Theorem 4.6.3 tell us that

$$D_{f,\vec{a}}(\vec{v}) = D_{\vec{v}}f(\vec{a}) \tag{4.16}$$

Notationally, the objects on either side of this equation look very similar; and of course, the equation asserts that they are equal. The reader should make sure to understand though that these objects are defined very differently.

On the left side of the equation we have a linear transformation acting on a vector. This linear transformation, once computed for a given function f at a given input point $\vec{a}$, is the same for all velocity vectors $\vec{v}$. We think of the linear transformation $D_{f,\vec{a}}$ as a single thing, separate from the vectors $\vec{v}$ to which it is applied.

On the right side of the equation, we have a directional derivative, which can be computed separately for individual vectors, directly from the definition.

Note also that for a function that is not differentiable, the left side of Equation 4.16 does not exist. However, the right side of the equation could still exist for some, or even all velocity vectors $\vec{v}$. We will see an example of such a function later in this section.

Given Equation 4.16, we get the same interpretations that we had for the directional derivative:

1. $\Delta f \approx D_{f,\vec{a}}(\Delta x)$
2. $\frac{df}{dt} = D_{f,\vec{a}}\left(\frac{d\vec{x}}{dt}\right)$

Because the derivative transformation can be thought of as a separate object from the velocity vector $\vec{v}$ to which it is applied, we can take a point of view of the above two interpretations that we could not take for the corresponding interpretations of the directional derivative. That is, as was the case for single variable derivatives, we now have a separate object that we can think of as doing something to the change in the input to compute the change in the output.

With this in mind we can phrase the above two interpretations in words very similarly to how we did for the interpretations of single variable derivatives, by saying that the derivative transformation $D_{f,\vec{a}}$ is:

1. the thing you do to a discrete change in x to approximate the discrete change in f; or
2. the thing you do to the rate of change of x to compute the rate of change of f

---

Example 4.6.2. Consider the polynomial function p from Example 4.6.1, at that same input point $\vec{a} = (1, 2)$. **What is the derivative transformation $D_{p,\vec{a}}$?**

In that example we computed a general formula for the directional derivatives for that function at that point.

$$D_{\vec{v}}p(\vec{a}) = 2v_1 + 5v_2$$

This is a linear transformation on the velocity vector $\vec{v}$, and since we know that all polynomials are differentiable, we can then write the derivative transformation by

$$D_{p,\vec{a}}(\vec{v}) = D_{\vec{v}}p(\vec{a}) = 2v_1 + 5v_2$$

We could also write this linear transformation in its matrix shorthand with the matrix

$$\begin{bmatrix} 2 & 5 \end{bmatrix}$$

4.6.3 Continuous differentiability

Of course as it stands, the only functions we have seen to have this very desirable property of differentiability are polynomials; and certainly there are many more functions that we need to make use of in multivariable calculus.

Recall however that in single variable calculus, we found that most functions can be approximated very closely by their Taylor polynomials, at least near the point from which that polynomial is computed. So, since single variable polynomials do such a good job of modeling the behavior of single variable functions locally, it would not seem like much of a stretch to expect multivariable polynomials to do a good job of modeling multivariable functions, at least locally.

In fact this is true for a large category of multivariable functions, though certainly not all. Since as we just mentioned all polynomials are differentiable, we would expect a large category of other multivariable functions also to be differentiable.

A very convenient category of multivariable functions for which this is true is the collection of "continuously differentiable" functions.

Definition 4.6.4. *A function $f : \mathbb{R}^n \to \mathbb{R}^m$ with component functions $f_1, \ldots, f_m$ is "continuously differentiable at the point $\vec{a}$" iff all of the partial derivatives $\partial f_i / \partial x_j$ of all of the component functions exist and are continuous at the point $\vec{a}$.*

Theorem 4.6.4. *If a function $f : \mathbb{R}^n \to \mathbb{R}^m$ is continuously differentiable at the point $\vec{a}$, then it is differentiable at the point $\vec{a}$.*

This gives us a much larger category of functions to which we can apply the idea of differentiability, which we will soon see is extremely useful in computations.

Note that Theorem 4.6.4 is not stated as an "if and only if" statement. In fact the converse of this theorem is not true. That is, there are functions that are differentiable which are not continuously differentiable.

Example 4.6.3. Consider the function $f : \mathbb{R}^2 \to \mathbb{R}^2$ defined by

$$f(x, y) = (\sin x, e^{xy})$$

Is this function differentiable at every point in its domain?

Aside from computing the directional derivatives directly, the only theorem we have which allows us to conclude differentiability for such a function is Theorem 4.6.4. So, we hope we can show this function to be continuously differentiable.

We compute the partial derivatives of both output variables with respect to both input variables, giving us

$$\frac{\partial f_1}{\partial x} = \cos x \quad \text{and} \quad \frac{\partial f_1}{\partial y} = 0$$

$$\frac{\partial f_2}{\partial x} = y e^{xy} \quad \text{and} \quad \frac{\partial f_2}{\partial y} = x e^{xy}$$

All four of these partial derivative functions are continuous functions. So, the function f is continuously differentiable, and thus Theorem 4.6.4 tells us that f is differentiable.

4.6.4 A hierarchy of regularity conditions

In single variable calculus, the notion of differentiability is fairly simple. That is, a function is differentiable at a point if and only if the derivative exists at that point. There is also the related notion of continuous differentiability, which adds the further requirement that the derivative function must be continuous.

4.6. DERIVATIVE TRANSFORMATIONS

In multivariable calculus there is a more complicated set of such ideas, all of which are somewhat similar to each other, but still different. In this subsection we will classify a total of five such notions. We have already seen three of them – earlier in this section we defined "directional linearity", "differentiability", and "continuous differentiability".

As it turns out, this notion of differentiability fits "in between" the notions of continuous differentiability and directional linearity. In other words:

Theorem 4.6.5. *If a function is continuously differentiable, then it is differentiable; and if the function is differentiable, then it is directional linear.*

It is also true that no two of these three conditions are equivalent – there are directional linear functions that are not differentiable, and there are differentiable functions that are not continuously differentiable. However, the differences between the conditions are subtle and the counterexamples distinguishing them are anomalous and rare. Students should be aware that these distinctions exist and should make a point of using these terms correctly based on their definitions – but in this text we will not discuss or see any of the distinguishing counterexamples.

The above theorem is the beginning of a "hierarchy" of conditions that a multivariable function might satisfy. We will add to that list two more conditions.

Definition 4.6.5. *A function $f : \mathbb{R}^n \to \mathbb{R}^m$ with component functions $f_1, \ldots, f_m$ is "partial differentiable at the point $\vec{a}$" iff all of the partial derivatives $\partial f_i / \partial x_j$ of all of the component functions exist at the point $\vec{a}$.*

Note that the only difference between partial differentiability and continuous differentiability is the continuity requirement for the latter.

Definition 4.6.6. *A function $f : \mathbb{R}^n \to \mathbb{R}^m$ is "directionally differentiable at the point $\vec{a}$" iff the directional derivatives $D_{\vec{v}} f(\vec{a})$ exist at the point $\vec{a}$ for all velocity vectors $\vec{v}$.*

Note that directional differentiability requires only that directional derivatives exist, while directional linearity requires both that they exist and that they satisfy a linearity condition.

These last two conditions are weaker than the previous three; and of course it is not hard to see that every directionally differentiable function must be partial differentiable.

We summarize and organize all of these conditions and relationships in the following diagram:

$$
\begin{array}{rcl}
f \text{ is continuously differentiable} & \iff & \text{all partial derivatives of } f \text{ exist and are continuous} \\
\Downarrow & & \\
f \text{ is differentiable} & \iff & \text{there is a "good linear approximation" to the nearby values of } f \\
\Downarrow & & \\
f \text{ is directional linear} & \iff & \exists \text{ a linear transformation } T \text{ such that } D_{\vec{v}} f(\vec{a}) = T(\vec{v}) \\
\Downarrow & & \\
f \text{ is directionally differentiable} & \iff & D_{\vec{v}} f(\vec{a}) \text{ exists for all vectors } \vec{v} \\
\Downarrow & & \\
f \text{ is partial differentiable} & \iff & \text{all partial derivatives of } f \text{ exist}
\end{array}
$$

All five of these notions are different. There are partial differentiable functions that are not directionally differentiable, directionally differentiable functions that are not directional linear, and as we have already stated there are directional linear functions that are not differentiable and differentiable functions that are not continuously differentiable.

The hierarchy diagram illustrates the importance of the difference between the mere existence of the partial derivatives, and the continuity of the partial derivatives.

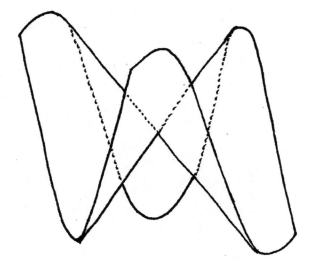

Figure 4.15:

Note also that this diagram shows a major difference between single variable calculus and multivariable calculus. That is, in single variable calculus if a function has derivatives that exist, then it is differentiable – but this is not true of multivariable functions.

We will present now examples of the first two distinctions above.

Example 4.6.4. We define the function $f : \mathbb{R}^2 \to \mathbb{R}^1$ by saying that $f(x,y) = 0$ if either x or y is zero, and $f(x,y) = 1$ if neither x nor y is zero.

We can compute the partial derivatives $\partial f/\partial x(\vec{0})$ and $\partial f/\partial y(\vec{0})$ very easily when we notice that all of the values of the function involved in either computation are all zero. So, those partial derivatives both exist at the origin.

However, the directional derivative $D_{(1,1)}f(\vec{0})$ is more of a problem. That directional derivative is defined as an instantaneous derivative of the values of f while moving in the direction $(1,1)$ from the origin – but those values of f are not continuous when moving in that direction, as they instantly jump up to 1 from 0 at the origin. Since these values are not continuous, the derivative is certainly not defined.

So this function is partial differentiable, but not directionally differentiable.

Example 4.6.5. Consider the function

$$f\left(\begin{bmatrix} x \\ y \end{bmatrix}\right) = \frac{3x^2y - y^3}{x^2 + y^2} = r\sin(3\theta)$$

where θ is the usual angle from the positive part of the x-axis to the vector (x,y). We further define f to be zero at the origin. The graph of this function is pictured in Figure 4.15

The polar interpretation of this function allows for the very easy computation of all of the directional derivatives at the origin. The only inconvenience we will encounter is the need to simplify the expression $f(t\vec{v})$ in the case when

4.6. DERIVATIVE TRANSFORMATIONS

t is either positive or negative. Of course the angle for $-\vec{v}$ is π greater than that of $\vec{v}$, so we have

$$f(t\vec{v}) = \begin{cases} \|t\vec{v}\|\sin(3\theta) & \text{when } t \geq 0 \\ \|t\vec{v}\|\sin(3(\theta + \pi)) & \text{when } t < 0 \end{cases}$$

$$= \begin{cases} |t|\|\vec{v}\|\sin(3\theta) & \text{when } t \geq 0 \\ |t|\|\vec{v}\|\sin(3\theta + 3\pi) & \text{when } t < 0 \end{cases}$$

$$= \begin{cases} t\|\vec{v}\|\sin(3\theta) & \text{when } t \geq 0 \\ (-t)\|\vec{v}\|(-\sin(3\theta)) & \text{when } t < 0 \end{cases}$$

$$= \begin{cases} t\|\vec{v}\|\sin(3\theta) & \text{when } t \geq 0 \\ t\|\vec{v}\|\sin(3\theta) & \text{when } t < 0 \end{cases}$$

Using this convenient result, we can compute the directional derivative by direct computation:

$$D_{\vec{v}}f(\vec{0}) = \frac{d}{dt}\bigg|_{t=0} f(\vec{0} + t\vec{v})$$

$$= \frac{d}{dt}\bigg|_{t=0} t\|\vec{v}\|\sin(3\theta)$$

$$= \|\vec{v}\|\sin(3\theta)$$

Now consider the following consequences of the above computation, and the observation of the angles that the vectors $(1, 0)$ and $(1/2, \sqrt{3}/2)$ make with the positive part of the x-axis:

$$D_{(1,0)}f(\vec{0}) = 0 \qquad D_{(1/2,\sqrt{3}/2)}f(\vec{0}) = 0$$

Notice that any vector in $\mathbb{R}^2$ can be written as a linear combination of the vectors $(1,0)$ and $(1/2, \sqrt{3}/2)$... So if the function were directional linear, then we would expect for all directional derivatives of this function to be zero. However we have already shown that this is not the case.

So, this function is directionally differentiable, but not directional linear.

Exercises

Exercise 4.6.1 (exer-Ders-DT-1). Find a general formula (in terms of the coordinates v_1 and v_2 of the vector $\vec{v}$) for the directional derivative $D_{\vec{v}}f(\vec{a})$ of the function $f(x,y) = x^2 + y^2$ at the point $(3,5)$. Write your answer as a matrix-vector product.

Exercise 4.6.2 (exer-Ders-DT-2). Find a general formula (in terms of the coordinates v_1 and v_2 of the vector $\vec{v}$) for the directional derivative $D_{\vec{v}}f(\vec{a})$ of the function $f(x,y) = (x^2y, xy^2)$ at the point $(2,0)$. Write your answer as a matrix-vector product.

Exercise 4.6.3 (exer-Ders-DT-3). Find a general formula (in terms of the coordinates v_1 and v_2 of the vector $\vec{v}$ and the coordinates x_0 and y_0 of the point $\vec{a} = (x_0, y_0)$) for the directional derivative $D_{\vec{v}}f(\vec{a})$ of the function $f(x,y) = 3xy$ at the point $\vec{a}$. Write your answer as a matrix-vector product.

Exercise 4.6.4 (exer-Ders-DT-4). Find a general formula (in terms of the coordinates v_1 and v_2 of the vector $\vec{v}$ and the coordinates x_0 and y_0 of the point $\vec{a} = (x_0, y_0)$) for the directional derivative $D_{\vec{v}}f(\vec{a})$ of the function $f(x,y) = e^{x-y}$ at the point $\vec{a}$. Write your answer as a matrix-vector product.

Exercise 4.6.5 (**exer-Ders-DT-5**). The function f is known to be differentiable at $\vec{a}$, and it is also known that at that point the directional derivative with velocity $(3, 4)$ is 7, and the directional derivative with velocity $(4, 1)$ is -3. Use this information to compute the directional derivative of this function at this point with velocity $(-1, 3)$.

Exercise 4.6.6 (**exer-Ders-DT-6**). Use Theorem 4.6.4 to show that the function $f : \mathbb{R}^3 \to \mathbb{R}^1$ given by $f(x, y, z) = xye^{x^2+y^2+z^2} - \sin(xy - z^3)$ is differentiable at every point in $\mathbb{R}^3$.

4.7 Jacobian Matrices

The main point of the last section is that if a function is continuously differentiable (and usually this is an easy condition to check), then there is a linear transformation (called the derivative transformation) that can be used to compute directional derivatives.

Of course matrices are the way we represent linear transformations in computations, so we make the following definition.

Definition 4.7.1. *The matrix representing the derivative transformation for a differentiable function $f : \mathbb{R}^n \to \mathbb{R}^m$ at the point $\vec{a}$ is called "the Jacobian matrix for f at the point $\vec{a}$", and is written as $J_{f,\vec{a}}$. We represent this with the equation*

$$D_{f,\vec{a}}(\vec{v}) = J_{f,\vec{a}}\vec{v} \tag{4.17}$$

The natural next question to ask is, "How can we compute the Jacobian matrix?"

Computing the Jacobian matrix for a differentiable function turns out to be a wonderfully elegant computation, combining several of the ideas we have used in this course.

We will perform the computation one column at a time. In particular, let's compute the jth column of $J_{f,\vec{a}}$. We begin by recalling from Definition 3.2.1 that the columns of a matrix are the images of the corresponding standard basis vectors. So we have

$$(j^{th} \text{ column of } J_{f,\vec{a}}) = D_{f,\vec{a}}(\vec{e}_j)$$

We can then complete the computation by rewriting this as a directional derivative, and then noting that since the vector is a standard basis vector we can further rewrite it as a partial derivative:

$$\begin{aligned}
(j^{th} \text{ column of } J_{f,\vec{a}}) &= D_{f,\vec{a}}(\vec{e}_j) \\
&= D_{\vec{e}_j} f(\vec{a}) \\
&= \frac{\partial f}{\partial x_j}(\vec{a})
\end{aligned}$$

This is a tremendous result! This means that we can compute the matrix for the derivative transformation simply by writing for each column the corresponding partial derivatives of the function f – each of which of course is just the vector of partial derivatives of the individual components $\{f_i\}$ of f, with respect to x_j.

Theorem 4.7.1. *Each column of the Jacobian matrix for a differentiable function f is the vector of partial derivatives of the components of f with respect to the input variable corresponding to that column.*

We can write this Jacobian matrix algebraically in several ways. First, interpreting directly from Theorem 4.7.1, we have

$$J_{f,\vec{a}} = \begin{pmatrix} | & & | \\ \frac{\partial f}{\partial x_1} & \cdots & \frac{\partial f}{\partial x_n} \\ | & & | \end{pmatrix}$$

or we can expand each column with its individual components to get

$$J_{f,\vec{a}} = \begin{pmatrix} \frac{\partial f_1}{\partial x_1} & \frac{\partial f_1}{\partial x_2} & \cdots & \frac{\partial f_1}{\partial x_n} \\ \frac{\partial f_2}{\partial x_1} & \cdots & & \frac{\partial f_2}{\partial x_n} \\ \vdots & \vdots & & \vdots \\ \frac{\partial f_m}{\partial x_1} & \frac{\partial f_m}{\partial x_2} & \cdots & \frac{\partial f_m}{\partial x_n} \end{pmatrix}$$

or for a more compact notation, we can simply list the ij-th term in the matrix by

$$(J_{f,\vec{a}})_{ij} = \frac{\partial f_i}{\partial x_j}$$

Recall of course that partial derivatives are generally easy to compute, and so because of the above result the Jacobian matrix can also generally be computed without much difficulty. Then using Definition 4.7.1, we can compute directional derivatives with relative ease.

Of course, this entire computation relies on the critical assumption that we made at the very beginning – that the function is differentiable in the first place! In practice, how do we know when a function is differentiable?

Our main tool for showing this important condition is Theorem 4.6.5 from the last section, stating that if a function is continuously differentiable, then it is differentiable. If the function is not continuously differentiable, then we have no other convenient option but to compute the directional derivatives directly from the definition.

Be sure to note that if a function is not differentiable, then the matrix of partial derivatives does not necessarily have any meaningful interpretation at all! As we discussed in the previous section, mere partial differentiability does NOT imply differentiability (or even directional linearity). For example, consider the function in Example 4.6.5. As we showed in that example this function is directionally differentiable and therefore partial differentiable, so we could assemble a matrix of partial derivatives – but the function is NOT differentiable (or even directional linear), so that matrix of partial derivatives has no meaningful significance.

Students must be careful not to be too casual when using Jacobian matrices. Remember that merely being able to compute the partial derivatives only shows the function is partial differentiable. Differentiability is required in order to define the Jacobian matrix and make interpretations from it about directional derivatives.

4.7.1 Computations

Here are some explicit computations using the Jacobian matrix.

Example 4.7.1. Suppose that

$$f\begin{bmatrix} x \\ y \end{bmatrix} = \begin{bmatrix} x \sin y \\ y \cos x \\ x^2 y^3 \end{bmatrix}$$

What is the velocity vector for the output of the function if we are at the point $\begin{bmatrix} \pi \\ \pi/2 \end{bmatrix}$ **in the domain, and moving in the domain with velocity given by the vector** $\begin{bmatrix} 3 \\ 5 \end{bmatrix}$**?**

The question is asking us to compute

$$D_{\begin{bmatrix} 3 \\ 5 \end{bmatrix}} f\left(\begin{bmatrix} \pi \\ \pi/2 \end{bmatrix}\right)$$

This function is continuously differentiable and thus differentiable, so we can use the derivative transformation and the Jacobian matrix.

4.7. JACOBIAN MATRICES

$$
\begin{aligned}
D_{\begin{bmatrix}3\\5\end{bmatrix}} f\left(\begin{bmatrix}\pi\\\pi/2\end{bmatrix}\right) &= D_{f,(\pi,\pi/2)}\left(\begin{bmatrix}3\\5\end{bmatrix}\right) \\
&= J_{f,(\pi,\pi/2)}\begin{bmatrix}3\\5\end{bmatrix} \\
&= \begin{pmatrix}\sin y & x\cos y \\ -y\sin x & \cos x \\ 2xy^3 & 3x^2y^2\end{pmatrix}\bigg|_{(\pi,\pi/2)}\begin{bmatrix}3\\5\end{bmatrix} \\
&= \begin{pmatrix}1 & 0 \\ 0 & -1 \\ \pi^4/4 & 3\pi^4/4\end{pmatrix}\begin{bmatrix}3\\5\end{bmatrix} \\
&= \begin{bmatrix}3\\-5\\9\pi^4/2\end{bmatrix}
\end{aligned}
$$

Example 4.7.2. For the same function f as in the previous example, find an estimate for $f(0.01, 0.03)$.

We know that $f(0,0) = \begin{bmatrix}0\\0\\0\end{bmatrix}$, and that we can write $(0.01, 0.03) = (0,0) + \begin{bmatrix}0.01\\0.03\end{bmatrix}$. **We use these observations and the derivative transformation to make the desired estimate.**

$$
\begin{aligned}
f(0.01, 0.03) &\approx f(0,0) + D_{f,(0,0)}\left(\begin{bmatrix}0.01\\0.03\end{bmatrix}\right) \\
&\approx f(0,0) + J_{f,(0,0)}\begin{bmatrix}0.01\\0.03\end{bmatrix} \\
&\approx f(0,0) + \begin{pmatrix}\sin y & x\cos y \\ -y\sin x & \cos x \\ 2xy^3 & 3x^2y^2\end{pmatrix}\bigg|_{(0,0)}\begin{bmatrix}0.01\\0.03\end{bmatrix} \\
&\approx \begin{bmatrix}0\\0\\0\end{bmatrix} + \begin{pmatrix}0 & 0 \\ 0 & 1 \\ 0 & 0\end{pmatrix}\begin{bmatrix}0.01\\0.03\end{bmatrix} \\
&\approx \begin{bmatrix}0\\0.03\\0\end{bmatrix}
\end{aligned}
$$

Example 4.7.3. Let's reconsider Example 4.3.2. We have the function $f : \mathbb{R}^3 \to \mathbb{R}^2$ given by

$$
f\left(\begin{bmatrix}x\\y\\z\end{bmatrix}\right) = \begin{bmatrix}xy\\yz\end{bmatrix}
$$

The directional derivative can be computed very conveniently by using the Jacobian matrix.

$$\begin{aligned} D_{\vec{v}} f(\vec{a}) &= D_{f,\vec{a}}(\vec{v}) \\ &= J_{f,\vec{a}}\vec{v} \\ &= \begin{pmatrix} y & x & 0 \\ 0 & z & y \end{pmatrix}\bigg|_{\vec{a}} \begin{bmatrix} 2 \\ 1 \\ 3 \end{bmatrix} \\ &= \begin{pmatrix} 2 & 1 & 0 \\ 0 & 0 & 2 \end{pmatrix} \begin{bmatrix} 2 \\ 1 \\ 3 \end{bmatrix} \\ &= \begin{bmatrix} 5 \\ 6 \end{bmatrix} \end{aligned}$$

Example 4.7.4. Let's reconsider Example 4.3.3. Again we can compute the directional derivative with the Jacobian, giving us

$$\begin{aligned} D_{\vec{v}} f(\vec{a}) &= D_{f,\vec{a}}(\vec{v}) \\ &= J_{f,\vec{a}}\vec{v} \\ &= \begin{pmatrix} 1 & 0 & -3 \\ 0 & 1 & 2 \\ 0 & 0 & 0 \end{pmatrix} \begin{bmatrix} 0 \\ -1 \\ 1 \end{bmatrix} \\ &= \begin{bmatrix} -3 \\ 1 \\ 0 \end{bmatrix} \end{aligned}$$

Example 4.7.5. Let's reconsider Example 4.4.1. This time we are computing with a unit directional derivative, but of course the same computation still works.

$$\begin{aligned} (v)D_{\vec{u}} f(\vec{a}) &= (v)D_{f,\vec{a}}(\vec{u}) \\ &= (v)J_{f,\vec{a}}\vec{u} \\ &= (v)\left((-2x)e^{-x^2-y^2} \quad (-2y)e^{-x^2-y^2}\right)\bigg|_{\vec{a}} \begin{bmatrix} -1 \\ 0 \end{bmatrix} \\ &= (3.5)\left((-4)e^{-5} \quad (-2)e^{-5}\right)\begin{bmatrix} -1 \\ 0 \end{bmatrix} \\ &= 14e^{-5} \end{aligned}$$

Example 4.7.6. We reconsider now Example 4.4.2. Again we can rephrase the computation with the Jacobian matrix.

4.7. JACOBIAN MATRICES

$$\begin{aligned}
D_{\vec{u}} f(\vec{a}) &= \frac{1}{\sqrt{2}} D_{\vec{v}} f(\vec{a}) \\
&= \frac{1}{\sqrt{2}} D_{f,\vec{a}}(\vec{v}) \\
&= \frac{1}{\sqrt{2}} J_{f,\vec{a}} \vec{v} \\
&= \frac{1}{\sqrt{2}} \begin{pmatrix} 2x & -4y \end{pmatrix} \Big|_{\vec{a}} \begin{bmatrix} 1 \\ 1 \end{bmatrix} \\
&= \frac{1}{\sqrt{2}} \begin{pmatrix} 4 & 0 \end{pmatrix} \begin{bmatrix} 1 \\ 1 \end{bmatrix} \\
&= \frac{4}{\sqrt{2}}
\end{aligned}$$

Exercises

Exercise 4.7.1 (**exer-Ders-JM-1**). Compute the Jacobian matrix for the function $f : \mathbb{R}^3 \to \mathbb{R}^2$ defined by $f(x, y, z) = (xyz, x - y^2 z + z^3)$.

Exercise 4.7.2 (**exer-Ders-JM-2**). Compute the Jacobian matrix for the function $f : \mathbb{R}^1 \to \mathbb{R}^2$ defined by $f(t) = (t^3 - t, e^t)$. How does this appear to relate to the parametric derivative?

Exercise 4.7.3 (**exer-Ders-JM-3**). Suppose that the Jacobian matrix for a function f at some point in the domain has four rows and seven columns. How many input variables must this function have? How many output variables must this function have?

Exercise 4.7.4 (**exer-Ders-JM-4**). Suppose that the Jacobian matrix for a function f at some point $\vec{a}$ in the domain is given by
$$\begin{pmatrix} 3 & 4 & 0 \\ 7 & -2 & -3 \end{pmatrix}$$
For each of the following vectors $\vec{v}$, compute the directional derivative $D_{\vec{v}} f(\vec{a})$.

1. $\vec{v} = (2, 1, 3)$
2. $\vec{v} = (-2, 3, 0)$
3. $\vec{v} = (5, 1, 1)$

Exercise 4.7.5 (**exer-Ders-JM-5**). Consider the function f given by $f(x, y, z) = x^2 \sin(x - 2y) - z e^{xy}$. Compute the directional derivative $D_{\vec{v}} f(\vec{a})$ for each of the following points $\vec{a}$ and velocities $\vec{v}$ given below.

1. $\vec{a} = (0, \pi/3, 2)$, $\vec{v} = (3, 2, 5)$
2. $\vec{a} = (0, \pi/3, 2)$, $\vec{v} = (-3, 1, -6)$
3. $\vec{a} = (0, \pi/3, 2)$, $\vec{v} = (1, 0, 0)$
4. $\vec{a} = (\pi/4, 0, 5)$, $\vec{v} = (-3, 3, 8)$
5. $\vec{a} = (\pi/4, 0, 5)$, $\vec{v} = (7, 1, -2)$

6. $\vec{a} = (\pi/4, 0, 5)$, $\vec{v} = (1, 0, 0)$

Exercise 4.7.6 (exer-Ders-JM-6). Consider the function

$$f\left(\begin{bmatrix} x \\ y \end{bmatrix}\right) = \frac{3x^2y - y^3}{x^2 + y^2} = r\sin(3\theta)$$

from Example 4.6.5. Using the results of that example we know that $\frac{\partial f}{\partial x}(\vec{0}) = D_{\vec{e}_1}f(\vec{0}) = 0$ and $\frac{\partial f}{\partial y}(\vec{0}) = D_{\vec{e}_2}f(\vec{0}) = -1$.

If we were to be interested in computing $D_{(1,2)}f(\vec{0})$, is the argument below valid or not? Explain.

$$D_{(1,2)}f(\vec{0}) = J_{f,\vec{0}}\begin{bmatrix} 1 \\ 2 \end{bmatrix} = \begin{bmatrix} \frac{\partial f}{\partial x} & \frac{\partial f}{\partial y} \end{bmatrix}\begin{bmatrix} 1 \\ 2 \end{bmatrix} = \begin{bmatrix} 0 & -1 \end{bmatrix}\begin{bmatrix} 1 \\ 2 \end{bmatrix} = -2$$

What is the actual correct value of $D_{(1,2)}f(\vec{0})$?

4.8 Gradients

4.8.1 Definition

If we have a differentiable function $f : \mathbb{R}^n \to \mathbb{R}^1$, then its Jacobian matrix consists of a single row.

$$J_{f,\vec{a}} = \begin{bmatrix} \frac{\partial f}{\partial x_1} & \frac{\partial f}{\partial x_2} & \cdots & \frac{\partial f}{\partial x_n} \end{bmatrix}$$

Directional derivatives are thus computed as

$$D_{\vec{v}}f(\vec{a}) = D_{f,\vec{a}}(\vec{v}) = J_{f,\vec{a}}\vec{v}$$

$$= \begin{bmatrix} \frac{\partial f}{\partial x_1} & \frac{\partial f}{\partial x_2} & \cdots & \frac{\partial f}{\partial x_n} \end{bmatrix} \begin{bmatrix} v_1 \\ v_2 \\ \vdots \\ v_n \end{bmatrix}$$

By definition of matrix vector products, we can rewrite this as a dot product of two vectors as

$$D_{\vec{v}}f(\vec{a}) = \begin{bmatrix} \frac{\partial f}{\partial x_1} \\ \frac{\partial f}{\partial x_2} \\ \cdots \\ \frac{\partial f}{\partial x_n} \end{bmatrix} \cdot \begin{bmatrix} v_1 \\ v_2 \\ \vdots \\ v_n \end{bmatrix}$$

The conclusion we can draw from this is that for a real valued differentiable function, we can compute $D_{\vec{v}}f(\vec{a})$ either by multiplying by a matrix, or by a dot product with a vector.

While on a computational level these are practically identical, there is an advantage to viewing this as a dot product with a vector. The reason is that while matrices and vectors are both thought of as arrays of numbers, vectors have the additional geometric interpretations of direction and length – and the dot product has properties that relate strongly to those geometric notions.

In order to investigate this point of view, we must first give this vector a name. We call it the "gradient of f at $\vec{a}$".

Definition 4.8.1. *For a differentiable function $f : \mathbb{R}^n \to \mathbb{R}^1$, we define the "gradient of f at the point $\vec{a}$" as the vector*

$$\nabla f(\vec{a}) = \begin{bmatrix} \frac{\partial f}{\partial x_1} \\ \frac{\partial f}{\partial x_2} \\ \vdots \\ \frac{\partial f}{\partial x_n} \end{bmatrix}$$

We summarize the above computation of directional derivatives then in the following theorem. Note that this theorem only applies to real valued functions, not vector valued functions.

Theorem 4.8.1. *If $f : \mathbb{R}^n \to \mathbb{R}^1$ is differentiable, then*

$$D_{\vec{v}}f(\vec{a}) = \nabla f(\vec{a}) \cdot \vec{v}$$

All of the above discussion applies only to functions that are real valued. But we can make a connection between gradients and vector valued functions. In particular, a vector valued function $f : \mathbb{R}^n \to \mathbb{R}^m$ has m component functions f_i, each of which is itself a real valued function. Each of these component functions has a gradient vector which can be used to compute its directional derivatives.

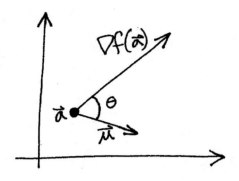

Figure 4.16:

Looking at the form of the Jacobian matrix,

$$J_{f,\vec{a}} = \begin{pmatrix} \frac{\partial f_1}{\partial x_1} & \frac{\partial f_1}{\partial x_2} & \cdots & \frac{\partial f_1}{\partial x_n} \\ \frac{\partial f_2}{\partial x_1} & \cdots & & \frac{\partial f_2}{\partial x_n} \\ \vdots & \vdots & & \vdots \\ \frac{\partial f_m}{\partial x_1} & \frac{\partial f_m}{\partial x_2} & \cdots & \frac{\partial f_m}{\partial x_n} \end{pmatrix}$$

we can see directly that each row vector of the Jacobian is just the gradient vector of the corresponding component function. So, we have yet another way of writing the Jacobian matrix, with rows given by the gradients of the individual component functions.

$$J_{f,\vec{a}} = \begin{pmatrix} -\!\!\!- & \nabla f_1 & -\!\!\!- \\ & \vdots & \\ -\!\!\!- & \nabla f_m & -\!\!\!- \end{pmatrix}$$

4.8.2 Geometric interpretations

As we hoped and suspected, the gradient vector of a real valued function has several interesting geometric properties:

Theorem 4.8.2. *Of all unit vectors $\vec{u}$, the one that maximizes the directional derivative $D_{\vec{u}} f(\vec{a})$ is the unit vector that points in the same direction as $\nabla f(\vec{a})$.*

Said differently, the direction of the gradient vector is the direction one should move in the domain in order to realize the fastest increase in the values of f. So we say that the gradient points in the "direction of fastest increase".

This is an easy result of Theorem 4.8.1, by writing that dot product in terms of lengths and angles. We consider an arbitrary unit vector $\vec{u}$ in the domain, and draw it at the point $\vec{a}$ along with the gradient vector. Let θ be the angle between these two vectors. See Figure 4.16.

We can then compute the directional derivative by

$$D_{\vec{u}} f(\vec{a}) = \nabla f(\vec{a}) \cdot \vec{u} = \|\nabla f(\vec{a})\| \|\vec{u}\| \cos(\theta) = \|\nabla f(\vec{a})\| \cos(\theta)$$

Since $\|\nabla f(\vec{a})\|$ is positive and completely independent of the choice of $\vec{u}$, we maximize the expression above simply by maximizing $\cos(\theta)$. This of course is accomplished by having $\theta = 0$. Since θ is the angle between the two vectors ∇f and $\vec{u}$, we see that this directional derivative is maximized when $\vec{u}$ points in the same direction as $\nabla f(\vec{a})$.

(To say this algebraically, we could write $\vec{u} = \frac{\nabla f(\vec{a})}{\|\nabla f(\vec{a})\|}$.)

Given this property, and thinking in terms of graphs (where the value of the function is a "height"), the direction of the gradient vector is called the "direction of steepest ascent". See Figure 4.17.

From the same setup, we can conclude another theorem.

4.8. GRADIENTS

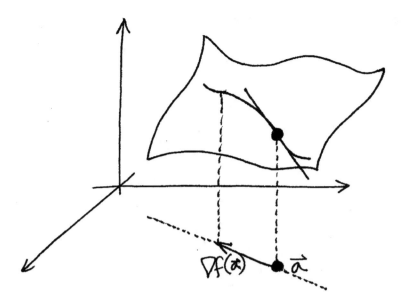

Figure 4.17:

Theorem 4.8.3. *(2) If $\vec{u}$ is the direction computed above, then the value of that directional derivative is $D_{\vec{u}} f(\vec{a}) = \|\nabla f(\vec{a})\|$.*

This is seen with a simple computation using Theorem 4.8.1.

$$D_{\vec{u}} f(\vec{a}) = \nabla f(\vec{a}) \cdot \vec{u} = \nabla f(\vec{a}) \cdot \frac{\nabla f(\vec{a})}{\|\nabla f(\vec{a})\|} = \frac{\|\nabla f(\vec{a})\|^2}{\|\nabla f(\vec{a})\|} = \|\nabla f(\vec{a})\|$$

The interpretations of Theorems 4.8.2 and 4.8.3 can be restated as follows, emphasizing these natural uses of each of the two geometric traits (direction and length) of the gradient vector:

(1) The direction of $\nabla f(\vec{a})$ is the direction of fastest increase.

(2) The length of $\nabla f(\vec{a})$ is the (unit) directional derivative in that direction.

4.8.3 Level sets

The gradient vector of a function turns out also to have a close relationship with the level sets of that function.

Theorem 4.8.4. *Suppose we have a differentiable function $f : \mathbb{R}^n \to \mathbb{R}^1$ and a point $\vec{a}$ in a given level set S in the domain. If $\vec{v}$ is tangent to the level set S at $\vec{a}$, then $\vec{v}$ must be perpendicular to $\nabla f(\vec{a})$.*

Stated more casually, we say that at any point in the domain of a differentiable function $f : \mathbb{R}^n \to \mathbb{R}^1$, the gradient vector at that point is perpendicular to the level set through that point. See Figure 4.18.

The idea here is that if $\vec{v}$ is tangent to the level set S, then as we move in the direction given by $\vec{v}$ we are moving "along" the level set, and so the value of the function should not be changing.

We can make this precise by imagining a point $\vec{x}$ moving through the level set S, with position given as a function of time by $\vec{x}(t)$, and with $\vec{x}(0) = \vec{a}$ and $\vec{x}'(0) = \vec{v}$. See Figure 4.19.

Since the point is moving through the level set, the value $f(\vec{x}(t))$ is not changing. So, we should have

$$\left.\frac{d}{dt}\right|_{t=0} f(\vec{x}(t)) = 0$$

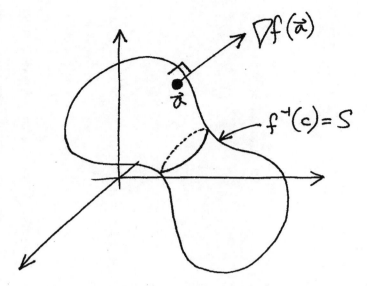

Figure 4.18:

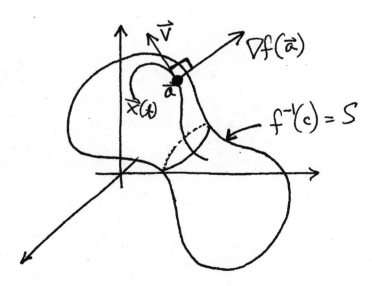

Figure 4.19:

4.8. GRADIENTS

which we can rewrite as the directional derivative

$$D_{\vec{v}}f(\vec{a}) = 0$$

Of course applying Theorem 4.8.1 allows us to rewrite this as $\nabla f(\vec{a}) \cdot \vec{v} = 0$, and thus we conclude that $\nabla f(\vec{a})$ and $\vec{v}$ are perpendicular.

4.8.4 Examples

Example 4.8.1. Let's revisit Example 4.3.1, in which we computed the directional derivative $D_{\vec{v}}f(\vec{a})$ of the function $f : \mathbb{R}^3 \to \mathbb{R}^1$ given by

$$f\left(\begin{bmatrix} x \\ y \\ z \end{bmatrix}\right) = x^2 + yz$$

with $\vec{a} = (1, 2, 0)$ and $\vec{v} = (2, 1, 3)$.

Here we will do the computation using the gradient vector. The function is continuously differentiable and thus differentiable, so we compute the gradient as

$$\nabla f(\vec{a}) = \begin{bmatrix} 2x \\ z \\ y \end{bmatrix}\bigg|_{(1,2,0)} = \begin{bmatrix} 2 \\ 0 \\ 2 \end{bmatrix}$$

We can then compute the directional derivative by

$$D_{\vec{v}}f(\vec{a}) = \nabla f(\vec{a}) \cdot \vec{v} = \begin{bmatrix} 2 \\ 0 \\ 2 \end{bmatrix} \cdot \begin{bmatrix} 2 \\ 1 \\ 3 \end{bmatrix} = 10$$

Example 4.8.2. In what direction is the function from Example 4.8.1 increasing the fastest at the point $\vec{a} = (1, 2, 0)$?

We know from Theorem 4.8.2 that we need only compute the direction of the gradient vector, at the given point. That gradient vector has already been computed in Example 4.8.1 as

$$\nabla f(\vec{a}) = \begin{bmatrix} 2 \\ 0 \\ 2 \end{bmatrix}$$

So the direction of fastest increase is the direction of the vector $(2, 0, 2)$, which we can represent with the unit vector

$$\vec{u} = \begin{bmatrix} \sqrt{1/2} \\ 0 \\ \sqrt{1/2} \end{bmatrix}$$

Example 4.8.3. Let $f : \mathbb{R}^3 \to \mathbb{R}^1$ be defined by

$$f(x, y, z) = e^{xy} + xz^2$$

What is the equation of the tangent plane to the level set of f through the point $(1, 3, 2)$?

We know from Theorem 4.8.4 that $\nabla f(1, 3, 2)$ is orthogonal to all of the vectors parallel to the given level set. So, we can use that gradient as a normal vector for the tangent plane to that level set.

$$\vec{n} = \nabla f(1, 3, 2) = \begin{bmatrix} ye^{xy} + z^2 \\ xe^{xy} \\ 2xz \end{bmatrix} \bigg|_{(1,3,2)} = \begin{bmatrix} 3e^3 + 4 \\ e^3 \\ 4 \end{bmatrix}$$

Using the point $(1, 3, 2)$ as our $\vec{x}_0$, we can then write the equation of the plane as

$$(3e^3 + 4)x + e^3 y + 4z = 6e^3 + 12$$

Exercises

Exercise 4.8.1 (**exer-Ders-G-1**). It was noted in this section that each row of the Jacobian matrix of a function $f : \mathbb{R}^n \to \mathbb{R}^m$ can be thought of as the gradient of the corresponding component function of f. How can we instead interpret the columns of J_f? (Hint: Note that if we fix all but one of the values of the input variables, then the resulting function can be thought of as a parametric curve.)

Exercise 4.8.2 (**exer-Ders-G-2**). Use the gradient vector to recompute the directional derivatives listed in Exercise 4.7.5

Exercise 4.8.3 (**exer-Ders-G-3**). Consider the temperature function $T : \mathbb{R}^3 \to \mathbb{R}^1$ given by $T(x, y, z) = 75 + 10(x - 3y + z)e^{-x^2 - y^2 - z^2}$. At the origin, in what direction is the temperature increasing the fastest? What is the rate of change of temperature with respect to distance traveled in that direction?

Exercise 4.8.4 (**exer-Ders-G-4**). Suppose that at the point $\vec{a}$ in the domain of the continuously differentiable function $f : \mathbb{R}^2 \to \mathbb{R}^1$, we know that the unit directional derivative in the x-direction is 4 and the unit directional derivative in the y-direction is 8. What is the direction of fastest increase of this function, and what is the unit directional derivative in that direction?

Exercise 4.8.5 (**exer-Ders-G-5**). Suppose that at the point $\vec{a}$ in the domain of the continuously differentiable function $f : \mathbb{R}^3 \to \mathbb{R}^1$, we know that the direction of fastest increase is in the direction of the vector $(3, 4, 12)$, and that the rate of change of f with respect to distance traveled in the x-direction is 7. What is the largest unit directional derivative of f at that point?

Exercise 4.8.6 (**exer-Ders-G-6**). A mountain has a shape which is equivalent to the graph of the function f defined by $f(x, y) = e^{x^2 - x^4 - y^2}$. How steep is the side of the mountain at the point above $(1, 0)$?

Exercise 4.8.7 (**exer-Ders-G-7**). Find the equation of the plane tangent to the level set of the function g defined by $g(x, y, z) = x^2 y - ye^z + y^3$ at the point $(3, 2, 1)$.

Exercise 4.8.8 (**exer-Ders-G-8**). Find the equation of the plane tangent to the graph of the function f defined by $f(x, y) = x^2 y - ye^x$ at the point above $(1, -1)$. (Hint: Recall that the graph of a function can be written as a level set of a different function.)

Exercise 4.8.9 (**exer-Ders-G-9**). Use properties of the gradient vector to show that the tangent plane approximation to the continuously differentiable function $f : \mathbb{R}^2 \to \mathbb{R}^1$ at the point (a, b) in the domain is given by

$$P(x, y) = f(a, b) + \left(\frac{\partial f}{\partial x}\right)(x - a) + \left(\frac{\partial f}{\partial y}\right)(y - b)$$

(Hint: Find the equation for the tangent plane to the graph of f at the point $(a, b, f(a, b))$, and then solve for z.)

4.9 The Chain Rule

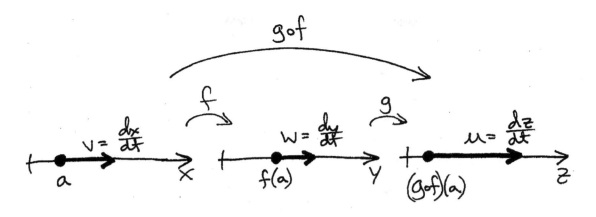

Figure 4.20:

Thinking of the derivative of a multivariable function as a linear transformation, it turns out that there is a very strong analogy between the multivariable chain rule and the single variable chain rule. In order to take that approach to understanding the multivariable chain rule, we will first review the single variable chain rule from that point of view.

4.9.1 Single variable version

Suppose that we have two differentiable functions, $f : \mathbb{R}^1 \to \mathbb{R}^1$ and $g : \mathbb{R}^1 \to \mathbb{R}^1$, and we are interested in the composition $(g \circ f)$. Let's define the variables x, y, z such that $y = f(x)$ and $z = g(y)$, as indicated in Figure 4.20

For an arbitrary point a on the x-axis, there is its image $f(a)$ on the y-axis, and its further image $(g \circ f)(a)$ on the z-axis.

If as usual we think of a point x being momentarily at a and moving with velocity $v = \frac{dx}{dt}$, then the image $y = f(x)$ is moving with some velocity $w = \frac{dy}{dt}$ and $z = g(y) = (g \circ f)(x)$ is moving with some velocity $u = \frac{dz}{dt}$.

We recall from Section 4.1 that one interpretation of the derivative of a single variable function at a specific point in its domain gives us a relationship between the velocity of its input and the velocity of its output. For example, the derivative of f at the point a gives us a relationship between $v = \frac{dx}{dt}$ and $w = \frac{dy}{dt}$; specifically, we have

$$w = f'(a)v \tag{4.18}$$

We can use similar reasoning to write down similar equations concerning $u = \frac{dz}{dt}$. But, notice that we can do this two different ways – (1) we can view u as being the velocity of the output of $(g \circ f)$, when the input velocity is v; but also (2) we can view u as being the velocity of the output of g, when the input velocity is w.

From the point of view of (1), we get

$$u = (g \circ f)'(a)v \tag{4.19}$$

From the point of view of (2), we get

$$u = g'(f(a))w \tag{4.20}$$

We must be sure to note that in Equation 4.20, the derivative of g is evaluated at the point $f(a)$, because of course that is the point that is the input to the function g.

Combining equations 4.18 and 4.20, we get

$$u = g'(f(a))w = g'(f(a))f'(a)v \tag{4.21}$$

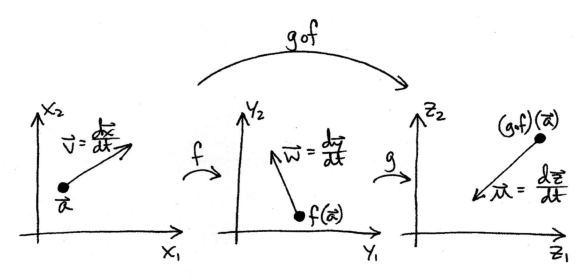

Figure 4.21:

Equating the results in equations 4.19 and 4.21, we get

$$(g \circ f)'(a)v = g'(f(a))f'(a)v \tag{4.22}$$

and thus

$$(g \circ f)'(a) = g'(f(a))f'(a) \tag{4.23}$$

This is our single variable chain rule.

4.9.2 Multivariable version

Let's now take a similar approach to thinking about a composition of multivariable functions, and see what sort of a conclusion we come to.

Suppose that we have two differentiable functions, $f : \mathbb{R}^n \to \mathbb{R}^m$ and $g : \mathbb{R}^m \to \mathbb{R}^p$, and we are interested in the composition $(g \circ f) : \mathbb{R}^n \to \mathbb{R}^p$. Let's define the variables $\{x_1, \ldots, x_n\}$, $\{y_1, \ldots, y_m\}$, $\{z_1, \ldots, z_p\}$ such that $y_i = f_i(\vec{x})$ and $z_j = g_j(\vec{y})$, as indicated in Figure 4.21. (Note that in the figure, we draw each of the multi-dimensional domains as being two dimensional, simply because it is easiest to draw.)

For an arbitrary point $\vec{a}$ in the $\vec{x}$-space, there is its image $f(\vec{a})$ in the $\vec{y}$-space, and its further image $(g \circ f)(\vec{a})$ in the $\vec{z}$-space.

If as usual we think of a point $\vec{x}$ being momentarily at $\vec{a}$ and moving with velocity $\vec{v} = \frac{d\vec{x}}{dt}$, then the image $\vec{y} = f(\vec{x})$ is moving with some velocity $\vec{w} = \frac{d\vec{y}}{dt}$ and $\vec{z} = g(\vec{y}) = (g \circ f)(\vec{x})$ is moving with some velocity $\vec{u} = \frac{d\vec{z}}{dt}$.

Similarly to the interpretation that we previously took for single variable derivatives, the derivative transformation of a function at a point in its domain gives us a relationship between the velocity of its input and the velocity of its output. For example, the derivative of f at the point $\vec{a}$ gives us a relationship between $\vec{v} = \frac{d\vec{x}}{dt}$ and $\vec{w} = \frac{d\vec{y}}{dt}$; specifically, we have

$$\vec{w} = D_{f,\vec{a}}(\vec{v}) \tag{4.24}$$

We can use similar reasoning to write down similar equations concerning $\vec{u} = \frac{d\vec{z}}{dt}$. But, notice that we can again do this two different ways – (1) we can view $\vec{u}$ as being the velocity of the output of $(g \circ f)$, when the input velocity is $\vec{v}$; but also (2) we can view $\vec{u}$ as being the velocity of the output of g, when the input velocity is $\vec{w}$.

4.9. THE CHAIN RULE

From the point of view of (1), we get

$$\vec{u} = D_{(g \circ f), \vec{a}}(\vec{v}) \tag{4.25}$$

From the point of view of (2),

$$\vec{u} = D_{g, f(\vec{a})}(\vec{w}) \tag{4.26}$$

As in the single variable case, we must be sure to note that in Equation 4.26, the derivative of g is evaluated at the point $f(\vec{a})$, because of course that is the point that is the input to the function g.

Combining equations 4.24 and 4.26, we get

$$\vec{u} = D_{g, f(\vec{a})}(\vec{w}) = D_{g, f(\vec{a})}\left(D_{f, \vec{a}}(\vec{v})\right) \tag{4.27}$$

Equating the results in equations 4.25 and 4.27, we get

$$D_{(g \circ f), \vec{a}}(\vec{v}) = D_{g, f(\vec{a})}\left(D_{f, \vec{a}}(\vec{v})\right) \tag{4.28}$$

Since this is true for all vectors $\vec{v}$, we have an equality of linear transformations,

$$D_{(g \circ f), \vec{a}} = D_{g, f(\vec{a})} \circ D_{f, \vec{a}} \tag{4.29}$$

This is our multivariable chain rule.

We can write this in matrix form by recalling that the matrix for a derivative transformation is the Jacobian matrix, and that the matrix for a composition of linear transformations is the matrix product of the corresponding matrices. This gives us

$$J_{(g \circ f), \vec{a}} = J_{g, f(\vec{a})} J_{f, \vec{a}} \tag{4.30}$$

Note that in this form, the multivariable chain rule says exactly the same thing as the single variable chain rule – the derivative of a composition is the product of the derivatives of the individial functions, evaluated at the appropriate points in their domains.

Example 4.9.1. Suppose that we have $f: \mathbb{R}^2 \to \mathbb{R}^3$ and $g: \mathbb{R}^3 \to \mathbb{R}^3$; and the only other facts we are given about these functions are

$$f\left(\begin{bmatrix} 1 \\ 2 \end{bmatrix}\right) = \begin{bmatrix} 3 \\ 1 \\ 0 \end{bmatrix}$$

$$J_{f,(1,2)} = \begin{pmatrix} 1 & 0 \\ 3 & -1 \\ 5 & -1 \end{pmatrix}$$

$$J_{g,(3,1,0)} = \begin{pmatrix} 0 & 1 & 0 \\ 1 & 0 & 1 \\ 0 & 1 & 0 \end{pmatrix}$$

What is the Jacobian matrix of the composition $(g \circ f)$ at the point $(1, 2)$, and also what is the value of $\frac{\partial g_3}{\partial x_1}$ at the point $(1, 2)$?

We can compute the Jacobian as the product of the given Jacobians:

$$\begin{aligned}
J_{(g\circ f),(1,2)} &= J_{g,f(1,2)} J_{f,(1,2)} \\
&= J_{g,(3,1,0)} J_{f,(1,2)} \\
&= \begin{pmatrix} 0 & 1 & 0 \\ 1 & 0 & 1 \\ 0 & 1 & 0 \end{pmatrix} \begin{pmatrix} 1 & 0 \\ 3 & -1 \\ 5 & -1 \end{pmatrix} \\
&= \begin{pmatrix} 3 & -1 \\ 6 & -1 \\ 3 & -1 \end{pmatrix}
\end{aligned}$$

To determine the value of $\frac{\partial g_3}{\partial x_1}$**, simply observe that this is the third element of the first column of** $J_{(g\circ f),(1,2)}$**; so, we have**

$$\frac{\partial g_3}{\partial x_1} = 3$$

4.9.3 Intermediate variables

The above form is the simplest and most elegant way think of the chain rule. For a different point of view that is more specific to the variables involved instead of the functions themselves, let's write out each of the above Jacobian matrices:

$$\begin{pmatrix} \frac{\partial z_1}{\partial x_1} & \frac{\partial z_1}{\partial x_2} & \cdots & \frac{\partial z_1}{\partial x_n} \\ \frac{\partial z_2}{\partial x_1} & \cdots & & \frac{\partial z_2}{\partial x_n} \\ \vdots & \vdots & & \vdots \\ \frac{\partial z_p}{\partial x_1} & \frac{\partial z_p}{\partial x_2} & \cdots & \frac{\partial z_p}{\partial x_n} \end{pmatrix} = \begin{pmatrix} \frac{\partial z_1}{\partial y_1} & \frac{\partial z_1}{\partial y_2} & \cdots & \frac{\partial z_1}{\partial y_m} \\ \frac{\partial z_2}{\partial y_1} & \cdots & & \frac{\partial z_2}{\partial y_m} \\ \vdots & \vdots & & \vdots \\ \frac{\partial z_p}{\partial y_1} & \frac{\partial z_p}{\partial y_2} & \cdots & \frac{\partial z_p}{\partial y_m} \end{pmatrix} \begin{pmatrix} \frac{\partial y_1}{\partial x_1} & \frac{\partial y_1}{\partial x_2} & \cdots & \frac{\partial y_1}{\partial x_n} \\ \frac{\partial y_2}{\partial x_1} & & \cdots & \frac{\partial y_2}{\partial x_n} \\ \vdots & \vdots & & \vdots \\ \frac{\partial y_m}{\partial x_1} & \frac{\partial y_m}{\partial x_2} & \cdots & \frac{\partial y_m}{\partial x_n} \end{pmatrix}$$

Written in this form, we can conclude specific formulas for the individual partial derivatives of the composition variables in terms of the partial derivatives of the variables for f and g. For example, suppose we are interested in computing

$$\frac{\partial z_i}{\partial x_j}$$

We notice immediately that this is simply the element of $J_{(g\circ f),\vec{a}}$ (on the left in the above equation) in the ith row and jth column. Since that matrix is the product of the two matrices on the right, we can write this element as the dot product of the corresponding row of $J_{g,f(\vec{a})}$ and the corresponding column of $J_{f,\vec{a}}$. Thus, we have

$$\frac{\partial z_i}{\partial x_j} = \frac{\partial z_i}{\partial y_1} \frac{\partial y_1}{\partial x_j} + \frac{\partial z_i}{\partial y_2} \frac{\partial y_2}{\partial x_j} + \cdots + \frac{\partial z_i}{\partial y_m} \frac{\partial y_m}{\partial x_j} \tag{4.31}$$

Another way to motivate Equation 4.31 is to think of the role that the variables $y_1, \ldots, y_m$ play in this setup. These are the variables that define the "intermediate" space, between the domain and the target of the composition, and so we call those variables "intermediate variables".

We can represent these variables symbolically, along with the variables x_j and z_i we are interested in, by the figure below.

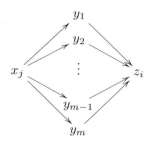

4.9. THE CHAIN RULE

These intermediate variables are the input variables to the function g_i, whose output is the variable z_i that we are interested in. So, we could argue that changes in z_i are caused by changes in those intermediate variables; these are represented by the partial derivatives $\frac{\partial z_i}{\partial y_k}$, where k takes values from 1 to m.

Of course each of those intermediate variables changes by an amount related to the amount that x_j changes, represented by the partial derivatives $\frac{\partial y_k}{\partial x_j}$, where again k takes values from 1 to m.

(The arrows in the above diagram represent the direction that these changes propagate through these variables.)

So, for each intermediate variable y_k, we have a product $\frac{\partial z_i}{\partial y_k} \frac{\partial y_k}{\partial x_j}$ representing the contribution that the intermediate variable y_k makes to the way changes in x_j cause changes in z_i. Adding these up for all of the intermediate variables gives Equation 4.31.

Example 4.9.2. Let's revisit the setup of Example 4.9.1, and again let's compute $\frac{\partial g_3}{\partial x_1}$, but this time using Equation 4.31 to do the computation instead of computing the entire Jacobian.

We are given the Jacobians of the functions f and g, so we can interpret the needed partial derivatives simply as the appropriate elements of those matrices. For example, the partial derivatives $\frac{\partial g_3}{\partial y_k}$ are simply the entries in the third row of the matrix $J_{g,(3,1,0)}$. And the partial derivatives $\frac{\partial y_k}{\partial x_1}$ are simply the entries in the first column of the matrix $J_{f,(1,2)}$. Plugging these values into Equation 4.31, we get

$$\frac{\partial g_3}{\partial x_1} = \frac{\partial g_3}{\partial y_1}\frac{\partial y_1}{\partial x_1} + \frac{\partial g_3}{\partial y_2}\frac{\partial y_2}{\partial x_1} + \frac{\partial g_3}{\partial y_3}\frac{\partial y_3}{\partial x_1}$$

$$= (0)(1) + (1)(3) + (0)(5)$$

$$= 3$$

Note of course that from a computational standpoint, this is identical to simply computing the single element of the Jacobian of the composition, as a dot product of the corresponding row of the one matrix and the corresponding column of the other.

Recall that another way to write the chain rule for single variable derivatives is

$$\frac{dz}{dx} = \frac{dz}{dy}\frac{dy}{dx} \tag{4.32}$$

Students may have been tempted in a single variable calculus course to view this equation as a simple consequence of "canceling" the "dy" in each "fraction" on the right side. Of course this is not a strictly valid point of view because actually these are not fractions – rather, they are limits. Still, having proved the single variable chain rule by more appropriate means many students still think of it as a cancelation in practice.

Equation 4.31 makes it very clear that this sort of thinking is not only technically incorrect, but completely inappropriate in any sense in the multivariable setting. Of course this is because if we did allow such "cancelation", even in a casual sense, the equation simply does not work.

Students should make sure to remember that the use of the chain rule in a multivariable setting has nothing to do with any sort of cancelation, and that any application of the chain rule must consider all of the intermediate variables, not just one or some.

4.9.4 Confusion of variables

One nice way to phrase the chain rule is to say, "the derivative of a composition is the product of the individual derivatives"; this is true, but it must be noted that those individual derivatives must be evaluated at the appropriate points. If the variables are set up correctly this will be no problem – but if not, it is easy to make a mistake, as in the following:

Example 4.9.3. Suppose we have functions f and g with $f(x) = x^2$, and $g(x) = x^3$. What is $(g \circ f)'(x)$?

Incorrect: We know that $f' = 2x$ and $g' = 3x^2$, so the above interpretation of the chain rule seems to tell us

$$(g \circ f)' = g'f' = (3x^2)(2x) = 6x^3$$

But clearly this is the wrong answer, since we know we that $(g \circ f)' = \left((x^2)^3\right)' = (x^6)' = 6x^5$

So, what went wrong? Well, we were sloppy about where we evaluated g; in particular the chain rule says

$$(g \circ f)'(x) = g'(f(x))f'(x)$$

So g' must be evaluated at $f(x)$, not at x as we inadvertently did in our careless computation above. We should have computed as follows.

Better: We know that $f' = 2x$ and $g' = 3x^2$, so

$$(g \circ f)' = g'(f(x))f'(x) = (3(f(x))^2)(2x) = (3(x^2)^2)(2x) = 6x^5$$

There is an easy way to avoid this sort of problem – don't confuse variables! The mistake we made in the above computation is that even though we are composing the functions f and g, we are using the same variable (x) to define each; in other words, we are using x to represent the input to g, which is the output of f... but we are also using x to represent the input to f. This is what creates the problem.

We can avoid that problem simply by rewriting one of the functions using a different dummy variable.

Example 4.9.4. Suppose we again consider functions f and g with $f(x) = x^2$, and $g(x) = x^3$. What is $(g \circ f)'(x)$?

Best: Let's write $f(x) = x^2 = y$. Since our new variable y is the output of f and we will be composing these as $(g \circ f)$, we should use y also as the input to g; so we write $g(y) = y^3$. Note that this is still the same function g – we are just expressing it with a different dummy variable.

Then, our computation becomes worry-free – the new variable y forces us to remember that the derivative of g must be evaluated at $f(x)$.

$$(g \circ f)' = g'f' = (3y^2)(2x) = 6xy^2 = 6x(x^2)^2 = 6x^5$$

This same idea can be put to good use in the multivariable situation.

Example 4.9.5. Suppose we have multivariable functions f and g with

$$f\left(\begin{bmatrix} x \\ y \end{bmatrix}\right) = \begin{bmatrix} x^2 + y^2 \\ x^2 - y^2 \end{bmatrix} \quad \text{and} \quad g\left(\begin{bmatrix} x \\ y \end{bmatrix}\right) = \begin{bmatrix} 2y^2 \\ 3x^2 + 1 \end{bmatrix}$$

What is the Jacobian matrix for the composition $(g \circ f)$?

4.9. THE CHAIN RULE

Incorrect: We might carelessly just compute the Jacobians for each function as written and multiply them, which gives us the wrong answer:

$$J_{(g \circ f)} = J_g J_f$$
$$= \begin{pmatrix} 0 & 4y \\ 6x & 0 \end{pmatrix} \begin{pmatrix} 2x & 2y \\ 2x & -2y \end{pmatrix}$$

As in our previous example of single variable functions, we have failed here to evaluate the Jacobian of g at $f(\vec{x})$.

Best: We can fix this by avoiding the confusion of variables. Let's write

$$f\left(\begin{bmatrix} x \\ y \end{bmatrix}\right) = \begin{bmatrix} x^2 + y^2 \\ x^2 - y^2 \end{bmatrix} = \begin{bmatrix} u \\ v \end{bmatrix}$$

Since our new variables u and v are the outputs of f, and since we will be composing the two functions as $(g \circ f)$, we should use u and v as the inputs for g. So, we rewrite g in terms of these variables:

$$g\left(\begin{bmatrix} u \\ v \end{bmatrix}\right) = \begin{bmatrix} 2v^2 \\ 3u^2 + 1 \end{bmatrix}$$

Now, as in the single variable case, we can proceed worry-free, since the new variables u and v will force us to remember to evaluate J_g at the appropriate point:

$$J_{(g \circ f)} = J_g J_f$$
$$= \begin{pmatrix} 0 & 4v \\ 6u & 0 \end{pmatrix} \begin{pmatrix} 2x & 2y \\ 2x & -2y \end{pmatrix}$$
$$= \begin{pmatrix} 0 & 4(x^2 - y^2) \\ 6(x^2 + y^2) & 0 \end{pmatrix} \begin{pmatrix} 2x & 2y \\ 2x & -2y \end{pmatrix}$$

4.9.5 Second derivatives

Suppose we consider the chain rule for single variable derivatives, for the situation

$$x \longrightarrow y \longrightarrow z$$

As in Equation 4.32, we have

$$\frac{dz}{dx} = \frac{dz}{dy}\frac{dy}{dx} \tag{4.33}$$

which we could instead write as

$$z_x = z_y y_x \tag{4.34}$$

Suppose we want to compute the second derivative, $\frac{d^2z}{dx^2}$, in terms of the derivatives of the functions f and g. How would we do this?

Naturally, we would want to compute the derivative with respect to x of the expression in Equation 4.34. But we must note that in this equation, there are new variables that we have not yet listed in the above diagram. That is, we must acknowledge that the derivatives $z_y = \frac{dz}{dy}$ and $y_x = \frac{dy}{dx}$ are themselves both functions, of y and x respectively.

In order to recognize this, we can insert them into the appropriate places in the above diagram to get

$$x \longrightarrow y \longrightarrow z$$
$$ y_x z_y$$

We can then just take the derivative, making sure to obey all of the appropriate derivative rules.

First, note that the expression we are differentiating is a product, so we must use the product rule.

$$\frac{d^2z}{dx^2} = \frac{d}{dx}(z_y\, y_x) \tag{4.35}$$

$$= \left[\frac{d}{dx}z_y\right] y_x + z_y \left[\frac{d}{dx}y_x\right] \tag{4.36}$$

The last item in parentheses in Equation 4.36 is of course just the second derivative of y with respect to x. But the first item in parentheses in that equation requires the use of the chain rule – because, as is indicated in the diagram, z_y is a function of y, and y is a function of x. So we get

$$\frac{d^2z}{dx^2} = \left[\left(\frac{d}{dy}z_y\right)\left(\frac{dy}{dx}\right)\right] y_x + z_y\, y_{xx} \tag{4.37}$$

$$= z_{yy}\,(y_x)^2 + z_y\, y_{xx} \tag{4.38}$$

We could call this a second derivative chain rule for single variable derivatives.

The same sort of approach can be used to compute second partials of composed multivariable functions. We will not write out a general form for that here; but in order to do such computations one needs only to remember that after taking one composed partial derivative, the individual partials in that formula are themselves functions of some of the variables.

We demonstrate this with an important example relating to polar coordinates.

Example 4.9.6. Suppose we have a function $g : \mathbb{R}^2 \to \mathbb{R}^1$, defined by a given formula $z = g(x,y)$. Suppose also that we consider x and y to be function of r and θ, in the usual polar coordinates way with $x = r\cos\theta$ and $y = r\sin\theta$. How can we compute the derivatives of z with respect to these polar coordinates?

We have the following setup of variables.

$$r \longrightarrow x \longrightarrow z$$
$$\theta \longrightarrow y$$

We can compute the first partials of z with respect to r and θ simply by applying the chain rule. We get

$$\frac{\partial z}{\partial r} = \frac{\partial z}{\partial x}\frac{\partial x}{\partial r} + \frac{\partial z}{\partial y}\frac{\partial y}{\partial r} \tag{4.39}$$

$$= z_x\cos\theta + z_y\sin\theta \tag{4.40}$$

and

$$\frac{\partial z}{\partial \theta} = \frac{\partial z}{\partial x}\frac{\partial x}{\partial \theta} + \frac{\partial z}{\partial y}\frac{\partial y}{\partial \theta}$$

$$= z_x(-r\sin\theta) + z_y\, r\cos\theta$$

Let's now compute one of the second partials, $\frac{\partial^2 z}{\partial r^2}$. Of course we will simply be taking a partial derivative of Equation 4.40 with respect to r. But we must first acknowledge that the two partials remaining in that equation, z_x and z_y, are functions of x and y, giving us the updated setup

4.9. THE CHAIN RULE

[Diagram showing dependency: $r, \theta \to x, y \to z, z_x, z_y$]

Now we can proceed, first writing down the derivative and applying the product rule.

$$\frac{\partial^2 z}{\partial r^2} = \frac{\partial}{\partial r}(z_x \cos\theta + z_y \sin\theta) \tag{4.41}$$

$$= \left(\frac{\partial}{\partial r} z_x\right)\cos\theta + z_x\left(\frac{\partial}{\partial r}\cos\theta\right) + \left(\frac{\partial}{\partial r} z_y\right)\sin\theta + z_y\left(\frac{\partial}{\partial r}\sin\theta\right) \tag{4.42}$$

Conveniently, the second and fourth terms in Equation 4.42 are zero, because θ and r are independent variables.

The two second partials in Equation 4.42 must be computed with the chain rule, following the appended diagram above noting that x and y are intermediate variables for each of those second partials; this gives us

$$\frac{\partial^2 z}{\partial r^2} = \left[\frac{\partial}{\partial r} z_x\right]\cos\theta + \left[\frac{\partial}{\partial r} z_y\right]\sin\theta$$

$$= \left[\left(\frac{\partial}{\partial x} z_x\right)\frac{\partial x}{\partial r} + \left(\frac{\partial}{\partial y} z_x\right)\frac{\partial y}{\partial r}\right]\cos\theta + \left[\left(\frac{\partial}{\partial x} z_y\right)\frac{\partial x}{\partial r} + \left(\frac{\partial}{\partial y} z_y\right)\frac{\partial y}{\partial r}\right]\sin\theta$$

$$= [(z_{xx})\cos\theta + (z_{yx})\sin\theta]\cos\theta + [(z_{xy})\cos\theta + (z_{yy})\sin\theta]\sin\theta$$

$$= (z_{xx})\cos^2\theta + 2(z_{xy})\sin\theta\cos\theta + (z_{yy})\sin^2\theta$$

Exercises

Exercise 4.9.1 (exer-Ders-CR-1). The function $f : \mathbb{R}^3 \to \mathbb{R}^3$ is defined by $f(x, y, z) = (x^2 y, xz, y^2 - x^2) = (r, s, t)$, and we also have $(x, y, z) = g(u, v) = (u - v, u^2 - v^2, uv^3)$. Use the matrix form of the chain rule to compute the Jacobian matrix of the composition $f \circ g$ in terms of u and v.

Exercise 4.9.2 (exer-Ders-CR-2). Use the result of Exercise 4.9.1 to compute $\frac{\partial s}{\partial u}(1, 2)$

Exercise 4.9.3 (exer-Ders-CR-3). Consider the functions $f, g, h : \mathbb{R}^2 \to \mathbb{R}^2$ given by $f(x, y) = (x^2 - y^2, 2xy)$, $g(x, y) = (xy, 2x - 3y)$, $h(x, y) = (x + y, x - y)$. Compute the derivative matrices for each of the following composition functions, making sure in each case not to confuse variables.

1. $f \circ g$
2. $g \circ f$
3. $g \circ h$
4. $h \circ f \circ g$

Exercise 4.9.4 (exer-Ders-CR-4). Let $f(u, v, w) = (x, y, z)$ and $g(x, y, z) = (r, s, t)$ be differentiable functions, with $f(\vec{u}_0) = \vec{x}_0$. Suppose we are given the following elements of the matrices below:

$$J_{f, \vec{u}_0} = \begin{pmatrix} 1 & 0 & 3 \\ 2 & 4 & \\ -1 & & \end{pmatrix} \quad \text{and} \quad J_{g, \vec{x}_0} = \begin{pmatrix} 1 & 0 & \\ 4 & -2 & 2 \\ 1 & 0 & \end{pmatrix}$$

Compute the partial derivatives below if possible, or explain why there is not enough information.

1. $\frac{\partial s}{\partial u}(\vec{u}_0)$

2. $\frac{\partial r}{\partial u}(\vec{u}_0)$

3. $\frac{\partial r}{\partial v}(\vec{u}_0)$

Exercise 4.9.5 (**exer-Ders-CR-5**). Suppose we have the following information about the variables p, q, u, v, w, x, and y: $p = u - v$, $v = x - y$, $w = 3x + y$, $q = v - w^2$, and $u = 4x + 3y$. Compute $\frac{\partial q}{\partial y}$.

Exercise 4.9.6 (**exer-Ders-CR-6**). With variables defined as in Example 4.9.6, write out and simplify a formula for $\frac{\partial^2 z}{\partial r \partial \theta}$ in terms of the derivatives of z with respect to x and y.

Exercise 4.9.7 (**exer-Ders-CR-7**). With variables defined as in Example 4.9.6, write out and simplify a formula for $\frac{\partial^2 z}{\partial \theta^2}$ in terms of the derivatives of z with respect to x and y.

Exercise 4.9.8 (**exer-Ders-CR-8**). Suppose we have a parametric curve $\vec{x}(t)$ in $\mathbb{R}^3$, and a differentiable function $f : \mathbb{R}^3 \to \mathbb{R}^1$. Use the chain rule to write out a formula for $\frac{df}{dt}$ in terms of the derivatives of the individual functions f and $\vec{x}$.

Now use the gradient vector ∇f and the velocity vector $\vec{x}'$ to compute the directional derivative of f, again giving us $\frac{df}{dt}$. How does this compare to the chain rule formula in this case?

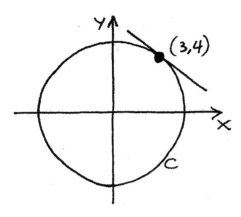

Figure 4.22:

4.10 Implicit Differentiation

Single variable calculus courses usually cover implicit differentiation as it applies to curves in the plane. We will begin this section with a review of that technique and a discussion of the theorem that makes it work. We will then move on to discussing an extension of that theorem to higher dimensional problems.

4.10.1 Single variable computations

Suppose we consider the solution set C in the xy-plane to the equation $x^2 + y^2 = 25$, and the particular point $(3, 4)$ on that curve C. How would we find the slope of the tangent line to C at that point? See Figure 4.22.

Of course we know that the derivative is a very convenient way to find the slope of a tangent line. But, critically, we must recognize that the derivative gives us the slope of the tangent line to the *graph* of a function. In this case, the curve C is not presented as the graph of a function; rather, it is presented simply as the solution set to an equation.

Further complicating the issue is the fact that C is a circle and thus fails the vertical line test. So it would seem that not only is the curve not presented as the graph of a function, but it cannot be expressed as the graph of a function.

Given these two impediments it would appear that we will not be able to use the derivative to find the desired slope. Fortunately though, there is a clever trick we can play that will allow us to proceed.

Note that even though we are given an entire curve, all we are really interested in is the tangent line at the given point. And the tangent line is entirely determined by the portion of the curve *near* the given point, so most of the curve C is irrelevant to the problem. In fact we can throw away most of C, and retain only a very small piece of curve which we will call C'. See Figure 4.23.

We can then state an equivalent problem – how can we find the slope of the tangent line to the curve C' at the given point $(3, 4)$?

The good news here is that the curve C' passes the vertical line test, and so we know that we can view C' as being the graph of a function. Because C' represents a portion of the curve C near $(3, 4)$, we will sometimes refer to this as saying that the curve C "is locally a graph near $(3, 4)$".

For future notational convenience we will call this function y. For this function, x is the input variable and y is the output variable. (We must acknowledge however that it is clumsy to use the same letter to denote both the function and the output variable.)

Of course we still do not have an explicit formula for this function and so we cannot take an explicit derivative. But, we know that the values for this function are determined *implicitly* by the given equation, $x^2 + y^2 = 25$. That is, for any

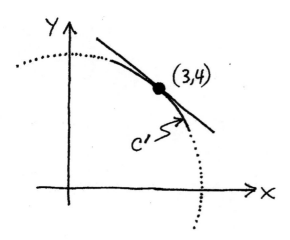

Figure 4.23:

value of x near 3 there is a unique corresponding point on C'; the y-coordinate of this point is $y(x)$.

Because the point $(x, y(x))$ is on the curve C', and every point on C' is also on C, we know that the point $(x, y(x))$ must satisfy the equation defining C. So for any value of x near 3, we have

$$x^2 + \Big(y(x)\Big)^2 = 25 \tag{4.43}$$

We can interpret each side of the above equation then as being entirely a function of x. We can then take the derivative, making sure to note that we must apply the chain rule. This gives us

$$2x + 2\Big(y(x)\Big)y'(x) = 0 \tag{4.44}$$

Of course simply having taken the derivative does not immediately give us the answer to the question, because of course the two sides of Equation 4.43 do not represent the function y whose graph is the curve C'. However, we note that having taken the above derivative, the desired y' ends up appearing in the resulting equation – and we can then simply solve for y'. We get

$$y'(x) = -\frac{x}{y(x)} \tag{4.45}$$

It is interesting to note that we now have an expression for the derivative of the function y, even though we never did find an expression for the function y itself.

Equation 4.45 gives us the slope that we were looking for in the first place, simply by evaluating y' when $x = 3$. This gives us

$$y'(3) = -\frac{3}{y(3)} \tag{4.46}$$

Even though we do not have an expression that allows us to compute $y(3)$, we do not need such an expression because by definition we know that $y(3) = 4$. So, we conclude that the desired slope is -3/4.

For notational convenience, the above computation can be performed without explicitly denoting y as a function of x, resulting in a computation that looks like

$$\begin{aligned} x^2 + y^2 &= 25 \\ 2x + 2yy' &= 0 \\ y' &= -\frac{x}{y} \\ y'\Big|_{(3,4)} &= -\frac{3}{4} \end{aligned}$$

Of course in order to do this computation correctly it is critical to remember when taking the derivative that y is considered a function of x, and that it is not an independent variable.

4.10. IMPLICIT DIFFERENTIATION

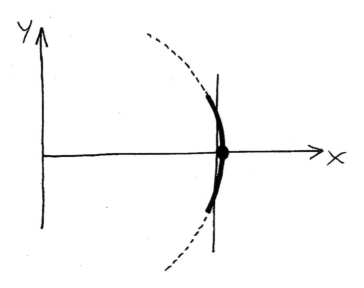

Figure 4.24:

4.10.2 Single variable implicit function theorem

In the above computation we made an assertion on which the computation depends critically, but for which we did not give a full justification.

That is, we asserted that even though the curve C' is not given explicitly as the graph of a function, it can nevertheless be viewed as the graph $y = y(x)$ of some function. We argued that this is true because the curve satisfies the vertical line test, but this claim depended on our incidental knowledge of this specific curve. Certainly we will not always be so lucky.

This is a weak point in the above computation. Not only is it not obvious that we can always do this, in fact it is not even true that we can always do this!

For example, suppose we consider instead the point $(5,0)$ on the curve C. The curve C is not locally a graph near this point. That is, no matter how small of a piece of C we take near this point, that piece of curve will *always* fail the vertical line test. See Figure 4.24.

In order for the computational strategy in Subsection 4.10.1 to be useful in situations where we are not already very familiar with the curve in question, we need a test that will allow us to determine when in fact we can view a curve, defined by some given equation, as being locally a graph near a given point.

The tool that allows us to answer this question is called the implicit function theorem. We will not prove this theorem in this text, but we will motivate it with the following line of reasoning.

We consider the following situation. Suppose we have an equation with a solution set C, and a point (x_0, y_0) on C.

First, note that Observation 2.4.1 tells us that whatever equation of two variables x and y we might be given, we can write that equation as

$$F(x, y) = 0 \qquad (4.47)$$

for some function $F : \mathbb{R}^2 \to \mathbb{R}^1$, where x and y are both input variables for this function. Of course then the solution set to this equation can be viewed as the level set $F^{-1}(0)$ for this function F.

(Make sure to note that this function F is NOT the same as the function y from Subsection 4.10.1. Clearly they are related to each other, but they are entirely different functions. For the function F, both x and y are input variables, whereas for the function y the variable x is an input and y is the output. The curve is a *level set* of F, while it is locally a *graph* of y.)

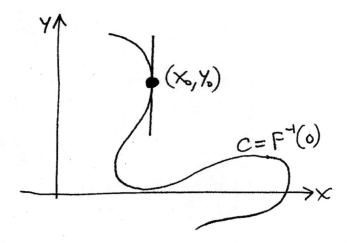

Figure 4.25:

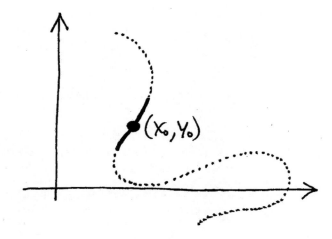

Figure 4.26:

Suppose now that the tangent line to C at (x_0, y_0) is vertical, as shown in Figure 4.25.

As we have the curve C drawn in that figure, note that the curve is not locally a graph. The problem here is that the curve "wraps over itself", loosely speaking – the same problem that we encountered with the point $(5, 0)$ on the circle. This is made possible because at the point (x_0, y_0) we are looking at, the tangent line is vertical.

If the tangent line were not vertical at the point (x_0, y_0), and if the curve were smooth, then the curve would simply not be able to wrap over or under itself immediately. So, over some sufficiently short distance, the curve would have to satisfy the vertical line test, thus making the curve locally a graph. See Figure 4.26.

This problem of having a vertical tangent line can be quickly associated to something that can be tested algebraically.

We know that level sets are always perpendicular to the gradient vector. So, a vertical tangent line at the point (x_0, y_0) for the level set $F^{-1}(0)$ is equivalent to saying that the gradient vector $\nabla F(x_0, y_0)$ must be horizontal, as is pictured in Figure 4.27.

This means that the y-component of the gradient vector is zero. Of course that y-component is $\frac{\partial F}{\partial y}(x_0, y_0)$.

So we conclude that if the curve has a vertical tangent line, then $\frac{\partial F}{\partial y}(x_0, y_0) = 0$. Said differently, if that partial derivative is not zero, then the curve does not have a vertical tangent line, and so we would expect for the curve to be locally a graph.

4.10. IMPLICIT DIFFERENTIATION

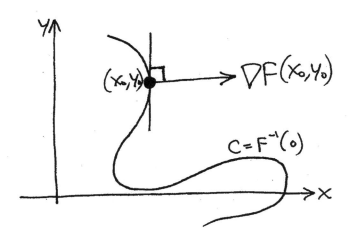

Figure 4.27:

In fact, for continuously differentiable functions F this is a correct result. We now state this as a theorem.

Theorem 4.10.1 (Implicit Function Theorem). *Suppose $F : \mathbb{R}^2 \to \mathbb{R}^1$ is continuously differentiable, with $F(x_0, y_0) = 0$ and $\frac{\partial F}{\partial y}(x_0, y_0) \neq 0$. Then near (x_0, y_0) on the curve C defined by $F(x, y) = 0$, the variable y can be thought of implicitly as a function $y = y(x)$ for values of x near x_0.*

This theorem allows us to give a more rigorous presentation of the example from Subsection 4.10.1.

Example 4.10.1. Let C be the curve in the xy-plane defined by the equation $x^2 + y^2 = 25$. What is the slope of the tangent line to this curve at the point $(3, 4)$?

We observe that the given equation can be written as $F(x, y) = 0$ where the function F is defined by $F(x, y) = x^2 + y^2 - 25$. We easily compute $\frac{\partial F}{\partial y} = 2y$, and so at the point in question we have

$$\left.\frac{\partial F}{\partial y}\right|_{(3,4)} = 8 \neq 0$$

This means that near the point $(3, 4)$, we can think of the curve C as being the graph of a function; that is, we can treat y as a function of x.

We can then compute as before.

$$\begin{aligned} x^2 + y^2 &= 25 \\ 2x + 2yy' &= 0 \\ y' &= -\frac{x}{y} \\ \left.y'\right|_{(3,4)} &= -\frac{3}{4} \end{aligned}$$

4.10.3 Multivariable implicit function theorem

We can generalize the ideas from the previous sections to the multivariable setting.

Suppose we have an equation $F(x_1, \ldots, x_n, y) = 0$ for a continuously differentiable function F. In the case where $n = 1$ of course we would expect this to define a curve in the plane, and similarly when $n = 2$ we expect this to define a surface in space. The multivariable implicit function theorem gives us a condition to test when we can view this level set as being the graph of a function.

Theorem 4.10.2 (Implicit Function Theorem). *Suppose $F : \mathbb{R}^{n+1} \to \mathbb{R}^1$ is continuously differentiable, with $F(a_1, \ldots, a_n, a_{n+1}) = 0$ and $\frac{\partial F}{\partial y}(a_1, \ldots, a_n, a_{n+1}) \neq 0$. Then near $(a_1, \ldots, a_n, a_{n+1})$, the equation $F(x_1, \ldots, x_n, y) = 0$ implicitly defines y as a function $y = y(x_1, \ldots, x_n)$ for values of $x_1, \ldots, x_n$ near $a_1, \ldots, a_n$.*

Here is a way to think of the multivariable implicit function theorem. Suppose we are given an equation of several variables, representing some sort of surface in the appropriate dimension space. Of course we can always view the equation as a level set of a function F, for which all of the given variables are input variables.

Suppose however that for one reason or another we wish to view one of those variables as being the *output* variable for a function of the other variables, near a given point on the level set. The implicit function theorem says that it is with respect to *that* variable that we must confirm the partial derivative of F is not zero.

We show the use of this theorem in the following example.

Example 4.10.2. Consider the surface in xyz-space defined by the equation $x^2 + y^4 + z^6 = 3$. At the point $(1, 1, 1)$ on this surface, what is the slope of the tangent line to the cross section of this surface perpendicular to the y-axis?

We quickly note that the answer to the question should be $\frac{\partial z}{\partial x}$, but as of the moment we cannot make sense of that because we do not know that z can in fact be written as a function of x and y near the given point.

Theorem 4.10.2 allows us to resolve this matter. We write the given equation as $F(x, y, z) = 0$ where $F(x, y, z) = x^2 + y^4 + z^6 - 3$. This function is continuously differentiable everywhere. Since z is the variable that we would like to view as a function of the other two variables, we must compute $\frac{\partial F}{\partial z}$ at the given point.

$$\frac{\partial F}{\partial z} = 6z^5$$

$$\left.\frac{\partial F}{\partial z}\right|_{(1,1,1)} = 6 \neq 0$$

This means that we can now view z as being a function of x and y near the given point.

Since we are interested in computing $\frac{\partial z}{\partial x}$, we will instead take the partial derivative with respect to x of the equation $x^2 + y^4 + z^6 = 3$ that we know the function z satisfies, and then solve for the derivative we are interested in. We get

$$2x + 0 + 6z^5 \frac{\partial z}{\partial x} = 0$$

$$\frac{\partial z}{\partial x} = -\frac{2x}{6z^5}$$

$$= -\frac{x}{3z^5}$$

$$\left.\frac{\partial z}{\partial x}\right|_{(1,1,1)} = -\frac{1}{3}$$

This is the desired slope.

Exercises

4.10. IMPLICIT DIFFERENTIATION

Exercise 4.10.1 (exer-Ders-ID-1). Consider the point $(3/2, 3/2)$ on the "Folium of Descartes", defined by the equation $x^3 + y^3 = 3xy$. Near this point on the curve, can we view y locally as a function of x? Near this point on the curve, can we view x locally as a function of y?

Exercise 4.10.2 (exer-Ders-ID-2). Consider the point $(\frac{1}{2}, \frac{1}{2\sqrt{3}})$ on the "Conchoid of de Sluze", defined by the equation $(x+1)(x^2+y^2) = 2x^2$. Near this point on the curve, can we view y locally as a function of x?

Exercise 4.10.3 (exer-Ders-ID-3). Find all of the points on the Conchoid of de Sluze from Exercise 4.10.2 where y cannot be viewed locally as a function of x.

Exercise 4.10.4 (exer-Ders-ID-4). Compute $\frac{dy}{dx}$ at the point in Exercise 4.10.2.

Exercise 4.10.5 (exer-Ders-ID-5). Show that at every point on the curve $2(x+y) = \sin(x-y)$ we can view y locally as a function of x, and find a formula for $\frac{dy}{dx}$ in terms of x and y.

Exercise 4.10.6 (exer-Ders-ID-6). Show that at every point on the surface $e^{z(x^2+y^2)} + e^{3z} = 5$ we can view z as being locally a function of x and y. Find formulas for $\frac{\partial z}{\partial x}$ and $\frac{\partial z}{\partial y}$ in terms of x, y, and z.

Exercise 4.10.7 (exer-Ders-ID-7). Near the point $(1, 3, 2)$ on the surface $x^2 y - yz^3 + 21 = 0$, determine if x can be viewed as a function of y and z. If so, compute $\frac{\partial x}{\partial z}$ at that point.

Exercise 4.10.8 (exer-Ders-ID-8). Near the point $(0, 5, 2)$ on the surface $3x^3 - 5z^2 + 2xyz = -20$, determine if y can be viewed as a function of x and z. If so, compute $\frac{\partial y}{\partial z}$ at that point.

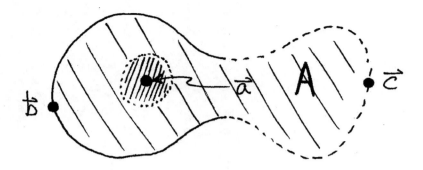

Figure 4.28:

4.11 Unconstrained Local Extrema

Optimization is the name given to the process of finding points in the domain that realize the extreme values of a function (these points are called "extrema", or more specifically, "maxima" or "minima"). This is one of the most natural and powerful applications of differentiation. Students have already seen many natural applications of optimization techniques in single variable calculus courses. Many of those same ideas can be adapted to handle multivariable optimization problems.

For example, suppose we are given a function $f : (D \subset \mathbb{R}^n) \to \mathbb{R}^1$. Very often, we will want to find a point in D that maximizes the value of this function f. Such a point is called an "absolute maximizer" or "global maximizer" of f. There is a process by which we can find the answer to this question, and we will discuss this process in Section 4.13.

In this section and the next, we consider a related question that is simpler to answer. Specifically, we will discuss how to find "local maximizers" for such a function f.

(Note that the techniques we will use to find maxima can be adapted very easily to find minimizers instead. We will omit these observations in most of these sections on optimization, leaving those extensions to the reader.)

4.11.1 Terminology

We first need to define several terms.

Definition 4.11.1. *Given a point $\vec{a} \in \mathbb{R}^n$ and a positive number r, we define the "open ball of radius r centered at $\vec{a}$" by*

$$B_r(\vec{a}) = \{\vec{x} \mid \|\vec{x} - \vec{a}\| < r\}$$

These open balls (which we will often casually refer to simply as "balls") will be the basic tool from which we will make several more definitions.

Definition 4.11.2. *Given a point $\vec{a}$ in a set A, we say that the point is an "interior point" of A if there exists a ball $B_r(\vec{a})$ around $\vec{a}$ which is entirely contained in A. The collection of all interior points of A is called the "interior of A", denoted by "int(A)".*

Definition 4.11.3. *Given a point $\vec{a}$, we say that the point is a "boundary point" of A if every ball $B_r(\vec{a})$ around $\vec{a}$ contains at least one point not in A <u>and</u> at least one point that is in A. (Note, a boundary point does NOT have to be in the set A itself.) The collection of all boundary points of A is called the "boundary of A", denoted by "bdry(A)".*

As examples of the above, Figure 4.28 shows a set A and points $\vec{a}$, $\vec{b}$ and $\vec{c}$. The point $\vec{a}$ is in the interior of A, and the points $\vec{b}$ and $\vec{c}$ are on the boundary of A. Note however that the point $\vec{c}$ is not actually in the set A.

Definition 4.11.4. *A set A is "open" if every point in A is an interior point.*

4.11. UNCONSTRAINED LOCAL EXTREMA

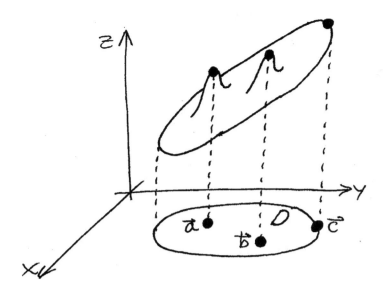

Figure 4.29:

Definition 4.11.5. *A set A is "closed" if it contains its entire boundary.*

Note that a connection between the above two definitions can be made by noting that a set is open if and only if it contains none of its boundary.

Students should be sure to note that these terms are not complementary. That is, it is not the case that every set must be either open or closed – rather, there are many sets that are neither open nor closed. For example, in light of the observation in the previous paragraph, easy examples can be generated simply by defining sets that contain only certain parts of their boundaries, such as the set A in Figure 4.28.

It turns out that each of these notions is important in the discussion of optimization.

Definition 4.11.6. *A set is "bounded" if there exists some ball of finite radius that entirely contains the set.*

The notion of a bounded set will be needed in Section 4.13.

Definition 4.11.7. *Given a set $D \subset \mathbb{R}^n$, a function $f : D \to \mathbb{R}^1$ and a point $\vec{a} \in D$, we say that $\vec{a}$ is a "local maximum point of f on D" if there exists a ball $B_r(\vec{a})$ such that for any $\vec{x} \in (B_r(\vec{a}) \cap D)$, we have $f(\vec{x}) \leq f(\vec{a})$.*

More casually, a local maximum point is a point in the domain for which the value of f is at least as big as that of any nearby point in the set – though not necessarily at least as big as values of the function at more distant points. In Figure 4.29, we show the graph of a function $f : (D \subset \mathbb{R}^2) \to \mathbb{R}^1$ and a few points in the domain D. The points $\vec{a}$, $\vec{b}$ and $\vec{c}$ are all local maximum points.

Note that the above definition does not make any distinction between points that are in the interior and points that are on the boundary. As it turns out, the tools that we will use to find local maxima are different in these two separate cases.

To distinguish these cases from each other, and to allow for a more convenient discussion of the tools we will use to find them, we make the following definition.

Definition 4.11.8. *Given a set $D \in \mathbb{R}^n$ and a point $\vec{a} \in D$, we say that the vector $\vec{v}$ is a "tangent vector to D at $\vec{a}$" if there is a differentiable parametric curve $p : (-1, 1) \to D$ for which $p(0) = \vec{a}$ and $p'(0) = \vec{v}$.*

We illustrate two examples of this in Figure 4.30. At the point $\vec{a}$, the indicated curve shows that the vector $\vec{v}$ is a tangent vector to D at $\vec{a}$. In fact, it is easily seen that from the point $\vec{a}$, all vectors are tangent to D. At the point $\vec{b}$, the indicated

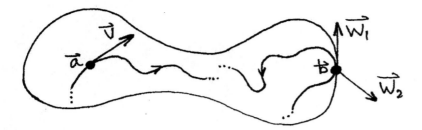

Figure 4.30:

curve shows that the vector $\vec{w}_1$ is tangent to D at $\vec{b}$. But there is no curve through $\vec{b}$ within the set D that is tangent to the vector $\vec{w}_2$, so the vector $\vec{w}_2$ is not tangent to D at $\vec{b}$.

It is fairly easy to see that at any interior point, all vectors are tangent. But for points on the boundary, only some vectors are tangent. That is, there are constraints on the tangent vectors from such points.

In this section we will consider points at which all vectors are tangent. We will deal with other kinds of points later.

Definition 4.11.9. *A point $\vec{a} \in D$ is an "unconstrained point" if all vectors are tangent at $\vec{a}$. An unconstrained point that is also a local maximum is called an "unconstrained local maximum".*

In Figure 4.29, the points $\vec{a}$ and $\vec{b}$ are unconstrained local maxima, but the point $\vec{c}$ is not.

Students might note that all interior points are unconstrained. However, not every unconstrained point is an interior point! (Consider the set D that is the union of the solid unit disk with the half plane $\{x \geq 1\}$; the reader can check that the point $(1, 0)$ is unconstrained, and that it is *not* in the interior.)

4.11.2 Critical points

Outline

The process by which we find unconstrained (interior) local maxima is a process of elimination. That is, we don't actually solve directly for such points – rather, we use particular criteria to rule out all but a small collection of candidates, and then consider each of these individually. These candidates which we cannot rule out with our criteria are referred to as "critical points".

It is important for the student to recognize that the most interesting thing about critical points is, very simply, that our active criteria fail at such points. So, upon finding a critical point, in fact we know very little about it. It is certainly not the case that a critical point must be an extremum. Further consideration will need to be given to draw any actual conclusions about these critical points.

With that in mind, let us proceed to consider the active criteria that we will use to rule out points in the domain as possible extrema.

Criteria for ruling out points

The criteria we use to rule out points as possible unconstrained extrema come from a very unsophisticated observation. In fact the same basic observation will be used to determine criteria in Section 4.12 for points that are not in the interior, though of course the details will turn out to be different.

Observation 4.11.1. *If at a point $\vec{a} \in D$ there is a tangent vector $\vec{v}$ such that $D_{\vec{v}} f(\vec{a}) \neq 0$, then the point $\vec{a}$ could not possibly be a local maximum.*

4.11. UNCONSTRAINED LOCAL EXTREMA

The point here is that if $D_{\vec{v}}f(\vec{a}) > 0$, then the function is increasing in the direction of $\vec{v}$, so that greater values of the function will certainly be found arbitrarily close to $\vec{a}$, thus showing it could not possibly be any sort of maximum. Similarly, if $D_{\vec{v}}f(\vec{a}) < 0$, then the function is increasing in the direction of $-\vec{v}$, so again $\vec{a}$ could not possibly be any sort of maximum.

To turn this observation into algebraic criteria that we can use in our search for unconstrained local maxima, we use Theorem 4.8.1, which tells us that for a differentiable function f, we have $D_{\vec{v}}f(\vec{a}) = \nabla f(\vec{a}) \cdot \vec{v}$. Because in this section we are considering only unconstrained points, we know all vectors are tangent at $\vec{a}$, and we can choose $\vec{v} = \nabla f(\vec{a})$, for which we then have $D_{\vec{v}}f(\vec{a}) = \|\nabla f(\vec{a})\|^2$. And of course if the gradient vector is itself not the zero vector, this result will not be zero.

(Alternatively, we could view this simply as a reminder that the gradient vector points in the direction in which the function is increasing the most quickly.)

So, we have the following tool for ruling out points as possible unconstrained local maxima.

Theorem 4.11.1. *If both*

1. *f is differentiable at $\vec{a} \in D$, and*
2. *$\nabla f(\vec{a}) \neq \vec{0}$,*

then $\vec{a}$ is not an unconstrained local maximum point of $f : D \to \mathbb{R}^1$

The great news is that for most problems that we will encounter in this course, the above two criteria are met for the overwhelming majority of points in the interior of the domain D. So, this theorem allows us to rule out most of those points.

The only points we cannot rule out with the above theorem are those for which at least one of the criteria fails. As we discussed at the beginning of this subsection, we label these remaining points as "critical points", to be analyzed individually.

We emphasize that in order to be a critical point, only one of the criteria in Theorem 4.11.1 needs to fail – because BOTH criteria are needed in order for the theorem to rule that point out as a possible local maximum.

Definition 4.11.10. *An "unconstrained critical point" of the function f on the domain D is any unconstrained point in D (any point in int(D)) for which either*

1. *f is not differentiable at $\vec{a} \in D$, or*
2. *$\nabla f(\vec{a}) = \vec{0}$,*

Using this language, Theorem 4.11.1 states simply that the unconstrained critical points are the only possible unconstrained local maxima in D.

Example 4.11.1. **We consider now the function $f(x, y) = x^3 - 3x + y^2$, defined on all of $\mathbb{R}^2$, and want to find the unconstrained local maxima. Of course since the domain is all of $\mathbb{R}^2$, all points are unconstrained.**

Of course we will not be able at this point to draw any confident conclusions about local maxima. All we can do is use Theorem 4.11.1 hopefully to rule out most points, and note that the remaining points are candidates that will require further study. Said differently, all we can do is find the unconstrained critical points.

To do this, first we must find all points for which f is not differentiable. Of course f is a polynomial, and we know that all polynomials are differentiable everywhere. So, this possibility does not lead to any critical points.

Next we must find all points for which $\nabla f = 0$. Simple computation gives us

$$\nabla f = \begin{bmatrix} 3x^2 - 3 \\ 2y \end{bmatrix}$$

In order for this gradient vector to be the zero vector, we must have both $3x^2 - 3 = 0$ and $2y = 0$. Fortunately, this system of equations is easy to solve, giving us $x = \pm 1$ and $y = 0$.

So, there are exactly two critical points:

$$\vec{a}_1 = \begin{bmatrix} 1 \\ 0 \end{bmatrix} \quad \text{and} \quad \vec{a}_2 = \begin{bmatrix} -1 \\ 0 \end{bmatrix}$$

We will discuss in Subsection 4.11.3 how to identify what is actually happening with the function at these critical points.

Example 4.11.2. Now let us consider the function $f(x,y,z) = \cos(\pi x) + \cos(\pi y) + \cos(\pi z)$, defined only on the closed unit ball in $\mathbb{R}^3$, defined by

$$B = \{\vec{x} \mid \|\vec{x}\| \leq 1\}$$

Again we wish to find the unconstrained local maxima, but for the moment have to be satisfied with finding only unconstrained critical points.

First we try to find points where f is not differentiable, but again there are no such points for this function.

Next, we again look for points where $\nabla f = 0$. Simple computation gives us

$$\nabla f = \begin{bmatrix} -\pi \sin(\pi x) \\ -\pi \sin(\pi y) \\ -\pi \sin(\pi z) \end{bmatrix}$$

Setting this equal to zero tells us that x, y and z must all be integers.

At first glance, it might appear that we have found an infinite number of unconstrained critical points for this function. But we must remember that of course an unconstrained critical point must actually be in the domain, which most of the above points are not. Furthermore, an unconstrained critical point must also in fact be unconstrained – that is, it must be in the interior of the given domain.

With these observations, we find that in fact there is exactly one unconstrained critical point for this function on the given domain, namely

$$\vec{a} = \begin{bmatrix} 0 \\ 0 \\ 0 \end{bmatrix}$$

(The point $(1,0,0)$ (and five other similar points) satisfies $\nabla f = 0$ and is in the domain – but it is not in the interior of the domain, and therefore it is not unconstrained.)

In order to identify what the function is actually doing at this point, we make a crafty observation. That is, we note that the value of f at this critical point is $f(0,0,0) = 3$, and furthermore that 3 is the largest possible value of f anywhere because we know that the cosine function is always ≤ 1. So this point must be a local maximum. (Note, we will usually not get this lucky when working optimization problems, but it is always a good idea to keep one's eyes open for easy tricks like this.)

Types of critical points

Having found a critical point for a multivariable function, we must acknowledge that, just as in the case of single variable functions, there are several possibilities of what the function might be doing near that point.

4.11. UNCONSTRAINED LOCAL EXTREMA

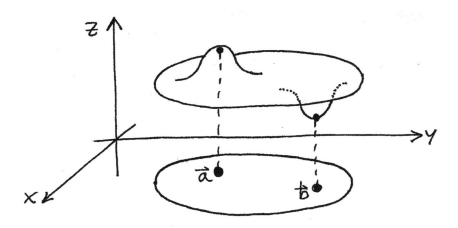

Figure 4.31:

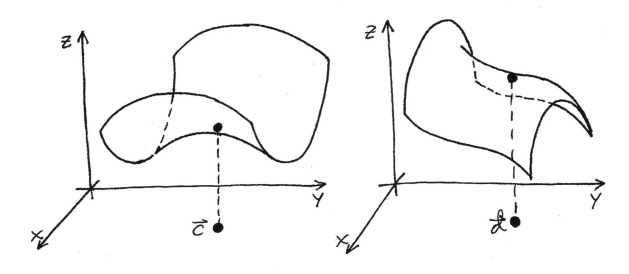

Figure 4.32:

Of course it is possible that the critical point might be a local maximum or a local minimum, as are the points $\vec{a}$ and $\vec{b}$ respectively in Figure 4.31.

But these are not the only possibilities. In single variable calculus, a third possibility was that the point might be an inflection point, and therefore neither a maximum or a minimum. There are several such analogues in the case of multivariable functions. In Figure 4.32, the first graph shows what is called a "saddle point", because the nature of the curvature is similar to that of a saddle, curving up in one direction but down in the perpendicular direction. The second shows what one might call a "shoulder point", because the graph resembles the shape of a person's shoulder.

There are many other such possibilities which we cannot enumerate. But they all have the simple property that they are neither maxima nor minima, and so we lump them all together with the term "saddle point".

We will see in Subsection 4.11.3 how to classify unconstrained critical points as one of the above.

4.11.3 Second derivative test

Readers will have noticed at this point that there are many strong analogies between multivariable optimization and single variable optimization. Along these lines, just as in the case of single variable optimization, there is a second derivative test for multivariable unconstrained critical points.

It should be noted however that the multivariable second derivative test is quite different from the single variable test that students are familiar with. Specifically, it is just not as simple as taking a second derivative and noting whether it is positive, negative or zero. In fact, the full statement of the second derivative test for multivariable functions will not be given in this text. Rather we will state the test only for functions of two variables.

We will however give a bit of an outline for the motivation for the test, and note the continued analogy with single variable functions.

Recall that for single variable functions there is a second order Taylor polynomial, which, for twice differentiable functions, provides good approximations to the function near some point a. That second order Taylor polynomial is

$$T_2(x) = f(a) + f'(a)(x-a) + \tfrac{1}{2}f''(a)(x-a)^2$$

Students can directly confirm that this function T_2 shares the same value, derivative, and second derivative as the original function f at the point a, so it is not surprising that this function gives good approximations near a.

This second order Taylor polynomial gives a plausibility argument for the single variable second derivative test. That is, suppose we have a critical point for f at the point a. That means that $f'(a) = 0$, so the second order Taylor polynomial at this critical point becomes

$$T_2(x) = f(a) + \tfrac{1}{2}f''(a)(x-a)^2$$

Of course $(x-a)^2$ is always positive, so if the second derivative $f''(a)$ is positive then this estimate suggests that the function is always larger than $f(a)$ near the point a, and thus that the critical point is a local minimum. Similarly, if $f''(a)$ is negative then the second term above is negative, suggesting that the function is always smaller than $f(a)$ near the point a, and thus that the critical point is a local maximum.

This same sort of motivational argument can be made for a multivariable function. The form of the second order Taylor polynomial is slightly different, but still highly analogous in certain ways. For a twice continuously differentiable function f, near a point $\vec{a}$, the multivariable second order Taylor polynomial is

$$T_2(\vec{x}) = f(\vec{a}) + \nabla f(\vec{a}) \cdot (\vec{x} - \vec{a}) + \tfrac{1}{2}(\vec{x} - \vec{a}) \cdot H_{\vec{a}}(\vec{x} - \vec{a})$$

where H is the matrix of second derivatives of f, called the "Hessian matrix", given by

$$H_{\vec{a}} = \begin{pmatrix} \frac{\partial^2 f}{\partial x_1^2}(\vec{a}) & \cdots & \frac{\partial^2 f}{\partial x_1 \partial x_n}(\vec{a}) \\ \vdots & & \vdots \\ \frac{\partial^2 f}{\partial x_n \partial x_1}(\vec{a}) & \cdots & \frac{\partial^2 f}{\partial x_n^2}(\vec{a}) \end{pmatrix}$$

(The Hessian matrix is always symmetric, because of the equality of mixed partials for these functions.) To clarify the notation, in the second order term, the Hessian matrix is being evaluated at the point $\vec{a}$, and then it is matrix-vector multiplied by the vector $(\vec{x} - \vec{a})$, before finally a dot product with $(\vec{x} - \vec{a})$.

Note that in comparison to the single variable version, the first terms are practically identical; in the second terms the gradient plays a role very similar to that of the single variable derivative; and in each of the third terms there is a quadratic polynomial that involves the second derivative(s) of f as coefficients.

Again, students can check directly that the above multivariable second order Taylor polynomial has the same value, derivatives and second derivatives as f at the point $\vec{a}$.

As in the single variable case, this second order Taylor polynomial gives a plausibility argument for what we will eventually call the multivariable second derivative test. Suppose we have a critical point for f at the point $\vec{a}$. That means that $\nabla f(\vec{a}) = \vec{0}$, so the second order Taylor polynomial at this critical point becomes

$$T_2(\vec{x}) = f(\vec{a}) + \tfrac{1}{2}(\vec{x} - \vec{a}) \cdot H_{\vec{a}}(\vec{x} - \vec{a})$$

4.11. UNCONSTRAINED LOCAL EXTREMA

Just as in the single variable case then, the question comes down to determining under what circumstances that second order term is either always positive or always negative.

This is where the analogy breaks down. In the single variable case we were lucky that the second order term consisted actually of a single term, and so a simple consideration of the signs of the factors in that term determined the answer to the question. In the multivariable case, there a multiple terms generated by the dot product and the Hessian matrix, so such a simple analysis could not possibly work.

As it happens, the second order term above is an example of something called a "quadratic form", which can be analyzed with techniques of linear algebra slightly beyond what we have presented in Chapter 3. In fact there is a precise set of conditions with which we can determine whether a quadratic form is always positive, or always negative, or neither. The full statement of these conditions is tedious, and will not be presented in this text.

The theorem about quadratic forms allows us to come up with a multivariable second derivative test. We will state it here for functions of two variables only. (We will not state the result for functions of more than two variables.)

Theorem 4.11.2. *Suppose $f : \mathbb{R}^2 \to \mathbb{R}^1$ is a twice differentiable function of x and y, and $\nabla f(\vec{a}) = \vec{0}$. Then*

1. *If $\det H > 0$ and $\frac{\partial^2 f}{\partial x^2}(\vec{a}) < 0$, then $\vec{a}$ is a local maximum point for f.*

2. *If $\det H > 0$ and $\frac{\partial^2 f}{\partial x^2}(\vec{a}) > 0$, then $\vec{a}$ is a local minimum point for f.*

3. *If $\det H < 0$, then $\vec{a}$ is a saddle point for f.*

4. *If none of the above conditions holds, then this test fails to draw any conclusion.*

Example 4.11.3. Let's go back to Example 4.11.1, where we were considering the function $f(x,y) = x^3 - 3x + y^2$, defined on all of $\mathbb{R}^2$, and we had found the two critical points

$$\vec{a}_1 = \begin{bmatrix} 1 \\ 0 \end{bmatrix} \quad \text{and} \quad \vec{a}_2 = \begin{bmatrix} -1 \\ 0 \end{bmatrix}$$

Using Theorem 4.11.2, we can now classify these critical points. First we compute the determinant of the Hessian matrix,

$$H = \begin{pmatrix} 6x & 0 \\ 0 & 2 \end{pmatrix} \quad \Longrightarrow \quad \det H = 12x$$

We can evaluate this at each of the above critical points, giving us

$$\det H(\vec{a}_1) = 12 \quad \text{and} \quad \det H(\vec{a}_2) = -12$$

Theorem 4.11.2 then immediately tells us that $\vec{a}_2$ is a saddle point. To classify $\vec{a}_1$, we compute $\frac{\partial^2 f}{\partial x^2} = 6x$, giving us $\frac{\partial^2 f}{\partial x^2}(\vec{a}_1) = 6 > 0$, so Theorem 4.11.2 tells us that $\vec{a}_1$ is a local minimum.

In Example 4.11.3, Theorem 4.11.2 gave us the desired classification. However, we will not always be so fortunate.

Example 4.11.4. Suppose we consider the function $f(x,y) = y^2 - x^2 y$. The gradient is

$$\nabla f = \begin{bmatrix} -2xy \\ 2y - x^2 \end{bmatrix}$$

and it is easily seen that there is a critical point at the origin. Computing the Hessian gives us

$$H = \begin{pmatrix} -2y & -2x \\ -2x & 2 \end{pmatrix} \quad \text{and} \quad \det H(\vec{0}) = 0$$

Since this is neither positive nor negative, Theorem 4.11.2 fails to give us any information about the classification of this critical point.

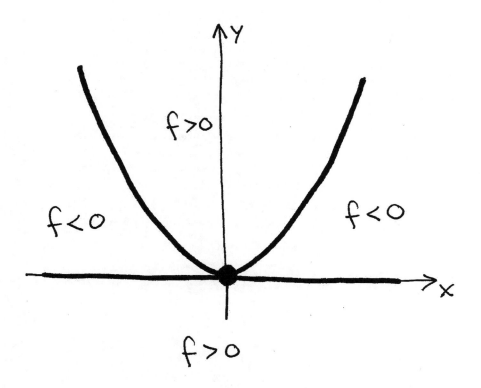

Figure 4.33:

When the second derivative test fails, sometimes another method can be employed.

Example 4.11.5. The value of the function f from Example 4.11.4 is zero at the critical point at the origin. Noting that f factors as $f(x,y) = (y)(y - x^2)$ and that f is equal to zero when either of these factors is zero, we conclude that the level set $f = 0$ includes both the line $y = 0$ and the curve $y = x^2$, both of which pass through the origin.

These curves divide the remainder of the plane into four distinct regions, as shown in Figure 4.33.

In the region above the parabola, both of the factors (y) and $(y-x^2)$ are positive, so their product $f(x,y) = (y)(y-x^2)$ is positive. Similarly, in the region below the x-axis both of the factors (y) and $(y-x^2)$ are negative, so their product $f(x,y) = (y)(y-x^2)$ is positive.

However, in the two other regions, the factor (y) is positive while the factor $(y - x^2)$ is negative – making the function $f(x,y) = (y)(y - x^2)$ negative in those regions.

Since the critical point at the origin is contiguous with regions where the function has greater values and also with regions where the function has lesser values, it cannot be either a local maximum or a local minimum. Therefore it must be a saddle point.

It is tempting to try to turn the multivariable classification problem discussed in Examples 4.11.4 and 4.11.5 into a single variable problem by considering the value of the function along all straight lines through the critical point at the origin; the temptation is to believe that if the function along each of these lines has the same concavity, then surely that must suggest the correct classification. Unfortunately this is NOT true. As is the case with multivariable limits, it is not possible to determine the behavior of a multivariable function merely by looking along all straight lines through the point.

Example 4.11.6. Considering again the function from Examples 4.11.4 and 4.11.5, we have already seen that on the line $y = 0$ the value of the function is $f = 0$. It is easily checked that on the line $x = 0$, the function is $f = y^2$, which is concave up at the origin. We can look along all other lines through the origin by plugging in $y = mx$ with $m \neq 0$, which gives us
$$f = m^2 x^2 - mx^3$$
and compute that the second derivative at the origin along such a line is $f''(0) = 2m^2 \geq 0$.

So, along every line through the origin, the function is either identically zero or concave up. It is tempting to conclude then that the critical point at the origin is a local minimum.

Of course we know from the analysis in Example 4.11.5 that this is not true, and that the critical point at the origin is a saddle point. Using the idea of approaching along a curve, this can be detected by considering the function on the curve $y = x^2/2$, on which the function simplifies as $f = -x^4/4$.

The lesson here then is that it is simply not sufficient to consider the behavior of the function along lines. Students will note a strong similarity between this lesson and the analogous lesson from the discussion of multivariable limits in Section 2.7.

Exercises

Exercise 4.11.1 (exer-Ders-ULE-1). For each of the following points, determine if the point is in the interior of the given set. If yes, then find an open ball around that point that satisfies Definition 4.11.2.

1. $\vec{a} = (1/2, 0)$, S is the closed unit ball centered at the origin.
2. $\vec{a} = (1, 0)$, S is the closed unit ball centered at the origin.
3. $\vec{a} = (1/2, 0)$, S is the part of the closed unit ball centered at the origin that is to the left of the y-axis.
4. $\vec{a} = (0.4, 0.9)$, S is the unit square (defined by $0 \leq x \leq 1$, $0 \leq y \leq 1$).

Exercise 4.11.2 (exer-Ders-ULE-2). For each of the items in Exercise 4.11.1, determine if the point is on the boundary of the given set.

Exercise 4.11.3 (exer-Ders-ULE-3). For each of the following sets, determine (a) if it is closed, and (b) if it is bounded.

1. The part of the xy-plane with $x \geq 0$.
2. The part of the xy-plane with $x > 0$.
3. The part of the xy-plane with $x \geq 0$, $y \geq 0$, and $xy \leq 1$.
4. The part of the xy-plane with $x \geq 0$, $y \geq 0$, and $x^2 + y^2 \leq 1$.
5. The part of the xyz-space with $x \geq 0$, $y \geq 0$, and $x^2 + y^2 \leq 1$.
6. The part of the xyz-space with $e^x + e^y + e^z \leq 1$.

Exercise 4.11.4 (exer-Ders-ULE-4). For each of the following, determine if the given vector $\vec{v}$ is tangent at the given point $\vec{a}$ in the given set S. If it is tangent, indicate a curve that satisfies the condition described in Definition 4.11.8.

1. S is the xy-plane, $\vec{a} = (1, 2)$, $\vec{v} = (6, 5)$.

2. S is the part of the xy-plane with $y \geq 0$, $\vec{a} = (3, 0)$, $\vec{v} = (1, 0)$.

3. S is the part of the xy-plane with $y \geq 0$, $\vec{a} = (3, 0)$, $\vec{v} = (0, 1)$.

4. S is the closed unit disk, $\vec{a} = (1, 0)$, $\vec{v} = (0, 1)$.

5. S is the closed unit disk, $\vec{a} = (1, 0)$, $\vec{v} = (1, 1)$.

6. S is the closed unit disk, $\vec{a} = (1, 0)$, $\vec{v} = (-1, 1)$.

7. S is the closed unit ball in $\mathbb{R}^3$, $\vec{a} = (1, 0, 0)$, $\vec{v} = (0, 1, 1)$. (Hint: Think about using spherical coordinates with $\rho = 1$, and θ and ϕ being well-chosen functions of t.)

Exercise 4.11.5 (**exer-Ders-ULE-5**). For each of the functions below, find all of the unconstrained critical points.

1. $f(x, y) = 3x^2 - 2y^2$.
2. $f(x, y) = xy - y^3$.
3. $f(x, y) = x^2 y - y^4$.
4. $f(x, y) = \sin(2\pi x) + \cos(2\pi y)$.
5. $f(x, y, z) = x^3 - 4y^3 + z$.
6. $f(x, y) = \sqrt{x^{2/3} + y^{2/3}}$.

Exercise 4.11.6 (**exer-Ders-ULE-6**). For each of the items below, find the second order Taylor polynomial for the given function f at the given point $\vec{a}$. Also, determine if the given point $\vec{a}$ is an unconstrained critical point of the given function f, and if it is, classify the critical point as being either a local max, a local min, or a saddle point.

1. $f(x, y) = x^2 - y^2$, $\vec{a} = (0, 0)$.
2. $f(x, y) = x^2 - y^2$, $\vec{a} = (3, 2)$.
3. $f(x, y) = 2x^3 - 6y^3 + 3xy$, $\vec{a} = (0, 0)$.
4. $f(x, y) = x^4 - x^2 - 2x - 3y^2 + x^2 y^2$, $\vec{a} = (-1, 0)$.

Exercise 4.11.7 (**exer-Ders-ULE-7**). In this problem we consider the function $f(x, y) = (x^2 + y^2 - 1)(y - x^2 - 1)$ and the point $\vec{a} = (0, 1)$.

1. Show that $\vec{a}$ is an unconstrained critical point of f.

2. Show that the second derivative test fails to classify this critical point.

3. Use the given factored form of f to show that the set where $f = 0$ is the union of a circle and a parabola, and draw this set.

4. For points in the xy-plane that are above the parabola, what can you say about the sign of $(y - x^2 - 1)$? For those same points, what can you say about the sign of $(x^2 + y^2 - 1)$, and therefore what can you say about the sign of f?

5. What can you say about the sign of f inside the circle?

6. What can you say about the sign of f at points that are below the parabola but outside of the circle?

7. Using the results above, decide what is the correct classification of this critical point.

4.12 Constrained Local Extrema and Lagrange Multiplier Theorems

In this section we consider points in the domain that are not unconstrained.

Definition 4.12.1. *A point $\vec{a} \in D$ is a "constrained point" if there is a vector that is not tangent to D at $\vec{a}$. A constrained point that is also a local maximum is called an "constrained local maximum".*

The search for constrained local extrema has the same basic outline as that for unconstrained local extrema. That is, it is a process of elimination, where we establish criteria by which we can rule out the majority of the points in the domain, and label the remaining points as (constrained) critical points.

(However, we will not discuss any sort of second derivative test for constrained extrema, so we will be left to classify the extrema by other means, if at all. We will see in Section 4.13 a strong use for these constrained critical points even if we do not classify them.)

The criteria we use to rule points out in this case is based on the same observation that was used for the unconstrained case, namely Observation 4.11.1. That is, if at a point $\vec{a} \in D$ there is a tangent vector $\vec{v}$ such that $D_{\vec{v}} f(\vec{a}) \neq 0$, then the point $\vec{a}$ could not possibly be a local maximum.

In the unconstrained case, we turned this into an algebraic form by making the convenient choice $\vec{v} = \nabla f$. We were free to do this because the point under consideration was by assumption unconstrained, so we could choose $\vec{v}$ to be any vector. But in this case we are not so fortunate. We are considering constrained points in the domain, and the gradient vector might not be tangent at a given point. We must make sure to choose a tangent vector, and yet we also want to choose a vector that will let us draw a useful algebraic conclusion.

There is a crafty approach by which we can make such a choice. The nature of the details depends on the specifics of the constraints, so we will have to consider separate cases.

4.12.1 Single constraint critical points

We first turn our attention to finding constrained local maximum points where there is a single equation that defines the constraint on the given point. Specifically then, we wish to find local extrema of the function f, where the domain D is defined by a single equation $g = 0$.

Students should note very carefully at this point that there are now two functions under consideration, and they will both play important roles in the techniques in this section. First, most obviously, there is the function f which we are trying to optimize. We call this the "objective function".

However there is also the function g that is used to write down the equation defining the domain D. We are not trying to optimize g at all, nor would this make any sense given that it is by definition zero on the entire domain. Still, the function g will have an important role. We call it the "constraint function".

Before we get started, let's note that the domain D is in fact a level set of this constraint function g that is used to write down the equation. Geometrically then, if the problem involves two variables, we expect D to be a curve in $\mathbb{R}^2$; and if the problem involves three variables, we expect D to be a surface in $\mathbb{R}^3$. In either case, as a level set of g, we recall from Section 4.8 that at any point $\vec{a} \in D$, the gradient of the g (if it is not the zero vector, of course) must be perpendicular to the domain D. This is represented in Figure 4.34.

In the derivation below we will specifically address the question of functions of three variables, where the constraint set is a surface. Students should note that these arguments can readily be adapted to two variable problems, and are encouraged subsequently to work through those details.

We now begin to describe our strategy for choosing a tangent vector $\vec{v}$ which can be used with Observation 4.11.1 to rule out the point $\vec{a}$ as a possible maximum.

We have bemoaned already the fact that we cannot choose $\vec{v} = \nabla f$ because it need not be tangent; instead, we will in a very meaningful sense choose the tangent vector that is the closest to that we can get. Specifically, as indicated in Figure

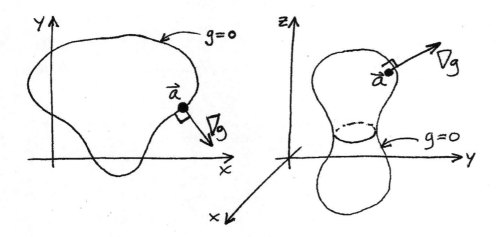

Figure 4.34:

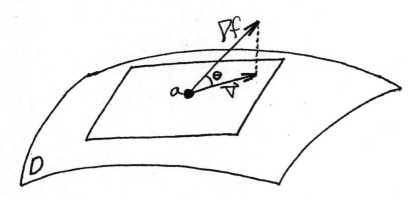

Figure 4.35:

4.35, we choose $\vec{v}$ to be the projection of ∇f to the plane tangent to the surface at $\vec{a}$. Note that this vector $\vec{v}$ is tangent to the surface D.

Computing the derivative and applying trigonometry in Figure 4.35, this gives us

$$D_{\vec{v}}f(\vec{a}) = \nabla f \cdot \vec{v} = \|\nabla f\|\|\vec{v}\|\cos\theta = \|\nabla f\|^2 \cos^2\theta$$

This is nonzero whenever $\nabla f \neq \vec{0}$ and $\theta \neq \pi/2$. That is, whenever ∇f is nonzero and not perpendicular to the surface D. This gives us a theorem analogous to Theorem 4.11.1.

Theorem 4.12.1. *If all three of the below conditions hold,*

1. *f is differentiable at $\vec{a} \in D$, and*
2. *$\nabla f(\vec{a}) \neq \vec{0}$, and*
3. *$\nabla f(\vec{a}) \not\perp D$*

then $\vec{a}$ is not a constrained local maximum point of $f : D \to \mathbb{R}^1$

As in the unconstrained case, most points in the domain will satisfy these conditions, and so this theorem rules out most of the points in the domain.

4.12. CONSTRAINED LOCAL EXTREMA AND LAGRANGE MULTIPLIER THEOREMS

Again, any point that fails to satisfy any one of the above criteria cannot be ruled out by the theorem. Again, we call these critical points for f on D.

Definition 4.12.2. *A "constrained critical point" of the function f on the domain $D = \{g = 0\}$ is any point in D for which* <u>either</u>

1. *f is not differentiable at $\vec{a} \in D$, or*
2. *$\nabla f(\vec{a}) = \vec{0}$, or*
3. *$\nabla f(\vec{a}) \perp D$*

Using this language, Theorem 4.12.1 states simply that the constrained critical points are the only possible constrained local maxima in D.

The ideas we have developed for constrained maxima so far are mostly analogous to what we did in the unconstrained case. However, in this constrained case we have found ourselves with a description of critical points that is not algebraically convenient. That is, we do not have a particularly convenient way algebraically to write that $\nabla f(\vec{a}) \perp D$.

The good news is that, because there is only a single constraint defining D, we expect for there to be only a single direction that is perpendicular to D. So, saying $\nabla f(\vec{a}) \perp D$ is the same as saying that ∇f is parallel to that direction.

Even better, we already have a vector that we know points in that direction. As was observed near the beginning of this subsection and represented in Figure 4.34, the gradient of g itself is always perpendicular to D, because D is a level set of g.

Combining these two observations, we note that $\nabla f \perp D$ is basically the same as saying $\nabla f \parallel \nabla g$, which we represent algebraically with
$$\nabla f = \lambda \nabla g \tag{4.48}$$
where λ is some scalar, referred to commonly in this context as a "multiplier". We say "basically" here because there are a couple of special cases that we have to deal with, involving zero vectors.

First, if $\nabla f = \vec{0}$, then Equation 4.48 is clearly satisfied (simply by choosing $\lambda = 0$), even though the lack of a well defined direction would make it awkward to claim that $\nabla f \perp D$. But this is good news – because $\nabla f = \vec{0}$ is of course another one of the conditions from Definition 4.12.2 that identifies a critical point. So, it would appear that Equation 4.48 actually replaces two of the conditions in that definition.

But second, we must consider what happens if $\nabla g = \vec{0}$. In that case, Equation 4.48 is not possible – but, of course it is still entirely possible that we might have $\nabla f \perp D$. That is, a point $\vec{a}$ with $\nabla g(\vec{a}) = \vec{0}$ should still be considered as a critical point, even though such a case is not represented in Equation 4.48. So we have to add this in as a new separate condition if we plan to use Equation 4.48.

The following theorem then is effectively a restatement of Definition 4.12.2.

Theorem 4.12.2 (Lagrange Multiplier Theorem). *Any constrained critical point (and therefore any constrained local maximum) of the function f on the domain $D = \{g = 0\}$ must be a point $\vec{a}$ on D for which* <u>either</u>

1. *f is not differentiable at $\vec{a} \in D$, or*
2. *$\nabla g(\vec{a}) = \vec{0}$, or*
3. *$\nabla f(\vec{a}) = \lambda \nabla g(\vec{a})$*

Condition (2) of Theorem 4.12.2 is sometimes called the "degeneracy condition", and condition (3) is called the "Lagrange condition". The advantage of Theorem 4.12.2 over Definition 4.12.2 is that these last two conditions are algebraic, not geometric, and thus they are easier to work with.

Condition (3) of Theorem 4.12.2 is represented in Figure 4.36.

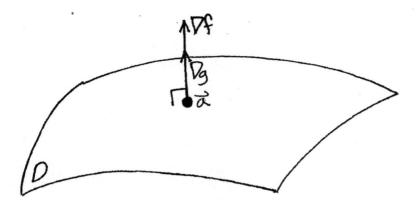

Figure 4.36:

Example 4.12.1. Suppose we want to find critical points for the function $f(x,y,z) = x^2 + y^2 + z^2$, on the domain D defined by $6x + 3y + 2z = 49$.

This domain is a level set of the function $g(x,y,z) = 6x + 3y + 2z - 49$.

We use Theorem 4.12.2 to check for critical points. First, we consider if there might be points where f is not differentiable, but since f is a polynomial we know such points cannot exist.

Second, we look at the degeneracy condition. But we have

$$\nabla g = \begin{bmatrix} 6 \\ 3 \\ 2 \end{bmatrix}$$

which is clearly never equal to zero. So again, this gives us no critical points.

Finally we consider the Lagrange condition, which turns into

$$\begin{bmatrix} 2x \\ 2y \\ 2z \end{bmatrix} = \lambda \begin{bmatrix} 6 \\ 3 \\ 2 \end{bmatrix}$$

and which we can then rewrite as three scalar equations by considering each component separately,

$$\begin{aligned} 2x &= 6\lambda \\ 2y &= 3\lambda \\ 2z &= 2\lambda \end{aligned}$$

This appears to be a system of three equations with four unknowns; in fact it appears that we could let λ be any value whatsoever, and then solve directly for x, y and z. But very importantly, we must recognize that this system is relevant ONLY for points which are, in fact, on the domain D. That is, we must include the equation $g = 0$ defining D with the above system, giving us

$$\begin{aligned} 2x &= 6\lambda \\ 2y &= 3\lambda \\ 2z &= 2\lambda \\ 6x + 3y + 2z &= 49 \end{aligned}$$

Now we have four equations and four unknowns, and we can solve this system with a variety of approaches. We could eliminate λ by solving for it in each of the first three equations and equating those results. Or, we could note

4.12. CONSTRAINED LOCAL EXTREMA AND LAGRANGE MULTIPLIER THEOREMS

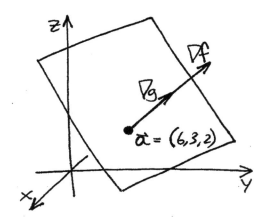

Figure 4.37:

that in this case we can solve each of the first three equations for x, y or z (respectively) in terms of λ and plug that into the fourth equation to solve for λ. Certainly there are many other variations.

Here we will take the former approach. Solving for λ in each of the first three equations above, we get

$$\lambda = x/3 = 2y/3 = z \implies y = x/2 \text{ and } z = x/3$$

Plugging these into the fourth equation above gives us

$$6x + 3(x/2) + 2(x/3) = 49 \implies x = 6 \implies y = 3 \text{ and } z = 2$$

So we have a single critical point, at $(6, 3, 2)$. We do not have any algorithmic means for identifying what kind of critical point this is, but since the function f approaches infinity as we move out in any direction on the plane, we can see that this is a local minimum. This is represented in Figure 4.37.

It does turn out often that the algebra in the Lagrange condition of constrained optimization problems can be a bit tricky, but the good news is that most of the time the tricks are similar. Students should generally try to eliminate a variable (as we did to λ in the above example), or if possible to solve explicitly for a variable if that opportunity presents itself.

Example 4.12.2. Suppose we want to find critical points for the function $f(x, y, z) = x^2 + y^2 + z^2$, on the ellipsoid D defined by $x^2 + 4y^2 + 9z^2 = 36$.

This domain is a level set of the function $g(x, y, z) = x^2 + 4y^2 + 9z^2 - 36$.

We use Theorem 4.12.2 to check for critical points. First, we consider if there might be points where f is not differentiable, but since again f is a polynomial we know such points cannot exist.

Second, we look at the degeneracy condition. We have

$$\nabla g = \begin{bmatrix} 2x \\ 8y \\ 18z \end{bmatrix}$$

This vector can be zero, at the point $(x, y, z) = (0, 0, 0)$. But that point does not satisfy the constraint $g = 0$, so it does not count as a constrained critical point. So the degeneracy condition again gives us no critical points.

Finally, the Lagrange condition turns into

$$\begin{bmatrix} 2x \\ 2y \\ 2z \end{bmatrix} = \lambda \begin{bmatrix} 2x \\ 8y \\ 18z \end{bmatrix} \qquad \Longrightarrow \qquad \begin{aligned} 2x &= \lambda(2x) \\ 2y &= \lambda(8y) \\ 2z &= \lambda(18z) \end{aligned}$$

We can deal with this system by cases. First, note that if $x \neq 0$, then the first equation above says we must have $\lambda = 1$, in which case the last two equations tell us we must have $y = 0$ and $z = 0$. Plugging this into our constraint, we get the two critical points $(6, 0, 0)$ and $(-6, 0, 0)$.

Similarly, if we consider the cases where $y \neq 0$ and $z \neq 0$, we get four more critical points, $(0, 3, 0)$, $(0, -3, 0)$, $(0, 0, 2)$ and $(0, 0, -2)$.

We will see in Section 4.13 how we can classify some of these critical points.

4.12.2 Two constraint critical points

We now consider the case of finding constrained local maximum points where there are two equations that define the constraints on the given point. That is, we wish to find local extrema of the function f, where the domain D is defined by the pair of equations $g = 0$ and $h = 0$. The functions g and h are both referred to as "constraint functions".

This problem is generally not interesting for functions (objective and constraint) of two variables, because then the domain is generically a zero dimensional set, consisting of a finite number of points. So we will consider this problem only for functions of three variables.

The picture for this problem is represented in Figure 4.38. The domain D is the curve that is the intersection of the surfaces represented by the constraints $g = 0$ and $h = 0$. This time then there are two constraint gradients, ∇g and ∇h, each of which is perpendicular to its own level set and thus also to the domain D.

As in the unconstrained and single constraint cases, the criteria we use to rule points out in this case is based on Observation 4.11.1. That is, if at a point $\vec{a} \in D$ there is a tangent vector $\vec{v}$ such that $D_{\vec{v}} f(\vec{a}) \neq 0$, then the point $\vec{a}$ could not possibly be a local maximum.

Again, as indicated in Figure 4.38, we choose $\vec{v}$ to be a projection of ∇f, this time to the line tangent to the curve at $\vec{a}$. Note that this vector $\vec{v}$ is tangent to the curve D.

As before, we compute
$$D_{\vec{v}} f(\vec{a}) = \nabla f \cdot \vec{v} = \|\nabla f\| \|\vec{v}\| \cos \theta = \|\nabla f\|^2 \cos^2 \theta$$

This is nonzero whenever $\nabla f \neq \vec{0}$ and $\theta \neq \pi/2$. That is, whenever ∇f is nonzero and not perpendicular to the curve D. This gives us a theorem again analogous to Theorem 4.11.1, and seemingly identical to Theorem 4.12.1, except of course that they apply in different contexts.

Theorem 4.12.3. *If all three of the below conditions hold,*

1. *f is differentiable at $\vec{a} \in D$, and*
2. *$\nabla f(\vec{a}) \neq \vec{0}$, and*
3. *$\nabla f(\vec{a}) \not\perp D$*

then $\vec{a}$ is not a constrained local maximum point of $f : D \to \mathbb{R}^1$

As in the unconstrained case, most points in the domain will satisfy these conditions, and so this theorem rules out most of the points in the domain.

4.12. CONSTRAINED LOCAL EXTREMA AND LAGRANGE MULTIPLIER THEOREMS

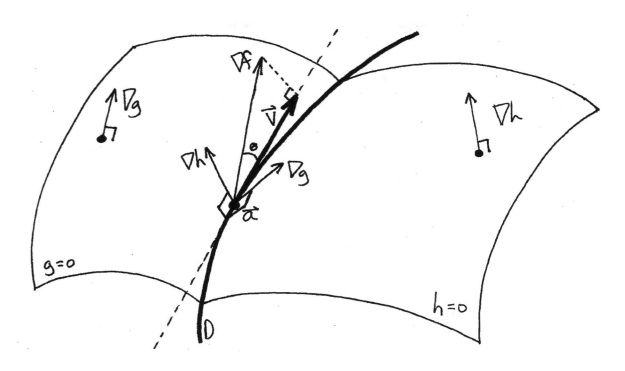

Figure 4.38:

Again, any point that fails to satisfy any one of the above criteria cannot be ruled out by the theorem. Again, we call these critical points for f on D.

Definition 4.12.3. *A "constrained critical point" of the function f on the domain $D = \{g = 0, h = 0\}$ is any point in D for which either*

1. *f is not differentiable at $\vec{a} \in D$, or*
2. *$\nabla f(\vec{a}) = \vec{0}$, or*
3. *$\nabla f(\vec{a}) \perp D$*

Using this language, Theorem 4.12.3 states simply that the constrained critical points are the only possible constrained local maxima in D.

Again we want to rephrase this in a way that will be more convenient algebraically. This time though, there is more than one direction that is perpendicular to D – in fact there is an entire plane perpendicular to D, and f need not point in any particular direction parallel to that plane.

But we have two vectors already (∇g, ∇h) that point perpendicular to D and therefore parallel to that plane. And if those vectors are not zero and not parallel, every other vector parallel to that plane will be some sort of linear combination of them. This is represented in Figure 4.39.

With this in mind, we can rephrase Definition 4.12.3 as a theorem.

Theorem 4.12.4 (Two Constraint Lagrange Multiplier Theorem). *Any constrained critical point (and therefore any constrained local maximum) of the function f on the domain $D = \{g = 0, h = 0\}$ must be a point $\vec{a}$ on D for which either*

1. *f is not differentiable at $\vec{a} \in D$, or*
2. *$\nabla g(\vec{a}) = \vec{0}$, or $\nabla h(\vec{a}) = \vec{0}$, or $\nabla g = k\nabla h$, or*
3. *$\nabla f(\vec{a}) = \lambda \nabla g(\vec{a}) + \mu \nabla h(\vec{a})$*

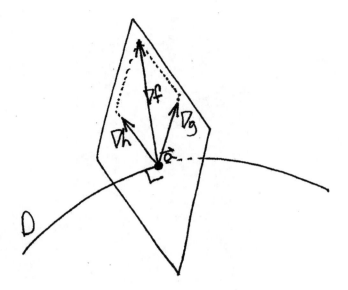

Figure 4.39:

Again, the conditions in (2) of Theorem 4.12.4 are called the "degeneracy conditions", and condition (3) is called the "Lagrange condition".

Example 4.12.3. Suppose we want to find critical points for the function $f(x,y,z) = 3x + 2y + z$, on the ellipse D that is the intersection of the cylinder $x^2 + y^2 = 1$ and the plane $x + y + z = 0$.

This domain is the intersection of level sets of the functions $g(x,y,z) = x^2 + y^2 - 1$ and $h(x,y,z) = x + y + z$.

We use Theorem 4.12.4 to check for critical points. First, we consider if there might be points where f is not differentiable, but since again f is a polynomial we know such points cannot exist.

Second, we look at the degeneracy conditions. We have

$$\nabla g = \begin{bmatrix} 2x \\ 2y \\ 0 \end{bmatrix} \quad \text{and} \quad \nabla h = \begin{bmatrix} 1 \\ 1 \\ 1 \end{bmatrix}$$

Clearly ∇h is never zero. On the other hand ∇g can be zero when x and y are both zero, which is the entire z-axis – but none of these points satisfies the constraint $g = 0$, so they do not count as constrained critical points. Finally we have to ask when one of these might be a multiple of the other; but in looking at the third components we quickly see this would require $\nabla g = \vec{0}$, and we have already determined that cannot happen on D.

So the degeneracy condition again gives us no critical points.

Third, the Lagrange condition combined with our constraints turns into

$$\begin{bmatrix} 3 \\ 2 \\ 1 \end{bmatrix} = \lambda \begin{bmatrix} 2x \\ 2y \\ 0 \end{bmatrix} + \mu \begin{bmatrix} 1 \\ 1 \\ 1 \end{bmatrix} \qquad \Longrightarrow \qquad \begin{aligned} 3 &= \lambda(2x) + \mu \\ 2 &= \lambda(2y) + \mu \\ 1 &= \lambda(0) + \mu \\ 1 &= x^2 + y^2 \\ 0 &= x + y + z \end{aligned}$$
$$1 = x^2 + y^2$$
$$0 = x + y + z$$

Here we have five equations and five unknowns. The convenient start to solving this system is to note that because of the zero in the third equation, we already know that $\mu = 1$. Plugging this in to the first two equations, we reduce

4.12. CONSTRAINED LOCAL EXTREMA AND LAGRANGE MULTIPLIER THEOREMS

the system to

$$\begin{aligned} 2 &= \lambda(2x) \\ 1 &= \lambda(2y) \\ 1 &= x^2 + y^2 \\ 0 &= x + y + z \end{aligned}$$

Looking at the first two equations, we can eliminate λ and conclude that $x = 2y$. Then the third equation above becomes

$$\begin{aligned} 1 &= (2y)^2 + y^2 \\ y &= \pm \frac{1}{\sqrt{5}} \end{aligned}$$

We can then solve directly for x and z, giving us two critical points, $\left(\frac{2}{\sqrt{5}}, \frac{1}{\sqrt{5}}, \frac{-3}{\sqrt{5}}\right)$ and $\left(\frac{-2}{\sqrt{5}}, \frac{-1}{\sqrt{5}}, \frac{3}{\sqrt{5}}\right)$.

Exercises

Exercise 4.12.1 (exer-Ders-CLE-1). Suppose we are considering a function f defined in $\mathbb{R}^3$, and suppose we know that at the point $\vec{a} = (1, 0, 0)$ the gradient vector for f is $\nabla f = (3, 2, 5)$. Show that $\vec{a}$ is not a constrained local maximum for f on the unit sphere by finding an explicit curve on the sphere on which f is increasing as it passes through this point.

Exercise 4.12.2 (exer-Ders-CLE-2). Suppose now we consider the same point $\vec{a}$ and function f from Exercise 4.12.1. Can we use a similar argument to rule out the possibility that $\vec{a}$ might be a constrained local maximum for f on the plane $3x + 2y + 5z = 3$? Explain why or why not.

Exercise 4.12.3 (exer-Ders-CLE-3). Use the method of Lagrange multipliers to find the highest point (greatest value of $f(x, y, z) = z$) on the "rotated ellipsoid" defined by $3x^2 - 2xy + y^2 + 4z^2 + xz = 3$.

Exercise 4.12.4 (exer-Ders-CLE-4). Find the point on the ellipsoid from Exercise 4.12.3 that maximizes the value of the function e^{x+y+z}.

Exercise 4.12.5 (exer-Ders-CLE-5). Find the point on the surface defined by $2x^2 + 4y^2 + 8z^2 = 1$ that maximizes the value of $x^3 + y^3 + z^3$.

Exercise 4.12.6 (exer-Ders-CLE-6). The Gotham City Police Department has been warned that the Joker will be attempting to rob one of three banks in the Greater Gotham area tonight. Given the clues, Commissioner Gordon thinks that there is a 60% chance that the First National Bank will be hit, a 10% chance for the Second National Bank, and a 30% chance for the Third National Bank.

For whichever bank is hit, the utility (in terms of the likelihood that they will be able to catch the Joker and protect the bank) of having x police stationed at that bank is estimated to be $\ln x$. So if police are stationed at the three banks above in numbers x_1, x_2, x_3, respectively, the expected total utility is

$$U(x_1, x_2, x_3) = (.6)\ln(x_1) + (.1)\ln(x_2) + (.3)\ln(x_3)$$

The GCPD has five hundred police available for Gordon to assign to protect these three banks, but he does not know how he should distribute them among the three banks in order to achieve maximum expected utility. As usual he consults Batman — but Batman never took a course in multivariable calculus, and so Batman consults you.

How many police should you instruct Batman to post at each of the three banks? Make sure to explain your reasoning clearly.

Exercise 4.12.7 (**exer-Ders-CLE-7**). Find the coordinates of the point (above the x-axis) on the ellipse $4x^2 + 9y^2 = 36$ that is closest to the point $(1, 0)$. *(Hint: Minimize the square of the distance.)*

Exercise 4.12.8 (**exer-Ders-CLE-8**). Find the absolute maximum value of the function $f(x, y) = x^2 - 6y$ on the circle in the xy-plane centered at the origin and with radius equal to 5.

Exercise 4.12.9 (**exer-Ders-CLE-9**). Find the absolute maximum value of the function $f(x, y, z) = z$ with constraints $h_1 = z - x^3 + 6y^2 = 0$ and $h_2 = x^2 + y^2 = 25$.

Exercise 4.12.10 (**exer-Ders-CLE-10**). In this problem we consider the function $f(x, y) = y - 3x^2$ on the domain defined by $y^2 - x^4 = 0$.

1. Show that there are no Lagrange critical points for f on this domain.

2. Use the fact that on this domain we must have either $y = x^2$ or $y = -x^2$ to show that f must attain a maximum value on this domain.

3. Explain how it is possible that f can attain a constrained local maximum at a point that is not a Lagrange critical point, and find where this happens on this domain.

The following exercises involve a type of problem that might appear on an exam. On this type of problem there is a question, followed by four options. It is known that exactly one of the options is the correct answer to the question. (Sort of like a multiple choice problem.)

Instead of choosing just one though, the student is asked to indicate next to each option his "confidence level" (x_i) that this option is the correct answer, in the form of a percentage. (Thus, the four indicated levels (x_1, x_2, x_3, x_4) must add up to 100.) So, for example, if the student is mostly sure the third option is correct, but thinks there is a chance it might be the others, he or she might respond with $x_1 = 10$, $x_2 = 10$, $x_3 = 70$, $x_4 = 10$. On another similar problem where the student is absolutely thoroughly confident in the first option and is certain the others are not correct, a reasonable response might be $x_1 = 94$, $x_2 = 2$, $x_3 = 2$, $x_4 = 2$.

The score that the student will receive is based on the confidence level indicated for the option that is correct, adjusted by a "scoring function" (f). Specifically, if option i is the correct option, the student will receive a score of $f(x_i)$.

Importantly, while the instructor can choose any scoring function, we must note that the student is free to answer any way he or she likes – and cannot be forced to answer honestly. That is, if the student's true confidence levels are p_1, p_2, p_3, p_4, he or she could answer with totally different values x_1, x_2, x_3, x_4 if it were somehow expedient to do so. We will assume for these problems that the student's goal is to choose values x_i that maximize the expected number of points he or she will receive on the problem, independent of whether those values are or are not indicative of the student's true confidence levels p_i.

Exercise 4.12.11 (**exer-Ders-CLE-11**). Show that the expected number of points the student will receive is

$$E = \frac{p_1}{100} f(x_1) + \frac{p_2}{100} f(x_2) + \frac{p_3}{100} f(x_3) + \frac{p_4}{100} f(x_4)$$

Exercise 4.12.12 (**exer-Ders-CLE-12**). Suppose that the scoring function is $f(x) = kx$, where k is a constant; in other words, that the number of points the student receives on that problem is just a multiple of his indicated confidence in the correct option. Suppose also that the student has absolutely no idea what the correct option is (and thus all p_i are 25). Show that in this case, his expected number of points on the problem does not depend on how he or she answers.

Exercise 4.12.13 (**exer-Ders-CLE-13**). With $f(x) = x/10$ (effectively making this a 10 point problem), now suppose that the student has some confidence in the first option, but no idea about the others, with $p_1 = 40$, $p_2 = 20$, $p_3 = 20$, $p_4 = 20$. What is the expected number of points for the following possible choices the student might make for his indicated levels?

1. $x_1 = 40$, $x_2 = 20$, $x_3 = 20$, $x_4 = 20$

2. $x_1 = 25$, $x_2 = 25$, $x_3 = 25$, $x_4 = 25$

3. $x_1 = 10$, $x_2 = 30$, $x_3 = 30$, $x_4 = 30$

4. $x_1 = 100$, $x_2 = 0$, $x_3 = 0$, $x_4 = 0$

In the interests of maximizing the expected number of points, should the student answer with true confidence levels, or not? Explain what the best strategy is for the student in this case.

Exercise 4.12.14 (exer-Ders-CLE-14). While the instructor cannot force the student to answer with true confidence levels, the instructor can choose a scoring function that makes it expedient for the student to do so.

Consider the scoring function $f(x) = k \ln(x) - c$, where k and c are both constants. (We observe that if any x_i is near zero, the expected number of points is very low (negative even!); from this we can conclude that the expected number of points then is maximized at a Lagrange critical point with all x_i nonzero.)

Show that the honest response ($x_i = p_i$) is the only Lagrange critical point for this optimization problem, and thus it must be the maximizer. (This means that, for this scoring function, the way that a student maximizes the expected number of points for the problem is actually by being honest, and representing his or her confidence levels accurately!)

Exercise 4.12.15 (exer-Ders-CLE-15). Suppose the scoring function is given as $f(x) = \ln(x)$, and the student has true confidence levels $p_1 = 80$, $p_2 = 10$, $p_3 = 5$, $p_4 = 5$.

The student is tempted to think that he might be able to increase his expected number of points by claiming greater confidence in the option that he thinks is most likely the right one, choosing to indicate his confidence levels as $x_1 = 97$, $x_2 = 1$, $x_3 = 1$, $x_4 = 1$.

Compute the expected number of points for this and for the honest response, and conclude that it is NOT to the student's advantage to do this.

Exercise 4.12.16 (exer-Ders-CLE-16). Suppose the scoring function is given as $f(x) = \ln(x)$, and the student has true confidence levels $p_1 = 80$, $p_2 = 10$, $p_3 = 5$, $p_4 = 5$.

The student notices that the log function does not increase quickly at higher values, but does at lower values; he is thus tempted to think that he might be able to increase his expected number of points by claiming lower confidence in the option where the scoring function increases slowly, and then being able to increase his claimed confidence in the other answers where the scoring function increases steeply. He is considering doing this by choosing to indicate his confidence levels as $x_1 = 55$, $x_2 = 15$, $x_3 = 15$, $x_4 = 15$.

Compute the expected number of points for this and for the honest response, and conclude that it is NOT to the student's advantage to do this.

4.13 Global Extrema

In Sections 4.11 and 4.12 we discussed methods for finding local extrema. These are worthwhile on their own, but they will also serve in the discussion in this section of finding global extrema.

Definition 4.13.1. *Given a set $D \in \mathbb{R}^n$, a function $f : D \to \mathbb{R}^1$ and a point $\vec{a} \in D$, we say that $\vec{a}$ is a "global maximum point of f on D" if for every $\vec{x} \in D$, we have $f(\vec{x}) \leq f(\vec{a})$.*

Students can readily check directly from the definitions that every global maximum point must also be a local maximum point.

Note that the definition of a global maximum is simpler than that in Definition 4.11.7 for a local maximum; that is, there is no discussion of the existence of balls – just a requirement that the function be at least as great at $\vec{a}$ as at any other point in the entire domain. Correspondingly, one could argue that global maxima are more intrinsically interesting things than local maxima, and certainly there are many applications in which this is the case. We will see in this section though that the two are closely linked.

4.13.1 Outline

The first step in the process of finding global extrema involves analyzing the domain. The observation we will outline here does not apply to every possible domain, but it does apply to every domain that we will consider in this textbook. That is, usually a domain D can be broken up into convenient pieces, which we will refer to here as "subdomains".

Definition 4.13.2. *Given a set $D \in \mathbb{R}^n$, a "subdomain" D_i of D is a subset of D with the property that all of the points in D_i have the same constraint equations.*

Example 4.13.1. Suppose we let D be the closed unit ball in $\mathbb{R}^3$, defined by $D = \{\vec{x} \mid \|\vec{x}\| \leq 1\}$. Note that we can write this set as the union $D = D_1 \cup D_2$, where

$$D_1 = \{\vec{x} \mid \|\vec{x}\| < 1\} \quad \text{and} \quad D_2 = \{\vec{x} \mid \|\vec{x}\| = 1\}$$

The set D_1 fits Definition 4.13.2 because every point in D_1 is unconstrained; that is, at every point in D_1, all vectors are tangent.

The set D_2 also fits because every point in D_2 is constrained by the same constraint function, $g(\vec{x}) = \|\vec{x}\| - 1$; that is, the tangent vectors at every point in D_2 are constrained to being tangent to the level set of the same function g.

Example 4.13.2. Let's consider now the unit square $D = [0,1] \times [0,1]$ in $\mathbb{R}^2$. We can break this up into nine pieces, as shown in Figure 4.40 – the interior, which we label as I; four edges (not including the vertices), which we label as E_1, E_2, E_3, E_4; and four vertices, labelled V_1, V_2, V_3, V_4. Points in the interior I are not subject to any constraints on vectors that are tangent. All points on any given edge are subject to the same single constraint on their tangent vectors; for example, on the top edge E_1, the constraint function is $g(x,y) = y - 1$. On each vertex, no vectors are tangent.

Having set up this idea, we can now describe the process by which we will find global maxima.

Suppose that the domain D is the union of the subdomains $D_1, \ldots, D_n$, that we have a global maximum of f on D at the point $\vec{a}$, and that this point $\vec{a}$ is in a particular one of the subdomains D_i.

4.13. GLOBAL EXTREMA

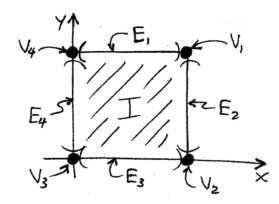

Figure 4.40:

As a global maximum of f on D, the point $\vec{a}$ must also be a global maximum of f on D_i. Therefore it must also be a local maximum of f on D_i, which then in turn means that it must be a critical point of f on D_i. And because of the way we defined subdomains, the process for finding critical points of f on each subdomain will simply be one of the methods discussed in Sections 4.11 or 4.12.

So if we have D written as a union of a finite number of subdomains, and if we can find all of the critical points of f on each of those subdomains, any global maximum point must be on that list. And since the global maximum must have a value of f greater than or equal to any other value, we can simply check the value of f at every such critical point, and rule out all but the greatest values.

Said differently, we can simply use the appropriate techniques on each subdomain to rule out most points, and then we can rule out most of the remaining critical points by direct comparison.

(The good news here is that there is no need to classify the critical points, because most of the remaining will be ruled out by the final direct comparison anyway.)

But there is a serious problem with this strategy. Note, all we have done is by one means or another rule out most of the points in the domain. The point(s) that remain might be global maxima, and they would certainly seem to be the only remaining options. But in order to make that final conclusion, we would need to use the assumption that global maxima exist in the first place.

As an example of how this might go wrong, suppose you wish to find the global maximum value of the function $f(x) = x^3$ where the domain D is the entire real line. The entire domain is unconstrained, and it is easily checked that the only critical point is $x = 0$. Thus every other point in the domain is eliminated; but this critical point is very clearly not a global maximum; in fact it is not even a local maximum.

It is important to recognize then that the above strategy simply will not work in some instances. There is however a very broad category of opimization problems where we can bypass this complication.

The powerful tool we will use is a very important theorem, often overlooked in many discussions of this topic.

Theorem 4.13.1. *Suppose that f is a <u>continuous</u> function, and D is a <u>closed</u> and <u>bounded</u> domain. Then the function f does achieve an absolute maximum value at some point $\vec{a} \in D$.*

(The proof of this theorem is slightly beyond the scope of this course.)

This theorem allows us to use the above strategy to find absolute maxima for continuous functions on closed and bounded domains. That is, knowing that absolute maximum points exist, it is true that the points remaining after the above process must be those absolute maximum points.

We will apply this method to some examples momentarily, but first we will emphasize the importance of the assumptions in Theorem 4.13.1. Note that in order to apply the theorem, the function must be continuous, and the domain must be both closed and bounded. Those terms were underlined in the statement of the theorem to emphasize their importance.

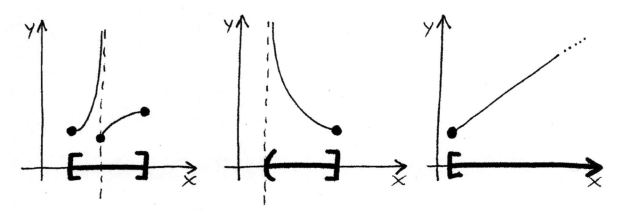

Figure 4.41:

In Figure 4.41 we show examples of how the theorem can fail when those assumptions are not satisfied. The first figure shows the graph of a function that is not continuous; the second shows a graph on a domain that is not closed; and the third shows a graph on a domain that is not bounded. Note that in none of these examples is an absolute maximum value attained at any point in the domain.

4.13.2 Examples

Example 4.13.3. Let's revisit Example 4.12.2, where we were considering the function $f(x,y,z) = x^2 + y^2 + z^2$ **on the ellipsoid** D **defined by** $x^2 + 4y^2 + 9z^2 = 36$.

Note that all of the points on this domain are constrained by the same constraint function, so we do not need to break it up into pieces. We already found six critical points: $(6,0,0)$, $(-6,0,0)$, $(0,3,0)$, $(0,-3,0)$, $(0,0,2)$ **and** $(0,0,-2)$.

The domain is closed and bounded, and the objective function is continuous, so Theorem 4.13.1 tells us that absolute maximum points must exist, and by the reasoning from Subsection 4.13.1 we know the six points above are the only possible candidates. Checking the values of the function at these six points, we get

$$\begin{aligned} f(6,0,0) &= 36 \\ f(-6,0,0) &= 36 \\ f(0,3,0) &= 9 \\ f(0,-3,0) &= 9 \\ f(0,0,2) &= 4 \\ f(0,0,-2) &= 4 \end{aligned}$$

We conclude that the absolute maximum value of this function on this domain is 36, achieved at the points $(6,0,0)$ **and** $(-6,0,0)$.

(Of course since these points are absolute maxima, they must also be local maxima; this is a conclusion that we could not draw when we saw this example previously.)

Example 4.13.4. Let's consider the function $f(x,y,z) = x + 2y + 3z$ **defined on the closed unit ball in** $\mathbb{R}^3$, $D =$

4.13. GLOBAL EXTREMA

$\{\vec{x} \mid \|\vec{x}\| \leq 1\}$, as in Example 4.13.1. As we did in that example, we write $D = D_1 \cup D_2$, where

$$D_1 = \{\vec{x} \mid \|\vec{x}\| < 1\} \quad \text{and} \quad D_2 = \{\vec{x} \mid \|\vec{x}\| = 1\}$$

We note first that this domain is both closed and bounded, and the objective function is continuous, so the absolute maximum points must exist and we can proceed with the method of Subsection 4.13.1.

Every point in D_1 is unconstrained; that is, at every point in D_1, all vectors are tangent. To find unconstrained critical points we look for points where the objective is not differentiable, but there are none because the objective is a polynomial; then we look for points where $\nabla f = \vec{0}$, which simplifies to $(1, 2, 3) = (0, 0, 0)$, and of course this does not yield any critical points either. So, there are no critical points in the interior subdomain D_1.

On D_2, we have a single constraint Lagrange problem. The constraint is given as $\|\vec{x}\| = 1$, but for algebraic convenience we will first square both sides to get the equivalent equation $\|\vec{x}\|^2 = 1$ (equivalent because lengths are positive), giving us the constraint function $g(x, y, z) = x^2 + y^2 + z^2 - 1$.

The degeneracy condition on D_2 then is $(2x, 2y, 2z) = (0, 0, 0)$, which can only happen at the origin, which of course is not on D_2. So, there are no degenerate critical points on D_2.

The Lagrange condition combined with the constraint on D_2 is

$$\begin{bmatrix} 1 \\ 2 \\ 3 \end{bmatrix} = \lambda \begin{bmatrix} 2x \\ 2y \\ 2z \end{bmatrix} \qquad \Longrightarrow \qquad \begin{aligned} 1 &= \lambda(2x) \\ 2 &= \lambda(2y) \\ 3 &= \lambda(2z) \\ 1 &= x^2 + y^2 + z^2 \end{aligned}$$
$$1 = x^2 + y^2 + z^2$$

From the three Lagrange equations we conclude that $y = 2x$ and $z = 3x$, which we can then plug into the constraint equation to get

$$1 = x^2 + y^2 + z^2 = x^2 + (2x)^2 + (3x)^2 = 14x^2$$

This gives us $x = \pm \frac{1}{\sqrt{14}}$, and thus two critical points, $\vec{a}_1 = \left(\frac{1}{\sqrt{14}}, \frac{2}{\sqrt{14}}, \frac{3}{\sqrt{14}}\right)$ and $\vec{a}_2 = \left(\frac{-1}{\sqrt{14}}, \frac{-2}{\sqrt{14}}, \frac{-3}{\sqrt{14}}\right)$.

These are the only two critical points from any of the subdomains, so we check the value of f at those points to get

$$\begin{aligned} f(\vec{a}_1) &= \sqrt{14} \\ f(\vec{a}_2) &= -\sqrt{14} \end{aligned}$$

So $\vec{a}_1$ is the unique absolute maximum point. (Similarly, $\vec{a}_2$ is the unique absolute minimum point.)

Example 4.13.5. Let's consider now the function $f(x, y) = x^2 - x + y^2$ defined on the unit square $D = [0, 1] \times [0, 1]$ in $\mathbb{R}^2$, as in Example 4.13.2. Again we break this up into the same nine pieces shown in Figure 4.40.

First, note that the objective function is continuous and the domain is closed and bounded, so Theorem 4.13.1 again tells us that absolute maximum points must exist and that we can therefore use the method of Subsection 4.13.1.

We consider the subdomains one at a time.

The interior, I: The objective again is clearly differentiable everywhere. The gradient is $\nabla f = (2x - 1, 2y)$, which can equal to zero at the point $\left(\frac{1}{2}, 0\right)$. But this point is not in the interior I, so it does not count as a critical point.

The top edge, E_1: This is a single constraint problem, with constraint $y = 1$. The constraint function is $g = y - 1$, and the constraint gradient is $\nabla g = (0, 1)$, which is clearly never zero, so there are no degenerate critical points.

The Lagrange condition and constraint equation together are

$$\begin{bmatrix} 2x - 1 \\ 2y \end{bmatrix} = \lambda \begin{bmatrix} 0 \\ 1 \end{bmatrix} \qquad \Longrightarrow \qquad \begin{aligned} 2x - 1 &= 0 \\ 2y &= \lambda \\ y &= 1 \end{aligned}$$
$$y = 1$$

These easily give us the critical point $\left(\frac{1}{2}, 1\right) \in E_1$.

The right edge, E_2: This is a single constraint problem, with constraint $x = 1$. The constraint function is $g = x - 1$, and the constraint gradient is $\nabla g = (1, 0)$, which is clearly never zero, so there are no degenerate critical points.

The Lagrange condition and constraint equation together are
$$\begin{bmatrix} 2x - 1 \\ 2y \end{bmatrix} = \lambda \begin{bmatrix} 1 \\ 0 \end{bmatrix} \qquad \Longrightarrow \qquad \begin{aligned} 2x - 1 &= \lambda \\ 2y &= 0 \\ x &= 1 \end{aligned}$$
$$x = 1$$

These give us the point $(1, 0)$, but this point is not in the set E_2 – so it does not count here as a critical point.

The bottom edge, E_3: This is a single constraint problem, with constraint $y = 0$. The constraint function is $g = y$, and the constraint gradient is $\nabla g = (0, 1)$, which is clearly never zero, so there are no degenerate critical points.

The Lagrange condition and constraint equation together are
$$\begin{bmatrix} 2x - 1 \\ 2y \end{bmatrix} = \lambda \begin{bmatrix} 0 \\ 1 \end{bmatrix} \qquad \Longrightarrow \qquad \begin{aligned} 2x - 1 &= 0 \\ 2y &= \lambda \\ y &= 0 \end{aligned}$$
$$y = 0$$

These easily give us the critical point $\left(\frac{1}{2}, 0\right) \in E_3$.

The left edge, E_4: This is a single constraint problem, with constraint $x = 0$. The constraint function is $g = x$, and the constraint gradient is $\nabla g = (1, 0)$, which is clearly never zero, so there are no degenerate critical points.

The Lagrange condition and constraint equation together are
$$\begin{bmatrix} 2x - 1 \\ 2y \end{bmatrix} = \lambda \begin{bmatrix} 1 \\ 0 \end{bmatrix} \qquad \Longrightarrow \qquad \begin{aligned} 2x - 1 &= \lambda \\ 2y &= 0 \\ x &= 0 \end{aligned}$$
$$x = 0$$

These give us the point $(0, 0)$, but this point is not in the set E_4 – so it does not count here as a critical point.

The vertices, V_1, V_2, V_3, V_4:

We deal with the vertices together because there is really nothing to do. For example, on V_1, no vectors are tangent so we cannot eliminate anything – but there is no need to eliminate anything, as the set itself is just a single point.

So we just call all four of these vertices critical points, to be considered with all of the others.

In total we have then six critical point, and we compute the value of the function at those six points:

$$\begin{aligned} f\left(\tfrac{1}{2}, 1\right) &= \tfrac{3}{4} \\ f\left(\tfrac{1}{2}, 0\right) &= -\tfrac{1}{4} \\ f(1, 1) &= 1 \\ f(1, 0) &= 0 \\ f(0, 0) &= 0 \\ f(0, 1) &= 1 \end{aligned}$$

The absolute maximum value of the function then is 1, achieved at the points $(0, 1)$ and $(1, 1)$.

Exercises

Exercise 4.13.1 (exer-Ders-GE-1). Using Definition 4.13.2, write each of the following as the union of subdomains and identify the constraint functions on each subdomain.

4.13. GLOBAL EXTREMA

1. The unit cube $[0,1] \times [0,1] \times [0,1]$ in $\mathbb{R}^3$.

2. The closed unit ball in $\mathbb{R}^3$.

3. The set of points (x,y) in the xy-plane for which $x \geq 0$, $y \geq 0$, $x+y \leq 2$, and $y - 2x \leq 1$.

4. The set of points in $\mathbb{R}^3$ for which $x^2 + y^2 + z^2 \leq 25$ and $(x-3)^2 + y^2 + z^2 \leq 16$.

5. The set of points in $\mathbb{R}^3$ for which $x \geq 0$, $y \geq 0$, $z \geq 0$, $x+y+z \leq 1$, and $2x + 3y + 6z = 6$.

Exercise 4.13.2 (exer-Ders-GE-2). Find the absolute maximum value of the function $f(x,y,z) = x+y+z$ subject to the constraints $z - x - y - 4 \leq 0$ and $x^2 + y^2 - z \leq 0$.

Exercise 4.13.3 (exer-Ders-GE-3). Find the absolute maximum value of the function $f(x,y) = xy$ subject to the constraints $x \geq 0$, $y \geq 0$, $x \leq 2$, $y \leq 2$, $x + y = 3$.

Exercise 4.13.4 (exer-Ders-GE-4). Find the absolute maximum value of the function $f(x,y,z) = x^2 - y^2$, subject to the constraints that $x + y \geq 0$, $x - y \geq 0$, $z \geq 0$, and $x + z = 1$.

Exercise 4.13.5 (exer-Ders-GE-5). Find the absolute maximum value of the function $f(x,y) = x^2 + y^2$ on the solid triangle with vertices at the points $(-2,1)$, $(1,-1)$ and $(1,1)$.

Exercise 4.13.6 (exer-Ders-GE-6). Find the absolute maximum value of the function $f(x,y) = x^3 + 3y^3 - 12x + 7y$ on the solid rectangle R in the xy-plane with corners at $(0,0)$, $(3,0)$, $(0,2)$, $(3,2)$.

Exercise 4.13.7 (exer-Ders-GE-7). Find the point (or points) in the unit disk in $\mathbb{R}^2$ that attains the maximum value on that domain of $f(x,y) = x^2 - x + y^2 - y$.

Exercise 4.13.8 (exer-Ders-GE-8). We have $f : (D \subset \mathbb{R}^3) \to \mathbb{R}^1$ with $f(x,y,z) = (x+y+z)^2$ and $D = \{x^2 + 4y^2 + 9z^2 \leq 1\}$. Find the point or points in the domain, if any exist, that attain the absolute maximum value of f on the domain D.

Exercise 4.13.9 (exer-Ders-GE-9). Find the absolute maximum value of the function $f(x,y,z) = x^2 - y^2 + z^2$ on the solid unit ball centered at the origin.

Exercise 4.13.10 (exer-Ders-GE-10). The solid D is bounded by the surface with equation $z^2 - \ln(2 - x^2 - y^2) = 0$. Find the absolute maximum value of the function $f(x,y,z) = x^2 + y^2 + z^2$ over the solid D.

Chapter 5

Integrals

5.1 Single Variable Integrals

Before we get into multivariable integrals, let's return for a moment to single variable integrals. In particular, let's ask the following question: "In the most fundamental sense, by definition, what is a single variable integral?"

Many if not most students who have completed a high school level course in single variable calculus would say that an integral is the area under a graph of a function. While this is a very useful geometric picture in dealing with integrals, this is not the correct answer.

Not only is it not correct, but knowing what we do about the applications of integrals, this could not possibly be the correct answer. In particular, we know that integrals can be used to compute many things other than areas – for example we can use an integral to compute the mass of a one-dimensional straight rod of non-constant density, or the volume obtained by rotating an area around an axis, or the total flow rate of a fluid through a tube, or any of a number of other such things. If integrals were defined as area, we would need to be able to relate these quantities directly to area... but what is the direct relationship between area and mass? Or between area and volume; or between area and flow rate?

So what is an integral then, in terms of its definition, if NOT the area under a graph?

The answer is that an integral is a Riemann sum.

$$\int_a^b f(x)\,dx = \lim_{n \to \infty} \sum_{i=1}^n f(x_i^*) \Delta x \tag{5.1}$$

This definition can be used to apply integrals to many different situations.

Example 5.1.1. For one, it can be used to compute the area under the graph of a positive, continuous function f, since that area can be cut into slices of width Δx and approximately constant height $f(x_i^*)$. Since area is computed as width times height, the Riemann sum then represents the area in this case. See Figure 5.1.

Example 5.1.2. Or, it can be used to compute the mass of a rod with non-constant density given by f, since that rod can be cut into pieces of width Δx and approximately constant density $f(x_i^*)$ on each piece. Since mass is size (length in this case) times density, the Riemann sum then represents mass in this case. See Figure 5.2.

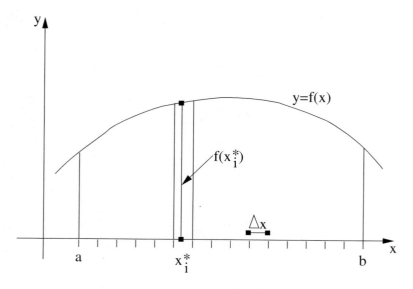

Figure 5.1: Computing area with a Riemann sum

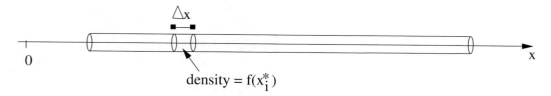

Figure 5.2: Computing mass with a Riemann sum

To be more explicit, all of these applications have in common that they satisfy the following four conditions, which allow the problem to be represented by a Riemann sum:

1. *There is a domain for the problem, and a function (called the "integrand") that is defined on the domain.*

    (a) In the first case above, the domain is the interval on the x-axis over which we want to compute the area, and the integrand is the function whose graph we are looking at.

    (b) In the second case, the domain is the interval on the x-axis corresponding to the rod's location, and the integrand is the function giving density as a function of location.

2. *The domain can be broken up into smaller pieces, and the quantity to be computed can be computed as the sum over all of the smaller pieces.*

    (a) In the first case above, the interval is broken up into subintervals, and of course the total area is the sum of the areas over all subintervals.

    (b) In the second case above, the rod is broken up into small chunks, of course the total mass is the sum of the masses of the chunks.

3. *The integrand is approximately constant over each smaller piece of the domain.*

    (a) In the first case, since the integrand is continuous, it is approximately constant over each subinterval.

    (b) In the second case, since density can only vary gradually, it is approximately constant over each chunk.

4. *The quantity to be computed, on the smaller pieces, can be approximated by the product of the size of the piece times the value of the integrand somewhere in that smaller piece.*

    (a) In the first case above, the value of the integrand is the height since in particular we are looking at the graph of the integrand; and area is computed as width times height.

5.1. SINGLE VARIABLE INTEGRALS

(b) In the second case above, the integrand represents density, and mass is computed as size times density.

Note that in every application of single variable integrals, thought of in the way described by the conditions above, the domain is one-dimensional; namely, an interval on a single axis such as the x-axis. In such a case, the problem in question turns into a Riemann sum such as that in Equation 5.1, and thus a single variable integral that we can compute with the fundamental theorem of calculus.

But note also that the four conditions above do not explicitly require that the domain be one dimensional. If we had a problem in which the domain were two dimensional, the only challenge in trying to fit the problem to the four conditions above is that we would have to have an appropriate notion of what we mean by the "size" of a small piece of a two-dimensional domain; and of course, area is that notion. Similarly, if we had a three-dimensional domain, then volume would be the appropriate notion of size.

There will be other details that will be different too. For one thing, the specifics of the Riemann sum in Equation 5.1 will not be appropriate to the new settings, and so we will have to find a new way to write down the new sum; and thus of course we will no longer have a single variable integral by the definition in Equation 5.1, so we will have to give the resulting expression a new name, with a new notation. Also, the fundamental theorem of calculus will no longer apply, and so we will have to come up with new ways to compute these expressions.

In this chapter we will take the ideas outlined in the four conditions above and try to apply them to new situations. We will try to apply them to situations where the domain is more than one-dimensional, and we will also try to apply them to situations where the integrand is not a scalar.

In all of these cases though, the resulting new types of integrals that we define will fit the same mold. We will call the four conditions below our four "integral conditions":

1. *There is a domain for the problem, and a function (the integrand) that is defined on the domain.*
2. *The domain can be broken up into smaller pieces, and the quantity to be computed can be computed as the sum over all of the smaller pieces.*
3. *The integrand is approximately constant over each smaller piece of the domain.*
4. *The quantity to be computed, on the smaller pieces, can be approximated by the product of the size of the piece times the value of the integrand somewhere in that smaller piece.*

The quantities in question in such problems can all be represented by Riemann sums as in equation 5.1. This can be written (imprecisely) in a more general form as

$$Q = \lim_{(\text{size} \to 0)} \sum_{D_i} \Big(f(x_i)\Big)\Big(\text{size of } D_i\Big) \tag{5.2}$$

where D is the domain for the problem in question, the subsets D_i form a partition of D into small pieces of a given size, and x_i is a point in the set D_i.

In this chapter we will apply this form to many different types of domains to define new types of integrals, and work out the details that will allow us to compute them.

Exercises

Exercise 5.1.1 (exer-Ints-SV-1). The number of termites per unit length at a position a distance x from the left end of a board of length 10 is given by the function $T(x) = 2 + \sin x$. If you were to use an integral to compute the total number of termites in the board, what would be the domain for this integral? What would be the integrand?

Exercise 5.1.2 (<u>exer-Ints-SV-2</u>). The amount of effort that it takes a colony of ants to add height to its bed is given by $E(z) = 2 + z$, where z represents the height above ground level, and the effort is measured in units of ant-hours per unit length. The bed is currently built up to height 10, but it needs to be built up to height 15. If you were to use an integral to compute the amount of effort it would take the colony to finish this task, what would be the domain for this integral? What would be the integrand?

Exercise 5.1.3 (<u>exer-Ints-SV-3</u>). Write out, but do not evaluate, explicit Riemann sums in terms of only i and n representing the quantities described in Exercises 5.1.1 and 5.1.2.

Exercise 5.1.4 (<u>exer-Ints-SV-4</u>). Evaluate directly (without using an integral) the Riemann sum from Exercise 5.1.3 referring to the quantity described in Exercise 5.1.2.

Exercise 5.1.5 (<u>exer-Ints-SV-5</u>). Use integrals and the fundamental theorem of calculus to compute the quantities described in Exercises 5.1.1 and 5.1.2.

5.2. DOUBLE INTEGRALS

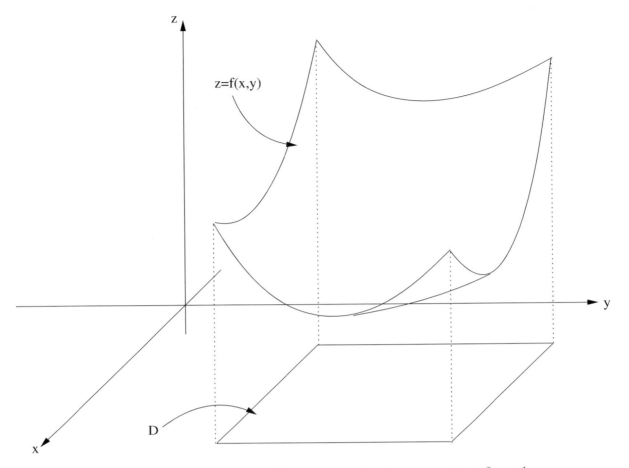

Figure 5.3: The volume under the graph of $f : \mathbb{R}^2 \to \mathbb{R}^1$

5.2 Double Integrals

We will begin by attempting to solve a problem that fits the four integral conditions described previously, where the domain is a two-dimensional subset of $\mathbb{R}^2$, instead of a one-dimensional subset of $\mathbb{R}^1$.

For example, suppose that we have a positive, continuous function $f : \mathbb{R}^2 \to \mathbb{R}$, and we wish to compute the volume under the graph of that function, over a rectangle D in $\mathbb{R}^2$, as in Figure 5.3.

This problem does fit the four integral conditions (See Figure 5.4):

1. The domain is the rectangle D, and the integrand is the function f whose graph we are considering.

2. The domain can be broken up into "sub-rectangles" by slicing D both perpendicular to the x-axis and perpendicular to the y-axis, and the volume under the graph over the whole rectangle is clearly the same as the sum of the volumes over all of the sub-rectangles.

3. Since the integrand is continuous, it is approximately constant over each sub-rectangle.

4. Volume can be computed as area times height; so over each sub-rectangle, the volume can be approximated by the product of the size (area, in this case) of the sub-rectangle times the value of the integrand (which is the height of the graph over that sub-rectangle).

Since these conditions are satisfied, we should be able to solve the given problem by writing it as a limit of a sum over these sub-rectangles. But we will have to develop some new notation in order to write this down explicitly.

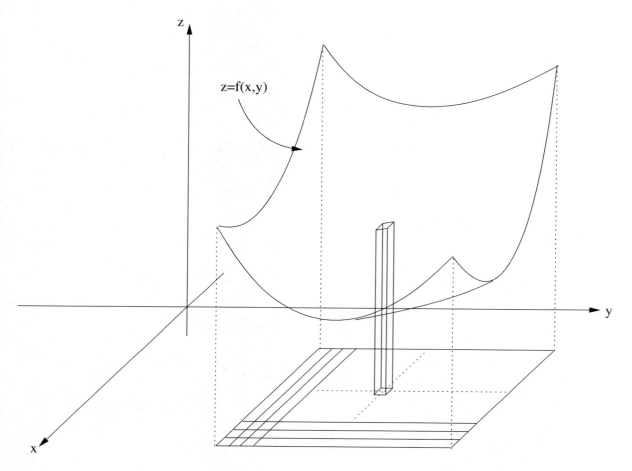

Figure 5.4: Computing volume with a sum

5.2. DOUBLE INTEGRALS

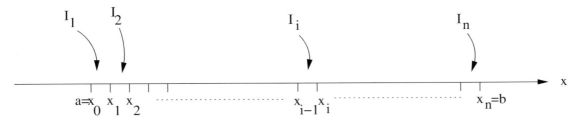

Figure 5.5: Indexing subintervals in $\mathbb{R}^1$

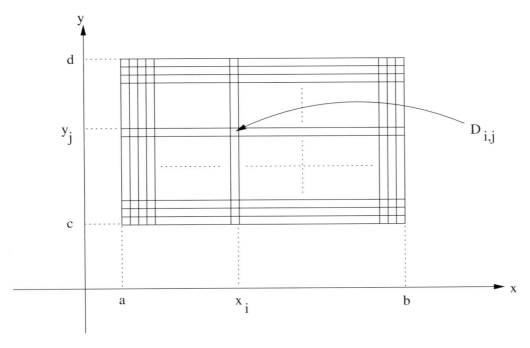

Figure 5.6: Indexing subrectangles in $\mathbb{R}^2$

5.2.1 Notation

In the one dimensional situation, illustrated in Figure 5.5, we used the index i to refer to specific subintervals. Specifically, the interval $I = [a, b]$ was broken up into n subintervals $I_i = [x_{i-1}, x_i]$, where $x_i = a + i\Delta x$ and $\Delta x = \frac{b-a}{n}$.

We will do something similar here in the two dimensional situation, but in this case we are subdividing along both the x- and y-axes, and so we will need a second index j for the y-axis.

Suppose that our rectangular domain is defined by the interval $[a, b]$ on the x-axis, and $[c, d]$ on the y-axis; we will use the notation $D = [a, b] \times [c, d]$ to represent this. We break up the interval $[a, b]$ as before, and similarly we break up the interval $[c, d]$ on the y-axis into n subintervals $[y_{j-1}, y_j]$, where $y_j = c + j\Delta y$ and $\Delta y = \frac{d-c}{n}$. See Figure 5.6.

These subintervals on the x- and y-axes then define "subrectangles" $D_{i,j}$, given by $D_{i,j} = [x_{i-1}, x_i] \times [y_{j-1}, y_j]$.

So when we do our summation, we are not summing over a list of subintervals I_i, indexed by a single index i, but rather over a list of subrectangles $D_{i,j}$, indexed by two indices i and j. So we will be writing down our Riemann sum in a form something like this:

$$\lim_{n \to \infty} \sum_{i,j=1}^{n} f(x_i^*, y_j^*)(\text{size of } D_{i,j})$$

The point (x_i^*, y_j^*) is just an arbitrary point chosen from the subrectangle $D_{i,j}$, and as before we evaluate the integrand function at such a point. Also as before, we multiply the value of the integrand by the size of the corresponding piece of the domain, and then we sum over all of those pieces. Note that in this summation, i and j each range over all integer values between 1 and n, so there are n^2 terms in this summation.

At this point then all we need is to quantify the notion of size for this new type of integral. But of course this is a very easy thing to do, namely, area is the way we measure size for a two dimensional object. In particular then, the area of the subrectangle $D_{i,j}$ is simply the product of its height and width, which is then just $\Delta A = \Delta x \Delta y$. So our expression becomes:

$$\lim_{n \to \infty} \sum_{i,j=1}^{n} f(x_i^*, y_j^*) \Delta A$$

5.2.2 Definition

Note that we arrived at this expression by a process almost identical to that by which we arrived at Riemann sums for single variable integrals. So, we call this a "two-index Riemann sum", since there are two indices i and j.

This two-index Riemann sum then represents the answer to our original question, namely, what is the volume under the surface which is the graph of the positive, continuous function $f(x, y)$, above a given rectangular domain D. (Of course we do not yet have a means for computing this expression, but that will come soon.) And in fact in any case where we have a problem that can be broken down with our initial four integral conditions and in which the corresponding domain is two dimensional, the answer to such a question can be represented by the same expression – just as was the case for one-dimensional domains. We will see soon that there are very many such questions.

Given this, we will give the above expression a name. By analogy with the single variable version, we call this a "double integral".

Definition 5.2.1. *For a continuous function f defined on a domain D in the xy-plane, "the double integral of the function f over the domain D" is*

$$\iint_D f(x, y)\, dA = \lim_{n \to \infty} \sum_{i,j=1}^{n} f(x_i^*, y_j^*) \Delta A$$

The "dA" is to be suggestive of the "ΔA" in the two-index Riemann sum, just like in single variable integrals the choice of notation "dx" is to be suggestive of the "Δx" in the single-index Riemann sum. Note that since $\Delta A = \Delta x \Delta y$, we sometimes write dA as $dx\, dy$.

The use of two integral symbols is just a reminder that this is a double integral and not a single integral.

Note that this fits the form from Equation 5.2.

(Note also – not all books use two integral symbols in this notation, since the fact that this is a double integral can also be deduced from the fact that the domain is two-dimensional.)

Of course the double integral only computes the volume under the graph in the case that the function is positive (as we noted when we went through the original derivation). The volume question was our original motivation for the double integral, but we will soon see that as with single variable integrals there are many other applications that can be addressed with the double integral, and as with single variable integrals not all of them require the integrand to be positive.

5.2.3 Examples

Here are a few examples of other applications of double integrals.

Example 5.2.1. Suppose we want to compute the mass of a sheet of metal represented by a rectangle $R = [a, b] \times [c, d]$ **in the** xy**-plane, with continuous density function given by** $\delta(x, y)$**.**

5.2. DOUBLE INTEGRALS

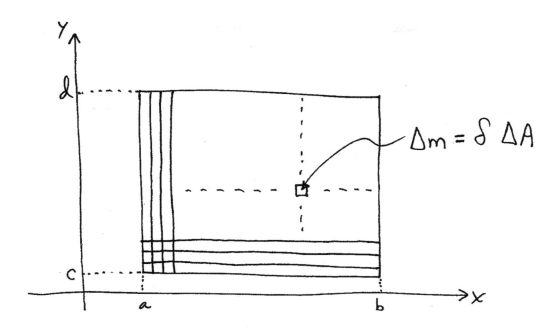

Figure 5.7: Example 5.2.1

This question fits the four integral conditions, by analogy with Example 5.1.2. The domain is the rectangle R, and the integrand is the density function δ. The domain can be broken up into pieces, indicated in Figure 5.7, each of which has a small mass that we might call Δm, and the total mass is certainly the sum of the masses of those little pieces. The density is given to be continuous and so it is approximately constant on each of the little pieces, and on each piece the mass is given by the area of the piece times the density on that piece.

So the mass can be written as a Riemann sum,

$$m = \lim \sum_{i,j} \Delta m = \lim \sum_{i,j} \delta(x,y) \Delta A \tag{5.3}$$

which in integral notation can be written as

$$m = \iint dm = \iint_R \delta(x,y)\, dA \tag{5.4}$$

(Note the similarity to Example 5.1.2)

Example 5.2.2. Suppose that we are given the population density function $f(x, y)$ in the state of Colorado, in units of thousands of people per square mile with x and y measuring position in miles east and north (respectively) of the southwest corner of the state; and we wish to compute from that the total population P of Colorado.

This question also fits our four conditions. The domain is the part of the xy-plane corresponding to the state of Colorado, and the integrand is the population density function f. The domain can again be broken up into pieces, each of which has a population we might call ΔP. Certainly the total population is the sum of those small populations. The population on each little piece is computed as $\Delta P = (1,000)f(x,y)\Delta A$, so we have

$$P = \lim \sum_{i,j} \Delta P = \lim \sum_{i,j} (1,000) f(x,y) \Delta A \tag{5.5}$$

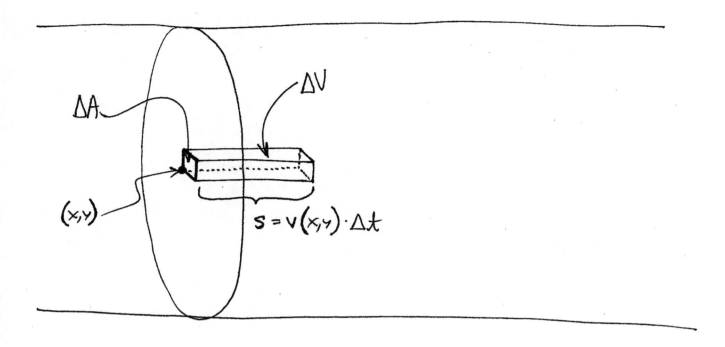

Figure 5.8:

which in integral notation can be written as

$$P = \iint dP = (1,000) \iint_R f(x,y)\,dA \tag{5.6}$$

Example 5.2.3. Suppose that a fluid is flowing through a tube with perpendicular cross-section D, and that the velocity of the fluid at a point (x, y) on that cross section is $v(x, y)$. What is the total flow rate F of fluid through the tube, in terms of volume per unit time?

Let's let the domain be the cross-section D. To begin with, it is not immediately clear what our integrand should be.

We can break up our domain D into small pieces of area, and certainly the total flow rate through the tube is equal to the sum of the flow rates through these individual pieces of area (because every particle of fluid must pass through one of those pieces of area). But, can we write the flow rate through such a piece of area as a product involving the area of that piece?

In Figure 5.8, we show a particular small piece of the cross-section at the point (x, y), and have represented the water that has flowed through that area in the last amount of time Δt. The volume of that water is $\Delta V = (\Delta A)(s) = (\Delta A)(v(x,y)\Delta t)$. So, the flow rate through that piece of area ΔA is thus

$$\Delta F = \frac{\Delta V}{\Delta t} = \frac{(\Delta A)(v(x,y)\Delta t)}{\Delta t} = v(x,y)\Delta A \tag{5.7}$$

So in fact it does turn out that the flow rate through a piece of area is a product involving the area of that piece. The other factor then is the integrand.

So we have

5.2. DOUBLE INTEGRALS

$$F = \lim \sum_{i,j} \Delta F = \lim \sum_{i,j} v(x,y) \Delta A \tag{5.8}$$

which in integral notation can be written as

$$F = \iint dF = \iint_R v(x,y)\, dA \tag{5.9}$$

Exercises

Exercise 5.2.1 (exer-Ints-DI-1). Suppose we are subdividing the rectangle $[2,5] \times [1,10]$ into subrectangles with $n=6$; what are the four vertices of the subrectangle $D_{2,5}$?

Exercise 5.2.2 (exer-Ints-DI-2). Suppose we are subdividing the rectangle $[2,5] \times [1,10]$ into subrectangles with $n=6$; write out a general expression for the subrectangle $D_{i,j}$.

Exercise 5.2.3 (exer-Ints-DI-3). Suppose we are subdividing the rectangle $[2,5] \times [1,10]$ into subrectangles. Write out a general expression for the subrectangle $D_{i,j}$ in terms of i, j, and n.

Exercise 5.2.4 (exer-Ints-DI-4). Suppose we wish to compute the volume under the graph of the positive continuous function $f(x,y) = 10 + 2x + 5y$ over the rectangle $[1,5] \times [-2,6]$. Use the form of a two-index Riemann sum to approximate this volume by considering the value of the summation when $n=4$, $x_i^* = x_i$, and $y_i^* = y_i$.

Exercise 5.2.5 (exer-Ints-DI-5). For each of the following, use $n=4$, $x_i^* = x_i$, and $y_i^* = y_i$ to approximate the value of the given integral.

1. $\iint_D 3x - y\, dA$, where $D = [2,5] \times [7,11]$
2. $\iint_D x^2 + 2y\, dA$, where $D = [1,5] \times [-1,2]$
3. $\iint_D xy\, dA$, where $D = [1,9] \times [-2,2]$

Exercise 5.2.6 (exer-Ints-DI-6). The mass per unit area on a rectangular sheet of width 2 and height 1 is given by the density function $\delta(x,y) = 2e^x - xe^y$, where x and y are the coordinates as measured from the bottom left corner of the rectangle. Write out, but do not evaluate, a double integral representing the mass of this sheet.

Exercise 5.2.7 (exer-Ints-DI-7). Suppose we wish to compute the volume under the graph of the function $\delta(x,y) = 2e^x - xe^y$ over the rectangle $[0,2] \times [0,1]$. Write out, but do not evaluate, a double integral representing this volume.

Exercise 5.2.8 (exer-Ints-DI-8). Noting that the integrals answering questions 5.2.6 and 5.2.7 are the same, does this suggest that there is an intrinsic relationship between volume and the mass of a sheet?

Exercise 5.2.9 (exer-Ints-DI-9). The number of termites per unit area in a rectangular board is given by $T(x,y) = 2 + \sin(\pi x) + \cos(\pi y)$, where x and y are the coordinates as measured from the bottom left corner of the board. The board has width 4 and height 7. Write out, but do not evaluate, a double integral representing the total number of termites in the board.

Exercise 5.2.10 (exer-Ints-DI-10). A canal (heading east) has a rectangular cross section, and is 50 feet wide and 10 feet deep. The speed of the water in the canal at a depth of y feet and x feet from the south edge is given by

$$s(x,y) = \frac{(11-y)}{100 + (x-25)^2}$$

where the speed is given in feet per second. Write out, but do not evaluate, a double integral representing the total flow rate (in cubic feet per second) through the canal.

Exercise 5.2.11 (exer-Ints-DI-11). Because of the shape of the roof and the direction of the wind, the accumulation of snow on a particular roof has depth (in feet) given by $d(x,y) = 3 - \sin(\pi x/2) - \cos(2\pi y)$, where x and y are the coordinates as measured from the chimney. This particular roof is rectangular with width (in the x direction) 30 feet and length (in the y direction) 50 feet, and the chimney in the middle of the roof. Write out, but do not evaluate, a double integral representing the total volume of snow accumulated on the roof.

Exercise 5.2.12 (exer-Ints-DI-12). Bob is putting a faux stone surface on an exterior wall of his house that is 20 feet wide and 9 feet tall. The pattern he is using requires the thickness (measured in inches) of the stone at a location (x,y) on the wall to be $T(x,y) = 3 - \sin(\pi(x+y)/3)$, where x represents the distance from the left side of the wall in feet, and y represents the height above the ground in feet. This faux stone weighs 20 lbs per cubic foot, and is all currently sitting at ground level. Write out, but do not evaluate, a double integral representing the total amount of work that it will take Bob to lift all of these stones up to their required height.

Exercise 5.2.13 (exer-Ints-DI-13). Roberta is throwing darts at a rectangular dart board. The probability that her dart will land in a tiny area at coordinates (x,y), where x and y are measured in inches from the center of the target, is given by the the product of that area and the function $P(x,y) = (3/\pi)e^{-3x^2 - 3y^2}$. (This function is called a "probability density function".) Write out, but do not evaluate, a double integral that represents the probability that her dart will land in a large rectangle with bottom left corner at $(-1,-1)$ and top right corner at $(3,3)$.

5.3 Nested Integrals

We have now successfully extended the ideas in the outlined four integral conditions to the situation where we have a two-dimensional domain. We have also adopted precise Riemann sum notation to represent this construction and a convenient integral notation to use as a shorthand for this Riemann sum.

But the very natural and important question that comes up now is, "How can we compute these things?". Certainly for single variable integrals this was an important question to consider, since the direct computation of a Riemann sum is extremely inconvenient at best, if even possible. And in the case of double integrals, the Riemann sum is even more inconvenient to compute directly.

In single variable calculus, there was a wonderful answer to this question, namely, the fundamental theorem of calculus. (In fact this theorem is such a powerful tool for computing integrals that many students get the mistaken impression that it is an alternative definition of integrals, along with the mistaken definition of "area under a curve".)

Will we be so fortunate when it comes to computing double integrals?

Well, not quite, unfortunately. The fundamental theorem of calculus relates the value of an integral to an expression computed over the boundary of the given interval. Conveniently in the one-dimensional case the boundary consists of only two points, thus making possible the extremely convenient answer "$F(b) - F(a)$". But for a two-dimensional domain the boundary is an entire curve of points, not just two points... So, we cannot hope for such a convenient answer in general for double integrals.

Still, we will find a computational approach that will be far, far more convenient to work with than attempting to compute the Riemann sums directly. This approach will be the topic of this section.

5.3.1 Another computation of volume

(In the calculation that follows we will give a geometric motivation for a formula that will allow us to compute a double integral. The rigorous proof of the result involves an algebraic computation that is beyond the scope of this course, and so it is not included in this text.)

Of course double integrals can be used to answer many different types of questions, but recall that we motivated our original definition of a double integral from our attempt to answer one specific question – how can we compute the volume under the graph of a positive, continuous function $f : \mathbb{R}^2 \to \mathbb{R}^1$ above a given rectangle $D = [a, b] \times [c, d]$ in $\mathbb{R}^2$, as in Figure 5.9?

So, rather than attempting to compute the Riemann sum definition of a double integral itself, directly, let's just try to find another alternative way to compute the volume under the graph of a function.

Let's recall that we do know of one instance in which single variable integrals were used to compute a volume – that is, in computing the volume obtained by rotating an area in the xy-plane around the x-axis. In that case, the strategy was to slice the given volume perpendicular to the x-axis, identify the volume of each slice as being the area of the cross-section times the width of the slice, and then use the single integral to add up those volumes. The integrand of the resulting integral then was the area of the cross-section, and since the volume was obtained by rotation, the areas of the cross-sections were easy to compute.

We can take a similar strategy in this case. We slice the volume perpendicular to the x-axis at an arbitrary place $x \in [a, b]$, with thickness Δx. The area of the cross-section will of course depend on the value of x, so we represent it as $A(x)$. Then the volume of that given slice will be $\Delta V = A(x) \Delta x$. See Figure 5.10

We can then write the volume as

$$V = \int dV = \int_a^b A(x)\, dx \tag{5.10}$$

The problem at this point is that our cross-section is not a simple geometric figure, and thus its area (our integrand) is not something that we can write down as a simple expression in terms of x. So, as it stands, we cannot compute this

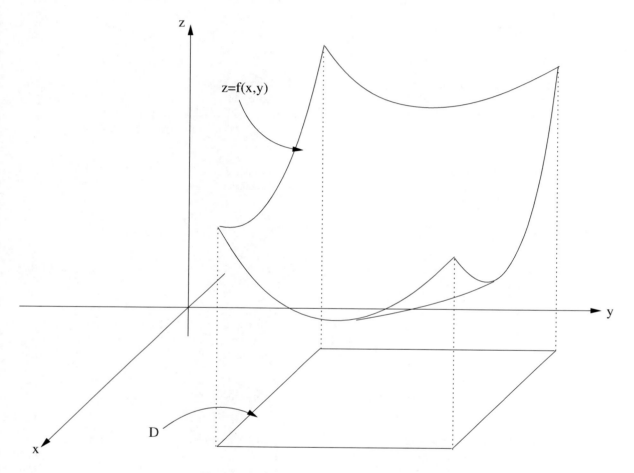

Figure 5.9: The volume under the graph of $f: \mathbb{R}^2 \to \mathbb{R}^1$

5.3. NESTED INTEGRALS

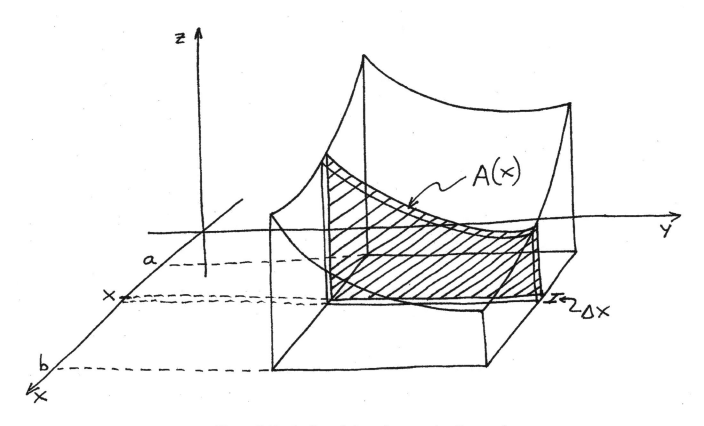

Figure 5.10: A slice of the volume under the graph

expression. In order to proceed, we need to be able to figure out how to compute the expression $A(x)$. So we fix the value of x for the moment, and turn our attention to computing this area for this fixed value of x.

But the area $A(x)$ is the area under a curve – in particular, the curve C which defines the top edge of the area we are interested in. Since we are accustomed to computing areas under curves with single variable integrals, let's try to do that in this situation.

Our domain will be the interval $[c, d]$ on the y-axis, and our integrand $h(y)$ will be defined as follows: For any given value of $y \in [c, d]$, $h(y)$ is the height of the curve C over the point (x, y). Of course, the curve C is the part of the graph of f corresponding to this fixed x, and so we can restate this more conveniently as $h(y) = f(x, y)$. See Figure 5.11.

The area $A(x)$ is the sum of the areas of the slices illustrated in Figure 5.12, and the area of each of those slices is the height $h(y)$ times the width Δy. Writing this as an integral, we get

$$A(x) = \int dA = \int_c^d h(y)\,dy = \int_c^d f(x, y)\,dy \tag{5.11}$$

This integral needs to be considered very carefully. While we are accustomed to single variable integrals, this is the first time that we have encountered an integral where it appears that there are two variables in the integrand. Each of these must be considered and treated differently, corresponding to the roles that each played in constructing this integral:

1. The variable y is the actual variable for this single variable integral; it represents the value moving from c to d along the y-axis, allowing us to refer to the areas in each of the slices of the desired area $A(x)$.

2. The variable x is a *constant* as far as this integral is concerned, because the area $A(x)$ we are trying to compute is the cross-section defined by a single, fixed value of x. Remember, we wrote $f(x, y)$ only because it represents the height function $h(y)$, which only depends on y. The resulting value of this integral depends on that choice of the fixed value of x.

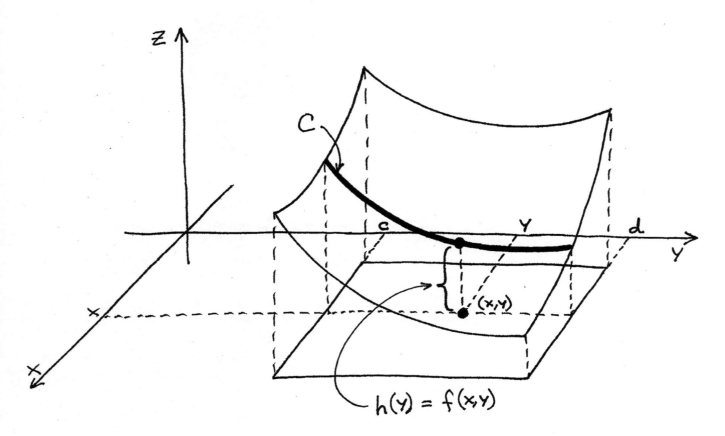

Figure 5.11: Height on C as a function of y

5.3. NESTED INTEGRALS

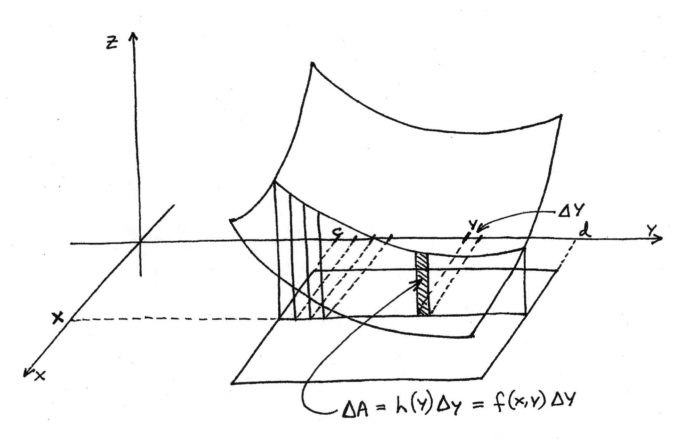

Figure 5.12: The area of the cross-section as a single variable integral

We can now combine equations 5.10 and 5.11 to get

$$V = \int_a^b A(x)\,dx = \int_a^b \left(\int_c^d f(x,y)\,dy \right) dx$$

We call this a "nested integral", because one integral takes place "inside" the other (at least notationally), much like nested parentheses. (The parentheses in the equation above serve to underscore that the inside integral is the integrand for the outside integral; but in most cases, those parentheses will be omitted.) It is also sometimes called an "iterated integral".

This is something that we can compute, because the computation of the inside integral yields a function of only x (since y is a dummy variable for that integral), and that resulting function of x can then be used as indicated as the integrand of the outside integral. The computation then turns out to be simply a succession of two single variable integral computations, and the result is the desired volume under the graph of the function.

Recall that of course we used the computation of the volume under the graph of a function as our motivation for defining the double integral as in Equation 5.2.1. So the nested integral above can be thought of as a computation of a double integral. So we have

$$\iint_D f(x,y)\,dA = \int_a^b \int_c^d f(x,y)\,dy\,dx \tag{5.12}$$

Note that while this equation seems very natural from a notational point of view, the two sides of this equation have very different definitions. The left side of the equation is a two-index Riemann sum, where the domain D is broken up into little pieces of area, and the sum is done over all of those pieces of area. The right side of the equation is a succession of two single variable integrals. The similarity in the notations is convenient and satisfying, but the distinction must be noted. In fact, this is a very elegant result and a very powerful computational tool.

Again, let's note the different roles that the variables x and y play at different points in the nested integral above:

1. For the inside integral, y is the variable and x is a constant.

2. For the outside integral, x is the variable, and there will be no y's present.

Let's try applying this result to solve a problem.

Example 5.3.1. Suppose we want to find the volume under the surface $z = x^2 + y^2$, and above the unit square $S = [0,1] \times [0,1]$.

First, we observe that the given surface is the graph $z = f(x,y)$ **of the function** $f(x,y) = x^2 + y^2$. **So, based on our arguments in Section 5.2, we conclude that this volume is**

$$V = \iint_S f(x,y)\,dA \tag{5.13}$$

Of course Equation 5.12 allows us to rewrite this as

$$V = \int_0^1 \int_0^1 f(x,y)\,dy\,dx \tag{5.14}$$

5.3. NESTED INTEGRALS

First we apply the fundamental theorem of calculus to the inside integral:

$$= \int_0^1 \left(\int_0^1 x^2 + y^2 \, dy \right) dx$$

$$= \int_0^1 \left(x^2 y + \frac{1}{3} y^3 \right) \Big]_{y=0}^{y=1} dx$$

$$= \int_0^1 \left((x^2(1) + \frac{1}{3}(1)^3) - (x^2(0) + \frac{1}{3}(0)^3) \right) dx$$

$$= \int_0^1 \left(x^2 + \frac{1}{3} \right) dx$$

(Note that when antidifferentiating for the inside integral, we treat x as a constant. Note also that the antiderivative still involves both x and y, so we clearly indicate the bounds as "$y = 0$" and "$y = 1$" to avoid inadvertently plugging in those values for x.)

At this point of course we have merely a single variable integral. Note that x is the only variable remaining (which is as it should be since the integrand represents the area of the cross-section, $A(x)$.) This integral can be computed by a second application of the fundamental theorem of calculus:

$$= \left(\frac{1}{3} x^3 + \frac{1}{3} x \right) \Big]_0^1$$

$$= \frac{2}{3}$$

5.3.2 Switching differentials

Now let's look back to the very beginning of this section, where we first decided to try to compute double integrals with nested single variable integrals. Notice that our first step was to slice perpendicular to the x-axis; after that, we then found it necessary to slice that resulting cross-section again, this time perpendicular to the y-axis.

But, did we need to slice in that order? Perhaps might we have sliced the volume first perpendicular to the y-axis, and then perpendicular to the x-axis? Would this have worked?

Let's give it a try. After slicing the volume perpendicular to the y-axis, we would have slices that look like that in Figure 5.13. The thickness would be Δy; and the area of the cross-section would depend on the value of y defining the cross-section, so we will write it as $A(y)$. So we have

$$V = \int dV = \int_c^d A(y) \, dy \tag{5.15}$$

Again we need to find a way to compute this integrand, and again it will be another single variable integral. This time the cross-section in question is the area under the graph for a fixed value of y, and x ranges over a set of values. The resulting integral has as its domain the interval $[a, b]$ on the x-axis, and the integrand is the height $h(x)$ of the surface above the point (x, y), where y is the given fixed value, and so $h(x) = f(x, y)$. See Figure 5.14. So we have

$$A(y) = \int_a^b h(x) \, dx = \int_a^b f(x, y) \, dx \tag{5.16}$$

Again combining these two results, we have

$$V = \int_c^d \int_a^b f(x, y) \, dx \, dy \tag{5.17}$$

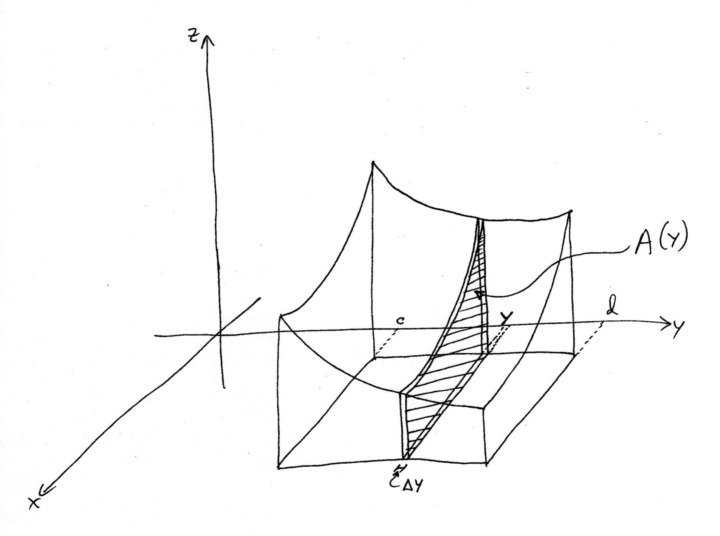

Figure 5.13: A slice taken perpendicular to the y-axis

5.3. NESTED INTEGRALS

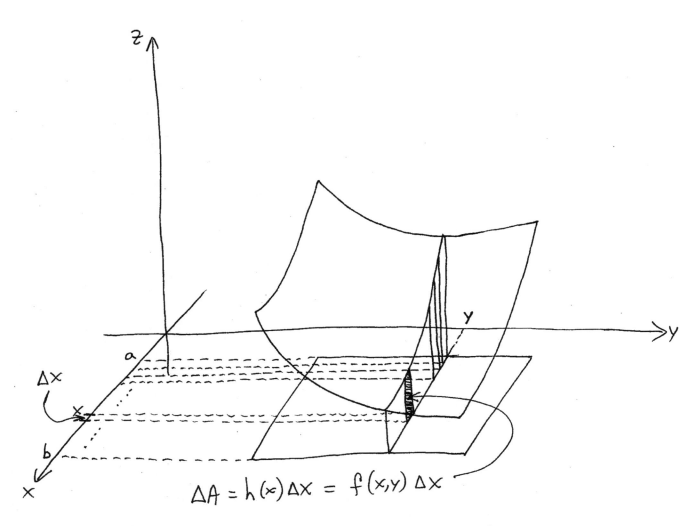

Figure 5.14: The area of a slice, as an integral

Nicely, we get an answer that is very similar to the one we got by slicing the other way. In each case, the differential for the outside integral corresponds to the axis we sliced perpendicular to first, and that variable is treated as a constant in the inside integral.

Fortunately, for continuous integrands, the results obtained by computing nested integrals in these two different orders will be the same. We state that here as a theorem.

Theorem 5.3.1. *For any continuous function $f : \mathbb{R}^2 \to \mathbb{R}^1$ and rectangle $R = [a,b] \times [c,d]$,*

$$\iint_R f(x,y)\, dA = \int_c^d \int_a^b f(x,y)\, dx\, dy = \int_a^b \int_c^d f(x,y)\, dy\, dx \tag{5.18}$$

Of course this theorem is highly believable, since each of these nested integrals was written down to compute the same thing.

Let's now revisit Example 5.3.1, this time solving the problem by slicing up the problem in the other order.

Example 5.3.2. Suppose we want to find the volume under the surface $z = x^2 + y^2$, and above the unit square $S = [0,1] \times [0,1]$.

We again note that this volume is

$$V = \iint_S f(x,y)\, dA \tag{5.19}$$

Theorem 5.3.1 allows us to write this as

$$V = \int_0^1 \int_0^1 f(x,y)\, dx\, dy \tag{5.20}$$

First we apply the fundamental theorem of calculus to the inside integral:

$$= \int_0^1 \left(\int_0^1 x^2 + y^2\, dx \right) dy$$

$$= \int_0^1 \left(\frac{1}{3}x^3 + xy^2 \right]_{x=0}^{x=1} dy$$

$$= \int_0^1 \left((\frac{1}{3}(1)^3 + (1)y^2) - (\frac{1}{3}(0)^3) + (0)y^2 \right) dy$$

$$= \int_0^1 \left(\frac{1}{3} + y^2 \right) dy$$

(Note that when antidifferentiating for the inside integral, we treat y as a constant. Note also that the antiderivative still involves both x and y, so we clearly indicate the bounds as "$x = 0$" and "$x = 1$" to avoid inadvertently plugging in those values for y.)

At this point of course we have merely a single variable integral, with y as the only variable remaining. This integral can be computed by a second application of the fundamental theorem of calculus:

$$= \left(\frac{1}{3}y + \frac{1}{3}y^3 \right]_0^1$$

$$= \frac{2}{3}$$

5.3.3 Examples other than volume

We established in Section 5.2 that double integrals can be used to represent many quantities other than the volume under the graph of a function.

But it was only in this present section that we have seen how actually to compute the volume under the graph of a function, which we did by applying a succession of single variable integrals, resulting in a nested integral. Of course the details of that argument were specific to the question of volume, and so as of the moment it would appear that our results about nested integrals apply only to that specific problem.

The question then arises, do we need to go through a similar process for every application of double integrals that we might run into?

For example, if we want to compute a mass as in Example 5.2.1, do we need to go through an argument in which we first slice up the rectangular sheet perpendicular to the x-axis, write down the mass of each slice as a product involving the width dx, and then continue by slicing perpendicular to the y-axis?

Fortunately, we do not!

The point here is that while we used the image of volume in deriving the method of nested integrals, our result is a formula for how to compute double integrals, and not just volume – volume is just the image that allowed us to arrive at the result.

(The volume question is particularly convenient for that purpose because of the fact that geometry is relatively easy to visualize and to draw, and the intermediate expression (area) is something for which we already have a name, and with which we already have experience and intuition. Having finished the computation though, we can apply the resulting formula relating double integrals to nested integrals to any other application of double integrals.)

So, any expression that we can write as a double integral, including all of those from the end of Section 5.2, can be computed by rewriting the corresponding double integral as a nested integral. There is no need to go through the breakdown process for every different application.

Let's do an example.

Example 5.3.3. Suppose we want to find the mass of a sheet of metal defined by the rectangle $R = [1, 2] \times [3, 4]$, where the density is given to be $\delta(x, y) = x + 2y$.

This problem can be approached with a double integral, as was done in Example 5.2.1.

We have

$$m = \iint dm = \iint_R \delta(x, y)\, dA \tag{5.21}$$

Using Theorem 5.3.1, we can choose to compute this double integral in either of two ways:

$$m = \int_3^4 \int_1^2 x + 2y\, dx\, dy = \int_1^2 \int_3^4 x + 2y\, dy\, dx \tag{5.22}$$

Arbitrarily, we will use the first of these options:

$$\begin{aligned} m &= \int_3^4 \int_1^2 x + 2y \, dx \, dy \\ &= \int_3^4 \left(\frac{1}{2}x^2 + 2xy\right]_{x=1}^{x=2} dy \\ &= \int_3^4 \frac{3}{2} + 2y \, dy \\ &= \left(\frac{3}{2}y + y^2\right]_3^4 \\ &= \frac{17}{2} \end{aligned}$$

The benefit of this versatility could be phrased differently, as the answer to the following reasonable question: If we can compute the volume under the graph of a function by nested integrals, what then is the purpose of the Riemann sum definition of a double integral?

The answer is that the Riemann sum is the connection that is shared by all of these applications, allowing us to use the same algebraic method to compute the answer to so many varied questions. Without Riemann sums, why would we have any reason to believe that there should be any connection at all between the mass of a sheet (as in the above example) and the volume under a graph? Or the population of Colorado? Without Riemann sums we might develop methods for computing each of these things which we would eventually notice have certain similarities, but we would have to work through the details of each of these methods individually for each different problem.

Note that this is also true for single variable integrals. Single variable integrals are used to compute the area under a curve, the mass of a rod, and many other quantities that seem to have nothing to do with each other. The same tool, integrals, could be used to compute all of them only because the quantities could all be expressed as Riemann sums.

Exercises

Exercise 5.3.1 (exer-Ints-NI-1). Consider the volume under the graph of the function $f(x, y) = x^3 + 3y^2$ over the rectangle $[2, 3] \times [1, 5]$ in the xy-plane. Suppose one would like to compute this volume with a single variable integral with respect to x. Evaluate the integrand $A(x)$ for this single variable integral as a function of x. What is the domain for this single variable integral?

Exercise 5.3.2 (exer-Ints-NI-2). Suppose that one would like to compute the volume from Exercise 5.3.1 with a single variable integral with respect to y. What is the integrand for this single variable integral (evaluate this as a function of y), and what is the domain?

Exercise 5.3.3 (exer-Ints-NI-3). Compute the integrals in Exercises 5.3.1 and 5.3.2.

Exercise 5.3.4 (exer-Ints-NI-4). Compute the volume under the graph of the function $f(x, y) = e^{2x-5y}$ over the rectangle $[0, 1] \times [0, 2]$.

Exercise 5.3.5 (exer-Ints-NI-5). Compute the mass described in Exercise 5.2.6.

Exercise 5.3.6 (exer-Ints-NI-6). Compute the number of termites described in Exercise 5.2.9.

Exercise 5.3.7 (exer-Ints-NI-7). Compute the total flow rate described in Exercise 5.2.10.

Exercise 5.3.8 (exer-Ints-NI-8). Compute the volume of snow described in Exercise 5.2.11.

Exercise 5.3.9 (<u>exer-Ints-NI-9</u>)**.** Compute the total amount of work described in Exercise 5.2.12.

Exercise 5.3.10 (<u>exer-Ints-NI-10</u>)**.** Compute the probability described in Exercise 5.2.13. *(Hint: You may use the approximation $\int_{-1}^{3} e^{-3u^2}\, du \approx 1.01625$.)*

5.4 Non-Rectangular Domains

We have a definition as a Riemann sum (Definition 5.2.1) for the double integral of a continuous function over a rectangle, and we also now have a means for computing such a double integral, using nested integrals. We have also seen many different types of applications of these tools.

But what if we wanted to compute an integral over a domain that is not a rectangle? The $\mathbb{R}^2$ setting makes it possible to have lots of different shapes as domains, while of course in $\mathbb{R}^1$ there is just not enough room for the domain to be anything other than an interval, or possibly a union of intervals.

For example, what if we wanted to compute the amount of volume under the graph of a function f, and over the unit circle in the xy-plane? Or what if we wanted to compute the mass of a triangular sheet of metal with a given non-constant density function δ?

We must acknowledge that the definition we have given for double integrals over a rectangle does not technically address these situations. Specifically, that definition relies on being able to compute $\Delta x = \frac{b-a}{n}$, and $\Delta y = \frac{d-c}{n}$, and if the domain is not a rectangle then there are no such defining values a, b, c and d.

As far as definitions go, however, this is just a matter of formality. It becomes more inconvenient to write down the formal definition, but this can be done. We will not discuss that in this book.

However, the intuition for non-rectangular domains is the same as for rectangular domains, coming straight from our original four integral conditions and resulting generalized Riemann sum in Equation 5.2 from page 229 – we chop up the domain into small pieces over which the integrand is approximately constant, write the sum (over all pieces) of the integrand times the size of the piece, and then take the limit.

Fortunately, when it comes to the computation of these integrals, the method of nested integrals still works very nicely. There are some minor differences in the implementation, but conceptually the method still works exactly the same as it did for rectangles.

5.4.1 Examples

Suppose we wish to compute the volume under the graph of the positive, continuous function $f(x, y)$, over the triangle in the xy-plane with vertices at the points $(0,0)$, $(1,0)$ and $(0,1)$. We will use nested integrals, and arbitrarily let's make our first slices perpendicular to the x-axis, as in Figure 5.15.

As before, the volume of this cross section is an area times the thickness Δx, and the area depends on the choice of x so we will write it as $A(x)$. Also as before we get such cross-sections for values of x between readily identifiable values (between $x = 0$ and $x = 1$ in this case), and so we can write down

$$V = \int_0^1 A(x)\,dx \tag{5.23}$$

Also as before, the area $A(x)$ is the area under a curve, and we would like to write it as an integral.

At this point we notice a difference, however. In particular, the upper and lower bounds for y are no longer constant values, as they were in the case of a rectangular domain. As we can see from Figure 5.15, the range of values for y depends on the value of x. For smaller values of x, the area $A(x)$ would correspond to values of y that range almost all the way up to $y = 1$, but for larger values of x, the relevant values of y would be smaller.

So not only will the areas $A(x)$ depend on x, but also the integrals that we write down to compute them. The good news is that, even though this is slightly different, it doesn't present any real problems for us. We can still just write down what we need to write down, and deal with this issue when it comes up.

As indicated in Figure 5.16, we can write, very similarly to what we have done already,

$$A(x) = \int_{y_1}^{y_2} h(y)\,dy = \int_{y_1}^{y_2} f(x,y)\,dy \tag{5.24}$$

5.4. NON-RECTANGULAR DOMAINS

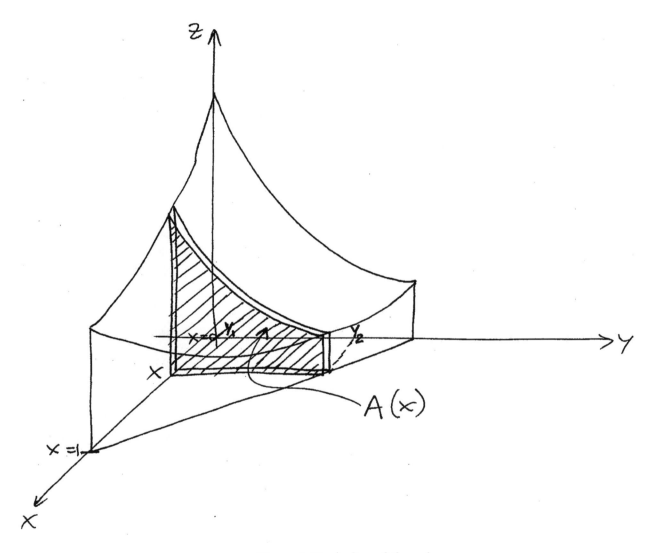

Figure 5.15: A slice of the volume

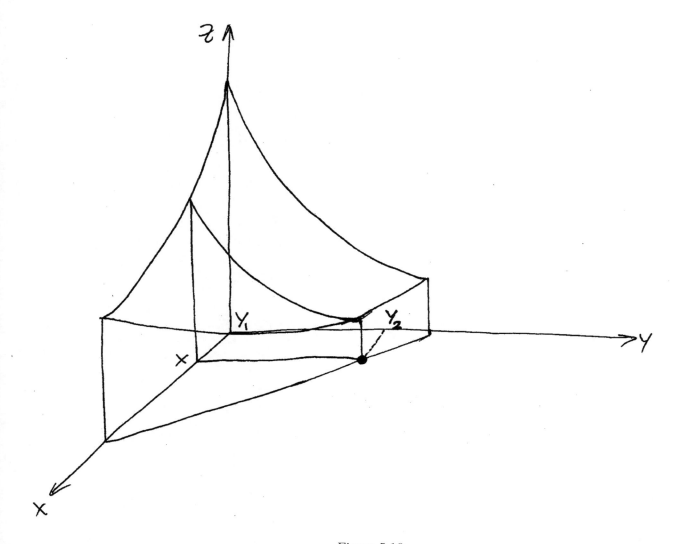

Figure 5.16:

Of course as we have already observed, in this case the bounds y_1 and y_2 are not constant, but depend on x. We will indicate that by rewriting the equation above as

$$A(x) = \int_{y_1(x)}^{y_2(x)} f(x,y)\,dy \tag{5.25}$$

It may look unusual to have the bounds for the integral depend on x, but remember that this is an integral with respect to y, and x is a constant as far as this integral is concerned. So, there really is no problem here.

Before we compute the integral, let's determine the functions y_1 and y_2 explicitly in terms of x.

The function $y_1(x)$ is easy to write down, because we can observe directly from Figure 5.16 that for every value of x, we have $y_1(x) = 0$.

The function $y_2(x)$ is not quite that easy to write down, but we can do so by again making an observation directly from the picture. In particular, the value y_2 relates to x by way of the line forming the hypoteneuse of the triangle. That is, the point (x, y_2) is a point on this hypoteneuse. Simple geometry allows us to note that the equation of this line is $x + y = 1$, and so the fact that (x, y_2) is a point on this line gives us $x + y_2 = 1$, and therefore that $y_2 = 1 - x$.

5.4. NON-RECTANGULAR DOMAINS

So we can rewrite our inside integral as

$$A(x) = \int_{y_1(x)}^{y_2(x)} f(x,y)\,dy = \int_0^{1-x} f(x,y)\,dy \tag{5.26}$$

Combining this with Equation 5.23, we have

$$V = \int_0^1 A(x)\,dx = \int_0^1 \int_0^{1-x} f(x,y)\,dy\,dx \tag{5.27}$$

So we have a nested integral, just like we did in the case of a rectangular domain. The only difference is that the bounds on the dy integral involve the variable x.

But remember that the variables play different roles in the two different integrals that appear here. In the outside integral, x is a variable, but in the inside integral, x is a constant and y is the only variable. So while it is correct to say that the bounds on the inside integral are variables from the point of view of the outside integral, they are constants from the point of view of the inside integral.

So this nested integral can be evaluated in the usual way.

Example 5.4.1. Suppose we want to find the volume under the graph of $f(x,y) = x+y$, over the triangle in the xy-plane with vertices at the points $(0,0)$, $(1,0)$ and $(0,1)$.

Noting that this is the same domain used in the derivation above, we can use Equation 5.27 to get

$$V = \int_0^1 \int_0^{1-x} x+y\,dy\,dx$$

The inside integral evaluates normally:

$$\begin{aligned} V &= \int_0^1 \left(xy + \frac{1}{2}y^2\right)\Big|_{y=0}^{y=1-x} dx \\ &= \int_0^1 \left(x(1-x) + \frac{1}{2}(1-x)^2\right) - \left(x(0) + \frac{1}{2}(0)^2\right) dx \\ &= \int_0^1 \frac{1}{2}(1-x^2)\,dx \end{aligned}$$

The remaining integrand for the outside integral depends on x, as it should, partly because the original function depended on x, and partly because in the inside integral we plugged in y values that depended on x, as mandated by the shape of the domain. Still, altogether, this is a single variable integral that can be worked out in the usual way, giving us $V = \frac{1}{3}$.

Of course the above example could also just as well have been done by slicing in the other order. For the outside integral y will range over $[0,1]$, and for the inside integral the range of values for x will depend on y. The reader is encouraged to work out these details as an exercise.

5.4.2 Simpler pictures

Looking at the above work, we notice eventually that even though the problem appears to be three-dimensional since we are computing volume, the work that went into writing down our explicit nested integral really did not involve knowing

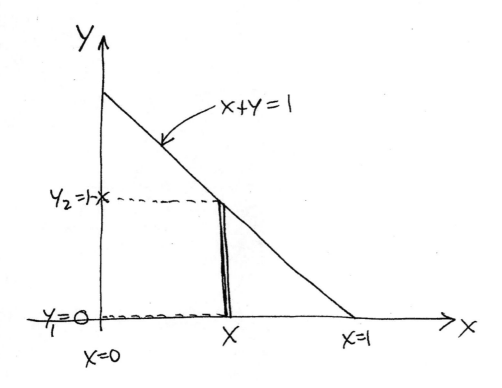

Figure 5.17:

anything about the integrand $f(x, y)$, nor did it involve using any of the three-dimensional aspects of the drawing. The cross-sections were three-dimensional, but as for determining the bounds for the inside and outside integrals all we needed to know about the cross-sections was contained in how those cross-sections intersected the xy-plane.

Of course the function $f(x, y)$ does make a difference in the final answer in that it is the integrand, and that will always be the case – the integrand for a double integral will always be the same as the integrand for the corresponding nested integral.

But the integrand does not make any difference in determining the bounds for the nested integral – only the domain determines the bounds for the nested integral. So, we could simplify our picture greatly by only drawing the domain in $\mathbb{R}^2$.

Let's go through the problem again, with that in mind.

Example 5.4.2. Suppose we want to find the volume under the graph of $f(x, y) = x + y$, over the triangle in the xy-plane with vertices at the points $(0, 0)$, $(1, 0)$ and $(0, 1)$.

The volume under the graph of a function is

$$V = \iint dV = \iint_D f(x, y)\, dA \tag{5.28}$$

We will write this as a nested integral. As we see in Figure 5.17, x ranges over $[0, 1]$, **and so we have**

$$= \int_0^1 \int f(x, y)\, dy\, dx \tag{5.29}$$

For a given fixed value of x, **we know that** y **ranges over some interval** $[y_1(x), y_2(x)]$, **and from the picture we can**

5.4. NON-RECTANGULAR DOMAINS

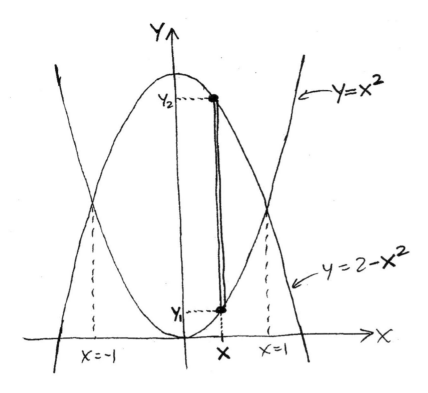

Figure 5.18:

see that $y_1 = 0$ and $y_2 = 1 - x$. So, we get

$$= \int_0^1 \int_0^{1-x} f(x,y)\,dy\,dx \tag{5.30}$$

This is the same integral we had before, and from here the evaluation is the same.

This point of view makes it much easier to write a given double integral problem as a nested integral, as we will see in the several following examples.

Example 5.4.3. Let's find the volume under the graph of $f(x,y) = 4 + x - y$ and above the region in the xy-plane bounded by $y = x^2$ and $y = 2 - x^2$.

The given region in the xy-plane is drawn in Figure 5.18.

We know that the volume in question is the double integral below, and that we can rewrite that as a nested integral. We only need to find the bounds on those individual integrals.

$$V = \iint f(x,y)\,dA = \int \left(\int (4 + x - y)\,dy \right) dx \tag{5.31}$$

We will first cut perpendicular to the x-axis, and note that we get non-empty slices between $x = -1$ and $x = 1$, since those are the points where the two given curves intersect. So these are the bounds on our outside integral.

$$V = \int_{-1}^1 \int (4 + x - y)\,dy\,dx \tag{5.32}$$

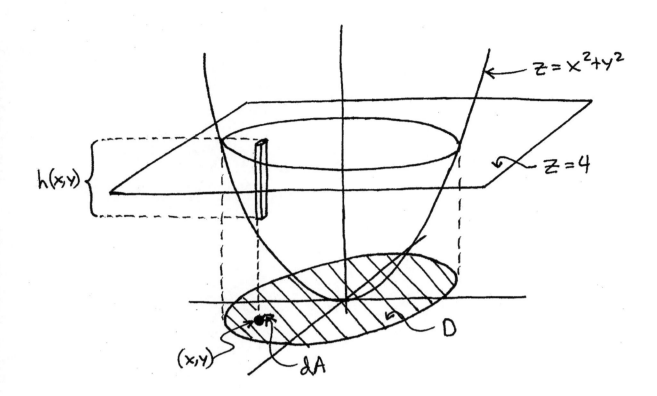

Figure 5.19:

To determine the bounds for the inside integral, we want to find y_1 and y_2 in terms of x, as indicated in Figure 5.18.

Note that for a given value of x, the point (x, y_1) is a point on the curve with equation $y = x^2$; so we conclude that $y_1 = x^2$. Similarly, we note that the point (x, y_2) is a point on the curve with equation $y = 2 - x^2$; so we conclude that $y_2 = 2 - x^2$. Our integral then becomes

$$V = \int_{-1}^{1} \int_{x^2}^{2-x^2} (4 + x - y) \, dy \, dx \qquad (5.33)$$

This nested integral can then be computed by the usual method.

Example 5.4.4. Suppose we want to find the volume bounded between the surfaces $z = 4$ and $z = x^2 + y^2$.

Looking at these two surfaces as in Figure 5.19, we can see that another way to describe this volume is as the volume between the surfaces over the set D in the xy-plane. We can write this as a double integral, adding up the volumes of the thin boxes as drawn. Such a volume dV is the area dA times the height $h(x, y) = (4) - (x^2 + y^2)$. So we have

$$V = \iint_D h(x, y) \, dA = \iint_D 4 - x^2 - y^2 \, dA \qquad (5.34)$$

To turn this into a nested integral, we need to consider the set D. From Figure 5.19, we can see that the boundary of D is the set of points (x, y) over which the two surfaces intersect – in other words, where they share the same z-value. This gives us $x^2 + y^2 = 4$, the equation for the circle which is the boundary of D.

5.4. NON-RECTANGULAR DOMAINS

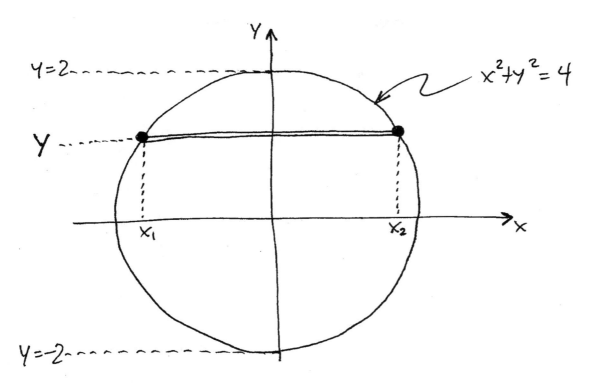

Figure 5.20:

At this point, we no longer need to consider the three-dimensional picture, since at this point all we need to do is determine the bounds for the nested integrals; and as previously noted these depend only on the domain itself. So, we turn our attention now to Figure 5.20.

We can slice in either direction, and this time arbitrarily we will slice perpendicular to the y-axis first. Given the equation $x^2 + y^2 = 4$ for the boundary of D, we easily note that y ranges from -2 to 2, giving us

$$V = \int_{-2}^{2} \int 4 - x^2 - y^2 \, dx \, dy \tag{5.35}$$

To find the bounds on the dx integral, we need to solve for the bound functions $x_1(y)$ and $x_2(y)$. The most immediate relationships we can find between these bounds and y are that the points (x_1, y) and (x_2, y) are both on the boundary circle, with equation $x^2 + y^2 = 4$, so we can conclude then that

$$x_1 = \pm\sqrt{4 - y^2} \quad \text{and} \quad x_2 = \pm\sqrt{4 - y^2} \tag{5.36}$$

Which one is which? Again looking at the picture, we notice that for all values of y we have $x_1 \leq x_2$, and so

$$x_1 = -\sqrt{4 - y^2} \quad \text{and} \quad x_2 = +\sqrt{4 - y^2} \tag{5.37}$$

So our integral becomes

$$V = \int_{-2}^{2} \int_{-\sqrt{4-y^2}}^{\sqrt{4-y^2}} 4 - x^2 - y^2 \, dx \, dy \tag{5.38}$$

5.4.3 Choosing the order

We have already observed that when writing a double integral as a nested integral, we have a choice as to the order in which we do the slicing – we can slice first perpendicular to the x-axis, resulting in a nested integral with differentials

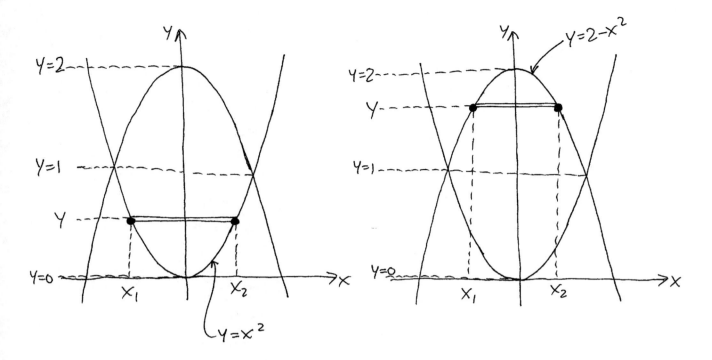

Figure 5.21:

"$dy\,dx$", or we can slice first perpendicular to the y-axis, resulting in a nested integral with differentials "$dx\,dy$". Either way, the theorem says that we will end up with the same numerical answer once the nested integrals are evaluated.

In the previous examples we made seemingly arbitrary choices as to which one of these options to use. And in many of those examples, the choice didn't really make any difference. But sometimes, even though both nested integral expressions represent the same value, one of them might be much more convenient to compute than the other.

To illustrate this point, let's revisit Example 5.4.3, in which we want to compute the volume under the graph of $f(x,y) = 4 + x - y$ and above the region in the xy-plane bounded by $y = x^2$ and $y = 2 - x^2$. This time, we will make our slices in the other order, slicing first perpendicular to the y-axis. From Figure 5.21, it is not hard to see that y goes from 0 to 2., giving us a nested integral of the form:

$$V = \int_0^2 \int (4 + x - y)\,dx\,dy \tag{5.39}$$

But we have a problem when we try to find the range of values for x for a given value of y. In particular, note that for values of y between 0 and 1, the points (x_1, y) and (x_2, y) are on the curve with equation $y = x^2$ – but for values of y between 1 and 2, the points (x_1, y) and (x_2, y) are on the curve with equation $y = 2 - x^2$.

So, depending on the value of y, the expressions for x_1 and x_2 will be different and thus the bounds that we put on the inside dx integral will be different.

In order to solve the problem with this slicing order, we will thus have to break the problem up into two separate pieces – first, the piece for which $y \in [0, 1]$, and second, the piece for which $y \in [1, 2]$.

On the first piece we have the points (x_1, y) and (x_2, y) on the curve with equation $y = x^2$, and so we conclude that $x_1 = -\sqrt{y}$ and $x_2 = \sqrt{y}$. On the second piece we have the points (x_1, y) and (x_2, y) on the curve with equation $y = 2 - x^2$, and so we conclude that $x_1 = -\sqrt{2-y}$ and $x_2 = \sqrt{2-y}$. So the total volume, which is the sum of the volumes computed over these two pieces, is

$$V = \int_0^1 \int_{-\sqrt{y}}^{\sqrt{y}} (4 + x - y)\,dx\,dy + \int_1^2 \int_{-\sqrt{2-y}}^{\sqrt{2-y}} (4 + x - y)\,dx\,dy \tag{5.40}$$

5.4. NON-RECTANGULAR DOMAINS

Clearly this will be a much less convenient computation than the one we arrived at in Example 5.4.3! First, there are two nested integrals here instead of one; second, there are square-roots here, and those are generally less convenient to deal with than polynomials. Of course this less convenient computation will still come out with the same numerical answer – nevertheless, it is still desirable to avoid unnecessary effort, given the opportunity.

Here are some observations that will allow the reader to make convenient choices about the order of slicing.

Note that the reason we ended up with two separate nested integrals in the above case is that the curves making up the boundary of our domain crossed – it was at that value of y ($y = 1$ for both of the crossing points $(-1,1)$ and $(1,1)$) that the bound points (x_1, y) and (x_2, y) switched from one curve to the other, with different equations, thus giving us different formulas for x_1 and x_2 and thus the need for more than one nested integral.

So ultimately, the problem is that our slices hit a corner on the boundary of our domain, and this caused the need for multiple nested integrals. We can summarize this as:

Observation 5.4.1. *When slicing, every time a slice hits a corner we must end the current nested integral and begin a new one.*

Note also that the reason we ended up with square roots in our answer above is that we had to solve for x in terms of y, while the equation was given in a form in which y was already solved for in terms of x. And the reason we had to solve for x is that we were slicing first perpendicular to the y axis. We can summarize this as:

Observation 5.4.2. *When slicing perpendicular to the axis of one variable, one will need to solve for the other variable in terms of that one.*

So roughly speaking, if the equations describing the domain are such that it is easier to solve for y, such as in the case above, there is a convenience to first slicing perpendicular to the x axis. If in another case it were easier somehow to solve for x, then there would be a convenience to first slicing perpendicular to the y axis.

Here are some examples.

Example 5.4.5. Suppose the domain D for a given double integral is the polygon in Figure 5.22. Should we slice first perpendicular to the x-axis, or to the y-axis?

Observation 5.4.1 tells us that every time we hit a corner, we will have to begin a new nested integral. There are four corners, but of course the corners at $(1,1)$ and $(8,4)$ will be the ends of the problem independent of which way we slice. So, we really only need to worry about the corners at $(6,2)$ and $(7,2)$. And since these have the same value of y, we note that we will hit them simultaneously if we slice first perpendicular to the y-axis.

So if we do our dy slices first, we will end up with two nested integrals; if we do our dx slices first, we will end up with three nested integrals.

Observation 5.4.2 doesn't seem too important here, since all of the boundary curves are straight lines. So, slicing perpendicular the y-axis first is probably preferred.

Example 5.4.6. Suppose the domain D for a given double integral is the shape in Figure 5.23. Should we slice first perpendicular to the x-axis, or to the y-axis?

Of course either way we slice, we will encounter a corner and have to make two nested integrals. But in this case, the top curve is described by an equation in which it is much more convenient to solve for y values in terms of x than it is to solve for x values in terms of y. Observation 5.4.2 thus suggests that it would be more convenient to first slice perpendicular to the x axis.

(Of course in this particular example there is even more reason to want to slice perpendicular to the x axis. Slicing perpendicular to the y axis, for any single value of $y \in [-1, 1]$ we would end up with cross-sections consisting of

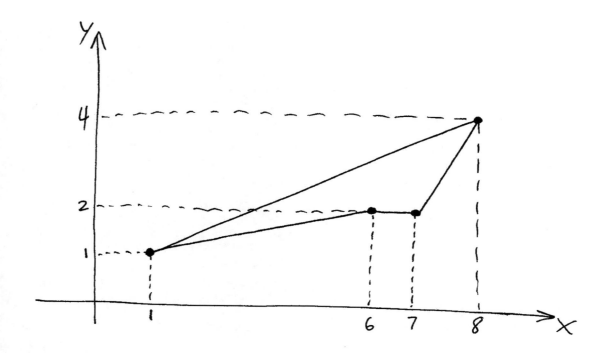

Figure 5.22:

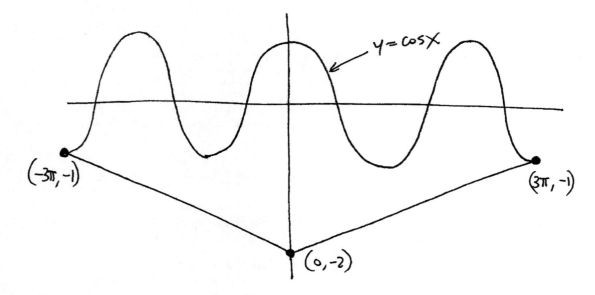

Figure 5.23:

5.4. NON-RECTANGULAR DOMAINS

three distinct intervals, instead of the usual one. So for that range of values of y we would have three nested integrals, plus the other for $y \in [-2, -1]$, making a total of four.)

Exercises

Exercise 5.4.1 (exer-Ints-NRD-1). Suppose we consider the domain that is the solid triangle with vertices at $(2, 2)$, $(4, 0)$, $(3, 5)$. If we fix $y = 3$, what is the range of values of x in this domain?

Exercise 5.4.2 (exer-Ints-NRD-2). Considering the same domain from Exercise 5.4.1, what is the range of values of x in the domain if we fix $y = 1$?

Exercise 5.4.3 (exer-Ints-NRD-3). Write down, but do not evaluate, a nested integral that could be used to compute the integral $\iint_D f(x, y)\, dA$, where D is the part of the unit disk that lies in the second quadrant of the xy-plane.

Exercise 5.4.4 (exer-Ints-NRD-4). Write down, but do not evaluate, a nested integral that could be used to compute the integral $\iint_D f(x, y)\, dA$, where D is the part of the unit disk that lies in the third quadrant of the xy-plane.

Exercise 5.4.5 (exer-Ints-NRD-5). Write down, but do not evaluate, a nested integral that could be used to compute the integral $\iint_D f(x, y)\, dA$, where D is the solid triangle with vertices at $(1, 2)$, $(6, 2)$, $(-1, 5)$.

Exercise 5.4.6 (exer-Ints-NRD-6). Write down, but do not evaluate, a nested integral that could be used to compute the integral $\iint_D f(x, y)\, dA$, where D is the region in the first quadrant bounded by the curves $y = x^2$ and $y = x^3$.

Exercise 5.4.7 (exer-Ints-NRD-7). Compute the volume under the graph of the function $f(x, y) = (x + y)^2$ above the region in the first quadrant bounded by the curves $y = x^2$ and $y = x^3$.

Exercise 5.4.8 (exer-Ints-NRD-8). The population density of the city of Triangula is given by $\delta(x, y) = 10 - x^2 - y^2$, and Triangula is contained in the part of the first quadrant of the xy-plane with $x + y \leq 2$. Compute the total population of Triangula.

Exercise 5.4.9 (exer-Ints-NRD-9). Compute the mass of the region below the curve $y = 3 - x^4$ and above the curve $y = 2x^2$, where the density is given by $\delta(x, y) = 6 + y$.

Exercise 5.4.10 (exer-Ints-NRD-10). The centroid (sometimes called the center of mass) of a two-dimensional sheet D with a given mass density function δ and total mass m is written as $\bar{\mathbf{x}} = (\bar{x}, \bar{y})$, where the coordinates are computed with the formulas

$$\bar{x} = \frac{1}{m} \iint_D x\delta(x, y)\, dA \quad \text{and} \quad \bar{y} = \frac{1}{m} \iint_D y\delta(x, y)\, dA$$

Use these formulas to compute the centroid of the region defined by $x^2 + y^2 \leq R^2$ and $x \geq 0$, with $\delta(x, y) = 1$.

Exercise 5.4.11 (exer-Ints-NRD-11). Show that if the density of a region is a nonzero constant, then the location of the centroid does not depend on what that constant is.

Exercise 5.4.12 (exer-Ints-NRD-12). When a small piece of area in the xy-plane is rotated around a line in the xy-plane, it sweeps out a thin band whose volume is that area times the distance it travels around that line (which of course is 2π times the distance from the small area to the line. Use this idea to write down and compute a double integral representing the volume obtained when the triangle with vertices at $(0, 0)$, $(1, 0)$, and $(0, 1)$ is rotated around the y-axis.

Exercise 5.4.13 (exer-Ints-NRD-13). Write down and compute a double integral representing the volume obtained when the triangle with vertices at $(0, 0)$, $(1, 0)$, and $(0, 1)$ is rotated around the x-axis.

Exercise 5.4.14 (exer-Ints-NRD-14). Write down and compute a double integral representing the volume obtained when the triangle with vertices at $(0, 0)$, $(1, 0)$, and $(0, 1)$ is rotated around the line $x = 2$.

Exercise 5.4.15 (exer-Ints-NRD-15). The Theorem of Pappus states that if L is a line in the xy-plane, and D is a region entirely on one side of that line, then the volume obtained by rotating D around L is equal to the area of D times the distance traveled by the centroid of D (using $\delta = 1$).

1. Assuming that D is a region with $x > 0$, write down a double integral representing the volume obtained by rotating D around the y-axis.

2. Compare this with the formula for $\bar{x}$ and show that this confirms the Theorem of Pappus in this case.

Exercise 5.4.16 (exer-Ints-NRD-16). Use the Theorem of Pappus to compute the value of $\bar{x}$ in Exercise 5.4.10 without using a double integral. *(Hint: Note that when you rotate that semidisk around the y-axis, you get a sphere, for which you already know the volume.)*

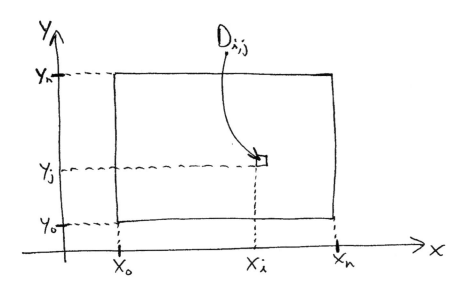

Figure 5.24:

5.5 Triple Integrals

At the beginning of this chapter in Section 5.1, we outlined a set of common conditions shared by integrals, and used those to make a general form for a Riemann sum in Equation 5.2. In Section 5.2 we applied that form to two-dimensional domains to come up with our definition of a double integral, and then in Sections 5.3 and 5.4 we developed tools for computing those double integrals.

In this chapter we will turn our attention to three-dimensional domains. Fortunately, most of the ideas from our development of double integrals extend very naturally to what we will momentarily define as "triple integrals".

5.5.1 Definition and interpretation

In Section 5.2 we motivated our definition of a double integral by a particular problem, namely the computation of the volume under the graph of a function $f : \mathbb{R}^2 \to \mathbb{R}^1$. This was a reasonable choice because of the convenient analogy to the similar and familiar application of single variable integrals, and because the figure corresponding to the question was three-dimensional (remember, the graph lives in a space of dimension equal to the sum of the dimension of the domain and the dimension of the target) and therefore something that we could visualize and draw.

If we try to do the same thing to motivate the definition of triple integrals, we run into a problem. In particular, a function $f : \mathbb{R}^3 \to \mathbb{R}^1$ has a graph which lives in $\mathbb{R}^4$, and we are unable to draw or even visualize this space. So, we will not be able to use this motivation for our definition.

But this is not a tragedy, because that particular application of double integrals was just one of many. For example, double integrals can be used to compute the mass of a two-dimensional sheet, or the population in a two-dimensional area; and each of these can be visualized using only two dimensions. We will be able to generalize these problems nicely for the case of three-dimensional domains, and we can have these problems in mind when we write down the definition of a triple integral.

In general, all we needed for the interpretation of a double integral is a two-dimensional domain, and an integrand that we can interpret as in the integral conditions on page 229 and the resulting general form for a Riemann sum, Equation 5.2. We can do the same thing for triple integrals, using volume as the natural notion of "size" in a three-dimensional setting. We will describe that below, leaving out some of the details that will end up not being used later in this book.

Suppose we have a solid rectangular box $D \subset \mathbb{R}^3$. We can break that region up into smaller pieces $D_{i,j,k}$ (note that

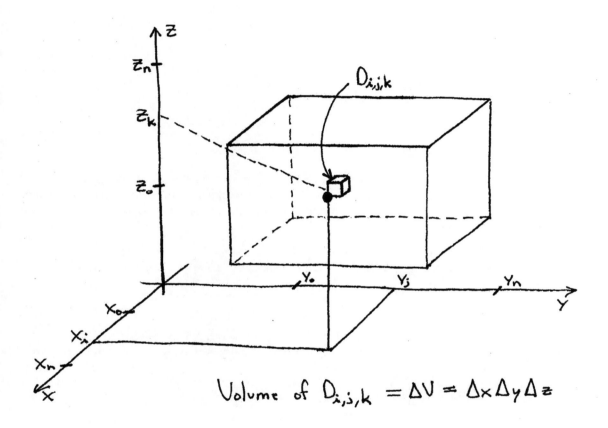

Figure 5.25:

5.5. TRIPLE INTEGRALS

we need three indices, because the set D is three-dimensional) over which our integrand $f : \mathbb{R}^3 \to \mathbb{R}^1$ is approximately constant, and for each piece compute the product of the value of f on that piece and the volume ΔV of that piece. We can then take the limit as the size of those pieces approaches zero. Note in Figures 5.24 and 5.25 the analogy with the two-dimensional picture.

Based on Equation 5.2 then, the triple integral of f over the domain D is defined with:

Definition 5.5.1.
$$\iiint_D f(x,y,z)\,dV = \lim_{n \to \infty} \sum_{i,j,k=1}^{n} f(x_i^*, y_j^*, z_k^*) \Delta V$$

As with double integrals, the "dV" is to be suggestive of the "ΔV" in the three-index Riemann sum. Note that since $\Delta V = \Delta x \Delta y \Delta z$, we sometimes write dV as $dx\,dy\,dz$. And again, the use of three integral symbols is just a reminder that this is a triple integral and not a single integral.

Note that this fits the form from Equation 5.2.

As before with double integrals, we have to modify this slightly to deal with the case of a domain that is not a rectangular box. Again, we will not present the details in this book, but suffice it to say that the intuition is precisely the same as for box domains – we chop up the domain into small pieces over which the integrand is approximately constant, write the sum (over all pieces) of the integrand times the size of the piece, and then take the limit.

Example 5.5.1. Suppose we wish to compute the mass of a solid box occupying the space defined by the solid rectangular box D. Suppose also that we know the density function $\delta(x, y, z)$.

We can break up D into little boxes, with masses Δm, and the total mass m is the sum of those masses. On each piece the mass is given by the volume of the piece times the density on that piece. So we can write

$$m = \lim \sum_{i,j,k} \Delta m = \lim \sum_{i,j,k} \delta(x,y,z) \Delta V \tag{5.41}$$

which in integral notation can be written as

$$m = \iiint dm = \iiint_D \delta(x,y,z)\,dV \tag{5.42}$$

(Note the similarity to Example 5.2.1)

Example 5.5.2. There is a population of bacteria in a large aquarium occupying the space defined by the solid rectangular box D. The number of bacteria per unit volume in the aquarium is known to be $f(x,y,z)$. What is the total population P of bacteria in the aquarium?

We can break up D into little boxes, with populations ΔP, and the total population P is the sum of those populations. On each piece the population is given by the volume of the piece times the populations density f on that piece. So we can write

$$P = \lim \sum_{i,j,k} \Delta P = \lim \sum_{i,j,k} f(x,y,z) \Delta V \tag{5.43}$$

which in integral notation can be written as

$$P = \iiint dP = \iiint_D f(x,y,z)\,dV \tag{5.44}$$

5.5.2 Nested triple integrals

Just as in the case of double integrals, we can compute triple integrals with nested integrals. In the three-dimensional case though we end up with three nested integrals instead of two, each corresponding to slicing up the domain perpendicular to one of the coordinate axes. And just as with double integrals, we can choose the order in which we make those slices – with three to choose from though, there are now six different possible orders. Again, for continuous functions, these will all yield the same answer.

Theorem 5.5.1. *For any continuous function $f : \mathbb{R}^3 \to \mathbb{R}^1$ and rectangular box $D = [x_1, x_2] \times [y_1, y_2] \times [z_1, z_2]$,*

$$\iiint_D f(x,y,z)\, dV = \int_{z_1}^{z_2} \int_{y_1}^{y_2} \int_{x_1}^{x_2} f(x,y,z)\, dx\, dy\, dz = \int_{y_1}^{y_2} \int_{x_1}^{x_2} \int_{z_1}^{z_2} f(x,y,z)\, dz\, dx\, dy = \cdots \quad (5.45)$$

(Note, as with double integrals, we will not present a proof of this theorem.)

This allows us to compute triple integrals almost as conveniently as double integrals.

Example 5.5.3. Suppose that a solid occupies the unit cube D, and has density $\delta(x,y,z) = 2x + 2y + 2z$. What is the mass of this cube?

Using Theorem 5.5.1 we can compute this directly.

$$\begin{aligned}
m &= \iiint_D dm = \iiint_D \delta\, dV = \iiint_D 2x + 2y + 2z\, dV \\
&= \int_0^1 \int_0^1 \int_0^1 2x + 2y + 2z\, dx\, dy\, dz \\
&= \int_0^1 \int_0^1 \left(x^2 + 2xy + 2xz\right)\Big]_{x=0}^{x=1} dy\, dz \\
&= \int_0^1 \int_0^1 1 + 2y + 2z\, dy\, dz \\
&= \int_0^1 \left(y + y^2 + 2yz\right)\Big]_{y=0}^{y=1} dz \\
&= \int_0^1 2 + 2z\, dz \\
&= \left(2z + z^2\right)\Big]_0^1 = 3
\end{aligned}$$

5.5.3 Non-rectangular domains

Just as was the case with double integrals, we can use nested integrals to compute triple integrals even when the domain is not a rectangular box. We only have to be careful when writing down the nested integral to make sure that we keep in mind the role that each of the three variables plays within each of the given integrals, and to write down correctly the bounds for each of the given integrals.

Recall that we showed that when writing a double integral as a nested integral, only the domain was needed to correctly determine the bounds on the nested integrals; so only the two-dimensional picture of the domain was necessary to draw. Similarly in the case of triple integrals, only the three-dimensional picture of the domain is necessary to draw.

And just as with double integrals, finding the correct bounds on triple nested integrals is just a matter of looking at the range of values for the appropriate variables as we continue to slice the domain.

5.5. TRIPLE INTEGRALS

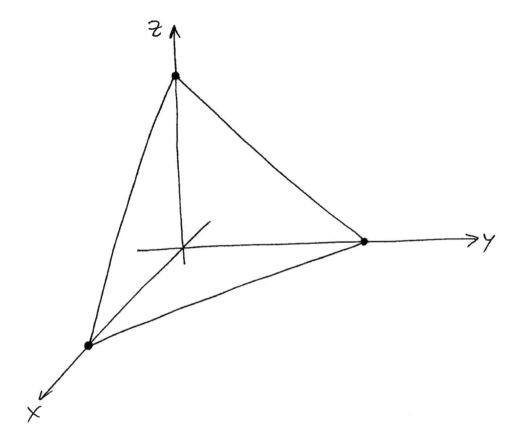

Figure 5.26:

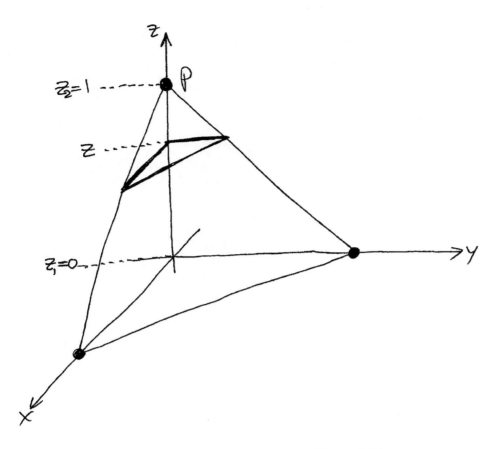

Figure 5.27:

For example, suppose that we wish to compute the triple integral $\iiint_D f \, dV$ for some function f, where D is the tetrahedron bounded by the three coordinate planes and the plane $x + y + z = 1$, as in Figure 5.26.

Arbitrarily, let's slice up the domain first perpendicular to the z-axis, then perpendicular to the x-axis, and lastly perpendicular to the y-axis. So our integral will be of the form

$$\iiint_D f \, dV = \int \int \int f(x, y, z) \, dy \, dx \, dz \tag{5.46}$$

When we slice perpendicular to the z axis as in Figure 5.27, we can easily observe that the lower bound z_1 for z is zero, since the xy-plane is the bottom face of our domain.

The upper bound z_2 for z is the z-coordinate of the point P in Figure 5.27. We see that this point P is the intersection of the plane $x + y + z = 1$ with the other two coordinate planes, which have equations $x = 0$ and $y = 0$. Given these three equations, we can quickly deduce that at the point P we have $z = 1$, and so the upper bound for z is $z_2 = 1$.

So our integral becomes

$$\iiint_D f \, dV = \int_0^1 \int \int f(x, y, z) \, dy \, dx \, dz \tag{5.47}$$

At this point, let's pause and compare our current situation with that in a comparable position in a double integral, as represented in Figure 5.17 from the main example in Section 5.4.1. In that double integal problem, taking that first slice yields a strip of the domain, corresponding to a set of allowed values for the remaining variable (y, in that case).

Here in this triple integral problem, taking our first slice yields an entire area – and this area corresponds to the allowed values for the two remaining variables (x and y, in this case).

5.5. TRIPLE INTEGRALS

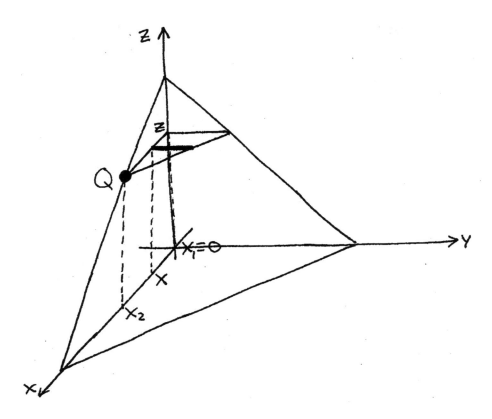

Figure 5.28:

And actually, we can view the remaining problem of finding the bounds for the dx and dy integrals as being just like finding the bounds on a double integral over the region of the slice in Figure 5.27, for a given fixed value of z.

So, to find the range of values for x and y, we fix a value for z and then look at that resulting triangular slice.

The next step we have chosen is to slice this triangular z-slice perpendicular to the x-axis, as in Figure 5.28, yielding strips perpendicular to both the z- and x-axes.

The range of values of x over the triangular z-slice (in other words, the range of x-coordinates of the strip slices in Figure 5.28) will be the bounds for the dx integral. Be careful to note however that this range of values can depend on z, since different values of z result in different sizes of triangular slices. So the bounds x_1 and x_2 for x will be functions of z, and thus could be written as $x_1(z)$ and $x_2(z)$.

Again, the lower bound x_1 for x is easily observed to be zero since, for any value of z, the strip slice with the smallest value of x is in the yz-plane.

The upper bound x_2 for x is the x-coordinate of the point Q in Figure 5.28. There are three pieces of information we can write down about Q from Figure 5.28 – first, it is on the plane $x + y + z = 1$; second, it is on the xz-plane and so $y = 0$; third, its z-coordinate is the given fixed value of z. Given this, we conclude that the x-coordinate of this point is $x = 1 - z$, and so the upper bound x_2 for x is $x_2 = 1 - z$.

So our integral becomes

$$\iiint_D f \, dV = \int_0^1 \int_0^{1-z} \int f(x, y, z) \, dy \, dx \, dz \tag{5.48}$$

Now, we fix a value of x, so that both z and x are now fixed, and look at the resulting single strip in Figure 5.29.

The next step we have chosen is to slice this strip perpendicular to the y-axis, as in Figure 5.29.

The range of values of y over this strip will be the bounds for the dy integral. At this point though note that this range

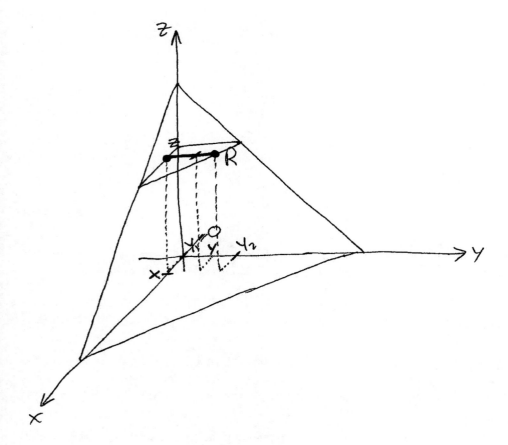

Figure 5.29:

5.5. TRIPLE INTEGRALS

of values will depend on both z and x, since different choices of those now fixed variables will result in different strips. So the bounds y_1 and y_2 for y will be functions of both z and x, and thus could be written as $y_1(x, z)$ and $y_2(x, z)$.

Again the lower bound y_1 for y is easily observed to be zero since for any values of x and z, the point on the strip with the lowest value of y is on the xz-plane.

The upper bound y_2 for y is the y-coordinate of the point R in Figure 5.29. There are three pieces of information we can write down about R from Figure 5.29 – first, it is on the plane $x + y + z = 1$; second, its z-coordinate is the given fixed value of z; third, its x-coordinate is the given fixed value of x. Given this, we conclude that the y-coordinate of this point is $y = 1 - x - z$, and so the upper bound y_2 for y is $y_2 = 1 - x - z$.

So our integral becomes

$$\iiint_D f\, dV = \int_0^1 \int_0^{1-z} \int_0^{1-x-z} f(x, y, z)\, dy\, dx\, dz \tag{5.49}$$

This can be evaluated in the usual way, starting with the inside integral and just proceeding one integral at a time.

Example 5.5.4. Suppose we want to find the mass of the region above, if the density is given to be $\delta(x, y, z) = 6x$.

By the above reasoning, we can evaluate the mass as

$$\begin{aligned}
m &= \iiint_D dm = \iiint_D \delta\, dV = \iiint_D 6x\, dV & (5.50) \\
&= \int_0^1 \int_0^{1-z} \int_0^{1-x-z} 6x\, dy\, dx\, dz & (5.51) \\
&= \int_0^1 \int_0^{1-z} (6xy]_{y=0}^{y=1-x-z}\, dx\, dz & (5.52) \\
&= \int_0^1 \int_0^{1-z} 6x - 6x^2 - 6xz\, dx\, dz & (5.53) \\
&= \int_0^1 \left(3x^2 - 2x^3 - 3x^2 z\right]_{x=0}^{x=1-z}\, dz & (5.54) \\
&= \int_0^1 \left(3(1-z)^2 - 2(1-z)^3 - 3(1-z)^2 z\right)\, dz & (5.55) \\
&= \int_0^1 (1-z)^3\, dz & (5.56) \\
&= \left(\frac{-1}{4}(1-z)^4\right]_0^1\, dz = \frac{1}{4} & (5.57) \\
& & (5.58)
\end{aligned}$$

Note that just like with double integrals, the variables play different roles at different points in the nested integral.

For the inner-most integral (corresponding to the last slicing), both other variables are fixed, and thus treated like constants; the bounds for that integral may depend on both of those other variables. For the second integral, the variable for the inner-most integral has disappeared, but the outer-most variable is still fixed and treated like a constant; the bounds for that integral may depend on the variable for the outer-most integral. For the outer-most integral, both other variables have disappeared, and the bounds must be actual numbers. For example, if we slice perpendicular to axes in order z first, then y, then x, we would have the following:

$$\iiint f\, dV = \int_{z_1}^{z_2} \left(\int_{y_1(z)}^{y_2(z)} \underbrace{\left(\int_{x_1(y,z)}^{x_2(y,z)} \underbrace{(f(x,y,z))}_{\text{fn of } x,y,z;\ x \text{ is dummy var., } y,z \text{ const.}} dx \right)}_{\text{fn of } y,z;\ y \text{ is dummy var., } z \text{ const.}} dy \right) dz \underbrace{}_{\text{fn of } z;\ z \text{ is dummy var.}} \tag{5.59}$$

5.5.4 Examples

Here are a few more examples.

Example 5.5.5. Suppose we have a triple integral over the unit ball $B \subset \mathbb{R}^3$, and we wish to write that as a nested integral. Arbitrarily, let's slice first perpendicular to the x-axis, then then the z-axis, and finally the y-axis. So our integral will take the form

$$\iiint_B f(x,y,z)\, dV = \int \int \int f(x,y,z)\, dy\, dz\, dx \tag{5.60}$$

To find the range of values of x for the bounds on the outside integral, we look at Figure 5.30 and see that the slices of B perpendicular to the x-axis start at $x = -1$ and end at $x = 1$.

For such a given fixed value of x then, what is the range of values of z? In Figure 5.31, we see that the range of values of z is defined by the z-coordinates of the points P_1 and P_2. These two points are on the boundary of the unit ball, and so we know that for each one, $x^2 + y^2 + z^2 = 1$. Note also that each one has the given fixed value of x, and that each one is on the xz-plane defined by $y = 0$. Solving, we get that z ranges from $z_1 = -\sqrt{1-x^2}$ to $z_2 = \sqrt{1-x^2}$. So at this point our integral has become

$$\iiint_B f(x,y,z)\, dV = \int_{-1}^{1} \int_{-\sqrt{1-x^2}}^{\sqrt{1-x^2}} \int f(x,y,z)\, dy\, dz\, dx \tag{5.61}$$

Now, taking fixed values of both x and z, the range of values of y is represented by the strip indicated in Figure 5.32.

The bounds for y then are determined by the y-coordinates of the points Q_1 and Q_2. Of course each of these points is on the boundary of the ball and thus $x^2 + y^2 + z^2 = 1$. And we also know that x and z are the given fixed values. This allows us to solve for the dy bounds, giving us

$$\iiint_B f(x,y,z)\, dV = \int_{-1}^{1} \int_{-\sqrt{1-x^2}}^{\sqrt{1-x^2}} \int_{-\sqrt{1-x^2-z^2}}^{\sqrt{1-x^2-z^2}} f(x,y,z)\, dy\, dz\, dx \tag{5.62}$$

Of course we made an arbitrary choice of the order in which to slice up the domain. By similar arguments we could have written the given triple integral as five different nested integrals.

Example 5.5.6. Suppose we have a triple integral over the region D bounded between the two paraboloids $z = 2x^2 + 2y^2$ and $z = 1 + x^2 + y^2$, and we wish to write that as a nested integral. As we see in Figure 5.33, the shape in question is somewhat like a bowl, in that the top surface is not flat, but concave.

5.5. TRIPLE INTEGRALS

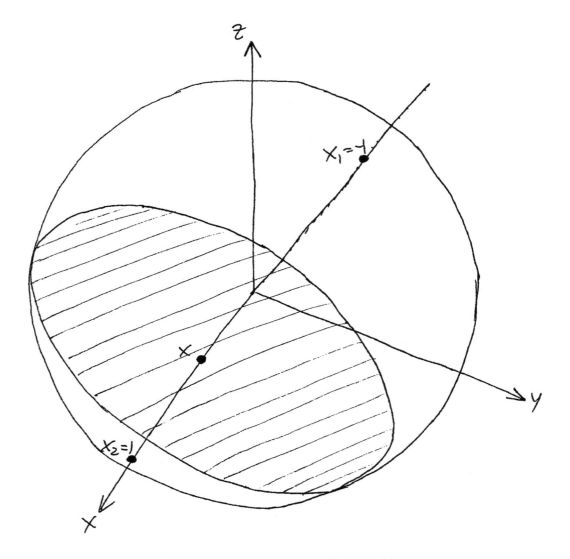

Figure 5.30:

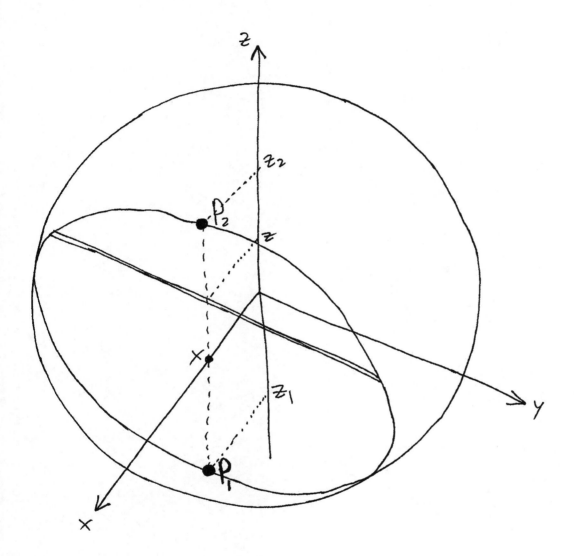

Figure 5.31:

5.5. TRIPLE INTEGRALS

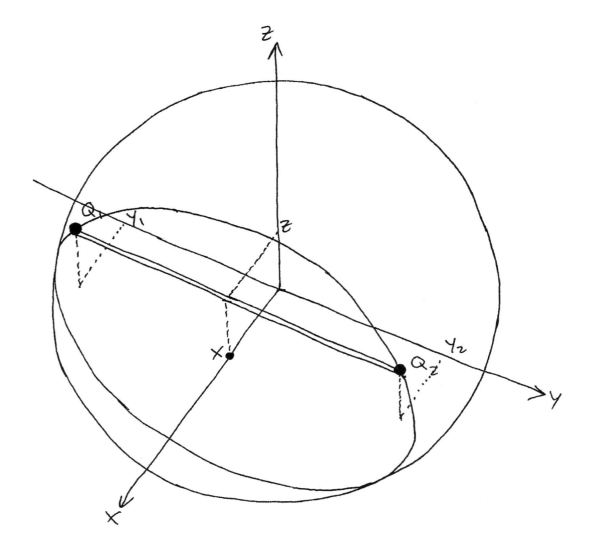

Figure 5.32:

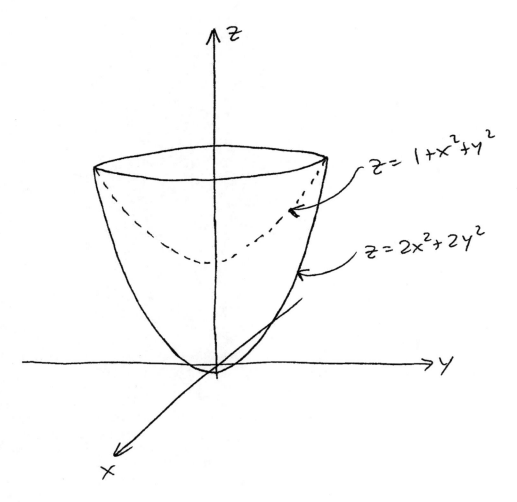

Figure 5.33:

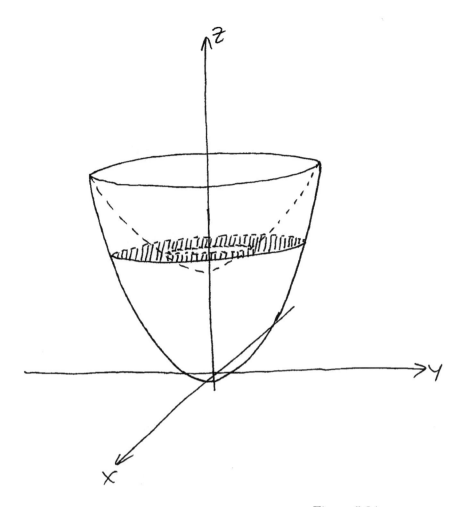

Figure 5.34:

In this example we will have to be very careful about the order in which we do our slicing, for reasons similar to those that we identified for double integrals. In particular, if possible we would like for the slices that we take to be simple enough that making further slices will not lead to writing down multiple nested integrals.

For example, what would happen if we were to make our first slice perpendicular to the z-axis? Well, as is indicated in Figure 5.34, some of those slices would be annular regions. And unfortunately once we slice one of those, in either direction, some of those resulting slices would be pairs of strips, not just single strips representing single intervals for a variable. So, if we made z slices first, we would have an inconvenient result.

As in Example 5.4.4, let's note that we can describe our domain D as the region above the surface $z = 2x^2 + 2y^2$ and below the surface $z = 1 + x^2 + y^2$, above the region in the xy-plane whose boundary is defined by the intersection of the two given surfaces, which is described by $2x^2 + 2y^2 = 1 + x^2 + y^2$, which is of course the unit circle $x^2 + y^2 = 1$.

If we choose to do x slices first and then y, we can easily note that the ranges of values of x and y over the domain D are the same as those over the unit circle. So, we can conveniently write

$$\iiint_D f(x, y, z)\, dV = \int_{-1}^{1} \int_{-\sqrt{1-x^2}}^{\sqrt{1-x^2}} \int f(x, y, z)\, dz\, dy\, dx \tag{5.63}$$

To determine the bounds for the dz integral, we need to find the range of values of z for given fixed values of x and y. As we see in Figure 5.35, that range is defined by the z-coordinates of the points P_1 and P_2; of course we can see from the figure that P_1 is on the surface with equation $z = 2x^2 + 2y^2$, and P_2 is on the surface with equation $z = 1 + x^2 + y^2$.

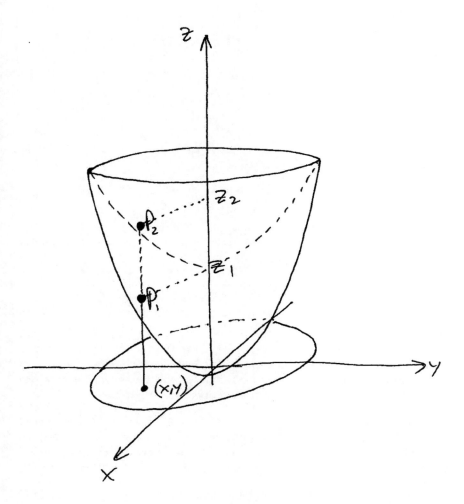

Figure 5.35:

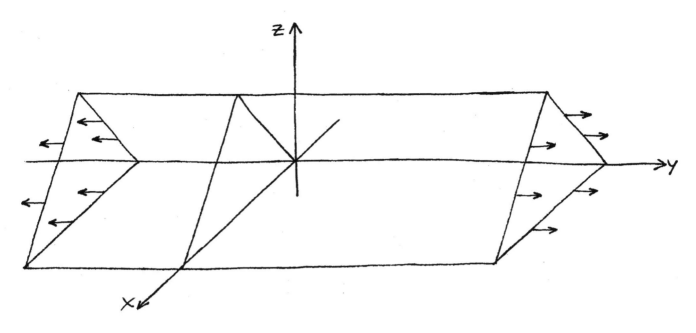

Figure 5.36:

So we can finish our integral as

$$\iiint_D f(x,y,z)\,dV = \int_{-1}^{1} \int_{-\sqrt{1-x^2}}^{\sqrt{1-x^2}} \int_{2x^2+2y^2}^{1+x^2+y^2} f(x,y,z)\,dz\,dy\,dx \tag{5.64}$$

Example 5.5.7. Suppose we have a triple integral over the region D bounded by the following five planes: $z = 0$, $z = x$, $z = 2 - x$, $y = 0$, and $y = 3 - z$. Let's rewrite this as a nested integral, again hopefully in such a way that we do not have to write down more than one. So again we will need to be careful about the slicing order.

First we need to draw a picture of this domain. There are five planes, and we could just draw all five planes first. But, three-dimensional pictures on a two-dimensional piece of paper are sometimes hard to visualize, especially when there are many intersections and overlaps.

In order to facilitate the creation of our diagram, let's first notice that the first three planes are completely devoid of mention of the variable y. So each one of these three planes is parallel to the y-axis, and in fact it is not hard to see that they form a triangular cylinder, drawn in **Figure 5.36**.

Then our complete picture of the domain D is formed by noting that the final two planes chop through the given triangular cylinder, giving us **Figure 5.37**.

With this figure we can now move on to choosing a first slicing direction. What would happen if we did x slices first? This does give us convenient cross-sections, but the problem is that the upper bound for z on such slices is defined by two different surfaces; between $x = 0$ and $x = 1$ the upper z bound is the plane $z = x$, but between $x = 1$ and $x = 2$ the upper z bound is the plane $z = 2 - x$. So, we would have to write this down as two separate nested integrals. Of course this is completely feasible, but we hope to find an even more convenient approach.

Note, the reason we encountered the problem above is that in the x slicing process we hit a "corner", namely the line that is the intersection of the two planes $z = x$ and $z = 2 - x$. As was the case with double integrals, slicing through corners generally causes a need for multiple nested integrals.

What about doing y slices first? Again, it is not hard to see that we will encounter a corner and thus the need for

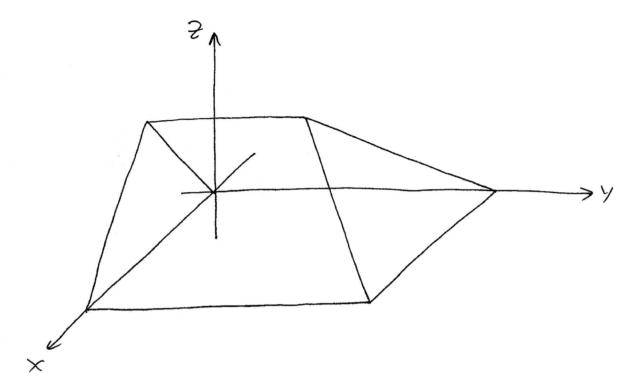

Figure 5.37:

multiple nested integrals. In this case the corner in question is the point at the intersection of the three planes $z = x$, $z = 2 - x$ and $y = 3 - z$.

Our only remaining choice then seems to be to do z slices first. We notice that we do not hit any corners of any kind while doing these slices except at the beginning and end, and so one single nested integral should do it. The diagram clearly shows that z starts at zero, and ends at the z-coordinate of the intersection of the planes $z = x$ and $z = 2 - x$, which is 1 (see Figure 5.38). So we have

$$\iiint_D f(x, y, z)\, dV = \int_0^1 \int \int f(x, y, z)\, dx\, dy\, dz \tag{5.65}$$

To determine the range of values for x and y, we fix z and look at that slice. For any fixed z we can see that the slice is bounded by the intersections of that z plane with the four planes $y = 0$, $y = 3 - z$, $x = z$, and $x = 2 - z$. Of course since z is fixed, the four equations above represent constant values of x and y, and so in fact this z slice is a rectangle, with the four equations above representing the ranges of values of x and y. So we can finish our nested integral in one step as

$$\iiint_D f(x, y, z)\, dV = \int_0^1 \int_0^{3-z} \int_z^{2-z} f(x, y, z)\, dx\, dy\, dz \tag{5.66}$$

or, equivalently

$$\iiint_D f(x, y, z)\, dV = \int_0^1 \int_z^{2-z} \int_0^{3-z} f(x, y, z)\, dy\, dx\, dz \tag{5.67}$$

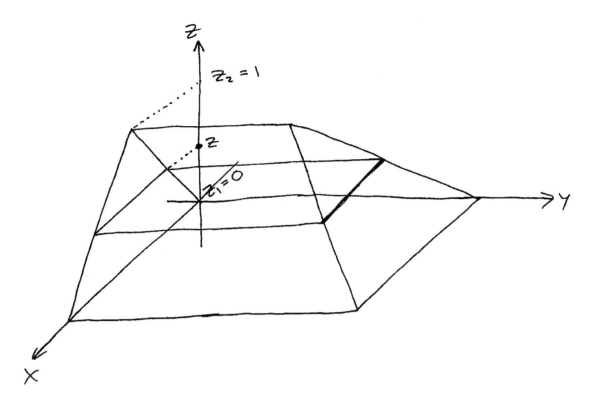

Figure 5.38:

Exercises

Exercise 5.5.1 (exer-Ints-TI-1). Suppose we are subdividing the rectangular box $[3, 6] \times [0, 7] \times [-4, 1]$ into sub-boxes. Write out a general expression for the sub-box $D_{i,j,k}$ in terms of i, j, k, and n.

Exercise 5.5.2 (exer-Ints-TI-2). Suppose we wish to compute the value of the triple integral $\iiint_D x + 3y - 2z \, dV$, where D is the box from Exercise 5.5.1. Use the form of a three-index Riemann sum to approximate this value by directly computing the value of the summation when $n = 2$, $x_i^* = x_i$, $y_j^* = y_j$, and $z_k^* = z_k$.

Exercise 5.5.3 (exer-Ints-TI-3). Write down, but do not evaluate, a triple nested integral representing the mass of the solid R bounded by the surfaces $x = 0$, $y = 0$, $z = 0$, and $3x + 4y + z = 12$, with density given by $\delta(x, y, z) = x^2 e^{yz}$.

Exercise 5.5.4 (exer-Ints-TI-4). Write down, but do not evaluate, a triple nested integral representing the number of molecules of methane in a spherical balloon of radius 7 centered at the origin, where the concentration of methane (in billions of molecules per unit volume) is given by $C(x, y, z) = 25e^{z-5}$.

Exercise 5.5.5 (exer-Ints-TI-5). The centroid (sometimes called the center of mass) of a three-dimensional region R with a given mass density function δ is written as $\bar{\vec{x}} = (\bar{x}, \bar{y}, \bar{z})$, where the coordinates are computed with the formulas

$$\bar{x} = \frac{1}{m} \iiint_R x\delta(x, y, z) \, dV \quad \text{and} \quad \bar{y} = \frac{1}{m} \iiint_R y\delta(x, y, z) \, dV \quad \text{and} \quad \bar{z} = \frac{1}{m} \iiint_R z\delta(x, y, z) \, dV$$

Use these formulas to compute the centroid of the region defined by $x^2 + y^2 + z^2 \leq k^2$ and $x \geq 0$, with $\delta(x, y, z) = 1$.

Exercise 5.5.6 (exer-Ints-TI-6). Students who have had a high school course in physics will recall that the kinetic energy of a mass m moving with linear speed v is given by the formula $KE = \frac{1}{2}mv^2$.

If the mass in a region R is thought of as moving in a circular orbit around an axis L, with an angular velocity ω, different parts of the mass are moving with different linear speeds, as represented by the equation $v = r\omega$, where r is the distance to the axis of rotation L. Writing the total kinetic energy then as

$$KE = \iiint_R d(KE) = \iiint_R \frac{1}{2}v^2 \, dm = \iiint_R \frac{1}{2}r^2\omega^2 \, dm = \frac{1}{2}\left(\iiint_R r^2 \, dm\right)\omega^2$$

we have a formula for rotational kinetic energy that has a similar form to that for linear kinetic energy. As ω is the angular analogue of velocity in this equation, similarly the expression

$$I = \iiint_R r^2 \, dm = \iiint_R r^2 \delta \, dV$$

is the angular analogue of mass. This expression I is called the *moment of inertia* of R around L.

Compute the moment of inertia around the z-axis of the solid unit box $[0,1] \times [0,1] \times [0,1]$ with density given by $\delta = x^2 + y^2 + z^2$.

Exercise 5.5.7 (**exer-Ints-TI-7**). Compute the moment of inertia around the x-axis of the solid cylinder defined by $y^2 + z^2 \leq 1$ and $0 \leq x \leq 1$, with $\delta = 1$.

Exercise 5.5.8 (**exer-Ints-TI-8**). Write out, but do not evaluate, a triple nested integral that represents the moment of inertia around the line defined by $x = 3$ and $z = 5$ of the region R bounded by the surfaces $z = x^2 + y^2$ and $z = 2 - x^2 - y^2$, with $\delta = 1$.

Exercise 5.5.9 (**exer-Ints-TI-9**). Write out, but do not evaluate, a triple nested integral that represents $\iiint_R f(x,y,z) \, dV$, where R is the region bounded by the surfaces $y = 0$, $z = 0$, $3y + 2z = 6$, $x = -y^2$, and $x = 1 + z^2$.

Exercise 5.5.10 (**exer-Ints-TI-10**). Write out, but do not evaluate, a triple nested integral that represents $\iiint_R f(x,y,z) \, dV$, where R is the region bounded by the surfaces $z = \left(x^2 + y^2 - 1\right)^2$ and $z = 64$.

Exercise 5.5.11 (**exer-Ints-TI-11**). Write out, but do not evaluate, a triple nested integral that represents $\iiint_R f(x,y,z) \, dV$, where R is the region inside of both of the spheres $x^2 + y^2 + z^2 = 25$ and $(x-6)^2 + y^2 + z^2 = 25$

Exercise 5.5.12 (**exer-Ints-TI-12**). Write out, but do not evaluate, a triple nested integral that represents $\iiint_R f(x,y,z) \, dV$, where R is the region inside the sphere $x^2 + y^2 + z^2 = 25$ and also inside the cylinder $y^2 + z^2 = 1$.

Exercise 5.5.13 (**exer-Ints-TI-13**). Write out, but do not evaluate, a triple nested integral that represents $\iiint_R f(x,y,z) \, dV$, where R is the region inside the sphere $x^2 + y^2 + (z-3)^2 = 16$ and outside of the sphere $x^2 + y^2 + z^2 = 25$.

5.6. CHANGE OF VARIABLES

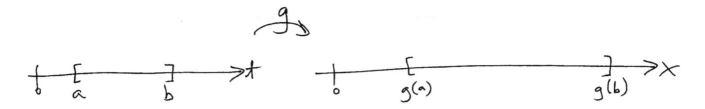

Figure 5.39:

5.6 Change of Variables

In this section we will use an idea that will be used again in several of the coming sections. Loosely speaking, the idea is the following: If for some reason the domain for a given integral is undesirable, you can rewrite the integral over a different domain, just so long as you are careful to make the appropriate modifications to the integrand. The details will depend on the setting, and will be discussed individually in this and several subsequent sections.

This will be a powerful tool, since as we have already seen it can sometimes be very inconvenient to write down multi-variable integrals over some types of domains.

The first example of this strategy will be a re-derivation of the substitution rule from single variable calculus.

(Of course, in practice the substitution rule is usually used with a different motivation than we will have when dealing with multivariable integrals; that is, the substitution rule is usually used for the purpose of changing the integrand in the hopes of facilitating the search for the antiderivative – changing the domain is only a detail that goes along with the result. With that point acknowledged, we will nevertheless present it here with the motivation of changing the domain, since our goal in revisiting this result is to outline such a method.)

5.6.1 The substitution rule

Suppose we have a single variable function $f(x)$, and we wish to compute an antiderivative $F(x)$ with $F'(x) = f(x)$. The substitution rule sometimes makes this more convenient, by allowing us to write $x = g(t)$, and then rewriting the antiderivative in terms of the new variable t. Algebraically we write the substitution rule for antiderivatives as

$$\int f(g(t))g'(t)\, dt = \int f(x)\, dx \tag{5.68}$$

This can be proved without too much trouble by using the chain rule, and the reader is encouraged to refer to a single variable calculus textbook to see that proof, even though it will not be used in this book.

Note that if we wish to change the above statement about antiderivatives into a statement about integrals, we must be careful about the bounds of those integrals. In particular, writing down bounds on the left integral suggests that these are values of t, so on the right integral we would have to write down the corresponding values of $x = g(t)$, which of course will not be the same numbers. Specifically then, the substitution rule for integrals is

$$\int_a^b f(g(t))g'(t)\, dt = \int_{g(a)}^{g(b)} f(x)\, dx \tag{5.69}$$

In this section, we will take a different point of view on Equation 5.69. Rather than thinking of this as an algebraic consequence of the chain rule, we will observe that this can be thought of as the result of taking a dx integral, and rewriting it with a different domain on the t-axis.

Let's represent the function g with separate domain and target, as in Figure 5.39. For convenience, we will take g to be monotonic; the interval $[a, b]$ on the t-axis then has as its image the interval $[g(a), g(b)]$ on the x-axis.

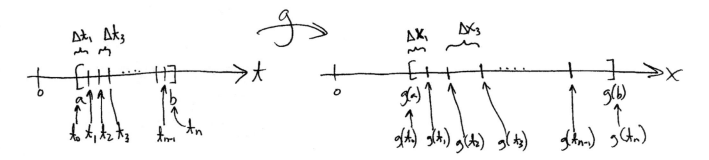

Figure 5.40:

Suppose that we wish now to represent an integral of the function $f(x)$ over the interval $[g(a), g(b)]$ on the x-axis. Recall that this means we will be chopping up that interval into small subintervals, and on each one we will multiply the value of the function there times the size of that subinterval.

First, let's recall that we can chop up the given interval $[g(a), g(b)]$ on the x-axis in any way we like – we do not have to chop using equally sized subintervals. With this slight modification then, we can write down the integral over the x-axis as a Riemann sum:

$$\int_{g(a)}^{g(b)} f(x)\,dx = \lim_{n\to\infty} \sum_{i=1}^{n} f(x_i)\Delta x_i \tag{5.70}$$

Note that since the subintervals on the x-axis are not all the same size, the Δx_i will be different for each subinterval.

In this case we will chop up the interval on the x-axis in the following way. First subdivide the interval $[a, b]$ on the t-axis, bounded by points $a = t_0, t_1, t_2, \ldots, t_n = b$. Each of these subintervals has an image on the x-axis, bounded by the image points $x_0 = g(t_0) = g(a), x_1 = g(t_1), x_2 = g(t_2), \ldots, x_n = g(t_n) = g(b)$. Those image intervals will not all be the same size, even if the t-subintervals were; but still, the collection of all of those images forms a subdivision of the interval $[g(a), g(b)]$ on the x-axis. See Figure 5.40.

(This might seem to be a pointlessly inconvenient way to form the subdivision of the interval on the x-axis, but it will have the advantage that it will allow us to manipulate the Riemann sum in Equation 5.70 in such a way that we can rewrite the given integral as a dt integral, instead of a dx integral.)

Note that there is a relationship between the size Δt_i of a t subinterval, and the size Δx_i of its image subinterval on the x-axis. In particular, recall that the derivative of a function gives a relationship between the change in the input and the change in the output of a function, and the above intervals represent changes in the input and output of the function g. So, we can write $\Delta x_i \approx g' \Delta t_i$ (the error in this approximation disappears in the limit), and thus rewrite the above Riemann sum as

$$\int_{g(a)}^{g(b)} f(x)\,dx = \lim_{n\to\infty} \sum_{i=1}^{n} f(x_i)\Delta x_i = \lim_{n\to\infty} \sum_{i=1}^{n} f(x_i)\Big(g'(t_i)\Delta t_i\Big) \tag{5.71}$$

Up to this point we have been viewing this Riemann sum as a representation of an integral that is taking place over the x-axis. And as it is written above, note that the summand is still a product of the value of the function f at a point x_i in an interval on the x-axis, and the size $\Delta x_i \approx g'\Delta t_i$ of that subinterval on the x-axis.

But simply by plugging in $x_i = g(t_i)$ and switching the parentheses around, we have

$$\int_{g(a)}^{g(b)} f(x)\,dx = \lim_{n\to\infty} \sum_{i=1}^{n} f(g(t_i))\Big(g'(t_i)\Delta t_i\Big) = \lim_{n\to\infty} \sum_{i=1}^{n} \Big(f(g(t_i))g'(t_i)\Big)\Delta t_i \tag{5.72}$$

The summand for this last Riemann sum is clearly written as a product of the value of a new function $f(g(t))g'(t)$ at a point t_i in an interval on the t-axis, and the size Δt_i of that subinterval on the t-axis. So this can now be thought of as

5.6. CHANGE OF VARIABLES

an integral on the t-axis, over the interval defined by the values over which the t_i range, namely $[a,b]$. So we have

$$\int_a^b f(g(t))g'(t)\,dt = \int_{g(a)}^{g(b)} f(x)\,dx \tag{5.73}$$

This of course is the same result that we had in Equation 5.69.

Despite the fact that the result we have proved was already known, this derivation we have just gone through serves to illustrate the idea that we referred to at the beginning of this section. We started with an integral over one domain, $[g(a), g(b)]$, and we have ended up with a different integral over a different domain, $[a,b]$.

As promised, we did have to change the integrand appropriately, and by going through this derivation we have derived those appropriate changes. For one thing, of course, we needed to get rid of all of the x's, which of course we did very simply by writing $x = g(t)$ in the given integrand.

More interestingly though, we have a new factor that has appeared in the integrand – namely, the factor $g'(t)$. In the derivation above, this factor is viewed as the relationship between the sizes of subintervals corresponding by the function g. Restating this part of the above derivation, the function g "stretches" pieces of the t-axis into pieces of different size on the x-axis, and the "stretching factor" of this action is the derivative $g'(t)$. So, using integral language, a piece of the t-axis of size dt is stretched by a factor $g'(t)$ to become a piece of the x-axis of size dx.

In this section we will be using pictures and arguments very similar to this one that we just went through to derive the idea of change of variables for multivariable integrals. In fact we will be doing the same in most of the remaining sections of this chapter, so this general strategy is very important.

Here is a symbolic summary of the general strategy. Suppose that we have some sort of integral, with a domain R and integrand f, which we will write here generally as $\int_R f\,dR$; this could be either a double integral or a triple integral, and we write dR to represent the size of the chunks of R as we subdivide it in defining the integral. To rewrite this integral, we must:

1. Find the given domain R as the image of another domain, D, by some one-to-one function g.

2. Subdivide D into chunks of size dD, and in so doing form a subdivision of R into chunks of size dR.

3. Find the "stretching factor" S_g by which the function g stretches out chunks of size dD into chunks of size dR, giving us $dR = S_g dD$.

4. Rewrite the original integral as

$$\int_D f \cdot S_g\,dD = \int_R f\,dR \tag{5.74}$$

In other words, if you want to rewrite an integral with a different domain, relating the given variables to new variables by way of a change of variables function g, then just make the substitution – and remember to insert the stretching factor S_g to account for the way the change of variables function distorted the domain.

We will refer to this as the "stretching factor strategy"; and we will derive the formula for the stretching factor S_g in each application as we apply this general strategy to several different situations throughout the rest of this chapter.

5.6.2 Change of variables in $\mathbb{R}^2$

We will begin by trying to apply this stretching factor strategy outlined above to rewrite a double integral over a different domain. Suppose for example that we wish to compute the integral $\iint_R f(x,y)\,dx\,dy$ over the domain R in the xy-plane, as shown in Figure 5.41. Note that if we did try to write such an integral as a nested integral, we would have to break the integral up into several pieces, making the task inconvenient at best.

As is indicated in the strategy outline, in order to change the domain we first need to find a one-to-one function g and a domain D in a new coordinate plane such that R is the image of D. Let us suppose that we can find such a function g,

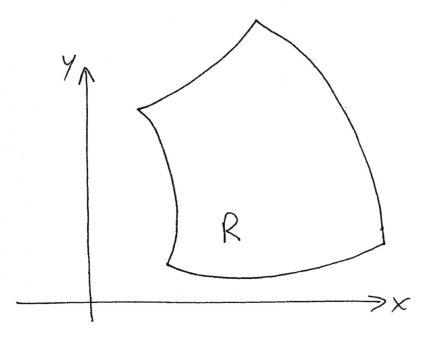

Figure 5.41:

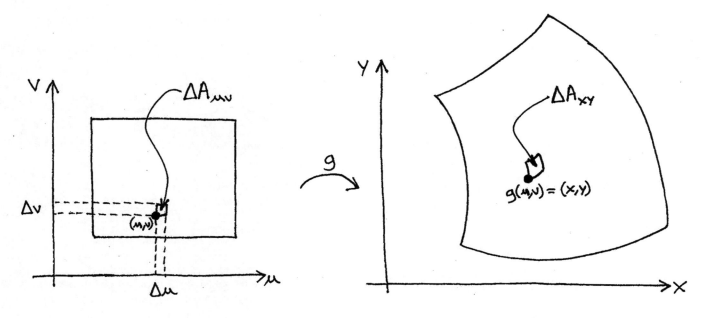

Figure 5.42:

5.6. CHANGE OF VARIABLES

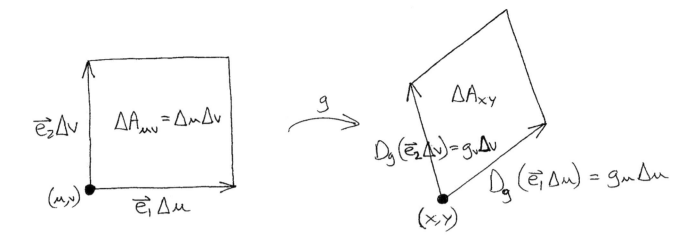

Figure 5.43:

from the uv-plane to the xy-plane, so that the domain D in the uv-plane is a more convenient shape (such as a rectangle), giving us Figure 5.42.

Continuing to follow the outlined strategy, we can subdivide D into rectangular pieces, of size $\Delta A_{uv} = \Delta u\, \Delta v$. The images of these rectangles have size which we will write as ΔA_{xy}, and form a partition of R, which can be used to represent the given integral over R with

$$\iint_R f(x,y)\, dx\, dy = \lim \sum f(x_i, y_i)\, \Delta A_{xy} \tag{5.75}$$

So we will write the given integral also as

$$\iint_R f(x,y)\, dx\, dy = \iint_R f(x,y)\, dA_{xy} \tag{5.76}$$

In order to rewrite this integral as an integral over D, we will again need to rewrite all of the components in terms of u and v. As before, we can of course write $f(x,y) = f(g(u,v))$ by simple substitution.

Changing the differential will again require us to measure how much the change of variables function g has stretched out the domain. In this case, we need to be able to measure the factor by which ΔA_{uv} is stretched out to form ΔA_{xy}.

In Figure 5.43, we draw expanded images of ΔA_{uv} and ΔA_{xy}. We know from Chapter 4 that differentiable functions are locally linear, and so the image of this rectangle represented by ΔA_{uv} is approximately some sort of parallelogram. (Of course it will not be exactly a parallelogram, but in the limit as we take smaller rectangles the approximation is valid. So we will draw the image with area ΔA_{xy} as an actual parallelogram.)

In order to compute ΔA_{xy} then, we need only find the two defining vectors and apply the determinant formula from Chapter 1.

Since ΔA_{uv} is a rectangle of width Δu and height Δv, we can represent its two defining edges by vectors $\vec{e}_1 \Delta u$ and $\vec{e}_2 \Delta v$. Thinking of these edges as vectors, they represent discrete changes in the input to the function g, so the derivative transformation $D_{g,(u,v)}$ applied to those vectors gives us the corresponding discrete changes in the output of g. These are the defining vectors for the parallelogram.

So we have as the two defining edge vectors of the parallelogram

$$D_g(\vec{e}_1 \Delta u) = g_u \Delta u = \begin{bmatrix} x_u \Delta u \\ y_u \Delta u \end{bmatrix} \quad \text{and} \quad D_g(\vec{e}_2 \Delta v) = g_v \Delta v = \begin{bmatrix} x_v \Delta v \\ y_v \Delta v \end{bmatrix} \tag{5.77}$$

The area ΔA_{xy} of the parallelogram is then

$$\begin{aligned} \Delta A_{xy} &= \left|\det\begin{pmatrix} x_u \Delta u & x_v \Delta v \\ y_u \Delta u & y_v \Delta v \end{pmatrix}\right| \\ &= \left|\det\begin{pmatrix} x_u & x_v \\ y_u & y_v \end{pmatrix}\right| \Delta u \Delta v \\ &= |\det J_g| \Delta A_{uv} \end{aligned}$$

So we have found our stretching factor – the rectangle of area ΔA_{uv} is stretched out by the function g into a parallelogram of area ΔA_{xy}, with area obtained by multiplying by a factor $|\det J_g|$.

We can put this all together now to rewrite everything in terms of u and v, just like we did in the case of the substitution rule for single variable integrals.

$$\begin{aligned} \iint_R f(x,y)\, dA_{xy} &= \lim \sum f(x_i, y_i)\, \Delta A_{xy} \\ &= \lim \sum \Big(f(g(u_i, v_i))\Big) \Big(|\det J_g| \Delta A_{uv}\Big) \\ &= \lim \sum \Big(f(g(u_i, v_i))\, |\det J_g|\Big) \Big(\Delta A_{uv}\Big) \\ \iint_R f(x,y)\, dA_{xy} &= \iint_D f(g(u,v))\, |\det J_g|\, dA_{uv} \end{aligned}$$

which we can also write as

$$\iint_R f(x,y)\, dx\, dy = \iint_D f(g(u,v))\, |\det J_g|\, du\, dv \tag{5.78}$$

This is the change of variables formula for double integrals.

Notice that we derived this result by referring directly to the Riemann sums representing the integrals in question. But, we can conclude the same result using integral shorthand. Here is that derivation again, using the shorthand.

We start with a piece of area dA_{uv} with edge vectors $\vec{e}_1 du$ and $\vec{e}_2 dv$, which is stretched into a parallelogram of area dA_{xy} with edge vectors

$$D_g(\vec{e}_1 du) = g_u du = \begin{bmatrix} x_u du \\ y_u du \end{bmatrix} \quad \text{and} \quad D_g(\vec{e}_2 dv) = g_v dv = \begin{bmatrix} x_v dv \\ y_v dv \end{bmatrix} \tag{5.79}$$

Again we can write this area as a determinant, from which we conclude

$$dA_{xy} = |\det J_g|\, dA_{uv} \tag{5.80}$$

and we plug this into the original given integral to get

$$\iint_R f(x,y)\, dA_{xy} = \iint_D f(g(u,v))\, |\det J_g|\, dA_{uv} \tag{5.81}$$

$$\iint_R f(x,y)\, dx\, dy = \iint_D f(g(u,v))\, |\det J_g|\, du\, dv \tag{5.82}$$

In future applications of the stretching factor strategy, we will just use this shorthand, with the understanding that there are Riemann sums that they refer to. Of course something similar could also be done with the derivation of the substitution rule in Subsection 5.6.1, and the reader is encouraged to try rewriting that derivation with such notation.

(Note that if we use the alternate notation $\det J_g = \frac{\partial(x,y)}{\partial(u,v)}$, Equation 5.80 becomes

$$\Delta A_{xy} = \left|\frac{\partial(x,y)}{\partial(u,v)}\right| \Delta A_{uv} \quad \text{or} \quad dx\, dy = \left|\frac{\partial(x,y)}{\partial(u,v)}\right| du\, dv \tag{5.83}$$

which appears to suggest some sort of cancellation. Of course this is not truly a cancellation, but the notational coincidence is somehow satisfying.)

5.6. CHANGE OF VARIABLES

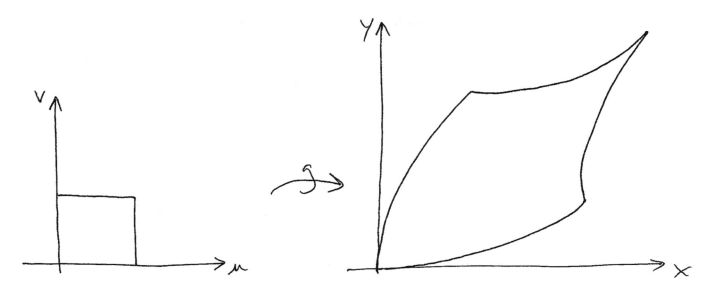

Figure 5.44:

5.6.3 Examples

Example 5.6.1. Suppose we wish to compute the integral $\iint_R x + y \, dx \, dy$, where R is the image of the unit square D in the uv-plane by the function $g(u, v) = (u^2 + 3v, v^2 + 3u) = (x, y)$, shown in Figure 5.44. Note that if we were to try to compute this directly, we would have to use at least three nested integrals.

Instead, let's use the change of variables formula. Conveniently the change of variables function is given, as is the new domain D. So we can apply Equation 5.82 directly once we compute the stretching factor $|\det J_g|$, which is

$$|\det J_g| = \left|\det \begin{pmatrix} 2u & 3 \\ 3 & 2v \end{pmatrix}\right| = |4uv - 9| \quad (5.84)$$

and since we know that u and v are both less than one, we have

$$|\det J_g| = 9 - 4uv \quad (5.85)$$

So our original integral becomes

$$\iint_R x + y \, dx \, dy = \iint_D ((u^2 + 3v) + (v^2 + 3u)) \, |\det J_g| \, du \, dv$$
$$= \int_0^1 \int_0^1 ((u^2 + 3v) + (v^2 + 3u)) \, (9 - 4uv) \, du \, dv$$

which can be computed directly.

Example 5.6.2. Suppose we wish to compute the integral $\iint_R x^2 + y^2 \, dx \, dy$, where R is the square with vertices $(0, 0)$, $(\sqrt{3}, 1)$, $(-1, \sqrt{3})$, $(\sqrt{(3)} - 1, \sqrt{(3)} + 1)$.

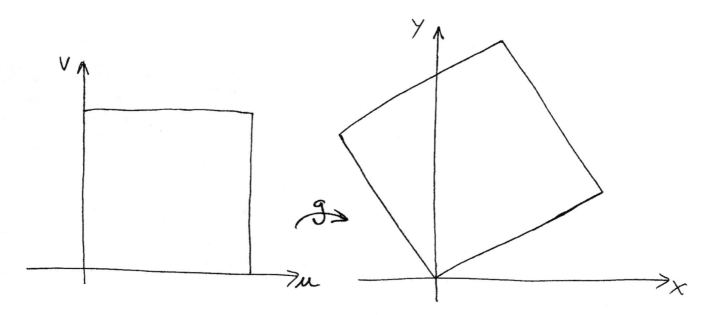

Figure 5.45:

We can first observe as shown in Figure 5.45 that R is the rotation around the origin of the square D with vertices at the points $(0,0)$, $(2,0)$, $(0,2)$, $(2,2)$, by the rotation matrix

$$T = \begin{pmatrix} \frac{\sqrt{3}}{2} & -\frac{1}{2} \\ \frac{1}{2} & \frac{\sqrt{3}}{2} \end{pmatrix} \tag{5.86}$$

Again, this would require at least three nested integrals if we tried to compute it directly; so instead let's use the change of variables formula.

Our change of variables function g is the rotation matrix T, giving us $g(u,v) = T(u,v) = (\frac{\sqrt{3}}{2}u - \frac{1}{2}v, \frac{1}{2}u + \frac{\sqrt{3}}{2}v) = (x,y)$; the new domain is the square D. Since g is a linear transformation, we know from Chapter 3 that the derivative transformation is the same linear transformation. So $J_g = T$, and thus

$$|\det J_g| = \left| \left(\frac{\sqrt{3}}{2}\right)\left(\frac{\sqrt{3}}{2}\right) - \left(-\frac{1}{2}\right)\left(\frac{1}{2}\right) \right| = 1 \tag{5.87}$$

(Of course it is not surprising that the stretching factor turned out to be one, since we know that rotations do not distort sizes at all.)

So we can then rewrite our integral as

$$\iint_R x^2 + y^2 \, dx\, dy = \iint_D \left(\frac{\sqrt{3}}{2}u - \frac{1}{2}v\right)^2 + \left(\frac{1}{2}u + \frac{\sqrt{3}}{2}v\right)^2 (1)\, du\, dv$$

$$= \int_0^2 \int_0^2 (u^2 + v^2)\, du\, dv$$

which can be computed directly.

5.6. CHANGE OF VARIABLES

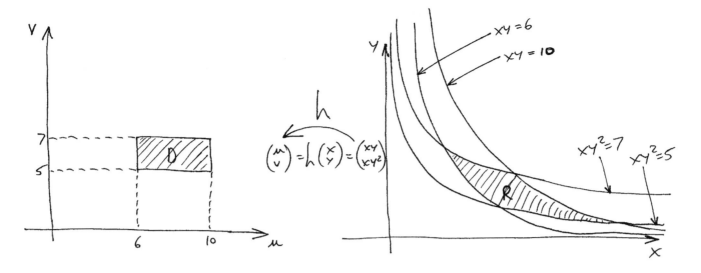

Figure 5.46:

Example 5.6.3. Suppose we wish to compute the integral $\iint_R x^2 y^3 \, dx\, dy$, where R is the region in the first quadrant of the xy-plane bounded by the four curves $xy = 6$, $xy = 10$, $xy^2 = 5$, $xy^2 = 7$.

We draw the given domain in Figure 5.46. Note however that the change of variables function is not given in the statement of the question, nor is it immediately clear how to form such a function from the given information.

We are tempted to form a function $h : \mathbb{R}^2 \to \mathbb{R}^2$ by making use of the two functions of x and y that are used to define the given curves. The problem is that in doing so the function is effectively backwards, in that x and y are inputs, not outputs — and our change of variables formula is not set up to deal with that arrangement.

Nevertheless, let's write that down and see if there is something we can do with it. We have

$$h\begin{pmatrix} x \\ y \end{pmatrix} = \begin{pmatrix} xy \\ xy^2 \end{pmatrix} = \begin{pmatrix} u \\ v \end{pmatrix} \tag{5.88}$$

Note that u is constant over each of the first two given curves, and v is constant over the each of the last two given curves, so the image of R by this function is a rectangle which we will call D.

One thing we could do in this situation is invert this function — that is, solve for x and y in terms of u and v, to form our usual function $g(u, v) = (x, y)$ so that we can apply the change of variables formulas that we already have.

Instead, let's try to deal with the situation as it is given.

Note that what we really need to do is to figure out some way to rewrite dA_{xy} in terms of dA_{uv}. We know that one way to do that is to use $dA_{xy} = |\det J_g| dA_{uv}$, since the function g stretches out pieces of area in the uv-plane to make pieces of area in the xy-plane. But equivalently, we could relate these two pieces of area by the action of the function h. Since h takes points the other direction, from the xy-plane to the uv-plane, we have

$$dA_{uv} = |\det J_h| dA_{xy} \quad \text{or} \quad dA_{xy} = \frac{1}{|\det J_h|} dA_{uv} \tag{5.89}$$

And of course we can compute $|\det J_h|$ directly, as

$$\det J_h = \det \begin{pmatrix} y & x \\ y^2 & 2xy \end{pmatrix} = xy^2 = v \tag{5.90}$$

So we can rewrite the given integral as

$$\iint_R x^2 y^3 \, dx\, dy = \iint_D x^2 y^3 \left(\frac{1}{v} du\, dv \right) = \iint_D (uv) \left(\frac{1}{v} du\, dv \right) = \iint_D u\, du\, dv \tag{5.91}$$

Note that the second of Equations 5.89 can be rewritten

$$dx\,dy = \frac{1}{\left|\frac{\partial(u,v)}{\partial(x,y)}\right|}du\,dv \qquad (5.92)$$

But of course we know from Equation 5.83 that

$$dx\,dy = \left|\frac{\partial(x,y)}{\partial(u,v)}\right|du\,dv \qquad (5.93)$$

which tells us that

$$\left|\frac{\partial(x,y)}{\partial(u,v)}\right| = \frac{1}{\left|\frac{\partial(u,v)}{\partial(x,y)}\right|} \qquad (5.94)$$

Of course we know that this notation is not actually a fraction – but the conclusion from the above is that despite this, it does sort of act like it. The proper interpretation of this statement would be that the Jacobian determinant of the inverse of a function is the reciprocal of the Jacobian determinant of the original function. But, for symbolic convenience, we can say that when working a change of variables problem, feel free to compute either $\frac{\partial(x,y)}{\partial(u,v)}$ or $\frac{\partial(u,v)}{\partial(x,y)}$, whichever happens to be more convenient – and then, use either to relate $dx\,dy$ to $du\,dv$ in the way that is suggested by the seemingly fractional notation.

5.6.4 Change of variables in $\mathbb{R}^3$

Having successfully applied our stretching factor strategy to double integrals, let's try to apply it to triple integrals. The solution will end up being remarkably similar to that for double integrals.

Suppose for example that we wish to compute the integral $\iiint_R f(x,y,z)\,dx\,dy\,dz$ over the domain R in the xyz-system, as shown in Figure 5.47. Note that if we did try to write such an integral as a nested integral, we would have to break the integral up into several pieces, making the task inconvenient at best.

As is indicated in the strategy outline, in order to change the domain we first need to find a one-to-one function g and a domain D in a new coordinate system such that R is the image of D. Let us suppose that we can find such a function g, from the uvw-system to the xyz-system, so that the domain D in the uvw-system is a more convenient shape (such as a rectangular box), giving us Figure 5.48.

In order to rewrite the given integral as an integral over D, we will again need to rewrite all of the components of the given integral in terms of u, v and w. As before, we can of course write $f(x,y,z) = f(g(u,v,w))$ by simple substitution.

Changing the differential will again require us to measure how much the change of variables function g has stretched out the domain. In this case, we need to be able to measure the factor by which the volume dV_{uvw} is stretched out to form the volume dV_{xyz}.

In Figure 5.49, we draw expanded images of dV_{uvw} and dV_{xyz}. We know from Chapter 4 that differentiable functions are locally linear, and so the image of this rectangular box represented by dV_{uvw} is approximately some sort of parallelopiped. (Of course it will not be exactly a parallelopiped, but in the limit as we take smaller boxes the approximation is valid. So we will draw the image with volume dV_{xyz} as an actual parallelopiped.)

In order to compute dV_{xyz} then, we need only find the three defining vectors and apply the determinant formula from

5.6. CHANGE OF VARIABLES

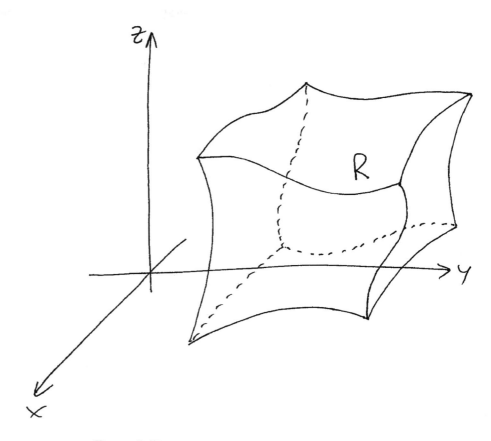

Figure 5.47:

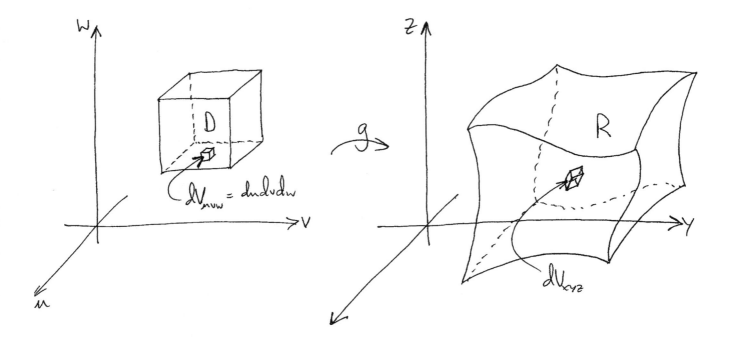

Figure 5.48:

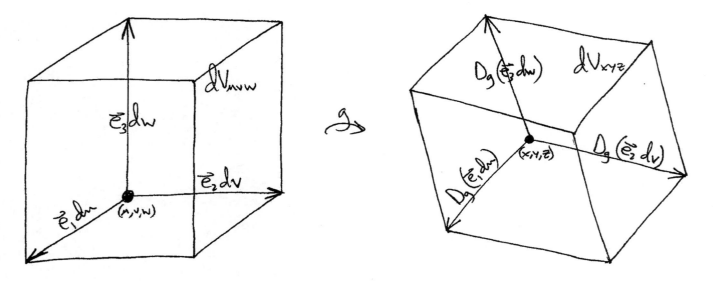

Figure 5.49:

Chapter 3. Just as in the case for double integrals, those defining vectors are

$$D_g(\vec{e}_1 du) = g_u du = \begin{bmatrix} x_u du \\ y_u du \\ z_u du \end{bmatrix}$$

$$D_g(\vec{e}_2 dv) = g_v dv = \begin{bmatrix} x_v dv \\ y_v dv \\ z_v dv \end{bmatrix}$$

$$D_g(\vec{e}_3 dw) = g_w dw = \begin{bmatrix} x_w dw \\ y_w dw \\ z_w dw \end{bmatrix}$$

Again we can write this volume as a determinant, from which we conclude

$$dV_{xyz} = \left| \det \begin{pmatrix} x_u du & x_v dv & x_w dw \\ y_u du & y_v dv & y_w dw \\ z_u du & z_v dv & z_w dw \end{pmatrix} \right|$$

$$= \left| \det \begin{pmatrix} x_u & x_v & x_w \\ y_u & y_v & y_w \\ z_u & z_v & z_w \end{pmatrix} \right| du\, dv\, dw$$

$$= |\det J_g|\, dV_{uvw}$$

and we plug this into the original given integral to get

$$\iiint_R f(x,y,z)\, dV_{xyz} = \iiint_D f(g(u,v,w))\, |\det J_g|\, dV_{uvw} \tag{5.95}$$

$$\iiint_R f(x,y,z)\, dx\, dy\, dz = \iiint_D f(g(u,v,w))\, |\det J_g|\, du\, dv\, dw \tag{5.96}$$

Note that this looks very similar to Equation 5.82. The only significant difference in the derivations is that we needed to compute the volume of a three-dimensional parallelopiped instead of the area of a two-dimensional parallelogram; and since those formulas are themselves very similar, the resulting 2- and 3-dimensional change of variables formulas are very similar. The stretching factor is represented by the same notation in fact, even though of course the Jacobian matrix in this case is now a 3×3 matrix.

Here is an example.

5.6. CHANGE OF VARIABLES

Example 5.6.4. Suppose we need to compute the integral of the function $f(x,y,z) = z$ over the domain R, which is the image of the box $[1,2] \times [2,3] \times [4,5]$ by the function $g(u,v,w) = (u^2 - v, v^2 - w, w^2 - u) = (x,y,z)$.

To use the change of variables formula, we need to know the Jacobian determinant, which we compute as

$$|\det J_g| = \left|\det \begin{pmatrix} 2u & -1 & 0 \\ 0 & 2v & -1 \\ -1 & 0 & 2w \end{pmatrix}\right| = |8uvw - 1| = 8uvw - 1 \tag{5.97}$$

since u, v and w are all ≥ 1.

We can then rewrite the given integral as

$$\iiint_R f(x,y,z)\,dx\,dy\,dz = \iiint_D (z)\left((8uvw - 1)\,du\,dv\,dw\right)$$
$$= \int_4^5 \int_2^3 \int_1^2 (w^2 - u)(8uvw - 1)\,du\,dv\,dw$$

5.6.5 Symmetry theorems

Students will recall from their single variable calculus courses that a function is called "odd" if $f(-x) = -f(x)$. (This terminology is motivated by the fact that the functions $f(x) = x^n$ have this property for all odd exponents n.)

An important theorem related to such functions is the symmetry theorem.

Theorem 5.6.1. *For any continuous odd function, we have*

$$\int_{-a}^{a} f(x)\,dx = 0 \tag{5.98}$$

This theorem can be proved by using substitutions.

In this section we will develop some analogous ideas for multivariable integrals. The tool that we will use to prove our results will also be analogous, namely, the method of change of variables.

Definitions

The single variable symmetry result in Theorem 5.6.1 uses the idea of symmetry over the origin. Here we will extend and generalize that idea to higher dimensional spaces.

We begin with three definitions in $\mathbb{R}^2$.

Definition 5.6.1. *The "reflection over the line L" in $\mathbb{R}^2$ is the function $R_L : \mathbb{R}^2 \to \mathbb{R}^2$ that reflects points over the given line L. Specifically, for any point $\vec{x}$, the line segment from $\vec{x}$ to $R_L(\vec{x})$ is perpendicularly bisected by the line L, as shown in Figure 5.50.*

Definition 5.6.2. *A set D in $\mathbb{R}^2$ is said to be "symmetric over the line L" if*

$$\left(\vec{x} \in D\right) \iff \left(R_L(\vec{x}) \in D\right) \tag{5.99}$$

Definition 5.6.3. *A function $f : \mathbb{R}^2 \to \mathbb{R}^1$ has "odd symmetry over the line L" if*

$$f\left(R_L(\vec{x})\right) = -f\left(\vec{x}\right) \tag{5.100}$$

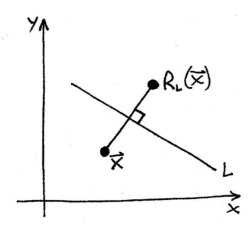

Figure 5.50:

We can make similar definitions in $\mathbb{R}^3$.

Definition 5.6.4. The "reflection over the plane P" in $\mathbb{R}^3$ is the function $R_P : \mathbb{R}^3 \to \mathbb{R}^3$ that reflects points over the given plane P. Specifically, for any point $\vec{x}$, the line segment from $\vec{x}$ to $R_P(\vec{x})$ is perpendicularly bisected by the plane P.

Definition 5.6.5. A set D in $\mathbb{R}^3$ is said to be "symmetric over the plane P" if

$$\left(\vec{x} \in D\right) \iff \left(R_P(\vec{x}) \in D\right) \tag{5.101}$$

Definition 5.6.6. A function $f : \mathbb{R}^3 \to \mathbb{R}^3$ has "odd symmetry over the plane P" if

$$f\left(R_P(\vec{x})\right) = -f\left(\vec{x}\right) \tag{5.102}$$

Symmetry theorems

It should not be too surprising that these reflection functions "preserve areas". That is, the area of any set is equal to the area of the reflection of that set. In fact, for any reflection R, in the plane or in space, we have that the stretching factor is always

$$|\det J_R| = 1 \tag{5.103}$$

(It turns out also that for any reflection, we have $\det J_R = -1$. The negative corresponds to the facts that reflections in the plane turn the plane "upside-down", and analogously, in space reflections change the "handed-ness" (right or left) of any triple of vectors.)

It should also be clear from the definition that any domain that is symmetric over a line or plane is its own image by that reflection function. That is, $R(D) = D$ for such domains D.

We can use these facts to prove the following symmetry theorems.

Theorem 5.6.2. If the domain D in $\mathbb{R}^2$ is symmetric over a line L and the function f has odd symmetry over the same line L, then

$$\iint_D f(\vec{x})\, dA = 0 \tag{5.104}$$

(Note that it is critical that the line of symmetry for the domain is the same as the line of symmetry for the function.)

5.6. CHANGE OF VARIABLES

Theorem 5.6.3. *If the domain D in $\mathbb{R}^3$ is symmetric over a plane P and the function f has odd symmetry over the same plane P, then*

$$\iiint_D f(\vec{x})\, dV = 0 \tag{5.105}$$

(Note that it is critical that the plane of symmetry for the domain is the same as the plane of symmetry for the function.)

Here we will prove Theorem 5.6.2. The proof of Theorem 5.6.3 is analogous, and left as an exercise to the reader.

Suppose we have a domain D and function f satisfying the hypotheses of the theorem. Then very simply, we consider the reflection function R_L as a change of variables function applying to the given integral.

First using our observation that $R(D) = D$, we have

$$\iint_D f(\vec{x})\, dA = \iint_{R(D)} f(\vec{x})\, dA \tag{5.106}$$

Then applying change of variables and using previously established properties, we get

$$= \iint_D f\big(R(\vec{x})\big) |\det J_{R_L}|\, dA$$
$$= \iint_D -f(\vec{x})\ (1)\, dA$$
$$= -\iint_D f(\vec{x})\, dA$$

Having just shown the the integral $\iint_D f(\vec{x})\, dA$ is its own negative, we conclude that the integral must be zero, as desired.

A very similar proof will work for Theorem 5.6.3.

Example 5.6.5. Suppose we wish to compute the double integral of the function $f(x,y) = x^2 y^3$ over the unit disk D in the xy-plane.

Of course the domain D is symmetric over many lines, including both the x-axis and the y-axis. But the integrand f has odd symmetry only over one line – the x-axis. We check that this symmetry is there by noting that the reflection over the x-axis is given by

$$R_x(x,y) = (x, -y) \tag{5.107}$$

and then computing

$$f(R_x(x,y)) = f(x, -y) = (x)^2(-y)^3 = -x^2 y^3 = -f(x,y) \tag{5.108}$$

We can then cite Theorem 5.6.2 and conclude that the integral must equal zero.

Example 5.6.6. Suppose we wish to compute the double integral of the function $f(x,y) = x^2 y^3$ over the triangle with vertices at $(0,1)$, $(-1,0)$ and $(1,0)$ in the xy-plane.

This domain is symmetric, but only over the y-axis. And the function f does NOT have odd symmetry over this line, as we see by computing

$$f(R_y(x,y)) = f(-x, y) = (-x)^2(y)^3 = x^2 y^3 \neq -f(x,y) \tag{5.109}$$

So we simply cannot apply the symmetry theorem here, and must resort to rewriting the integral as a nested integral.

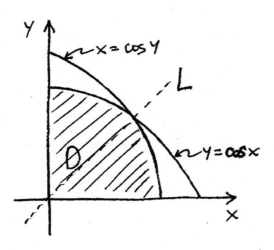

Figure 5.51:

Example 5.6.7. Suppose we wish to compute the integral of the function $f(x,y) = x^4 - y^4$ over the part of the first quadrant bounded by the curves $y = \cos x$ and $x = \cos y$ and the x- and y-axes.

As shown in Figure 5.51, this domain is symmetric over the line L with equation $y = x$. Reflections over that line are given by
$$R_L(x, y) = (y, x) \tag{5.110}$$
And the function f has odd symmetry over that line, as we see by computing
$$f(R_L(x, y)) = f(y, x) = (y)^4 - (x)^4 = y^4 - x^4 = -\Big(f(x, y)\Big) \tag{5.111}$$
So again, we can cite Theorem 5.6.2 and conclude that the integral must equal zero.

Example 5.6.8. Suppose we wish to compute the triple integral of the function $f(x, y, z) = x^2 y^3 z - z^3$ over the solid ellipsoid whose boundary has equation $x^2 + y^2 + 4z^2 = 4$.

The domain is symmetric over all three of the coordinate planes. However the integrand function does not have odd symmetry over all of those planes.

Fortunately it does have odd symmetry over one, the xy-plane. We see this by noting that the reflections through that plane are given by
$$R_{xy}(x, y, z) = (x, y, -z) \tag{5.112}$$
and computing
$$f(R_{xy}(x, y, z)) = f(x, y, -z) = (x)^2(y)^3(-z) - (-z)^3 = -x^2 y^3 z + z^3 = -\Big(f(x, y, z)\Big) \tag{5.113}$$
Citing Theorem 5.6.3 then, we have that the integral is zero.

5.6. CHANGE OF VARIABLES

Exercises

Exercise 5.6.1 (exer-Ints-CV-1). Compute the integral $\iint_D f(x,y)\,dx\,dy$, where $f = 1$ and D is the image of $[0,1] \times [1,2]$ by the function $g(u,v) = (v + \arctan u, uv) = (x,y)$.

Exercise 5.6.2 (exer-Ints-CV-2). Show that the image of the unit square by any function of the form $g(u,v) = (u, v + f(u))$, where f is a differentiable function, is a set with an area of 1.

Exercise 5.6.3 (exer-Ints-CV-3). Compute the centroid of the image D of the rectangle $[1,2] \times [2,3]$ by the function $g(u,v) = (\arctan u, uv + u^3 v) = (x,y)$.

Exercise 5.6.4 (exer-Ints-CV-4). Compute the mass of a sheet of metal in the shape of the region D from Exercise 5.6.3, whose density is given by $\delta(x,y) = x + y$.

Exercise 5.6.5 (exer-Ints-CV-5). Show that for any linear transformation $L : \mathbb{R}^2 \to \mathbb{R}^2$, the area of the image $L(D)$ of a region D is equal to the area of D times the area of the image of the unit square.

Exercise 5.6.6 (exer-Ints-CV-6). Compute the area of the region in the first quadrant of the xy-plane bounded by the curves $x^2 y = 1$, $x^2 y = 2$, $xy^2 = 3$, and $xy^2 = 4$.

Exercise 5.6.7 (exer-Ints-CV-7). Compute the centroid of the region in the first quadrant of the xy-plane bounded by the curves $y^2 = 1/x^3$, $y^2 = 2/x^3$, $y = 1/x$, and $y = 4/x$.

Exercise 5.6.8 (exer-Ints-CV-8). Compute the integral $\iint_D f(x,y)\,dx\,dy$, where $f(x,y) = x^3 y^4 - xy$ and D is the region bounded by the curves $y = e^x$, $y = e^{-x}$, and $y = x^2 - 1$.

Exercise 5.6.9 (exer-Ints-CV-9). Compute the integral $\iint_D f(x,y)\,dx\,dy$, where $f(x,y) = x^2 y^2 e^y - x^2 y^2 e^{-y}$ and D is the region bounded by the curve $y^2 = 1 - x^4 e^{x^2}$.

Exercise 5.6.10 (exer-Ints-CV-10). Compute the integral $\iint_D f(x,y)\,dx\,dy$, where $f(x,y) = x^2 y^3$ and D is the quadrilateral with vertices at $(0,-2)$, $(1,1)$, $(-1,1)$, and $(0,4)$.

Exercise 5.6.11 (exer-Ints-CV-11). Compute the integral $\iint_D f(x,y)\,dx\,dy$, where $f(x,y) = xe^y - ye^x$ and D is the region bounded by the curves $y = e^x$, $x = e^y$, $x + y = 0$, and $x + y = 1$.

Exercise 5.6.12 (exer-Ints-CV-12). Compute the integral $\iint_D f(x,y)\,dx\,dy$, where $f(x,y) = ye^x - ye^{4-x}$ and D is the quadrilateral with vertices (listed in order going clockwise around the interior) at $(1,5)$, $(2,1)$, $(3,5)$, and $(2,0)$.

Exercise 5.6.13 (exer-Ints-CV-13). Let L be a line in the xy-plane, and let R represent the reflection over that line. Show that for any set $D \subset \mathbb{R}^2$, the set $S = D \cup R(D)$ is symmetric over L.

Exercise 5.6.14 (exer-Ints-CV-14). Suppose that the function $f : \mathbb{R}^2 \to \mathbb{R}^1$ has odd symmetry over the line L, and that $\vec{p}$ is a point on L. Show that $f(\vec{p}) = 0$.

Exercise 5.6.15 (exer-Ints-CV-15). Suppose that the function $f : \mathbb{R}^3 \to \mathbb{R}^1$ has odd symmetry over the plane P, and that $\vec{p}$ is a point on P. Show that $f(\vec{p}) = 0$.

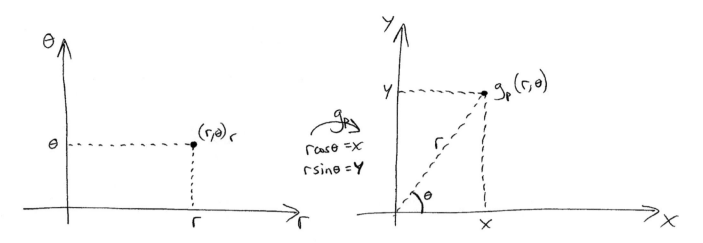

Figure 5.52:

5.7 Integrals in Coordinate Systems

5.7.1 Coordinate systems as change of variables functions

Most of the time we think about other coordinate systems in the way that they were presented in Section 1.6, as coordinate systems. In other words, we think of them as being an alternative to rectangular coordinates in terms of how we might represent a point in either $\mathbb{R}^2$ or $\mathbb{R}^3$ as an order pair or triple (resp.) of numbers.

At this point we need to take a different point of view of these coordinate systems.

Note that in our discussions of each of these coordinate systems, we found formulas for the individual rectangular coordinates in terms of these other sets of coordinates. These formulas can be put together to form a single function, which can be interpreted as a change of variables function.

The polar coordinates function

For example, in our discussion of polar coordinates we found formulas for x and y in terms of r and θ; namely, $x = r\cos\theta$ and $y = r\sin\theta$.

We could put these together to make a single function $g_p : \mathbb{R}^2 \to \mathbb{R}^2$, for the moment ignoring that these functions were motivated by a previously defined coordinate system. We get

$$g_p(r, \theta) = (r\cos\theta, r\sin\theta) = (x, y) \tag{5.114}$$

which we will now visualize as a function from the $r\theta$-plane (with rectangular coordinates r and θ) to the xy-plane, as in Figure 5.52.

Thought of in this way, we can view this function as being a change of variables function. In other words, with some domain D in the $r\theta$-plane and its image R in the xy-plane, we can use this function g_p to transform an integral over R into a different integral over D, using the ideas from Section 5.6.

For example, suppose we consider the rectangle $D = [0, 1] \times [0, \pi/2]$ in the $r\theta$-plane, as in Figure 5.53.

Its image R in the xy-plane by the given change of variables function g_p is the part of the unit disk that is in the first quadrant. We can see this by looking at the images of each of the edges of the given rectangle, remembering that what the function g_p is doing is simply to take the rectangular coordinates r and θ for points in the $r\theta$-plane and make those the polar coordinates for the image point in the xy-plane.

5.7. INTEGRALS IN COORDINATE SYSTEMS

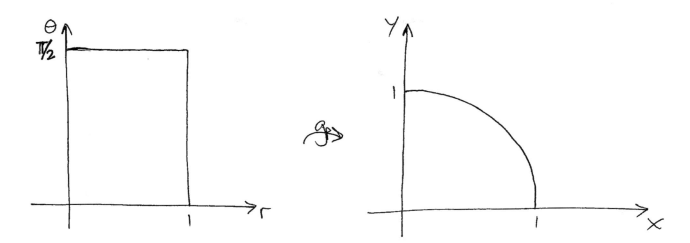

Figure 5.53:

For example, the right edge of the rectangle is on the line $r = 1$, going from $\theta = 0$ to $\theta = \pi/2$. In the xy-plane though, $r = 1$ is the unit circle, and going from $\theta = 0$ to $\theta = \pi/2$ simply traces out the part of the unit circle going counterclockwise from $(1,0)_r$ to $(0,1)_r$. Similarly the other three edges of the rectangle define the remainder of the border of the first quadrant part of the unit disk. (Note that one of the edges of the rectangle is on the line $r = 0$, and so that entire edge has its image at the origin in the xy-plane.)

Having noted this, suppose we wish to compute an integral over the first quadrant portion of the unit circle in the xy-plane. We could instead write this as a different integral over the rectangle above, using the change of variables formula. Of course we will need to know the Jacobian determinant for this change of variables function, which we can compute without too much trouble.

$$|\det J_{g_p}| = \left|\det \begin{pmatrix} \cos\theta & -r\sin\theta \\ \sin\theta & r\cos\theta \end{pmatrix}\right| = |r\cos^2\theta + r\sin^2\theta| = |r| = r \tag{5.115}$$

The absolute values in the above calculation go away at the end because, over this rectangle in the $r\theta$-plane, the r coordinate is always positive.

We can then apply the change of variables formula to get

$$\iint_R f(x,y)\,dx\,dy = \iint_D f(g_p(r,\theta))\left(|\det J_{g_p}|\,dr\,d\theta\right) = \int_0^{\pi/2}\int_0^1 f(r\cos\theta, r\sin\theta)r\,dr\,d\theta \tag{5.116}$$

Of course this could very well prove to be a nice alternative to the original integral, because writing the integral over a quarter circle in the usual way as a nested integral would require using square roots in the bounds; whereas in the result above, the bounds are constants.

This polar coordinates function is a useful change of variables function fairly often. So, it is worth making note of its Jacobian determinant $|\det J_{g_p}| = r$ for future use, rather than recomputing it every time we need it. Since of course we will always use that Jacobian determinant to relate the two area differentials, we can also note this as

$$dx\,dy = r\,dr\,d\theta \tag{5.117}$$

Note, in order to allow for eliminating the absolute values, using this formula requires that we represent our domain with only positive values of r; fortunately, this is usually the most convenient choice anyway.

The cylindrical coordinates function

Very similarly, we can define the cylindrical coordinates function $g_c : \mathbb{R}^3 \to \mathbb{R}^3$ by

$$g_c(r,\theta,z) = (r\cos\theta, r\sin\theta, z) = (x,y,z) \tag{5.118}$$

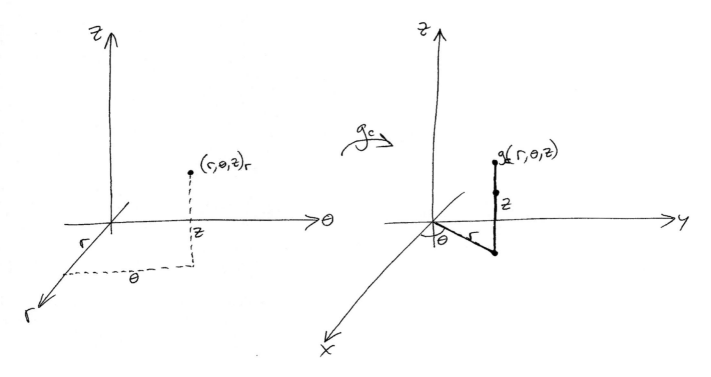

Figure 5.54:

which we will now visualize as a function from the $r\theta z$-system (with rectangular coordinates r, θ and z) to the xyz-system, as in Figure 5.54. (Note that here we are using z as both an input and output variable; technically this is a confusion of variables, but since z plays the same role in each coordinate system there is no harm in just leaving this as is.)

Thought of in this way, we can view this function as being a change of variables function too. In other words, with some domain D in the $r\theta z$-system and its image R in the xyz-system, we can similarly use this function g_c to transform an integral over R into a different integral over D, using the ideas from Section 5.6.

Of course to do this we will need to know the Jacobian determinant of g_c.

$$|\det J_{g_c}| = \left|\det \begin{pmatrix} \cos\theta & -r\sin\theta & 0 \\ \sin\theta & r\cos\theta & 0 \\ 0 & 0 & 1 \end{pmatrix}\right| = \left|0 + 0 + (r\cos^2\theta + r\sin^2\theta)\right| = |r| = r \quad (5.119)$$

(It should not be too surprising that we ended up with the same stretching factor as in the polar coordinates function, because remember that cylindrical coordinates is just polar coordinates for x and y, and the z is the same as in rectangular coordinates. Again, note that to use this result without the absolute values requires that r always be positive; and again, we are fortunate that this is usually convenient anyway.)

Just as with the polar coordinates function, we can use this as a change of variables function.

Suppose we consider the box $D = [0, 1] \times [0, \pi] \times [0, 3]$ in the $r\theta z$-system. Its image R by the cylindrical coordinates function g_c is half of a solid cylinder, as in Figure 5.55.

So we can compute an integral over this half cylinder with a different integral over the box.

$$\iiint_R f(x, y, z)\, dx\, dy\, dz = \iiint_D f(g_c(r, \theta, z))(|\det J_{g_c}|\, dr\, d\theta\, dz)$$
$$= \int_0^3 \int_0^\pi \int_0^1 f(r\cos\theta, r\sin\theta, z)\, r\, dr\, d\theta\, dz$$

Like the polar coordinates function, this cylindrical coordinates function is a useful change of variables function fairly often. So, it is worth making note of its Jacobian determinant $|J_{g_c}| = r$ for future use as well, rather than recomputing it

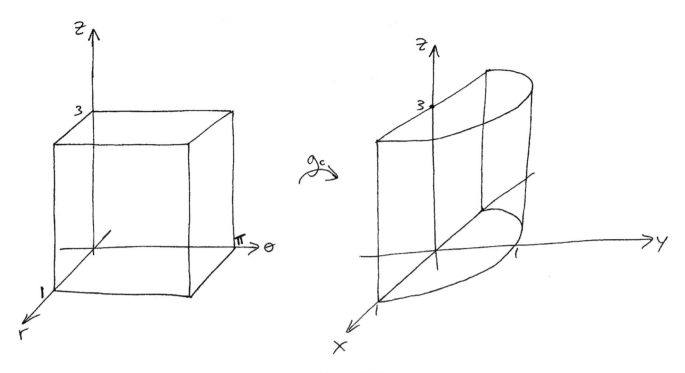

Figure 5.55:

every time we need it. Since of course we will always use this Jacobian determinant to relate the two volume differentials, we can also note this as

$$dx\,dy\,dz = r\,dr\,d\theta\,dz \tag{5.120}$$

The spherical coordinates function

Again very similarly, we can define the spherical coordinates function $g_s : \mathbb{R}^3 \to \mathbb{R}^3$ by

$$g_s(\rho, \phi, \theta) = (\rho \sin \phi \cos \theta, \rho \sin \phi \sin \theta, \rho \cos \phi) = (x, y, z) \tag{5.121}$$

which we will now visualize as a function from the $\rho\phi\theta$-system (with rectangular coordinates ρ, ϕ and θ) to the xyz-system, as in Figure 5.56.

Thought of in this way, we can view this function as being a change of variables function too. In other words, with some domain D in the $\rho\phi\theta$-system and its image R in the xyz-system, we can similarly use this function g_s to transform an integral over R into a different integral over D, using the ideas from Section 5.6.

Of course to do this we will need to know the Jacobian determinant of g_s.

$$\begin{aligned} |\det J_{g_s}| &= \left|\det \begin{pmatrix} \sin\phi\cos\theta & \rho\cos\phi\cos\theta & -\rho\sin\phi\sin\theta \\ \sin\phi\sin\theta & \rho\cos\phi\sin\theta & \rho\sin\phi\cos\theta \\ \cos\phi & -\rho\sin\phi & 0 \end{pmatrix}\right| \\ &= |\rho^2 \sin\phi| \end{aligned}$$

The details of the above computation are left as an exercise for the reader. Note at this point though that we can remove the absolute values if we know that $\sin\phi$ is positive; that is, if we require $\phi \in [0, \pi]$. As was the case in polar and cylindrical coordinates, we are again fortunate that this is usually the most convenient way to represent things in spherical coordinates. In fact we also usually choose to use $\rho \geq 0$, even though the square above makes this not necessary. Very often θ is viewed as being between 0 and 2π, but there are instances in which it can be convenient to view $\theta \in [-\pi, \pi]$ instead; and of course that does not affect the absolute values at all.

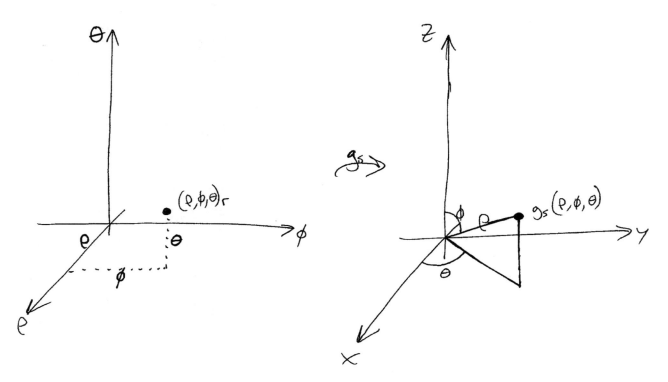

Figure 5.56:

With this constraint on ϕ, we then have

$$|\det J_{g_s}| = \rho^2 \sin \phi$$

Just as with the polar and cylindrical coordinates functions, we can use this as a change of variables function.

Suppose we consider the box $D = [0,1] \times [0, \pi/2] \times [0, 2\pi]$ in the $\rho\phi\theta$-system. Its image R by the spherical coordinates function g_s is the top half of the unit ball, as in Figure 5.57.

So we can compute an integral over this half of the unit ball with a different integral over the box.

$$\iiint_R f(x,y,z)\, dx\, dy\, dz = \iiint_D f(g_s(\rho, \phi, \theta))(|\det J_{g_s}|\, d\rho\, d\phi\, d\theta)$$
$$= \int_0^{2\pi} \int_0^{\pi/2} \int_0^1 f(\rho \sin \phi \cos \theta, \rho \sin \phi \sin \theta, \rho \cos \phi) \rho^2 \sin \phi\, d\rho\, d\phi\, d\theta$$

Like the polar and cylindrical coordinates functions, this spherical coordinates function is a useful change of variables function fairly often. So, it is worth making note of its Jacobian determinant $|\det J_{g_s}| = \rho^2 \sin \phi$ for future use as well, rather than recomputing it every time we need it. Since of course we will always use this Jacobian determinant to relate the two volume differentials, we can also note this as

$$dx\, dy\, dz = \rho^2 \sin \phi\, d\rho\, d\phi\, d\theta \qquad (5.122)$$

5.7.2 Integrals "in" coordinate systems

At this point we have all of the tools necessary to use these alternate coordinate systems to rewrite double and triple integrals. But there remains an observation which will allow us to rewrite integrals in this way without drawing pictures such as Figures 5.53, 5.55 and 5.57.

5.7. INTEGRALS IN COORDINATE SYSTEMS

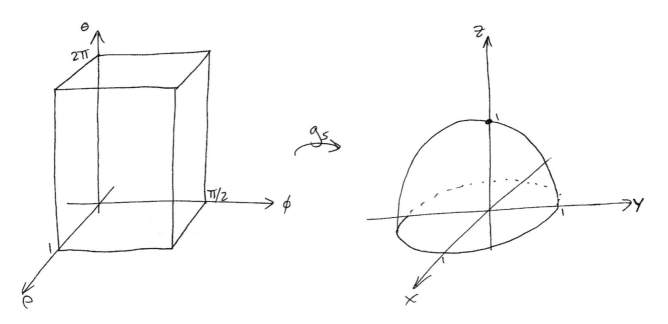

Figure 5.57:

To make this observation, let's revisit the example on page 302 in which we used the polar coordinates function as a change of variables function. We started with an integral over a domain R in the xy-plane, and the result was that we had a nested integral, where the differentials were $dr\, d\theta$ instead of the usual $dx\, dy$.

We got to this conclusion by thinking about the change of variables function and the resulting new domain D in the $r\theta$-plane explicitly, rewriting the original integral over R as an integral over D, and then writing that double integral over D as a nested integral in the usual way.

At this point let's remind ourselves of the details of how we turn the double integral over D into a nested integral. First, noting the order of the differentials, we look for the range of values of the variable corresponding to the outermost differential, in this case θ. In Figure 5.58, we draw in the domain D the first slice at $\theta_1 = 0$, the last slice at $\theta_2 = \pi/2$, and a sample slice for some arbitrary, fixed value of θ between those bounds. We know that these values $\theta_1 = 0$ and $\theta_2 = \pi/2$ are the bounds on the $d\theta$ integral.

Before we go on, let's pause and ask if there might be some way we could have arrived at those same bounds without having actually gone to the trouble of drawing the domain D in the $r\theta$-plane. Remember, the problem originally posed asked for an integral to be computed over the domain R in the xy-plane, and it would be nice if we could determine the bounds for the final $dr\, d\theta$ integral without having to actually draw an entirely new picture. In fact, we can determine these bounds very conveniently, without drawing the picture of D at all.

The point here is that the lines representing the slices we take of D have known images in the xy-plane by the change of variables function g_p. In particular, any horizontal line in the $r\theta$-plane, since it represents a constant value of θ, has as its image a line through the origin in the xy-plane.

So instead of looking at the actual horizontal slices of D, we can just look at their images in the given domain R. Those images are radial slices through the origin, which we might call "θ-slices".

As we see in Figure 5.58, the first θ-slice of R is on the line $\theta_1 = 0$, and the last θ-slice of R is on the line $\theta_2 = \pi/2$.

This is the same conclusion that we arrived at considering the actual new domain D, but note that we arrived at that

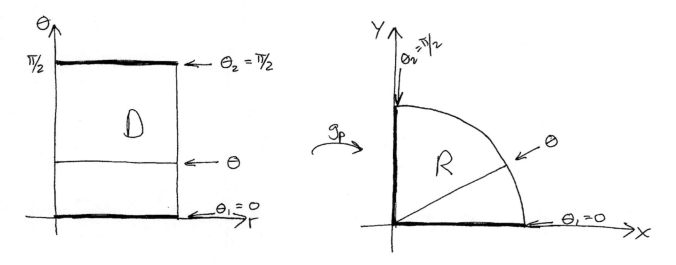

Figure 5.58:

conclusion by looking only at the given domain R. We thus have the bounds on the outermost $d\theta$ integral.

$$\iint_R f(x,y)\,dx\,dy = \iint_D f(g_p(r,\theta))(|\det J_{g_p}|\,dr\,d\theta)$$
$$= \int_0^{\pi/2} \int f(g_p(r,\theta))(|\det J_{g_p}|\,dr\,d\theta)$$

Now, on to finding the bounds for the inside integral. Looking at D, we consider the θ-slice for the arbitrary, fixed value of θ, and want to know the range of values that r takes on that given slice. From Figure 5.59 we see that on the given slice, r ranges from $r_1 = 0$ to $r_2 = 1$.

Again, we can come to this same conclusion without considering the new domain D directly. Again, we can just look at the images on the given domain R. As we see in Figure 5.59, on the arbitrary θ-slice, the variable r ranges from $r_1 = 0$ to $r_2 = 1$.

We thus have the bounds on the dr integral.

$$\iint_R f(x,y)\,dx\,dy = \iint_D f(g_p(r,\theta))(|\det J_{g_p}|\,dr\,d\theta)$$
$$= \int_0^{\pi/2} \int_0^1 f(g_p(r,\theta))(|\det J_{g_p}|\,dr\,d\theta)$$

When using the polar coordinates function as a change of variables function, note that the point of view we used above will always allow us to determine the bounds on the new $dr\,d\theta$ integral, without actually drawing the new domain D. And of course we know that the stretching factor will always be the same. So, in effect, we don't need to draw the $r\theta$-plane at all.

Here are a few more examples.

Example 5.7.1. Suppose I want to use the polar coordinates function to change variables for the integral $\iint_R f\,dx\,dy$, where $f(x,y) = x^2 + y^2$, and R is the top half of the unit disk, represented in Figure 5.60.

From the Figure we see that the angle θ goes from 0 to π; and for any given value of θ, r ranges from 0 to 1.

5.7. INTEGRALS IN COORDINATE SYSTEMS

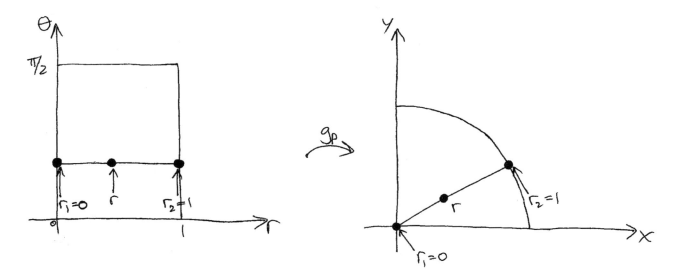

Figure 5.59:

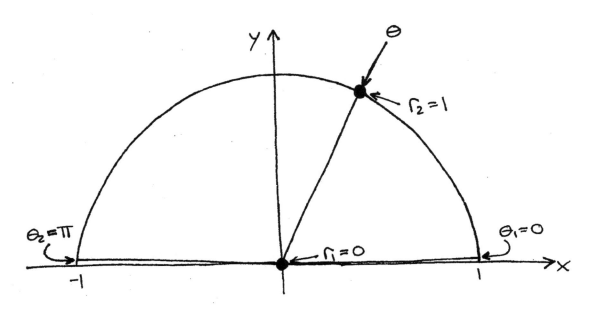

Figure 5.60:

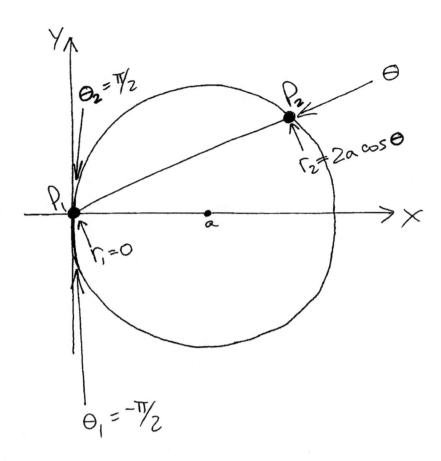

Figure 5.61:

Remembering to insert the stretching factor, we can then write

$$\iint_R x^2 + y^2 \, dx \, dy = \int_0^\pi \int_0^1 (r^2 \cos^2 \theta + r^2 \sin^2 \theta)(r \, dr \, d\theta)$$
$$= \int_0^\pi \int_0^1 r^3 \, dr \, d\theta$$

Of course in the above example we are using the change of variables formula, even though we don't explicitly represent it that way. Since the end result is that we write down the ranges of values of the polar coordinate variables, instead of the rectangular coordinate variables, we say that we wrote down the above integral "in polar coordinates".

Example 5.7.2. Suppose I want to use the polar coordinates function to change variables for the integral $\iint_R f \, dx \, dy$, where $f(x, y) = x^2 y$, and R is the disk of radius a with center at $(a, 0)$, represented in Figure 5.61.

From the Figure we see that the angle θ goes from $-\pi/2$ to $\pi/2$.

For a given, fixed value of θ, the range of values of r is determined by the points P_1 and P_2 in Figure 5.61. For point P_1 of course we have $r_1 = 0$. Point P_2 is on the boundary circle, and from Equation 1.29 we know that this means that $r_2 = 2a \cos \theta$.

5.7. INTEGRALS IN COORDINATE SYSTEMS

Remembering to insert the stretching factor, we can then write

$$\iint_R x^2 y \, dx \, dy = \int_{-\pi/2}^{\pi/2} \int_0^{2a \cos \theta} (r \cos \theta)^2 (r \sin \theta)(r \, dr \, d\theta)$$

$$= \int_{-\pi/2}^{\pi/2} \int_0^{2a \cos \theta} r^4 \cos^2 \theta \sin \theta \, dr \, d\theta$$

Coordinate slices

We noted early on in our discussion of double integrals that we could slice up a given domain in either order – in other words, we could do x-slices first and then y-slices, putting our differentials in the order $dy \, dx$; or we could do the y-slices first and then the x-slices, putting our differentials in the order $dx \, dy$. Very often neither option was significantly more convenient than the other; which is not surprising, since in each case the slices are lines.

Of course it is also the case for integrals in polar coordinates that we can do the slicing in either order we choose. However, in this case the slices do not look the same.

Remember, taking an x-slice means to look at a set of points that have the same value of the variable x, and that is a line perpendicular to the x-axis; similarly for a y-slice. Analogously a θ-slice is a set of points that have the same value of the variable θ, and that too is a line (though not perpendicular to any particular axis since there is no θ-axis).

So what then is an r-slice? Well of course it is a set of points in a given domain that have the same value of r. But note, that is not a line – rather, it is a circle centered at the origin. So, if we choose to do an integral in polar coordinates doing the r-slices first, our picture will be a bit different.

Let's revisit Example 5.7.1, this time doing the r-slices first.

Example 5.7.3. Suppose I want to use the polar coordinates function to change variables for the integral $\iint_R f \, dx \, dy$, where $f(x, y) = x^2 + y^2$, and R is the top half of the unit circle, represented in Figure 5.62.

Our r-slices begin with the slice $r = 0$, and continue out to the slice $r = 1$. An arbitrary fixed slice is also shown in the figure.

On that fixed r-slice, we need now to determine the range of values of θ, which is determined by the values of θ at the points Q_1 and Q_2. So we have $\theta_1 = 0$ and $\theta_2 = \pi$.

Remembering to insert the stretching factor, we can then write

$$\iint_R x^2 + y^2 \, dx \, dy = \int_0^1 \int_0^\pi (r^2 \cos^2 \theta + r^2 \sin^2 \theta)(r \, d\theta \, dr)$$

$$= \int_0^1 \int_0^\pi r^3 \, d\theta \, dr$$

Note that we ended up with a very similar answer to that from Example 5.7.1; this is not surprising, because of course we know that in fact this is a double integral over a rectangle in the $r\theta$-plane (indicated by the fact that the bounds are all constants), and when the bounds are all constants they do not change when switching the order of the differentials.

But when the bounds are not all constants, switching the order of the differentials in polar coordinates changes things around in the same way it did for rectangular coordinates. We will see that as we now revisit Example 5.7.2.

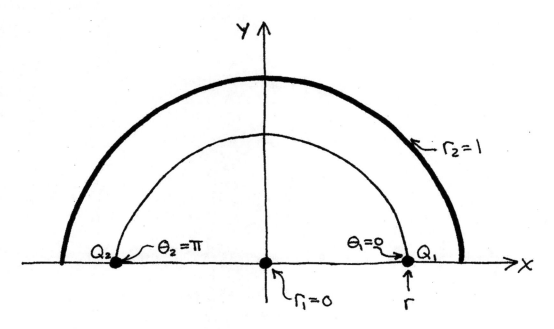

Figure 5.62:

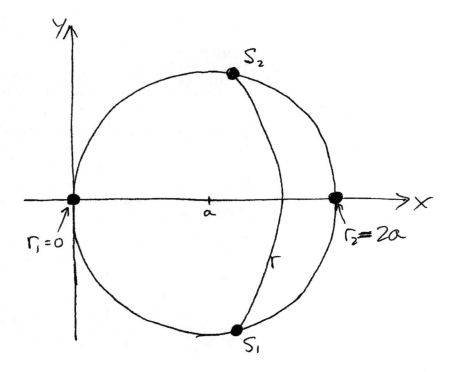

Figure 5.63:

5.7. INTEGRALS IN COORDINATE SYSTEMS

Example 5.7.4. Suppose I want to use the polar coordinates function to change variables for the integral $\iint_R f\,dx\,dy$, where $f(x,y) = x^2 y$, and R is the disk of radius a with center at $(a, 0)$, represented in **Figure 5.63**.

The r-slices begin with $r_1 = 0$, and continue along to the point furthest from the origin, which is $(2a, 0)_r$ at a distance from the origin $r_2 = 2a$.

For a given, fixed value of r, the range of values of θ is determined by the points S_1 and S_2 in **Figure 5.63**. Each of these is on the boundary circle, and from Equation 1.29 we know that this means that $r = 2a\cos\theta$ for both θ_1 and θ_2. Since $\theta_1 < 0$ and $\theta_2 > 0$, we conclude that $\theta_1 = -\arccos\left(\frac{r}{2a}\right)$ and $\theta_2 = \arccos\left(\frac{r}{2a}\right)$.

Remembering to insert the stretching factor, we can then write

$$\iint_R x^2 y\,dx\,dy = \int_0^{2a} \int_{-\arccos\left(\frac{r}{2a}\right)}^{\arccos\left(\frac{r}{2a}\right)} (r\cos\theta)^2 (r\sin\theta)(r\,d\theta\,dr)$$

$$= \int_0^{2a} \int_{-\arccos\left(\frac{r}{2a}\right)}^{\arccos\left(\frac{r}{2a}\right)} r^4 \cos^2\theta \sin\theta\,d\theta\,dr$$

Notice that in this case, the figure was a bit harder to draw, and the bounds on the $d\theta$ integral were inconvenient. So in this case doing the θ-slices first was certainly the more convenient option.

Both of these inconveniences were consequences of the nature of the coordinate system. The r-slices are not lines, and so they are a bit harder to draw; which tends to suggest that usually the θ-slices should come first. And most of the time curves turn out to be easiest to express with r as a function of θ, also suggesting that the θ-slices should come first in order to avoid needing to invert a function.

So most of the time, when doing integrals in polar coordinates, it is most convenient to do the θ-slices first, thus having the differentials in the order $dr\,d\theta$.

Cylindrical coordinate integrals

We have found that we can view polar coordinate change of variables integrals without actually looking at the new domain. In the same way, we can effectively do change of variables with the cylindrical and spherical coordinate functions also without looking at the new domain.

As was the case with polar coordinate functions, it all comes down to seeing what the appopriate coordinate slices look like for these coordinate variables.

In cylindrical coordinates, the three variables are r, θ and z. As with rectangular and polar coordinates, the slices corresponding to these variables are sets where these variables are constant. Thus, r-slices are cylinders centered on the z-axis; θ-slices are planes containing the z-axis; and z-slices are horizontal planes. See **Figure 5.64**.

Using these pictures and ideas similar to those we applied to polar coordinate integrals, we can do a triple integral "in cylindrical coordinates".

Example 5.7.5. Suppose that we are interested in the integral $\iiint_R f\,dx\,dy\,dz$, where R is the cone whose base is the unit disk in the xy-plane, and whose vertex is at the point $(0, 0, 1)$, as in **Figure 5.65**. And suppose that the function is $f(x, y, z) = x^2 + y^2 + z^2$

We can represent the side wall of the cone in cylindrical coordinates with $r = 1 - z$.

To change the integral into a cylindrical coordinates integral, keep in mind that what we are really doing is a change of variables using the cylindrical coordinates function, for which we have already computed the Jacobian determinant to be $|\det J_{g_c}| = r$. So all that remains is to determine the proper bounds for the nested integrals.

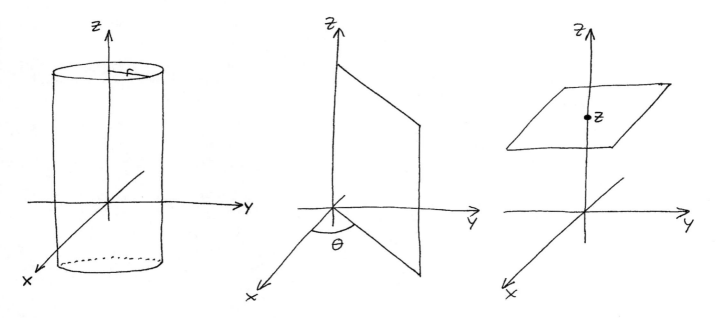

Figure 5.64:

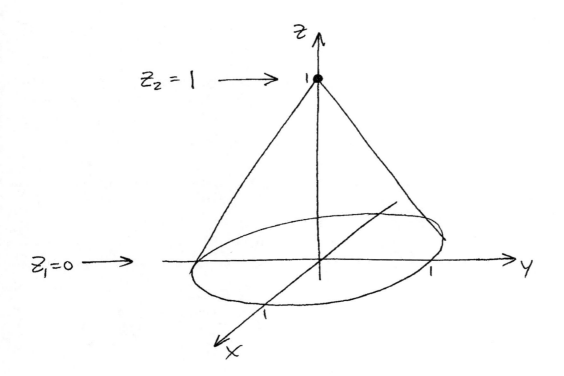

Figure 5.65:

5.7. INTEGRALS IN COORDINATE SYSTEMS

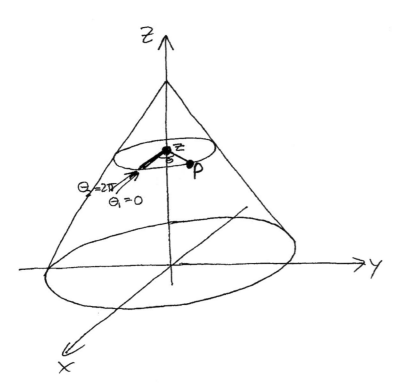

Figure 5.66:

Arbitrarily, let's slice in the order that makes our differentials $dr\,d\theta\,dz$. In Figure 5.65 we see that z starts at $z_1 = 0$ and ends at $z_2 = 1$.

Looking at Figure 5.66, we see that for any fixed value of z, the angle θ ranges all the way around from $\theta_1 = 0$ to $\theta_2 = 2\pi$. And for any given fixed values of both z and θ, we have that r begins at $r_1 = 0$ and ends at the point P; since P is on the surface with equation $r = 1 - z$, we then have $r_2 = 1 - z$.

So, the end result of applying change of variables with the cylindrical coordinates function is going to be the integral

$$\iiint_R f(x,y,z)\,dx\,dy\,dz = \int_0^1 \int_0^{2\pi} \int_0^{1-z} f(r\cos\theta, r\sin\theta, z)(r\,dr\,d\theta\,dz)$$
$$= \int_0^1 \int_0^{2\pi} \int_0^{1-z} r^3 + rz^2\,dr\,d\theta\,dz$$

This is one of the most often convenient orders in which to do the slices when doing an integral in cylindrical coordinates, for reasons similar to those for polar coordinates. That is, very often r is given as a function of θ and z making dr a natural choice for the inside integral; and dz slices are often the easiest to draw, since they are all horizontal planes.

Another order that is often convenient is $dz\,dr\,d\theta$. This would be useful in an instance where z is given as a function of r and θ (note that we still prefer not to do the r-slices first, since they are cylinders, and therefore curved and less convenient to draw.)

Spherical coordinate integrals

The same sort of reasoning can be applied to doing a triple integral "in spherical coordinates". Again we must note the shapes of the slices for these coordinate variables.

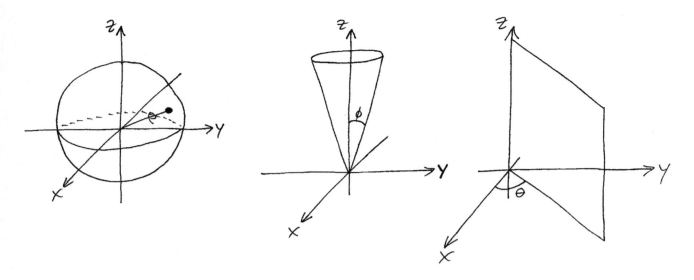

Figure 5.67:

In spherical coordinates, the three variables are ρ, ϕ and θ. As with the other coordinate systems, the slices corresponding to these variables are sets where these variables are constant. Thus, ρ-slices are spheres centered at the origin; ϕ-slices are cones with vertex at the origin and symmetric around the z-axis ; and θ-slices are planes containing the z-axis. See Figure 5.67.

Using these pictures and ideas similar to those we applied to other coordinate integrals, we can do a triple integral "in spherical coordinates".

In this case we note right away that the θ-slices are the only ones that are planes – both of the others are curved. Furthermore, ρ is usually given as a function of ϕ and θ. Given these factors, usually the most convenient order of differentials is $d\rho\, d\phi\, d\theta$.

Example 5.7.6. Suppose that we are interested in the integral $\iiint_R f\, dx\, dy\, dz$, where R is the ball of radius a with center at $(0, 0, a)$ on the z-axis (as in Figure 5.68), and the integrand is again $f(x, y, z) = x^2 + y^2 + z^2$.

We first note that we can write the spherical equation of the given boundary sphere by converting the rectangular equation. We get

$$\begin{aligned} x^2 + y^2 + (z-a)^2 &= a^2 \\ x^2 + y^2 + z^2 + a^2 - 2az &= a^2 \\ \rho^2 &= 2az \\ \rho^2 &= 2a(\rho \cos\phi) \\ \rho &= 2a \cos\phi \end{aligned}$$

(and of course also $\rho = 0$, noting that we divided by ρ in the last step above.)

To change the integral into a spherical coordinates integral, keep in mind again that what we are really doing is a change of variables using the spherical coordinates function, for which we have already computed the Jacobian determinant to be $|\det J_{g_s}| = \rho^2 \sin\phi$. So all that remains is to determine the proper bounds for the nested integrals.

Let's slice in the order that makes our differentials $d\rho\, d\phi\, d\theta$, as we discussed above. In Figure 5.69 we see that θ ranges all the way around from $\theta_1 = 0$ to $\theta_2 = 2\pi$.

In Figure 5.69 we draw an arbitrary, fixed θ-slice, and now move on to determining the range of values of ϕ. In order to make this more visible, we draw in Figure 5.70 just the given θ-slice – in that figure the vertical axis is the

5.7. INTEGRALS IN COORDINATE SYSTEMS

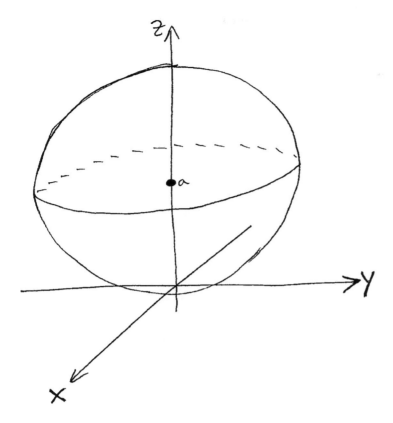

Figure 5.68:

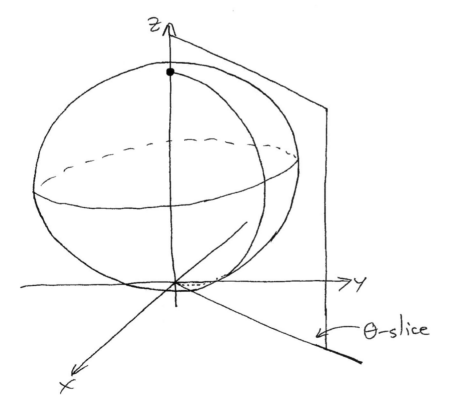

Figure 5.69:

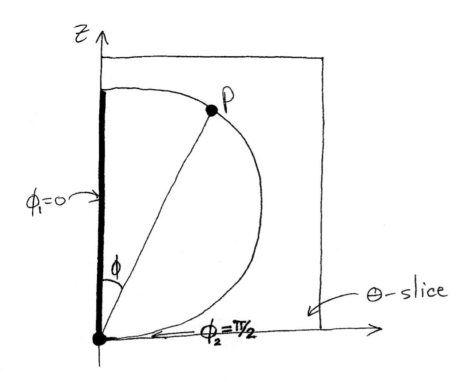

Figure 5.70:

z-axis, but note that the horizontal axis in neither the x- nor y-axis. In that figure we can then readily see that ϕ starts at $\phi_1 = 0$ and ends at $\phi_2 = \pi/2$.

For arbitrary fixed values of θ and ϕ then, we move on to finding the range of values for ρ. In Figure 5.70 we see that ρ starts at $\rho_1 = 0$, and the ending value is determined by the point P. That point P is on the circle which is a slice of the original sphere, so it is on that sphere, with equation $\rho = 2a\cos\phi$. So we have $\rho_2 = 2a\cos\phi$, and thus the end result of applying change of variables with the spherical coordinates function will be the integral

$$\iiint_R f(x,y,z)\,dx\,dy\,dz = \int_0^{2\pi}\int_0^{\pi/2}\int_0^{2a\cos\phi}(x^2+y^2+z^2)(\rho^2\sin\phi\,d\rho\,d\phi\,d\theta)$$
$$= \int_0^{2\pi}\int_0^{\pi/2}\int_0^{2a\cos\phi}\rho^4\sin\phi\,d\rho\,d\phi\,d\theta$$

Exercises

Exercise 5.7.1 (exer-Ints-ICS-1). Confirm by direct calculation that the Jacobian determinant for the spherical coordinates function is $\rho^2\sin\phi$.

Exercise 5.7.2 (exer-Ints-ICS-2). Set up, but do not evaluate, a double nested integral in polar coordinates that represents $\iint_D xe^y - y^2\,dA$, where D is the fourth quadrant portion of the solid disk of radius 3 centered at $(3,0)$.

Exercise 5.7.3 (exer-Ints-ICS-3). Set up, but do not evaluate, a double nested integral in polar coordinates that represents $\iint_D \ln(1+x^2+y^2)\,dA$, where D is the part of the disk D (of radius 1 centered at $(0,1)$) that is outside of the unit disk.

5.7. INTEGRALS IN COORDINATE SYSTEMS

Exercise 5.7.4 (exer-Ints-ICS-4). Set up, but do not evaluate, a double nested integral in polar coordinates that represents $\iint_D (x^2 + y^2)^4 \, dA$, where D is the second quadrant part of the disk D (of radius 3 centered at $(0,3)$) that is outside of the disk B (of radius 1 centered at $(0,1)$).

Exercise 5.7.5 (exer-Ints-ICS-5). Set up, but do not evaluate, a triple nested integral in cylindrical coordinates that represents $\iiint_D xyz \, dV$, where D is the unit ball.

Exercise 5.7.6 (exer-Ints-ICS-6). Set up, but do not evaluate, a triple nested integral in cylindrical coordinates that represents $\iiint_D x + z \, dV$, where D is the region satisfying $x^2 + y^2 + z^2 \leq 4$ and $x^2 + y^2 \leq 1$.

Exercise 5.7.7 (exer-Ints-ICS-7). Set up, but do not evaluate, a triple nested integral in cylindrical coordinates that represents $\iiint_D x + z \, dV$, where D is the region satisfying $x^2 + y^2 + z^2 \leq 4$ and $z^2 \leq 1$.

Exercise 5.7.8 (exer-Ints-ICS-8). Set up, but do not evaluate, a triple nested integral in spherical coordinates that represents $\iiint_D xyz \, dV$, where D is the unit ball.

Exercise 5.7.9 (exer-Ints-ICS-9). Set up, but do not evaluate, a triple nested integral in spherical coordinates that represents $\iiint_D x - 4yz \, dV$, where D is the part of the ball of radius 4 centered at $(0,0,4)$ with $z^2 \geq x^2 + y^2$.

Exercise 5.7.10 (exer-Ints-ICS-10). Set up, but do not evaluate, a triple nested integral in spherical coordinates that represents $\iiint_D 3x - 2y + e^z \, dV$, where D is the part of the solid ball of radius 1 centered at $(0,0,1)$ that is outside of the unit ball.

Exercise 5.7.11 (exer-Ints-ICS-11). Set up, but do not evaluate, a triple nested integral in spherical coordinates that represents the volume of the region in xyz-space defined by $\rho \leq 2 - \sin\phi\cos\theta$.

Exercise 5.7.12 (exer-Ints-ICS-12). Set up, but do not evaluate, a triple nested integral in spherical coordinates that represents the mass of the region in xyz-space defined by $x^2 + y^2 + z^2 \leq 1$, $z \geq \sqrt{x^2 + y^2}$, and $y \geq x$, where the density is given by $\delta(x,y,z) = y^2$.

Exercise 5.7.13 (exer-Ints-ICS-13). Show that the equation $\rho = \sin\phi$ represents the rotation around the z-axis of the circle of radius $1/2$ in the xz-plane centered at $(1/2, 0, 0)$.

Exercise 5.7.14 (exer-Ints-ICS-14). Use a spherical coordinates integral to compute the volume inside the surface described in Exercise 5.7.13, and then compute that same volume with the Theorem of Pappus described in Exercise 5.4.15.

Exercise 5.7.15 (exer-Ints-ICS-15). Compute the moment of inertia around the z-axis of the unit ball centered at the origin, with $\delta = 1$.

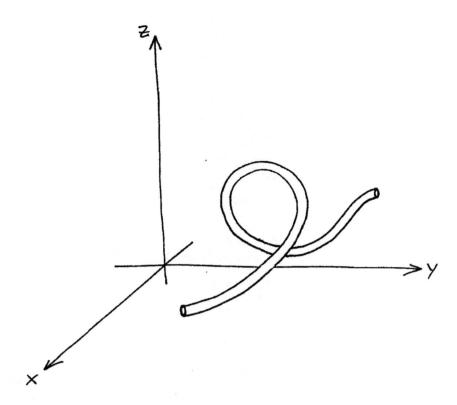

Figure 5.71:

5.8 Scalar Line Integrals

So far we have found ways to define integrals, from the point of view of Section 5.1, for two types of domains other than the familiar single variable integrals on intervals in $\mathbb{R}^1$ – we have defined double integrals for two-dimensional domains in $\mathbb{R}^2$, and triple integrals for three-dimensional domains in $\mathbb{R}^3$. Subsequent sections then gave us techniques allowing us to compute these integrals.

Note that in each case, the dimension of the domain for the integral was the same as the dimension of the space of which it was thought of as a subset. For example, an interval in $\mathbb{R}^1$ is one-dimensional, and $\mathbb{R}^1$ is itself one-dimensional. A rectangle or other type of area domain in $\mathbb{R}^2$ is two-dimensional, as is $\mathbb{R}^2$ itself; and a solid box or other type of volume domain in $\mathbb{R}^3$ is three-dimensional, as is $\mathbb{R}^3$ itself.

But how would we deal with a situation where this is not the case? What if we have a type of problem that fits the four integral conditions from Section 5.1, so that we would like to use some type of integral to solve the problem – but the domain is a curve in $\mathbb{R}^2$ or $\mathbb{R}^3$? Or a surface in $\mathbb{R}^3$?

For example, suppose there is a road following a known curved path in $\mathbb{R}^2$, and the depth of snow is a known, non-constant function h of location; how can we compute the total volume of snow on the road? This problem fits the four integral conditions, with the domain being the known curved path and the integrand being the product of the width w of the road and the snow height function (we will explain this in more detail momentarily). But this domain is not like any of the domains for which we have yet defined integrals.

Or suppose that we have a rod and wish to compute the mass, as in Example 5.1.2; but suppose that the rod is not straight, but rather that it follows a known curved path in $\mathbb{R}^3$, as in Figure 5.71. How can we compute the mass? This problem also fits the four integral conditions, with the domain being the known curved path and the integrand being the mass per unit length δ; but again this domain is not like any of the domains for which we have yet defined integrals.

Both of these problems feel very similar to single variable integrals, since they have one-dimensional domains. In fact,

5.8. SCALAR LINE INTEGRALS

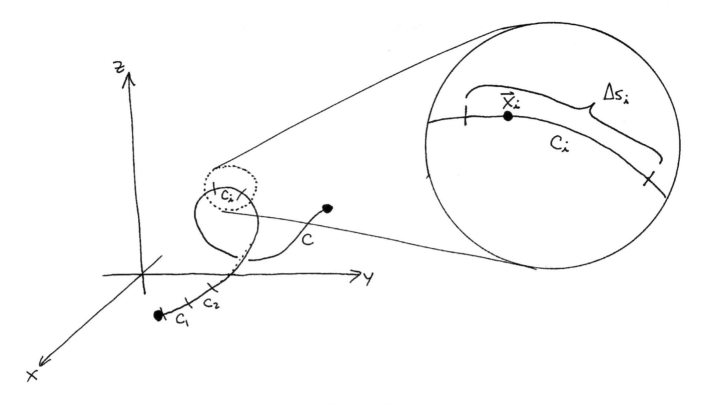

Figure 5.72:

ultimately we will use single variable integrals as part of the computation. But before we can make that connection, we must use the integral conditions on page 229 to find an appropriate definition for an integral over such domains to define the quantity we are actually computing.

5.8.1 Definition

In each of the above examples the domain is a path; let's call this path C. Following our usual model, we first need to break up the path C into smaller pieces C_i, with individual lengths Δs_i. In each piece we need to choose a sample point, and since these pieces are in $\mathbb{R}^3$ we will denote those points with position vectors $\vec{x}_i$. See Figure 5.72

Following the model of Equation 5.2, we define the "scalar line integral" of a given scalar (real-valued) function f over the path C as

Definition 5.8.1.

$$\int_C f(\vec{x})\,ds \;=\; \lim \sum \Big(f(\vec{x}_i)\Big)\Big(\Delta s_i\Big)$$

(Note: Even though all of the other integrals we have defined up to now also have scalar (real-valued) integrands, we explicitly note this for these integrals by defining the above as a "scalar line integral" because very shortly we will make a definition of something called a "vector line integral". In practice though, when the context is clear, each of these can be referred to as simply a "line integral".)

Let's apply this definition to the previous examples in more detail.

Example 5.8.1. Suppose there is a road following a known curved path C in $\mathbb{R}^2$, and the depth of snow is a known, non-constant function h of location; how can we represent the total volume of snow on the road?

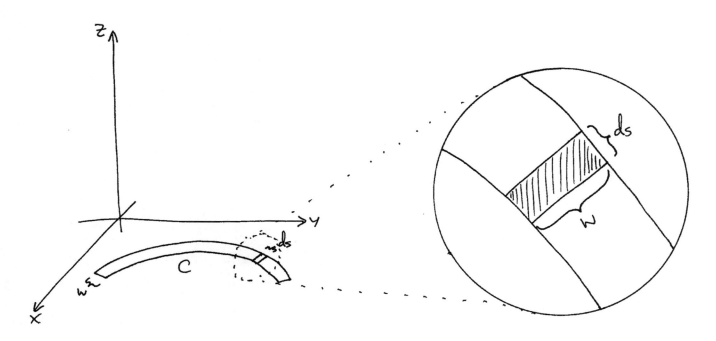

Figure 5.73:

We break up the path C into little pieces C_i of length ds (note that we are using the integral shorthand for the Riemann sum we are about to write down), as in Figure 5.73. Let dV represent the volume of snow over that piece of the road.

On each such piece, the road has width w and length ds, so the area of that piece of the road is $w\,ds$. Given that the depth of the snow over that area is $h(\vec{x})$, we can write the volume of snow over that piece of the road as $dV = \big(h(\vec{x})\big)\big(w\,ds\big)$.

Adding this up over the entire road as in Definition 5.8.1, we have that the total volume of snow is

$$V = \int_C dV = \int_C w\,h(\vec{x})\,ds \tag{5.123}$$

Example 5.8.2. Suppose that there is a bent rod following the curved path C, and that the density at a location $\vec{x}$ on the curve is $\delta(\vec{x})$. How can we represent the mass of this rod?

As before, we break up the path C into little pieces C_i, as in Figure 5.72. Each piece C_i has length ds and density $\delta(\vec{x})$, and thus has mass $dm = \delta(\vec{x})\,ds$. Then the total mass is

$$m = \int_C dm = \int_C \delta(\vec{x})\,ds \tag{5.124}$$

5.8.2 Strategy for computation

The definition we have above for the scalar line integral, like those for double and triple integrals, is a matter of breaking up a domain into little pieces, multiplying the integrand times the size of each piece, and then adding them all up. And as

5.8. SCALAR LINE INTEGRALS

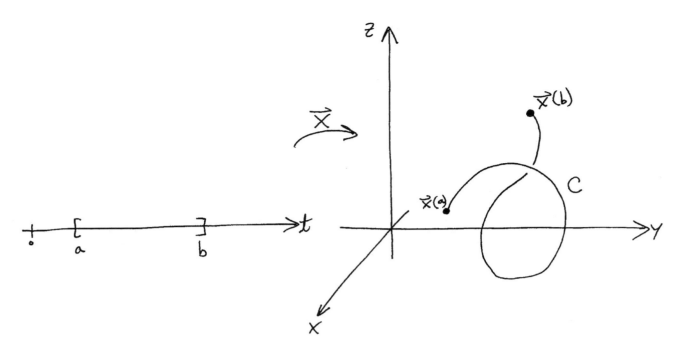

Figure 5.74:

was the case for double and triple integrals, we see that there are applications for which this construction is a convenient representation.

But, again just as with double and triple integrals, we still have the problem that having written down this Riemann sum we don't have any particularly convenient way to compute it. With double and triple integrals we found that nested integrals were a useful direct computational tool, and then of course there were the "stretching factor" methods – change of variables, and using different coordinate systems. But as of the moment we have not yet developed any convenient way of computing scalar line integrals.

Scalar line integrals clearly have much in common then with double and triple integrals. The most substantial difference though really is in the nature of the domain – it is of a dimension that is different than the space of which it is a subset. This makes the idea of nested integrals simply not work very well, at least not immediately.

So our problem then is really that we have an integral defined on a domain that is inconvenient. And this is a situation we have found ourselves in before, namely in the section about the change of variables method. In fact the way we will end up computing these scalar line integrals is completely analogous to the method we used to do change of variables; in particular, we will find the given domain as the image of a different, more convenient domain, and then use the stretching factor strategy from the beginning of Section 5.6.

Finding a path in either $\mathbb{R}^2$ or $\mathbb{R}^3$ as the image of a more convenient domain is actually something that we have done already; namely, it is a parametrization of that path. For example, if we have a scalar line integral whose domain is the path C which is upper half of the unit circle in $\mathbb{R}^2$, we can find that as the image of the interval $[0, \pi]$ on the t-axis, using the parametrization $\vec{x} : \mathbb{R}^1 \to \mathbb{R}^2$

$$\vec{x}(t) = \begin{pmatrix} \cos t \\ \sin t \end{pmatrix} \quad \text{where} \quad t \in [0, \pi] \tag{5.125}$$

In general, we can describe any curve in either $\mathbb{R}^2$ or $\mathbb{R}^3$ parametrically. We can geometrically represent such a parametrization of an arbitrary curve C by the function $\vec{x}$ with separate domain and target, by analogy with the way we geometrically represented change of variables functions, as in Figure 5.74. Note that in the figure we represent a curve in $\mathbb{R}^3$, though the picture is very similar for $\mathbb{R}^2$.

To apply the stretching factor strategy, we will want to rewrite a given integral defined on the curve C as an integral defined on the new domain, which is the interval $[a, b]$. Note of course that the interval $[a, b]$ is a one-dimensional domain

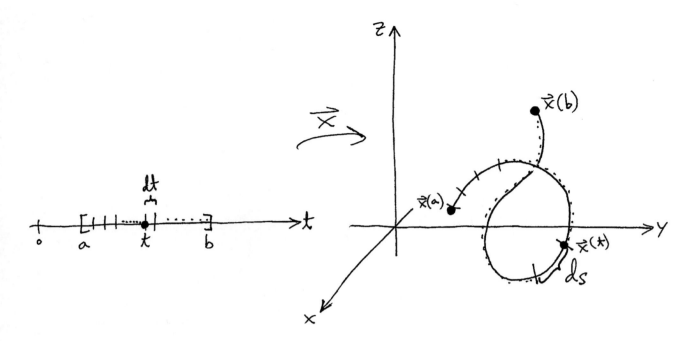

Figure 5.75:

on the one-dimensional t-axis, and so such an integral could then be computed by the fundamental theorem of calculus.

Just as was the case with the change of variables method, there now remain two steps in the process or rewriting the given integral over the new domain. First, everything must be rewritten in terms of t – but of course the parametrization function itself gives us an explicit representation of the variables x, y and z in terms of t, so that is no problem.

As with change of variables, the step that requires some consideration is in determining the stretching factor. Recall that the idea is that every subinterval of length dt on the t-axis will be stretched out to form a piece of length ds on the curve C; and the nature of the parametrization $\vec{x}$ will determine in some way how much the length dt is stretched to become ds. See Figure 5.75.

To determine this stretching factor, we will take an approach very similar to the one from the change of variables section; namely, we will use the derivative to compute the change in the output given a change in the input.

As in Figure 5.76, we can view the vector $\vec{e}_1 dt$ as representing the change in the input over the given interval in the domain. The corresponding change in the target then is well approximated by

$$d\vec{x} = D_{\vec{x},t}(\vec{e}_1 dt) = D_{\vec{x},t}(\vec{e}_1)\,dt = \vec{x}'(t)\,dt = \vec{v}(t)\,dt \tag{5.126}$$

where $\vec{v}$ is the velocity vector of the parametrization

(Of course as usual with linear approximations, this is not exact, but is sufficiently accurate in the limit as the pieces get small.)

The length ds is just the length of this image vector, so we have

$$\begin{aligned} ds &= \|d\vec{x}\| \\ &= \|\vec{v}(t)\|\,dt \end{aligned}$$

From this we can conclude that the stretching factor, by which dt is stretched to become ds, is simply the length of the velocity vector. (Actually this is no surprise, since we know that the length of the velocity vector is the speed, which is the amount of distance (ds) per unit time (dt) that the parametrized curve travels.)

Incorporating this into our general stretching factor strategy, we can conclude the following.

5.8. SCALAR LINE INTEGRALS

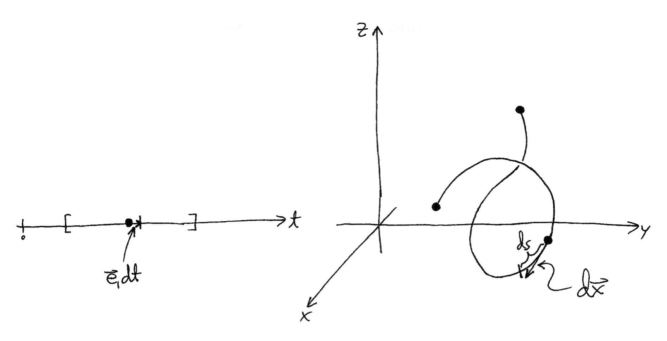

Figure 5.76:

Theorem 5.8.1. *If a path C is parametrized by the continuously differentiable function $\vec{x}(t)$ over the interval $[a,b]$ on the t-axis, with velocity vector given by $\vec{v}(t) = \vec{x}'(t)$, then a scalar line integral of a function $f(\vec{x})$ over C can be rewritten as*

$$\int_a^b f(\vec{x}(t)) \|\vec{v}(t)\| \, dt = \int_C f(\vec{x}) \, ds \tag{5.127}$$

Of course a curve can be parametrized in many different ways. It is interesting to notice that the left side of the above equation depends on the parametrization, while the right hand side does not. So, while the integral on the left side of the equation in Theorem 5.8.1 could take different forms depending on the parametrization, the resulting value is completely independent of that parametrization.

5.8.3 Examples

We now have a definition of the scalar line integral as a Riemann sum, and a result that allows us to convert that into a single variable integral which we can compute by the usual methods. So we are ready to try some example applications.

Example 5.8.3. Suppose that there is a bent rod following the curved path C, which is the top half of the unit circle in $\mathbb{R}^2$; and that the density at a location $\vec{x}$ on the curve is $\delta(\vec{x}) = x^2 + 2y^2$. What is the mass of this rod?

From Example 5.8.2 the total mass is

$$m = \int_C dm = \int_C \delta(\vec{x}) \, ds \tag{5.128}$$

In order to turn this integral into a single variable integral we can compute, note that we need to have a parametrization function $\vec{x}(t)$ of the path C; a convenient choice is

$$\vec{x}(t) = \begin{pmatrix} \cos t \\ \sin t \end{pmatrix} \quad \textbf{where} \quad t \in [0, \pi] \tag{5.129}$$

Of course we will need to know the length of the velocity vector as a function of t, so we compute

$$\|\vec{v}(t)\| = \|\vec{x}'(t)\| = \left\| \begin{pmatrix} \cos t \\ \sin t \end{pmatrix}' \right\| = \left\| \begin{pmatrix} -\sin t \\ \cos t \end{pmatrix} \right\| = 1 \tag{5.130}$$

Plugging this into Theorem 5.8.1, we get

$$\begin{aligned} m &= \int_C \delta(\vec{x})\, ds \\ &= \int_0^\pi \delta(\vec{x}(t))\|\vec{v}(t)\|\, dt \\ &= \int_0^\pi \delta\left(\begin{pmatrix} \cos t \\ \sin t \end{pmatrix} \right)(1)\, dt \\ &= \int_0^\pi \cos^2(t) + 2\sin^2(t)\, dt \\ &= \int_0^\pi 1 + \sin^2(t)\, dt \end{aligned}$$

This single variable integral can be computed the usual methods.

Note that in the above example, the very first thing we needed to do was to parametrize the given curve. The point here is that our computational tool, Theorem 5.8.1, uses the parametrization to convert the scalar line integral into a single variable integral we can compute, and so without that parametrization we cannot use the Theorem. Parametrizing the domain curve is then always the first thing that we need to do.

Example 5.8.4. Suppose there is a road following the path C in $\mathbb{R}^2$ which is the part of the graph of $y = 1000\sin(\pi x/1000)$ between $x = 0$ and $x = 1000$ (units are meters); the depth of snow is given by $h(\vec{x}) = \frac{x+y}{1000}$, and the road is three meters wide. What is the total volume of snow on the road?

From Example 5.8.1, we know that

$$V = \int_C w\, h(\vec{x})\, ds \tag{5.131}$$

The curve is given as a graph, so we can use the graph parametrization

$$\vec{x}(t) = \begin{pmatrix} t \\ 1000\sin(\pi t/1000) \end{pmatrix} \quad \text{where} \quad t \in [0, 1000] \tag{5.132}$$

The length of the velocity vector is

$$\|\vec{v}(t)\| = \|\vec{x}'(t)\| = \left\| \begin{pmatrix} t \\ 1000\sin(\pi t/1000) \end{pmatrix}' \right\| = \left\| \begin{pmatrix} 1 \\ \pi\cos(\pi t/1000) \end{pmatrix} \right\| = \sqrt{1 + \pi^2 \cos^2(\pi t/1000)} \tag{5.133}$$

Then Theorem 5.8.1 gives us

$$\begin{aligned} V &= \int_C w\, h(\vec{x})\, ds \\ &= \int_0^{1000} 3\, h(\vec{x}(t))\|\vec{v}(t)\|\, dt \\ &= \int_0^{1000} 3\left(\frac{t + 1000\sin(\pi t/1000)}{1000} \right) \sqrt{1 + \pi^2 \cos^2(\pi t/1000)}\, dt \end{aligned}$$

5.8. SCALAR LINE INTEGRALS

Again, this can be computed by single variable calculus methods.

Exercises

Exercise 5.8.1 (exer-Ints-SLI-1). A tree has a trunk which follows a curved path C. The thickness of the trunk is not constant though, but instead has cross-sections with areas (in square feet) at location (x, y, z) (all coordinates measured in feet) given by $A(x, y, z) = 1/(1+\sqrt{z})$. Use the notation of a scalar line integral to write down, but not evaluate, an expression representing the total volume of wood in the tree trunk.

Exercise 5.8.2 (exer-Ints-SLI-2). The tree in Exercise 5.8.1 has been invaded by a strain of bacteria from one side. The concentration of bacteria in the wood of the trunk (in thousands per cubic foot of wood) depends on the location by $C(x, y, z) = e^{3-x}$. Use the notation of a scalar line integral to write down, but not evaluate, an expression representing the total number of bacteria in the tree trunk.

Exercise 5.8.3 (exer-Ints-SLI-3). A coil spring is ten inches tall, two inches in radius, has seven full turns (moving counterclockwise as seen from above when moving up the spring), and the wire making up the spring has a radius of $1/10$ of an inch. It is oriented with its axis on the z-axis, and the bottom is resting on the xy-plane at the point $(2, 0, 0)$. The surface of the spring is rusting in a non-uniform way due to the presence of humidity near the bottom and near the right side; specifically, at any point (x, y, z), the thickness (in inches) of rust on the surface is given by $T(x, y, z) = (4+y)(10-z)/1000$. Compute the total volume of rust on this spring.

Exercise 5.8.4 (exer-Ints-SLI-4). The Enterprise follows the path of the graph of the function $y = 1 - (x^2/4)$ (between $x = -10$ and $x = 10$, all coordinates measured in millions of miles) in slingshot around the sun. Because of its varying speed and proximity to the sun, at any point (x, y) on the path the heat absorbed per unit distance is given by $H(x, y) = 1/(x^2 + y^2)$, in joules per mile. Compute the total amount of heat absorbed by the Enterprise.

Exercise 5.8.5 (exer-Ints-SLI-5). Compute the value of the line integral $\int_C f \, ds$, where $f(x, y, z) = yz$ and C is the curve parametrized by $\vec{x}(t) = (\sin^2 t, \sin t \cos t, \cos t)$ for $t \in [0, 2\pi]$.

Exercise 5.8.6 (exer-Ints-SLI-6). Compute the value of the line integral $\int_C f \, ds$, where $f(x, y, z) = xyz$ and C is the curve parametrized by $\vec{x}(t) = (3-t, 3+2t, 4t-5)$ for $t \in [0, 1]$.

Exercise 5.8.7 (exer-Ints-SLI-7). Suppose that a curve C in $\mathbb{R}^2$ is symmetric over a line L and that the function f has odd symmetry over L. Explain how you know that the line integral $\int_C f \, ds$ must be zero.

Exercise 5.8.8 (exer-Ints-SLI-8). Can the claim in Exercise 5.8.7 be used to compute the integral $\int_C f \, ds$, where C is the unit circle and $f(x, y) = x^2$? If yes, explain all of the details (identify the line of symmetry L, and explain how you know the domain and integrand have the required symmetries); if no, explain why the claim does not apply.

Exercise 5.8.9 (exer-Ints-SLI-9). Can the claim in Exercise 5.8.7 be used to compute the integral $\int_C f \, ds$, where C is the unit circle and $f(x, y) = x^2 - y^2$? If yes, explain all of the details; if no, explain why the claim does not apply.

Exercise 5.8.10 (exer-Ints-SLI-10). Suppose that a curve C in $\mathbb{R}^3$ is symmetric through a plane P and that the function f has odd symmetry through P. Explain how you know that the line integral $\int_C f \, ds$ must be zero.

Exercise 5.8.11 (exer-Ints-SLI-11). Can the claim in Exercise 5.8.10 be used to compute the integral $\int_C f \, ds$, where C is the intersection of the unit sphere and the plane $y = 0$ and $f(x, y, z) = x$? If yes, explain all of the details; if no, explain why the claim does not apply.

Exercise 5.8.12 (exer-Ints-SLI-12). The claim in Exercise 5.8.10 can be used to compute the integral $\int_C f \, ds$, where C is the intersection of the unit sphere and the plane $z = x$, and $f(x, y, z) = 3x - 4y + 3z$. Explain these details. (Hint: Think about a plane that is a level set of f.)

5.9 Scalar Surface Integrals

Scalar line integrals address the situation where the domain is a curve in either $\mathbb{R}^2$ or $\mathbb{R}^3$. Very similarly, we will define in this section "scalar surface integrals", which will address the situation where the domain is a surface in $\mathbb{R}^3$.

For example, suppose we wish to compute the mass of a curved sheet of metal, defined by a curved surface in $\mathbb{R}^3$, where the sheet has non-constant density $\delta(\vec{x})$ (mass per unit area, in this case). Certainly this fits our four integral conditions since we could cut the surface up into small pieces, and represent the total mass as the sum of the masses of the little pieces, each of which is density times area. Yet we cannot use any of our previously defined types of integrals because of the nature of the domain.

As we did with scalar line integrals, we must first make a definition that addresses this situation specifically. We will then move on to finding a method for computing such things.

5.9.1 Definition

In the situation we are considering here, the domain is a surface; let's call this surface S. Following our usual model, we first need to break up the surface S into smaller pieces S_i, with individual areas ΔS_i. In each piece we need to choose a sample point, and since these pieces are in $\mathbb{R}^3$ we will denote those points with position vectors $\vec{x}_i$. See Figure 5.77

(Note that we are using "ΔS" to represent the area of a piece of surface, while we used "Δs" to represent the length of a piece of curve in Section 5.8; students should make sure to be attuned to the notational difference.)

Following the model of Equation 5.2, we define the "scalar surface integral" of a given scalar (real-valued) function f over the surface S as

Definition 5.9.1.

$$\iint_S f(\vec{x})\, dS = \lim \sum \left(f(\vec{x}_i)\right)\left(\Delta S_i\right)$$

(Note: As with scalar line integrals, we specifically denote "scalar" in this definition because we will very soon define a different type of integral, called a "vector surface integral".)

Let's now revisit the above example in more detail.

Example 5.9.1. Suppose we wish to compute the mass of a curved sheet of metal, defined by the curved surface S in $\mathbb{R}^3$ which is the upper half of the unit sphere. The sheet has non-constant density $\delta(\vec{x}) = 3 + z$.

We break up the surface S into little pieces S_i, as in Figure 5.77. Each piece S_i has area dS (note that we are again using the integral shorthand for the Riemann sum we are about to write down) and density $\delta(\vec{x})$, and thus has mass $dm = \delta(\vec{x})\, dS$. Then the total mass is

$$m = \iint_S dm = \iint_S \delta(\vec{x})\, dS \qquad (5.134)$$

5.9.2 Strategy for computation

As was the case at this point with scalar line integrals, we now have a definition of a scalar surface integral as a Riemann sum, and applications for which this construction is a convenient representation. But we do not yet have any convenient ways to compute such things.

5.9. SCALAR SURFACE INTEGRALS

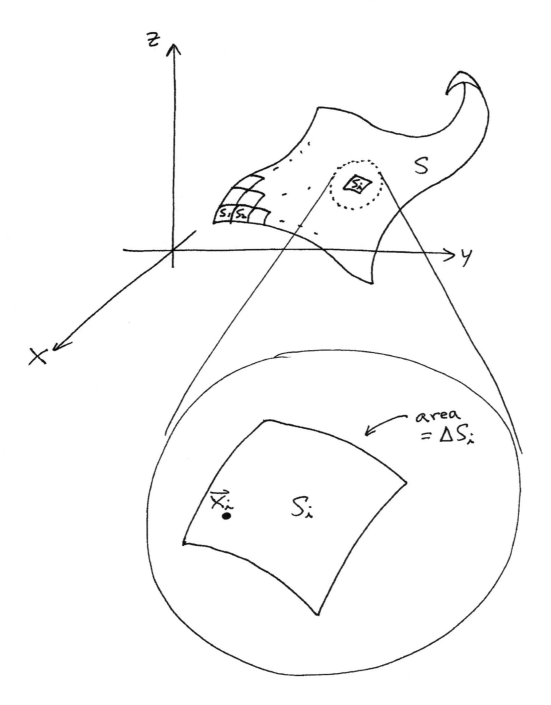

Figure 5.77:

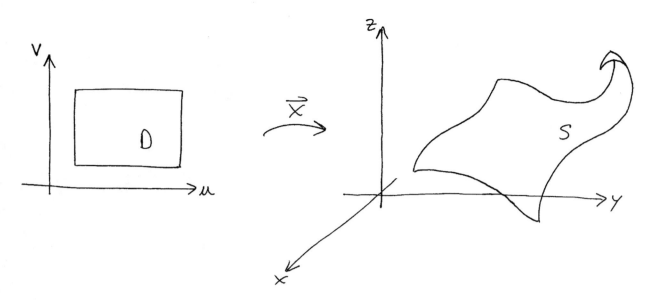

Figure 5.78:

Again as was the case with scalar line integrals, we will approach this problem with the stretching factor strategy using a parametrization.

We can geometrically represent such a parametrization of an arbitrary surface S by the function $\vec{x}$ with separate domain and target, by analogy with the way we geometrically represented scalar line integals, as in Figure 5.78.

To apply the stretching factor strategy, we will want to rewrite a given integral defined on the surface S as an integral defined on the new domain D in the uv-plane (we draw D here as a rectangle, though it need not be). Note of course that D is a two-dimensional domain in the two-dimensional uv-plane, and so such an integral could then be computed by double integral techniques from earlier in this chapter.

As was the case with scalar line integrals, there now remain two steps in the process of rewriting the given integral over the new domain. First, everything must be rewritten in terms of u and v – but of course the parametrization function itself gives us an explicit representation of the variables x, y and z in terms of u and v, so that is no problem.

Again, the step that requires some consideration is in determining the stretching factor. Recall that the idea is that every piece of area $dA = du\,dv$ in the uv-plane will be stretched out to form a piece of area dS on the surface S, which will be approximately a parallelogram (of course this is not exact, but is sufficiently accurate in the limit as the pieces get small.) And the nature of the parametrization $\vec{x}$ will determine in some way just how much the area dA is stretched to become dS. See Figure 5.79.

The computation of this stretching factor is highly analogous to that for scalar line integrals. The sides of the sub-rectangle of area $dA = du\,dv$ can be thought of as being defined by the two vectors $\vec{e}_1 du$ and $\vec{e}_2 dv$, as in Figure 5.80. As before, we can apply the derivative transformation to these two vectors to find the defining edges of the parallelogram that approximates the image surface of area dS. We have

$$\begin{aligned} D_{\vec{x},(u,v)}(\vec{e}_1 du) &= D_{\vec{x},(u,v)}(\vec{e}_1)\,du = \vec{x}_u\,du \\ D_{\vec{x},(u,v)}(\vec{e}_2 dv) &= D_{\vec{x},(u,v)}(\vec{e}_2)\,dv = \vec{x}_v\,dv \end{aligned}$$

In order to find the area dS of the parallelogram, we compute the length of the cross product of these two vectors as in Chapter 3. See Figure 5.81

5.9. SCALAR SURFACE INTEGRALS

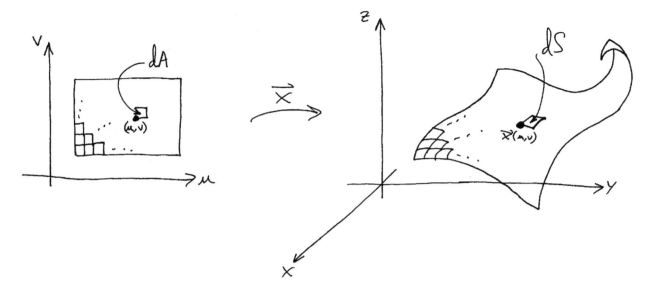

Figure 5.79:

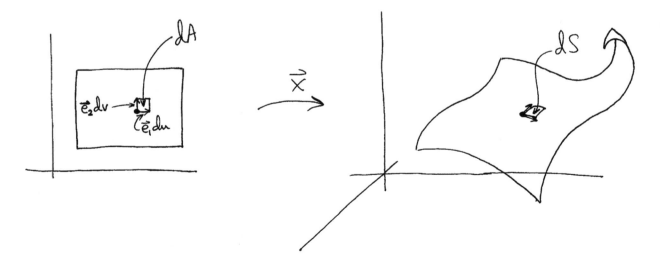

Figure 5.80:

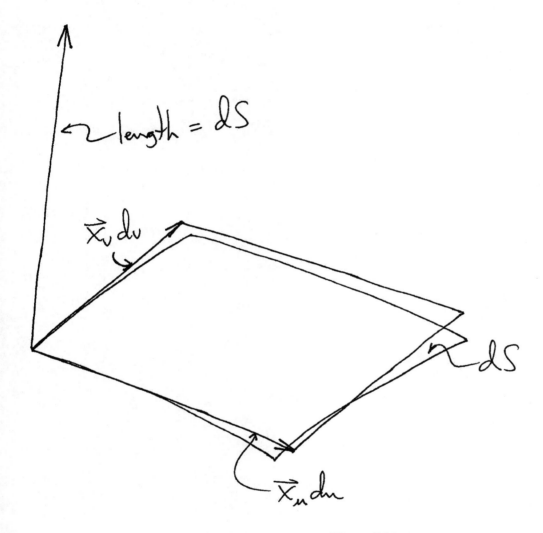

Figure 5.81:

5.9. SCALAR SURFACE INTEGRALS

So we have

$$\begin{aligned} dS &= \left\| \left(\vec{x}_u \, du \right) \times \left(\vec{x}_v \, dv \right) \right\| \\ &= \| \vec{x}_u \times \vec{x}_v \| \, du \, dv \\ &= \| \vec{x}_u \times \vec{x}_v \| \, dA \end{aligned}$$

So, we have that our stretching factor is $\| \vec{x}_u \times \vec{x}_v \|$.

Since this cross product $\vec{x}_u \times \vec{x}_v$ will come up often in this and other situations, we will give it a name. Note that it is perpendicular to the surface, since it is perpendicular to each of the two tangent vectors $\vec{x}_u$ and $\vec{x}_v$. So it is a normal vector, motivating our choice

$$\vec{N} = \vec{x}_u \times \vec{x}_v \tag{5.135}$$

We can thus rewrite our stretching factor as $\|\vec{N}\|$.

Incorporating this into our general stretching factor strategy, we can conclude the following.

Theorem 5.9.1. *If a surface S is parametrized by the continuously differentiable function $\vec{x}(u,v)$ over the domain D in the uv-plane, and if we define $\vec{N}(u,v) = \vec{x}_u \times \vec{x}_v$, then the scalar surface integral of the function $f(\vec{x})$ over S can be rewritten as*

$$\iint_D f(\vec{x}(u,v)) \, \|\vec{N}(u,v)\| \, du \, dv = \iint_S f(\vec{x}) \, dS \tag{5.136}$$

As was the case with scalar line integrals, note that the form of the integral on the left side of the above equation depends on the parametrization, but the right side does not. So, again, notice that the choice of parametrization does not affect the computed value of the scalar surface integral.

5.9.3 Examples

As with scalar line integrals, we now have a definition for scalar surface integrals and a theorem that allows us to transform such an integral into a more familiar integral – in this case a double integral on a plane region. Here we will combine these ideas to solve some problems.

Example 5.9.2. Suppose we have a metal sheet in the shape of the graph of the function $f(x,y) = x^2 + y^2$ over the unit disk in the xy-plane, and the density is given by the function $\delta(\vec{x}) = z^2$. Let's compute the mass of this sheet.

The sheet can be broken up into little pieces, and on each piece the mass is the density times the area; so we have

$$m = \iint_S dm = \iint_S \delta(\vec{x}) \, dS \tag{5.137}$$

To use the previous theorem to compute this surface integral, we must first parametrize the surface. We use the graph parametrization, giving us

$$\vec{x}(u,v) = \begin{pmatrix} u \\ v \\ u^2 + v^2 \end{pmatrix} \quad \text{where} \quad (u,v) \in \{D = \text{unit disk}\} \tag{5.138}$$

We will need to know the stretching factor $\|\vec{N}\|$, so we compute

$$\begin{aligned}
\|\vec{N}\| &= \|\vec{x}_u \times \vec{x}_v\| \\
&= \left\|\frac{\partial}{\partial u}\begin{pmatrix} u \\ v \\ u^2+v^2 \end{pmatrix} \times \frac{\partial}{\partial v}\begin{pmatrix} u \\ v \\ u^2+v^2 \end{pmatrix}\right\| \\
&= \left\|\begin{pmatrix} 1 \\ 0 \\ 2u \end{pmatrix} \times \begin{pmatrix} 0 \\ 1 \\ 2v \end{pmatrix}\right\| \\
&= \left\|\begin{pmatrix} -2u \\ -2v \\ 1 \end{pmatrix}\right\| \\
&= \sqrt{1 + 4u^2 + 4v^2}
\end{aligned}$$

Applying Theorem 5.9.1, we have

$$\begin{aligned}
m &= \iint_S \delta(\vec{x})\, dS \\
&= \iint_D \delta(\vec{x}(u,v)) \|\vec{N}\|\, dA \\
&= \iint_D \delta\left(\begin{pmatrix} u \\ v \\ u^2+v^2 \end{pmatrix}\right) \sqrt{1+4u^2+4v^2}\, du\, dv \\
&= \iint_D (u^2+v^2)^2 \sqrt{1+4u^2+4v^2}\, du\, dv
\end{aligned}$$

This is now a double integral, which can be computed by standard double integral techniques.

Note that, as was the case with scalar line integrals, the first step in the computation of a scalar surface integral is to find a parametrization of the surface to allow for the use of Theorem 5.9.1.

Example 5.9.3. At an aquarium there is a tank of depth 20 feet, with a hallway underneath. On the bottom surface of the tank is a hemispherical plexiglas dome (S) of radius 1 meter that allows viewers in the hallway to see the entire tank from the bottom. Recalling that force is pressure times area, let's compute the total force of the water on the plexiglas dome.

The pressure of water at a depth of h meters is approximately $P = 9800h \frac{N}{m^2}$. If we set up our xyz-axis system so that the origin is at the center of the hemisphere, then the depth is $h = 20 - z$, so we have $P = 9800(20 - z)$, measured in units of $\frac{N}{m^2}$. Breaking up the area of the hemisphere into pieces as in Figure 5.82, we then add up the total force as

$$F = \iint_S dF = \iint_S P(\vec{x})\, dS = 9800 \iint_S (20-z)\, dS \qquad (5.139)$$

Of course in order to compute this scalar surface integral we must have a parametrization of the hemisphere; this can be done by using a parametrization motivated by spherical coordinates:

$$\vec{x}(\phi, \theta) = \begin{pmatrix} \sin\phi\cos\theta \\ \sin\phi\sin\theta \\ \cos\phi \end{pmatrix} \quad \text{where} \quad \phi \in [0, \pi/2], \theta \in [0, 2\pi] \qquad (5.140)$$

5.9. SCALAR SURFACE INTEGRALS

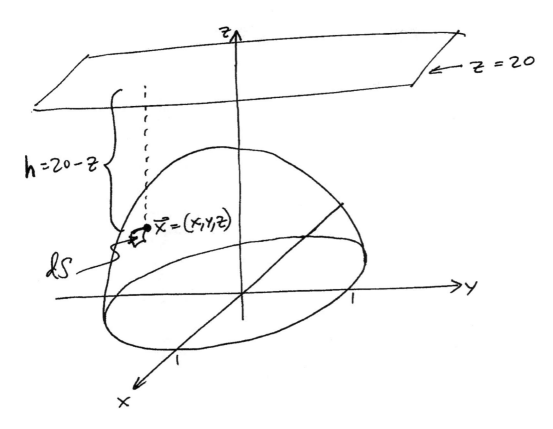

Figure 5.82:

(Note that while this is motivated by spherical coordinates, this is NOT the spherical coordinates function – the function above is a function of two variables, while the spherical coordinates function is a function of three variables.)

We compute the stretching factor in the usual way.

$$\begin{aligned}
\|\vec{N}\| &= \|\vec{x}_\phi \times \vec{x}_\theta\| \\
&= \left\| \frac{\partial}{\partial \phi} \begin{pmatrix} \sin\phi\cos\theta \\ \sin\phi\sin\theta \\ \cos\phi \end{pmatrix} \times \frac{\partial}{\partial \theta} \begin{pmatrix} \sin\phi\cos\theta \\ \sin\phi\sin\theta \\ \cos\phi \end{pmatrix} \right\| \\
&= \left\| \begin{pmatrix} \cos\phi\cos\theta \\ \cos\phi\sin\theta \\ -\sin\phi \end{pmatrix} \times \begin{pmatrix} -\sin\phi\sin\theta \\ \sin\phi\cos\theta \\ 0 \end{pmatrix} \right\| \\
&= \left\| \begin{pmatrix} \sin^2\phi\cos\theta \\ \sin^2\phi\sin\theta \\ \sin\phi\cos\phi \end{pmatrix} \right\| \\
&= |\sin\phi| = \sin\phi
\end{aligned}$$

Note that we can remove the absolute values at the last step because for all of the relevant values of ϕ, we know $\sin\phi \geq 0$.

With the above parametrization and stretching factor, we can then apply Theorem 5.9.1 to get

$$\begin{aligned} F &= 9800 \iint_S (20-z)\, dS \\ &= 9800 \int_0^{2\pi} \int_0^{\pi/2} (20-z) \|\vec{N}\| \, d\phi \, d\theta \\ &= 9800 \int_0^{2\pi} \int_0^{\pi/2} (20 - \cos\phi) \sin\phi \, d\phi \, d\theta \end{aligned}$$

This double integral can now be evaluated by standard techniques.

Exercises

Exercise 5.9.1 (**exer-Ints-SSI-1**). A net is submerged in the ocean, parametrized by $\vec{x}(u,v) = (2u-v, u+2v, u^2+v^2) = (x,y,z)$, with u and v ranging over the unit disk in the uv-plane and where z represents the depth in the ocean. The amount of heat absorbed in a given time period by a piece of the net is equal to the area times the temperature difference function, given by $T(x,y,z) = 3z$. Use the notation of a scalar surface integral to write down an expression representing the total amount of heat absorbed by the net in that amount of time, and compute that value.

Exercise 5.9.2 (**exer-Ints-SSI-2**). A screen lies over a garden, in the shape of half of a cylinder, with the radius equal to 10 feet and the length equal to 20 feet, and the axis of the cylinder is the x-axis. Locusts trap themselves in the screen at a rate of f per square foot per day, where the function f is given by $f(x,y,z) = x+z$ (x is the horizontal distance from one edge of the garden, and z is height above ground level). Use the notation of a scalar surface integral to write down an expression representing the total rate that locusts trap themselves in the screen, and compute that value. (Hint: Parametrize the screen using the angle θ measured at the x-axis from the top of the screen to the point in question.)

Exercise 5.9.3 (**exer-Ints-SSI-3**). Compute the value of $\iint_S f(x,y,z)\, dS$, where S is the graph of the function $g(x,y) = x^2 - y$ over the unit square in the xy-plane, and $f(x,y,z) = 3x$.

Exercise 5.9.4 (**exer-Ints-SSI-4**). The roof of a house is a surface over the foundation. This surface is made up of several faces, but each face has the same slope of m. Find a formula that gives the total square footage of the roof in terms of the slope m and the square footage A of the foundation. (Hint: Note that m is the length of the gradient, and that this can be used to compute the stretching factor in the graph parametrization of the roof.)

Exercise 5.9.5 (**exer-Ints-SSI-5**). A backyard deck is round, with a radius of 10 feet, thought of as being in the xy-plane and centered at the origin. The sunshade over the deck is in the shape of the part of the half cylinder described by $y^2 + z^2 = 100$ and $z \geq 0$ that is directly over the deck. Compute the area of the sunshade.

Exercise 5.9.6 (**exer-Ints-SSI-6**). Use a scalar surface integral and the spherical coordinates ($\rho = R$) parametrization to compute the area of a sphere of radius R.

Exercise 5.9.7 (**exer-Ints-SSI-7**). Suppose that a surface S in $\mathbb{R}^3$ is symmetric through a plane P and that the function f has odd symmetry through P. Explain how you know that the surface integral $\iint_S f\, dS$ must be zero.

Exercise 5.9.8 (**exer-Ints-SSI-8**). Can the claim in Exercise 5.9.7 be used to compute the integral $\iint_S f\, dS$, where S is the part of the unit sphere above the xy-plane and $f(x,y,z) = x$? If yes, explain all of the details; if no, explain why the claim does not apply.

Exercise 5.9.9 (**exer-Ints-SSI-9**). Can the claim in Exercise 5.9.7 be used to compute the integral $\iint_S f\, dS$, where S is the part of the unit sphere above the xy-plane and $f(x,y,z) = x + 3y$? If yes, explain all of the details; if no, explain why the claim does not apply. (Hint: Think about a vertical plane that is one of the level sets of f.)

5.9. SCALAR SURFACE INTEGRALS

Exercise 5.9.10 (**exer-Ints-SSI-10**). Compute the surface integral $\iint_S f \, dS$, where S is the part of the graph of the function $g(x,y) = x^2$ that is inside the unit sphere, and $f(x,y,z) = x^3yz - xy$.

Exercise 5.9.11 (**exer-Ints-SSI-11**). Compute the surface integral $\iint_S f \, dS$, where S is the part of the graph of the function $g(x,y) = x^2 + e^y$ sitting over the region in the xy-plane bounded by $y = 2+x^2$ and $y = e^{x^2}$, and $f(x,y,z) = x^5ye^z$.

5.10 Vector Fields

So far in this chapter we have discussed lots of different types of integrals (single variable integrals, double integrals, triple integrals, scalar line integrals, scalar surface integrals), all of which fit the integral conditions outlined in Section 5.1. These integrals differ only in the nature of the domain over which they are computed; but for each one, the domain is subdivided into smaller pieces, whose sizes are multiplied by a scalar integrand function, and then these products are summed over all of the pieces.

We have seen many applications for each of these, and we have found a variety of methods allowing us to compute such integrals.

At this point we will begin the discussion of a different kind of variation on our general picture of an integral.

Each of the previous integrals we have talked about has an integrand which is a scalar function, and the value of that scalar function for a particular piece of the domain is strictly a function of the position of that piece. We will see in the few sections after this one that there are other types of applications for which the integral conditions are satisfied and defining a new type of integral is appropriate – but the value of the integrand on a given piece of the domain will depend on both the position, *and the orientation*, of the given piece.

For each of these we will find an equivalent point of view on the nature of the integrand – rather than viewing the integrand as a scalar, we will be able to view it as being a vector; and the usual product will be replaced with a dot product. Details on this will be provided in those coming sections.

Before we can develop those ideas, however, we need to familiarize the reader with the idea of a function (called a "vector field") whose value is a vector instead of a scalar. That is the focus of this section.

5.10.1 Some motivating examples

We will begin by noting some examples that will motivate our eventual definition of a vector field.

Force fields

Any student who has taken an introductory course in physics is aware of the idea of a force. It is represented by a vector, because a force has both a magnitude and a direction.

In some cases, the force vector depends on position. With gravitational forces, for example, both the magnitude and the direction of the force depend on the position of the observer with respect to the mass creating the gravitational force – the magnitude is larger for points closer to the mass in question, and the direction is always directly toward the mass in question.

So, for every point in space, there is a corresponding vector representing the force at that point in space. We represent this situation by drawing the vectors in question at a selection of points in the domain. This is called a "force field", and is our first example of a vector field.

In Figure 5.83 we represent a gravitational force field for a single mass. (We draw this as a vector field in $\mathbb{R}^2$ so that it is easier to visualize, but of course a similar representation would be appropriate for a gravitational field in $\mathbb{R}^3$.)

Note of course that it would not be possible to draw the force vectors for all of the points in the domain, because there is simply not enough room. But drawing a representative sample will usually allow for a reasonably accurate idea of how the forces depend on position.

Of course gravity is not the only force that can be represented by a vector field. Electric and magnetic forces also give rise to a force vector that depends on location, and thus we can draw pictures for these vector fields just like we did for gravitational fields.

Students should be reminded that when dealing with forces, the direction of the force depends critically on the chosen

5.10. VECTOR FIELDS

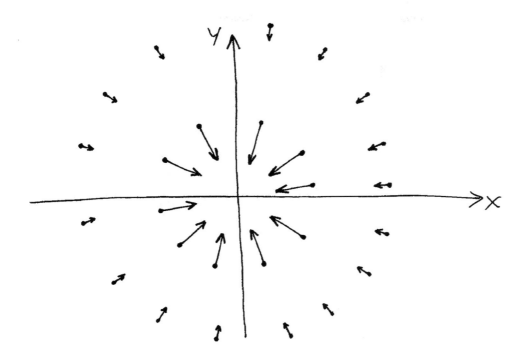

Figure 5.83:

perspective. For example, if we are holding a mass in the air with our hands, there are two forces acting on the mass – the gravitational force is pulling the mass down, while simultaneously the force that we personally are applying is pushing the mass up; the mass remains motionless because these two opposing forces cancel themselves out perfectly.

Flow fields

When a fluid is flowing in some region, then at a given moment in time and at a particular position, there is a vector representing the flow at that point. The length of the vector represents the flow rate, and the direction of the vector represents the direction the fluid is flowing.

Just as with forces, this then gives us a vector associated to every point location in a domain. We call this a "flow field", and is another example of a vector field. As with force fields we can represent flow fields by drawing the flow vectors at a sample of points in the domain.

For example, suppose we have a river through which water is flowing, and suppose that the flow rate is faster in the middle of the river than it is on the edges. We represent the flow field in Figure 5.84.

Similarly, we can represent the flow field for a fluid in a circular container that has been stirred in a clockwise direction, as in Figure 5.85.

Gradient fields

If we have a function $f : \mathbb{R}^n \to \mathbb{R}^1$, we know from Section 4.8 that for any given point in the domain, the gradient of f, which we write as ∇f, is a vector. Of course we recall the interpretations of the traits of this vector, namely that the direction of ∇f is the direction of fastest increase, and the magnitude is the rate of increase in that direction.

So, for every point in the domain the gradient provides a vector at that location. This gives us another example of a vector field, and for any given function we can represent this gradient field by drawing the gradient vectors at a sample of points in the domain.

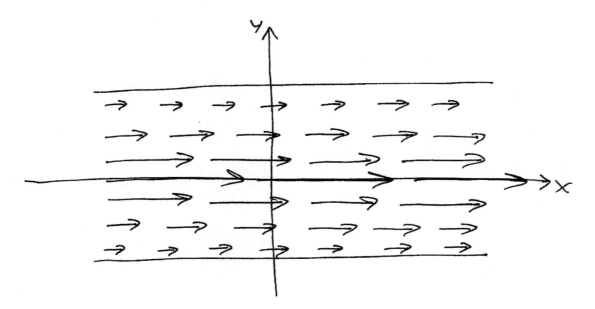

Figure 5.84:

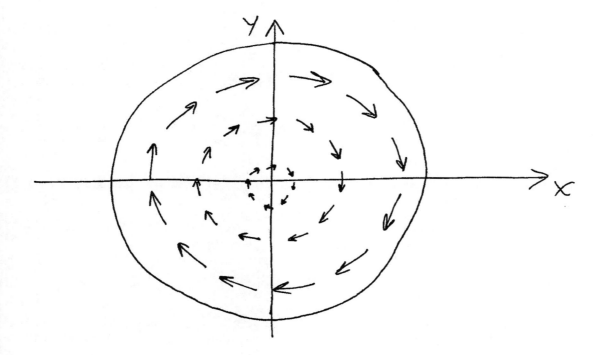

Figure 5.85:

5.10. VECTOR FIELDS

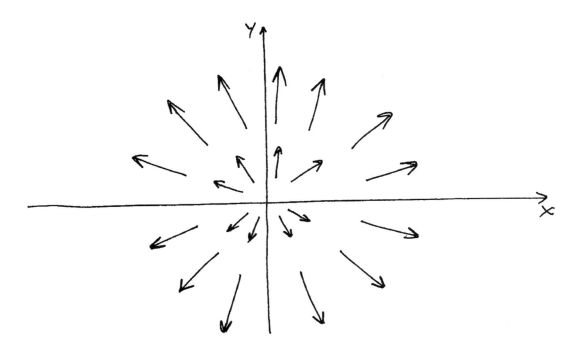

Figure 5.86:

For example, consider the function $f : \mathbb{R}^2 \to \mathbb{R}^1$, with $f(x, y) = x^2 + y^2$. We can compute the gradient easily as

$$\nabla f(\vec{x}) = \begin{bmatrix} 2x \\ 2y \end{bmatrix} \tag{5.141}$$

We can draw this gradient field by selecting a few points $\vec{x} = (x, y)$ in the domain at which to compute and draw the corresponding gradient vectors, giving us Figure 5.86.

Note that we can make interpretations from this vector field that are consistent with our prior knowledge of this function – the direction of fastest increase is always directly away from the origin, and the rate of increase in that direction is greater at greater distances from the origin.

Gradient fields will turn out to be particularly important in Chapter 6.

5.10.2 Definition

Based on the above motivating examples, we now make the following definition of a vector field.

Definition 5.10.1. *A "vector field on $\mathbb{R}^n$" is a function $\vec{F} : \mathbb{R}^n \to \mathbb{R}^n$.*

At first glance this definition seems more reminiscent of a change of variables function than the examples we just finished discussing; and certainly a change of variables function is also a function from $\mathbb{R}^n$ to $\mathbb{R}^n$.

The difference is just in the interpretation. For a change of variables function g, we interpret both the domain $\mathbb{R}^n$ and the target $\mathbb{R}^n$ as being sets of points – locations in space represented by their n coordinates. For any given input point in the domain, the output of g in the target $\mathbb{R}^n$ is thought of as a location in that target, which we think of as being the image of the input point. See Figure 5.87.

For a vector field, we simply make a different interpretation. We continue to think of the domain $\mathbb{R}^n$ as being a set of locations represented by their n coordinates; but in this case we interpret the output of the function $\vec{F}$ in the target $\mathbb{R}^n$ as being a vector that we associate to that point in the domain (the field vector at that point). See Figure 5.88.

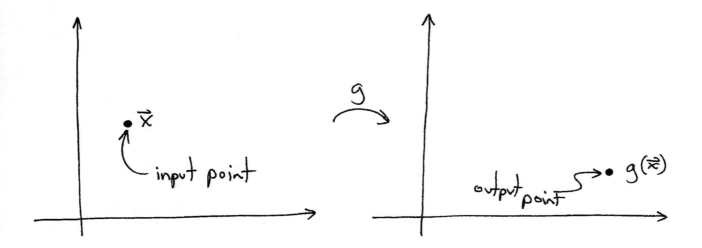

Figure 5.87: $g : \mathbb{R}^2 \to \mathbb{R}^2$, viewed as a change of variables function.

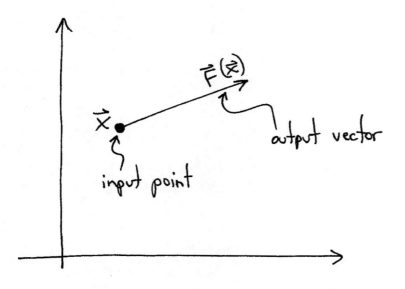

Figure 5.88: $\vec{F} : \mathbb{R}^2 \to \mathbb{R}^2$, viewed as a vector field.

5.10. VECTOR FIELDS

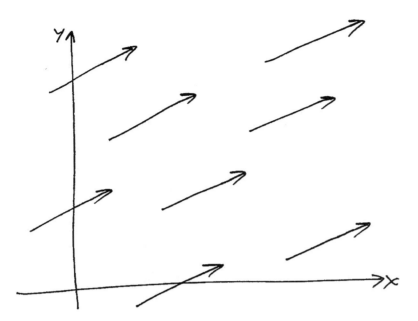

Figure 5.89:

Note, we require that the domain and target of $\vec{F}$ must be the same dimension; because for any point in $\mathbb{R}^n$ we allow the vector to point in any direction, and with any magnitude from that point in the domain.

5.10.3 More examples

Here we will list a few examples of explicit vector field functions, and draw their corresponding geometric representations.

Example 5.10.1. Consider the vector field $\vec{F}(x,y) = (2,1)$. What does that look like geometrically?

Thought of as a vector field, we interpret the above function as indicating that at every point in the xy-plane, the corresponding vector is the vector $(2,1)$. This is what we call a constant vector field, since the field vector is independent of location. See Figure 5.89.

Example 5.10.2. Consider the vector field $\vec{F}(x,y) = (x,y)$. What does that look like geometrically?

At every point, the field vector is the same as the position vector for that point. So the vector field looks like Figure 5.90.

Example 5.10.3. Consider the vector field $\vec{F}(x,y) = (-y,x)$. What does that look like geometrically?

For this vector field, we can notice that for every point (x,y), the field vector has the same length as the position vector, but is perpendicular to that position vector. Selecting a few specific points, we can see that this vector field points counter-clockwise around the origin; so the vector field looks like Figure 5.91.

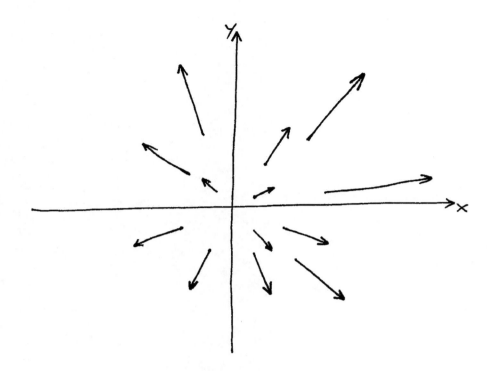

Figure 5.90:

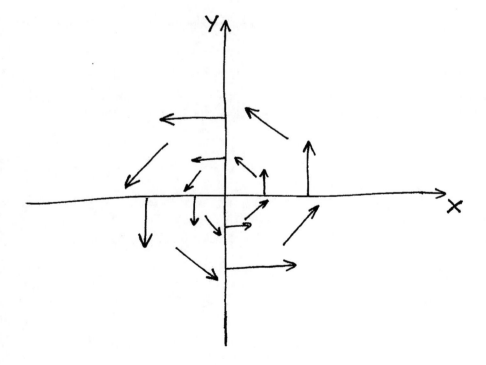

Figure 5.91:

5.10. VECTOR FIELDS

Exercises

Exercise 5.10.1 (exer-Ints-VF-1). A force field $\vec{F}$ acts on an object at position (x, y) in the plane by the force $\vec{F}(x, y) = (x, y)$. Draw a representative sample of vectors in this vector field and describe the force acting by this vector field.

Exercise 5.10.2 (exer-Ints-VF-2). A force field $\vec{F}$ acts on an object at position (x, y) in the plane by the force $\vec{F}(x, y) = (3, 1)$. Draw a representative sample of vectors in this vector field and describe the force acting by this vector field.

Exercise 5.10.3 (exer-Ints-VF-3). A force field in the plane acts to push points away from a single point. Which one of the following could possibly represent this vector field?

1. $\vec{F}(x, y) = (y, -x)$
2. $\vec{F}(x, y) = (e^x, x)$
3. $\vec{F}(x, y) = (x - 8, y + 2)$
4. $\vec{F}(x, y) = (x^2, y^2)$

Exercise 5.10.4 (exer-Ints-VF-4). A force field in the plane acts to push points in a single direction, with varying magnitudes. Which one of the following could possibly represent this vector field?

1. $\vec{F}(x, y) = (2, x)$
2. $\vec{F}(x, y) = (e^x, 3e^x)$
3. $\vec{F}(x, y) = (1, 4)$
4. $\vec{F}(x, y) = (3 - y, 4 + x)$

Exercise 5.10.5 (exer-Ints-VF-5). The flow of water on the surface of a river is given by $\vec{F}(x, y) = (e^{-y^2}, 0)$. Draw a representative sample of vectors in this vector field, and describe the flow.

Exercise 5.10.6 (exer-Ints-VF-6). The flow of water on the surface of a river is given by

$$\vec{F}(x, y) = \left(\frac{y}{\sqrt{x^2 + y^2}}, \frac{-x}{\sqrt{x^2 + y^2}} \right)$$

Draw a representative sample of vectors in this vector field, and describe the flow.

Exercise 5.10.7 (exer-Ints-VF-7). Compute and draw representative sample vectors of the gradient field for the function $f(x, y) = x^2 - y^2$. Draw the graph of this function, and note that the gradient field always points uphill on the graph.

Exercise 5.10.8 (exer-Ints-VF-8). Compute and draw representative sample vectors of the gradient field for the function $f(x, y) = \sin(x + y)$. Draw the graph of this function, and note that the gradient field always points uphill on the graph.

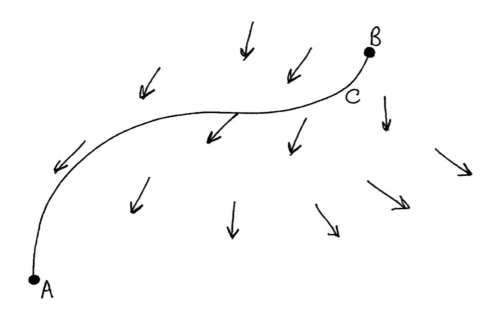

Figure 5.92:

5.11 Vector Line Integrals

In this section we address a type of problem that satisfies the four integral conditions almost exactly – but not quite exactly. Despite the imperfect fit, we will still use the word "integral" in representing its solution; but students should make sure to note that there is something qualitatively different about this particular type of integral.

5.11.1 Motivation and definition

Suppose that there is a fixed path C in $\mathbb{R}^2$ or $\mathbb{R}^3$, along which we need to move from point A to point B, as in Figure 5.92. Suppose also that we have a force field $\vec{F}$ over that domain; moving along the path through a force field then requires work to oppose the given force field. The problem we would like to address here is to find the total amount of work required to get from A to B.

Certainly we can break up the path C into pieces, and the total amount of work can be thought of as a sum of the amounts of work required to travel along each piece. So we have

$$W = \lim \sum \Delta W_i \tag{5.142}$$

Recall from Section 1.3 that with a constant force vector $\vec{F}$, the amount of work required to achieve a displacement by a vector $\vec{d}$ is $W = \vec{F} \cdot \vec{d}$. Over a small piece of curve C_i the force field is approximately constant at $\vec{F}(\vec{x}_i)$, and in order to oppose that force field we must exert a force $-\vec{F}(\vec{x}_i)$; and we can write the displacement as $\Delta \vec{x}_i$. See Figure 5.93.

So the work over a small piece of curve is given by

$$\Delta W_i = -\vec{F}(\vec{x}_i) \cdot \Delta \vec{x}_i \tag{5.143}$$

Certainly this is something that we could add up over all of the pieces in order to represent the total amount of work.

$$W = \lim \sum \Delta W_i = \lim \sum -\vec{F}(\vec{x}_i) \cdot \Delta \vec{x}_i \tag{5.144}$$

But, is this really something that we should call an integral? In order to fit the integral conditions, we would need for the work differential to be the product of a function and the length Δs of the piece of curve. As it stands, that is not the case; but there is something similar happening.

5.11. VECTOR LINE INTEGRALS

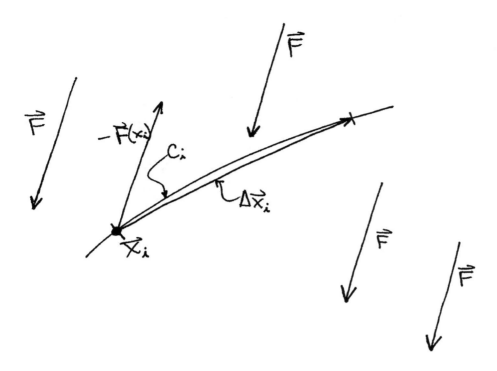

Figure 5.93:

In particular, note that $\Delta \vec{x}_i$ is a vector closely related to the piece of curve, in that the shaft of that vector closely approximates the piece of curve; and the length of the vector $\Delta \vec{x}_i$ is very close to the length Δs of the piece of curve. And of course a dot product is not the usual scalar product that we have seen at this point in previous integrals, but still it is a product of a sort.

So even though this does not fit the integral conditions perfectly, we will still use the word integral to denote the construction that we are about to define.

Definition 5.11.1. *Given a continuous vector field $\vec{F}$ defined over a smooth curve C, we define the vector line integral of $\vec{F}$ over C as*

$$\int_C \vec{F}(\vec{x}) \cdot d\vec{x} = \lim \sum \vec{F}(\vec{x}_i) \cdot \Delta \vec{x}_i \tag{5.145}$$

Using this notation then, we can write the work needed to oppose the force $\vec{F}$ over the curve C as

$$W = \int_C -\vec{F}(\vec{x}) \cdot d\vec{x} \tag{5.146}$$

Note that we write the differential as a vector – because it represents a vector in the Riemann sum; and we write the dot product in the integral notation for the same reason.

Let's now restate the construction using this integral notation, as in Figure 5.94. We break up the curve C into pieces over any one of which the position $\vec{x}$ changes by the differential vector $d\vec{x}$. Over any such piece, the quantity to be computed is the dot product of a given field vector $\vec{F}(\vec{x})$ with that differential $d\vec{x}$, and we add those up over the entire curve to get $\int_C \vec{F}(\vec{x}) \cdot d\vec{x}$.

5.11.2 Computation

At this point, as with all of the other previous types of integrals we have defined up to this point, we now have a definition for a vector line integral based on an application to a specific problem – but we have not yet determined any methods

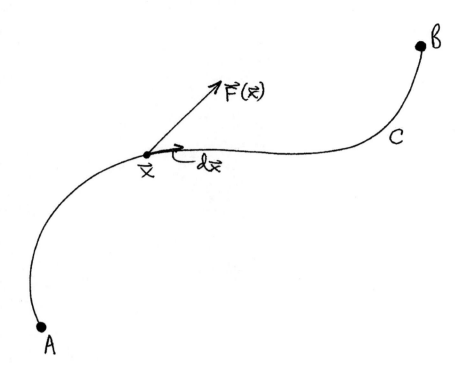

Figure 5.94:

by which we can compute it.

Conveniently, the method we will use to compute vector line integrals is based on the same setup as the method we used to compute scalar line integrals. In particular, we will begin with a parametrization $\vec{x}(t)$ of the curve C, and find a way to rephrase the given integral instead as an integral on the t-axis.

Suppose that the curve C, from the point A to the point B, is parametrized by

$$\vec{x}(t) = \begin{pmatrix} x(t) \\ y(t) \end{pmatrix} \quad \text{where} \quad t \in [a, b] \tag{5.147}$$

As in Figure 5.95, a sub-interval of length dt on the t-axis has as its image an interval on the curve C, closely approximated by the differential vector $d\vec{x}$.

Normally this is the point where we would begin an attempt to find the stretching factor. But in this case, we do not. The reason is that, as we have formulated the definition of the vector line integral, there is no ds explicitly written, and thus we don't need to rewrite it in terms of dt. (Nevertheless, it will still show up eventually; we will see that later in this section.)

What we need to do instead is to find the differential vector $d\vec{x}$ in terms of dt. Fortunately this is something that we have already done, in Section 5.8. Specifically, Equation 5.126 tells us that

$$d\vec{x} = \vec{v}(t)\, dt \tag{5.148}$$

where $\vec{v}$ is the velocity vector of the parametrization. See Figure 5.96.

So, very simply, we can rewrite the vector line integral over the curve C as an integral on the t-axis:

$$\begin{aligned} \int_C \vec{F}(\vec{x}) \cdot d\vec{x} &= \int_a^b \vec{F}(\vec{x}(t)) \cdot \left(\vec{v}(t)\, dt\right) \\ &= \int_a^b \left(\vec{F}(\vec{x}(t)) \cdot \vec{v}(t)\right) dt \end{aligned}$$

5.11. VECTOR LINE INTEGRALS

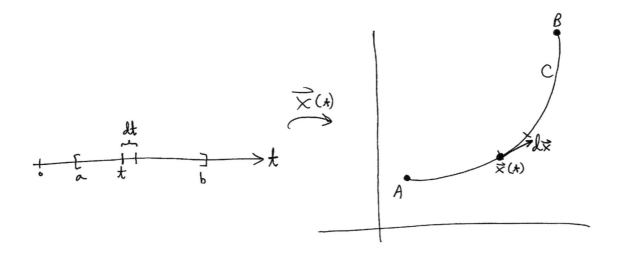

Figure 5.95:

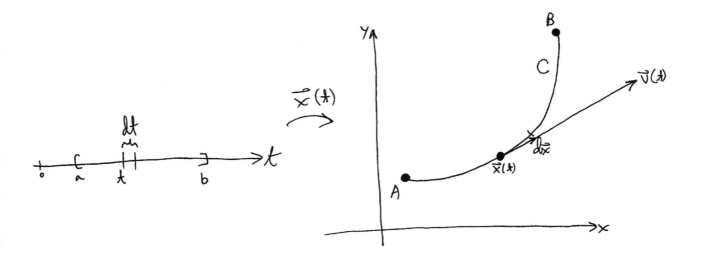

Figure 5.96:

Everything in this last integral is written in terms of t, and so we have a plain single variable integral at this point which can be computed by the usual methods.

Note that from a certain point of view, this computation is easier than the one for scalar line integrals, because we didn't have to go to the trouble of computing the length of the $d\vec{x}$ vector to find ds. Other than that, the computation is very similar.

We state this result here as a theorem:

Theorem 5.11.1. *Given a vector line integral $\int_C \vec{F}(\vec{x}) \cdot d\vec{x}$ of a continuous vector field $\vec{F}$ defined over a smooth curve C, we can compute the line integral with*

$$\int_C \vec{F}(\vec{x}) \cdot d\vec{x} = \int_a^b \vec{F}(\vec{x}(t)) \cdot \vec{v}(t) \, dt \tag{5.149}$$

5.11.3 Examples

Here are a few examples in which we phrase a problem as a vector line integral, and them compute that integral using Theorem 5.11.1.

Example 5.11.1. Suppose we wish to compute the amount of work that it takes to walk counterclockwise along the unit circle from the point $(1, 0)$ to the point $(-1, 0)$, where the wind at the point (x, y) necessitates pushing with an opposing force $\vec{F}(x, y) = (-y, x)$.

(Note that the vector field given is that of the opposing force required to counter the force of the wind — not the force of the wind itself. So, the negative sign from our original work computation will not be present here in this example.)

We know from our previous computations that the work is computed as a vector line integral over the curve C.

$$W = \int_C dW = \int_C \vec{F} \cdot d\vec{x} \tag{5.150}$$

In order to compute this line integral, we must first parametrize the curve C. We use

$$\vec{x}(t) = \begin{pmatrix} \cos t \\ \sin t \end{pmatrix} \quad \text{where} \quad t \in [0, \pi] \tag{5.151}$$

From this we can compute the velocity vector $\vec{v}(t)$ as

$$\vec{v}(t) = \vec{x}\,'(t) = \begin{bmatrix} -\sin t \\ \cos t \end{bmatrix} \tag{5.152}$$

Then Theorem 5.11.1 gives us

$$\begin{aligned} W &= \int_a^b \vec{F}(\vec{x}(t)) \cdot \vec{v}(t) \, dt \\ &= \int_0^\pi \vec{F}\left(\begin{pmatrix} \cos t \\ \sin t \end{pmatrix}\right) \cdot \begin{bmatrix} -\sin t \\ \cos t \end{bmatrix} dt \\ &= \int_0^\pi \begin{bmatrix} -\sin t \\ \cos t \end{bmatrix} \cdot \begin{bmatrix} -\sin t \\ \cos t \end{bmatrix} dt \\ &= \int_0^\pi 1 \, dt \\ &= \pi \end{aligned}$$

5.11. VECTOR LINE INTEGRALS

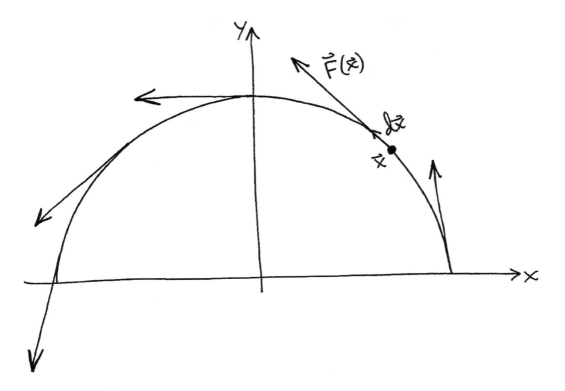

Figure 5.97:

This is a believable result, because we notice that the vector field given is exactly the vector field from Example 5.10.3, which we saw in that example always points perpendicular to the position vector, and thus is always tangent to the unit circle. So, as in Figure 5.97, the force $\vec{F}$ is always being pushed in exactly the same direction as the direction of motion and with magnitude 1.

Example 5.11.2. Suppose we wish to compute the amount of work that it takes to walk counterclockwise along the unit circle from the point $(1,0)$ to the point $(-1,0)$, where the wind at the point (x,y) necessitates pushing with an opposing force $\vec{F}(x,y) = (x,y)$.

As in the previous example, we have

$$W = \int_C dW = \int_C \vec{F} \cdot d\vec{x} \tag{5.153}$$

Since we have the same path C, we can use the same parametrization

$$\vec{x}(t) = \begin{pmatrix} \cos t \\ \sin t \end{pmatrix} \quad \text{where} \quad t \in [0, \pi] \tag{5.154}$$

and thus we have the same velocity vector $\vec{v}(t)$

$$\vec{v}(t) = \vec{x}'(t) = \begin{bmatrix} -\sin t \\ \cos t \end{bmatrix} \tag{5.155}$$

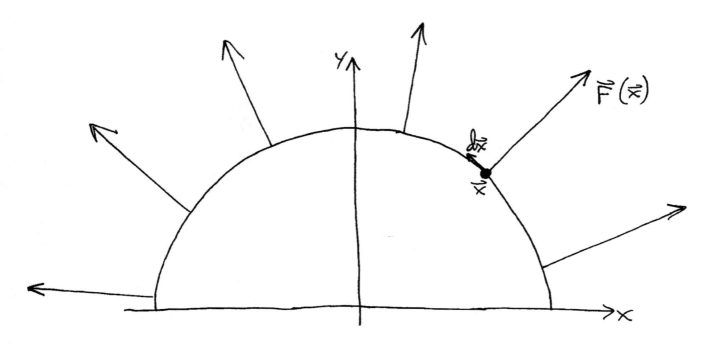

Figure 5.98:

Then Theorem 5.11.1 gives us

$$\begin{aligned} W &= \int_a^b \vec{F}(\vec{x}(t)) \cdot \vec{v}(t)\, dt \\ &= \int_0^\pi \vec{F}\left(\begin{pmatrix} \cos t \\ \sin t \end{pmatrix}\right) \cdot \begin{bmatrix} -\sin t \\ \cos t \end{bmatrix} dt \\ &= \int_0^\pi \begin{bmatrix} \cos t \\ \sin t \end{bmatrix} \cdot \begin{bmatrix} -\sin t \\ \cos t \end{bmatrix} dt \\ &= \int_0^\pi 0\, dt \\ &= 0 \end{aligned}$$

This too is a believable result, because we notice that the vector field given is exactly the vector field from Example 5.10.2, which we saw in that example always points the same as the position vector, and thus is always perpendicular to the unit circle. So, as in Figure 5.98, the force $\vec{F}$ is always being pushed perpendicular to the direction of motion, and thus there is no actual work being exerted.

Example 5.11.3. Suppose we wish to compute the amount of work that it takes to walk counterclockwise along the unit circle from the point $(1,0)$ to the point $(0,1)$, where the wind at the point (x,y) necessitates pushing with an opposing force $\vec{F}(x,y) = (2,5)$.

As in the previous example, we have

$$W = \int_C dW = \int_C \vec{F} \cdot d\vec{x} \tag{5.156}$$

We have a similar path C, which we parametrize with

$$\vec{x}(t) = \begin{pmatrix} \cos t \\ \sin t \end{pmatrix} \quad \text{where} \quad t \in [0, \pi/2] \tag{5.157}$$

5.11. VECTOR LINE INTEGRALS

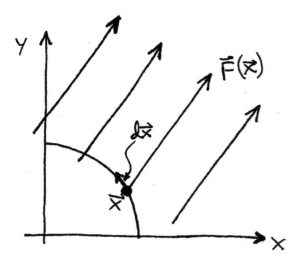

Figure 5.99:

and the velocity vector $\vec{v}(t)$ is still

$$\vec{v}(t) = \vec{x}'(t) = \begin{bmatrix} -\sin t \\ \cos t \end{bmatrix} \tag{5.158}$$

Then Theorem 5.11.1 gives us

$$\begin{aligned} W &= \int_a^b \vec{F}(\vec{x}(t)) \cdot \vec{v}(t)\, dt \\ &= \int_0^{\pi/2} \vec{F}\left(\begin{pmatrix} \cos t \\ \sin t \end{pmatrix}\right) \cdot \begin{bmatrix} -\sin t \\ \cos t \end{bmatrix} dt \\ &= \int_0^{\pi/2} \begin{bmatrix} 2 \\ 5 \end{bmatrix} \cdot \begin{bmatrix} -\sin t \\ \cos t \end{bmatrix} dt \\ &= \int_0^{\pi/2} -2\sin t + 5\cos t\, dt \\ &= 3 \end{aligned}$$

Here the final single variable integral does not have a constant integrand, even though the vector field itself is constant. Of course this is because the direction of motion changes as we move along the curve C, and of course the work differentials depend on the direction of motion. See Figure 5.99.

5.11.4 The unit tangent vector

The length differential ds did not appear in any of the previous computations involving vector line integrals. But the setup for the computation of vector line integrals closely resembles that for scalar line integrals, and so this seems unexpected.

In fact we will momentarily see that there is a close connection between these two types of integrals because their differentials, $d\vec{x}$ and ds respectively, are so closely related – remember, $ds = \|d\vec{x}\|$. We will rewrite a vector line integral in such a way that it looks more like a scalar line integral by considering the unit tangent vector.

Let $\vec{T}$ be the unit tangent vector to the curve C at the point $\vec{x}$. The position differential $d\vec{x}$ is of course also tangent to

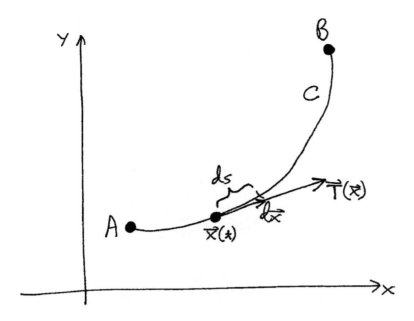

Figure 5.100:

the curve C at the point $\vec{x}$, with length ds, and since $\vec{T}$ is a unit vector we can then write (see Figure 5.100)

$$d\vec{x} = \vec{T}\, ds \tag{5.159}$$

Using this result, we can rewrite a vector line integral in a different way as

$$\int_C \vec{F}(\vec{x}) \cdot d\vec{x} = \int_C \vec{F}(\vec{x}) \cdot \left(\vec{T}(\vec{x})\, ds\right) \tag{5.160}$$
$$= \int_C \left(\vec{F}(\vec{x}) \cdot \vec{T}(\vec{x})\right) ds \tag{5.161}$$

At first glance this appears to be just a scalar line integral, but this is not quite true. While the integrand does depend on the position $\vec{x}$, note that the unit tangent vector also depends on the direction that the path C is going at that point – so, the integrand is not strictly a function of position.

Nevertheless, ignoring this point, we can still use the above as an alternative interpretation of a vector line integral. Recall that the component of a vector in the direction of a given unit vector is computed with a dot product (see Figure 5.101).

So we can rewrite the above integrand as $\vec{F}(\vec{x}) \cdot \vec{T}(\vec{x}) = \text{comp}_{\vec{T}}\vec{F}$, giving us another interpretation of a vector line integral as

$$\int_C \vec{F}(\vec{x}) \cdot d\vec{x} = \int_C \left(\text{comp}_{\vec{T}}\vec{F}\right) ds \tag{5.162}$$

Equation 5.162 tells us that the vector line integral computes effectively a scalar line integral, where the integrand at a given point on the curve is the component of the vector field $\vec{F}$ in the direction that the curve is moving at that point.

This allows us to make another interpretation of vector line integrals, very different from the interpretation as work. Recall from our motivation of the definition that vector line integrals represent work if the vector field in question is interpreted as representing a force. Instead, now suppose that we view the vector field $\vec{F}$ as representing the flow field of some fluid.

5.11. VECTOR LINE INTEGRALS

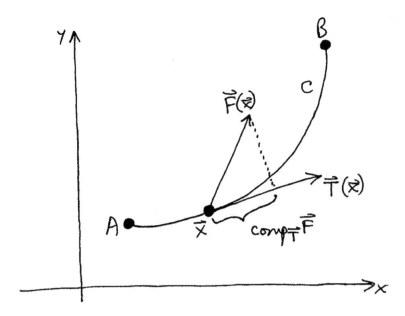

Figure 5.101:

The expression $\text{comp}_{\vec{T}}\vec{F}$ then would be interpreted as representing the amount of the fluid flow ($\vec{F}$) in the same direction ($\vec{T}$) as the curve, at a given point. The vector line integral $\int_C \vec{F} \cdot d\vec{x} = \int_C \left(\text{comp}_{\vec{T}}\vec{F}\right) ds$ could then be viewed as representing, in some sense, the total amount that the fluid is flowing along the entire curve.

While this does give us a new intuitive interpretation of a vector line integral, it is usually not particularly useful as a computational tool. The reason for this is that if we were to use Equation 5.161 to compute a vector line integral, the next step would be to compute the ensuing scalar line integral, which of course would involve computing the stretching factor $\|\vec{v}\| = \frac{ds}{dt}$. So we would compute

$$\int_C \vec{F}(\vec{x}) \cdot d\vec{x} = \int_C \left(\vec{F} \cdot \vec{T}\right) ds \tag{5.163}$$

$$= \int_a^b \left(\vec{F} \cdot \vec{T}\right) \|\vec{v}\| \, dt \tag{5.164}$$

$$= \int_a^b \vec{F} \cdot \left(\|\vec{v}\|\vec{T}\right) dt \tag{5.165}$$

But of course $\|\vec{v}\|\vec{T}$ is just the velocity vector $\vec{v}$, and so this computation has simply taken us back to the same integral we already had in Theorem 5.11.1; basically, all that has happened is that we divided by $\|\vec{v}\|$ in order to compute the unit tangent vector $\vec{T}$, only to then multiply back by the same $\|\vec{v}\|$ as the stretching factor for the ensuing scalar line integral.

5.11.5 Summary

We now have three fairly different ways of thinking about vector line integrals.

First, we write a vector line integral as

$$\int_C \vec{F} \cdot d\vec{x} \tag{5.166}$$

as is suggested by our motivation for the given definition – namely, the curve C is broken up into pieces over which the position differential is $d\vec{x}$, and the differential of the quantity to be computed is the dot product of the integrand vector field $\vec{F}$ with that position differential.

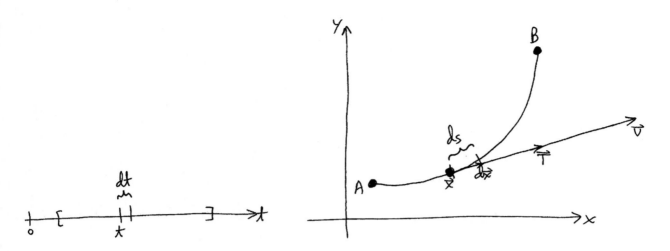

Figure 5.102:

This perspective is useful for relating the vector line integral to specific applications, since the notation is highly suggestive of the process by which such integrals are defined.

Second, we can write a vector line integral as

$$\int_a^b \vec{F} \cdot \vec{v}\, dt \tag{5.167}$$

where $\vec{v}$ is the velocity of the parametrization. This is a convenient notation when it comes to doing computations, since it is written as a single variable integral and can thus be computed directly.

Third, we can write a vector line integral as

$$\int_C \vec{F} \cdot \vec{T}\, ds \tag{5.168}$$

where $\vec{T}$ is the unit tangent vector for the curve C. This notation is nice in the sense that it relates the vector line integral to the previously established construction of the scalar line integral, which fits into the mold of all of the other types of integrals we have defined so far in this chapter.

Each of the above representations involves different notations representing different quantities related to the vector line integral. Of course all three of the above notations for the vector line integral represent the same thing, and so there are many relationships between the different quantities. Here are a few of those relationships, pictured in Figure 5.102.

There are three vectors tangent to the curve that of course all relate to each other – the position differential $d\vec{x}$, the unit tangent vector $\vec{T}$, and the velocity vector $\vec{v}$. We have

$$d\vec{x} = \vec{v}\, dt = \vec{T}\, ds \tag{5.169}$$

The vectors $\vec{T}$ and $\vec{v}$ are thus both related to $d\vec{x}$ in that they have the same direction as $d\vec{x}$; they differ only in their lengths, whose ratio is the scalar stretching factor $\|\vec{v}\|$.

The scalar differentials dt and ds are both related to $d\vec{x}$ in that they are all differentials; and they too differ only in their lengths, whose ratio is again the scalar stretching factor $\|\vec{v}\|$.

Exercises

5.11. VECTOR LINE INTEGRALS

Exercise 5.11.1 (exer-Ints-VLI-1). Express as a vector line integral and then compute the amount of work that it takes to move through the force field $\vec{F} = (x^2, y)$ along the curve parametrized by $\vec{x}(t) = (t, t^2 - 1)$ with $t \in [0, 3]$.

Exercise 5.11.2 (exer-Ints-VLI-2). Express as a vector line integral and then compute the amount of work that it takes to move through the force field $\vec{F} = (xy, e^x)$ along the curve parametrized by $\vec{x}(t) = (t^2, 2t^2 - 4)$ with $t \in [0, 1]$.

Exercise 5.11.3 (exer-Ints-VLI-3). Compute the line integral of the vector field $\vec{F} = (y^2, x - y)$ over the curve parametrized by $\vec{x}(t) = (t, t^2 + 3)$ with $t \in [0, 1]$. Compute also the line integral of the same vector field over the curve parametrized by $\vec{x}(t) = (t^3, t^6 + 3)$ with $t \in [0, 1]$. Why do these calculations give you the same result?

Exercise 5.11.4 (exer-Ints-VLI-4). Compute the line integral of the vector field $\vec{F} = (x + 5y, x - y)$ over the upper half of the unit circle, moving to the right. Compute the line integral of the same vector field over the same curve, but moving to the left. How do these results compare?

Exercise 5.11.5 (exer-Ints-VLI-5). Compute the line integral of the vector field $\vec{F} = (x + y, x - y)$ over the graph of the function $f(x) = x^2 - x$, starting at the point on the graph where $x = 2$ and ending where $x = 5$.

Exercise 5.11.6 (exer-Ints-VLI-6). Compute the line integral of the vector field $\vec{F} = (x + y, x - y)$ over the curve with equation $x - y - y^2 = 0$, starting at the point $(2, 1)$ and ending at $(20, 4)$.

Exercise 5.11.7 (exer-Ints-VLI-7). Compute the line integral of the vector field $\vec{F} = (x + y, x - y, 3x^2)$ over the curve that is the intersection of the cylinder $x^2 + y^2 = 4$ and the plane $x - y + z = 3$, moving clockwise as seen from above.

Exercise 5.11.8 (exer-Ints-VLI-8). We will say that a vector field $\vec{F}(x, y)$ has "mirror symmetry over the y-axis" if the x coordinate has odd symmetry over the y-axis and the y coordinate has even symmetry over the y-axis. Written precisely,

$$\vec{F}(x, y) = \begin{bmatrix} F_1(x, y) \\ F_2(x, y) \end{bmatrix} \quad \text{with} \quad F_1(-x, y) = -F_1(x, y) \quad \text{and} \quad F_2(-x, y) = F_2(x, y)$$

Draw a representative sample of vectors in each of the following vector fields, all of which have mirror symmetry over the y-axis.

1. $\vec{F}(x, y) = (x, y)$
2. $\vec{F}(x, y) = (x, -y)$
3. $\vec{F}(x, y) = (-x, y)$
4. $\vec{F}(x, y) = (x/(1 + x^2), 0)$

(Note then that the vectors in such a vector field appear to be mirror images of each other over the y-axis.)

Exercise 5.11.9 (exer-Ints-VLI-9). The curve C is the part of $y = x^2$ going from $(-3, 9)$ to $(3, 9)$. C is symmetric over the y-axis, and the vector field $\vec{F}(x, y) = (x, -y)$ has mirror symmetry over the y-axis. Draw the curve C, and for the points $(1, 1)$ and $(2, 4)$ on C do the following:

1. Draw the points and their reflection points over the y-axis.
2. Compute and draw the field vector $\vec{F}$ at each of these points and their reflections.
3. Compute and draw the unit tangent vector $\vec{T}$ to the curve C at each of these points and their reflections.
4. Compute $\vec{F} \cdot \vec{T}$ at each of these points and their reflections.
5. Confirm that the value of $\vec{F} \cdot \vec{T}$ at each of these points is exactly the negative of the value of $\vec{F} \cdot \vec{T}$ at their corresponding reflection points.

Exercise 5.11.10 (exer-Ints-VLI-10). As is suggested by the result of Exercise 5.11.9, if C is symmetric over the y-axis and $\vec{F}$ has mirror symmetry over the y-axis, then the value of the line integral $\int_C \vec{F} \cdot d\vec{x} = \int_C \vec{F} \cdot \vec{T}\, ds$ must be zero. Confirm by direct computation that this claim is true when the curve C is the unit circle going counterclockwise, and $\vec{F}(x, y) = (x^3, -y^2)$.

Exercise 5.11.11 (exer-Ints-VLI-11). Use the result in Exercise 5.11.10 to compute the value of $\int_C \vec{F} \cdot d\vec{x}$ where C is the curve $e^{x^2} + e^{y^2} = 4$, moving clockwise, and $\vec{F}(x,y) = (x^3 e^y - \sin(x^5), x^2 y^2)$.

Exercise 5.11.12 (exer-Ints-VLI-12). Use the ideas in Exercises 5.11.8 and 5.11.9 to define what it should mean to say that a vector field has mirror symmetry over the x-axis. Then state a corresponding symmetry result analogous to the one in Exercise 5.11.10.

Exercise 5.11.13 (exer-Ints-VLI-13). Use the result of Exercise 5.11.12 to compute the value of $\int_C \vec{F} \cdot d\vec{x}$ where C is the curve $e^{x^2} + e^{y^2} = 4$, moving clockwise, and $\vec{F}(x,y) = (x \cos y + x^2 y^4, x^2 y - x^3 y^3 e^x)$.

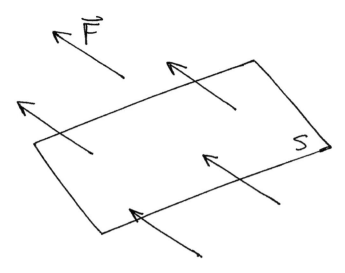

Figure 5.103:

5.12 Vector Surface Integrals

As scalar surface integrals turned out to be very similar to scalar line integrals, so will vector surface integrals have much in common with vector line integrals. In particular they are both motivated by a particular type of problem, and they both turn out to be a type of integral that does not fit our four integral conditions quite perfectly. They both end up involving a vector differential and a dot product, and they both have alternate interpretations involving components.

Aside from obvious dimensional differences, the main qualitative difference between these two types of integrals is in the applications which motivate them. Vector line integrals are most easily motivated by the application to computing work; vector surface integrals are most easily motivated by the application to computing something called "flux". Since most students in this course have probably not seen a presentation of flux before, we will start with a discussion of this notion.

5.12.1 Flux

Suppose we have a vector field $\vec{F}$, perhaps representing a flow of some sort. Given a surface S, it would be natural to ask for the rate of flow passing through the surface. For example, if the vector field is the flow field of water in a tank and the surface represents a net, we could ask how much volume of water per unit time is flowing through the net. This is called the "flux" of the vector field $\vec{F}$ through the surface S.

In this subsection we will consider flux only in the context of a constant vector field $\vec{F}$ and a flat surface S; the more general case will be the subject of the following subsections.

So, suppose we have a constant vector field $\vec{F}$ representing the uniform flow of some sort of fluid, and a flat surface S with area A through which we wish to compute the flux. See Figure 5.103.

Before we proceed we must make a choice about direction; in other words, which direction through the surface do we want to consider to be the "positive" direction through the surface? This is an arbitrary choice, but it is a choice that must be made in order for the question to be well-posed.

By analogy, when computing the work of a vector field along a curve, we must specify whether we are moving one direction along the curve, or the other. Each is itself a reasonable question to ask, but the two answers will be the negatives of each other. For example, lifting a mass up along a given curve requires some positive amount of work; if instead we were to ask how much work it takes to let it down along precisely the same path, we would get precisely the negative answer.

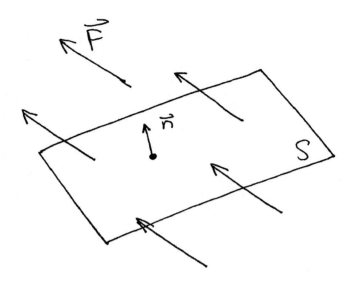

Figure 5.104:

With curves, the "positive" direction along the curve is indicated by the unit tangent vector $\vec{T}$. For a surface, we will indicate our choice of direction again by a unit vector, but since we are measuring flows through the surface and not along it, we choose a unit normal vector, which we will write as $\vec{n}$. Note that just as with unit tangent vectors, there are two choices, and they are precisely the negatives of each other.

We refer to the choice of unit normal vectors on the surface as the "orientation" of the surface.

We should note here that flux is a notion that is relevant for more than just fluid flows; we will see in Chapter 6 that there are other extremely important applications. Still, the problem of fluid flows is the easiest application with which to motivate the definition, and so that is the perspective we will take in this section.

Computation

With a given choice of unit normal vector then, as in Figure 5.104, we can now proceed to computing the flux through the given surface in the direction indicated by that normal vector.

Let's first write the flow field as the sum of two vectors – one perpendicular to the surface, and one parallel to the surface. We will write this as
$$\vec{F} = \vec{F}_\perp + \vec{F}_\parallel \tag{5.170}$$
To compute the flux of $\vec{F}$ we can compute the flux of each of these component vector fields individually.

Intuitively, we can see that the flux of $\vec{F}_\parallel$ through the surface is zero – because this vector field represent flow moving along the surface, and therefore not at all through the surface.

So, we are left only with computing the flux of $\vec{F}_\perp$. Since this vector is pointed directly perpendicular to the surface, it represents fluid flowing directly through the surface, and we can compute this flow simply as the product of the magnitude of the vector field times the area of the surface. Using the Greek letter Φ to represent flux, we then have
$$\Phi = \|\vec{F}_\perp\| A \tag{5.171}$$
Of course we know that $\|\vec{F}_\perp\| = \text{comp}_{\vec{n}} \vec{F} = \vec{F} \cdot \vec{n}$, and so we can rewrite this much more conveniently as
$$\Phi = \left(\vec{F} \cdot \vec{n}\right) A \tag{5.172}$$

5.12. VECTOR SURFACE INTEGRALS

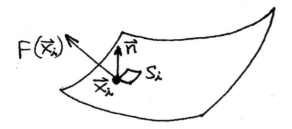

Figure 5.105:

Example 5.12.1. Suppose that the flat surface S in question is a parallelogram with edge vectors $\vec{v}$ and $\vec{w}$, and we wish to compute the flux of the constant vector field $\vec{F}$ through this parallelogram.

First, we can rewrite the above equation as

$$\Phi = \vec{F} \cdot \left(\vec{n} A\right) \tag{5.173}$$

Let's now consider the vector in parentheses – it is perpendicular to the surface S, and has length equal to the area A of S. Said differently – it is perpendicular to each of $\vec{v}$ and $\vec{w}$, and has length equal to the area of the parallelogram defined by $\vec{v}$ and $\vec{w}$. Of course we recall from Section 1.4 that the cross product $\vec{v} \times \vec{w}$ fits this description perfectly.

The only possible discrepancy involves the choice of the direction of the unit normal vector. But certainly if we know that the chosen unit normal vector points in the same direction as the cross product, then we can conclude that

$$\vec{n} A = \vec{v} \times \vec{w} \tag{5.174}$$

and therefore that

$$\Phi = \vec{F} \cdot \left(\vec{v} \times \vec{w}\right) \tag{5.175}$$

5.12.2 Definition

Suppose now that we are faced with the question of representing a flux, this time of a non-constant vector field, and through a surface that is not flat.

Analogously to the approach we have taken in many instances in the past, we approach this problem by breaking up the surface into small pieces. The total flux then is the sum of the flux through each of the pieces. If we write the pieces of surface as S_i, and the flux though such a piece as Φ_i, we get

$$\Phi = \lim \sum \Delta \Phi_i \tag{5.176}$$

Even though the surface as a whole is not flat, each small piece is approximately flat. And, even though the vector field is not constant, over each small piece the vector field is approximately constant. So if we refer to the area of S_i as ΔS_i, and choose a sample point $\vec{x}_i \in S_i$ as is represented in Figure 5.105, we can use Equation 5.172 to rewrite this as

$$\Phi = \lim \sum \Delta \Phi_i = \lim \sum \left(\vec{F}(\vec{x}_i) \cdot \vec{n}\right) \Delta S_i \tag{5.177}$$

Again we note that this construction does not look exactly like integrals we have previously defined, but, as was the case with vector line integrals, it will turn out to relate to previously defined integrals.

Definition 5.12.1. *Given a continuous vector field $\vec{F}$ defined over a smooth surface S, we define the vector surface integral of $\vec{F}$ over S as*

$$\iint_S \vec{F} \cdot \vec{n}\, dS = \lim \sum \left(\vec{F}(\vec{x}_i) \cdot \vec{n}\right) \Delta S_i \tag{5.178}$$

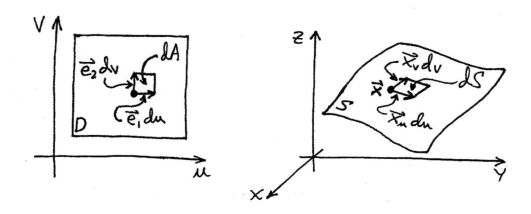

Figure 5.106:

Because this definition is motivated by flux, sometimes this is called a "flux integral".

Notation

In the notation of Definition 5.12.1, the area differential "dS" plays the role of "ΔS_i" in the Riemann sum, representing the area of a piece of the surface. We can combine the dS with the normal vector to rewrite the above integral as

$$\iint_S \vec{F} \cdot \vec{n}\, dS = \iint_S \vec{F} \cdot (\vec{n}\, dS) = \iint_S \vec{F} \cdot d\vec{S} \tag{5.179}$$

where we define the differential vector $d\vec{S}$ by $d\vec{S} = \vec{n}\, dS$.

This notation is useful in a few ways. Most obviously, it is slightly more terse than the original. More importantly though, the vector $d\vec{S}$ has features and plays a role very similar to that of the differential vector "$d\vec{x}$" used in vector line integrals. That is, both vectors have the properties that their directions define the orientations of the pieces of the domain they represent, and their lengths represent the sizes of those pieces. So the vector $d\vec{S}$ itself is a natural and interesting object, and deserves its own notation.

Finally, this notation distinguishes this as a different kind of integral than a scalar surface integral. As with line integrals, the point here is that the integrand in Equation 5.178 depends not only on the position, but also the orientation of each piece of the domain. Of course as that equation itself shows we can certainly think of it as a scalar surface integral; but it is useful to distinguish it as something different.

5.12.3 Computation

As in the case of all of the previous line and surface integrals we have seen, the computation of vector surface integrals involves the use of a parametrization of the surface. In fact we will proceed starting from the parametrization setup that was used to compute scalar surface integrals in Section 5.9.

Recall that in that section, we had already assumed a parametrization $\vec{x}$ for the surface S over the domain D. We broke up the domain D into pieces of size $dA = du\, dv$ with edges defined by the vectors $\vec{e}_1 du$ and $\vec{e}_2 dv$, and approximated the image of that piece of the domain by a parallelogram with edges we computed as $\vec{x}_u du$ and $\vec{x}_v dv$, respectively. This is represented in Figure 5.106.

We computed the area of that parallelogram as $dS = \|\vec{N}\|\, du\, dv$, where $\vec{N} = \vec{x}_u \times \vec{x}_v$. This vector $\vec{N}$ is perpendicular to the surface, as is of course the unit normal vector $\vec{n}$. The vector area differential $d\vec{S}$ defined in Equation 5.179 is also perpendicular to the surface. A closeup of the differential parallelogram, along with these vectors, is represented in Figure 5.107.

5.12. VECTOR SURFACE INTEGRALS

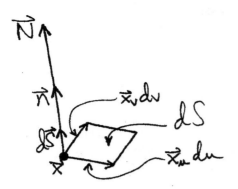

Figure 5.107:

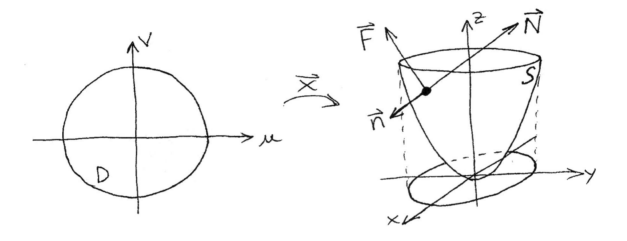

Figure 5.108:

Given this setup, we can easily rewrite the vector area differential $d\vec{S} = \vec{n}\, dS$ in terms of the parametrization. That is, we have

$$d\vec{S} = \vec{n}\, dS = \vec{n}\left(\|\vec{N}\|\, du\, dv\right) = \left(\vec{n}\|\vec{N}\|\right) du\, dv = \vec{N}\, du\, dv \tag{5.180}$$

So we can rewrite a vector surface integral as

$$\iint_S \vec{F}(\vec{x}) \cdot d\vec{S} = \iint_D \vec{F}(\vec{x}(u,v)) \cdot \vec{N}\, du\, dv \tag{5.181}$$

This domain of this integral is a two dimensional set D in the two dimensional uv-plane, so it can be computed as a nested integral.

5.12.4 Examples

Example 5.12.2. Let S be the part of the surface $z = x^2 + y^2$ directly above the unit disk in the xy-plane, oriented downward, and let $\vec{F}(x, y, z) = (x, y, z)$. Suppose we wish to compute the flux of the vector field $\vec{F}$ through the surface S. See Figure 5.108.

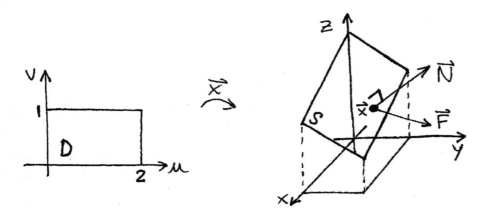

Figure 5.109:

We will use Equation 5.179. First we parametrize the surface with the graph parametrization, by $\vec{x}(u,v) = (u, v, u^2 + v^2)$, defined on the unit disk D in the uv-plane. We then compute the normal vector $\vec{N}$ as

$$\begin{aligned}\vec{N} &= \vec{x}_u \times \vec{x}_v \\ &= (1, 0, 2u) \times (0, 1, 2v) \\ &= (-2u, -2v, 1)\end{aligned}$$

Before we proceed, we pause to note that this normal vector is pointing upward because the z component is positive. But the surface S is defined as being oriented downward, meaning that this parametrization is "upside down" – that is, the vector $\vec{N}$ from our parametrization is pointing in the opposite direction of the orientation normal vector $\vec{n}$ (see Figure 5.108). Rather than reparametrize, we note that the result of the computation we are doing will be the negative of the correct answer. We thus fix the problem simply by inserting a minus sign.

With this extra minus sign, we can thus plug in to get

$$\begin{aligned}\iint_S \vec{F}(\vec{x}) \cdot d\vec{S} &= -\iint_D \vec{F}(\vec{x}(u,v)) \cdot \vec{N} \, du \, dv \\ &= -\iint_D \vec{F}(u, v, u^2+v^2) \cdot (-2u, -2v, 1) \, du \, dv \\ &= -\iint_D (u, v, u^2+v^2) \cdot (-2u, -2v, 1) \, du \, dv \\ &= -\iint_D -(u^2+v^2) \, du \, dv\end{aligned}$$

We can cancel the negatives and rewrite in polar coordinates to get

$$\begin{aligned} &= \int_0^{2\pi} \int_0^1 (r^2) r \, dr \, d\theta \\ &= \pi/2\end{aligned}$$

Example 5.12.3. Suppose the surface S is the part of the plane $x + y + z = 4$ that is directly above the rectangle in the xy-plane with corners at $(0,0,0)$, $(2,0,0)$, $(0,1,0)$, $(2,1,0)$, oriented upward, and let $\vec{F}(x, y, z) = (2y, z, -x)$. We will compute the flux of the vector field $\vec{F}$ through the surface S. See Figure 5.109.

Again we will use Equation 5.179. First we parametrize the surface, again with the graph parametrization, by $\vec{x}(u,v) = (u, v, 4 - u - v)$, defined on the rectangle D with corners at $(0,0)$, $(2,0)$, $(0,1)$, $(2,1)$ in the uv-plane. We

5.12. VECTOR SURFACE INTEGRALS

can then compute the normal vector $\vec{N}$ as

$$\begin{aligned} \vec{N} &= \vec{x}_u \times \vec{x}_v \\ &= (1, 0, -1) \times (0, 1, -1) \\ &= (1, 1, 1) \end{aligned}$$

(Of course a quick sanity check on this computation of $\vec{N}$ comes from looking at the given equation for the plane. Note however that this can NOT be used to compute $\vec{N}$ in the first place, because the length of $\vec{N}$ depends on the parametrization, and that is a critical part of the ensuing integral.)

Again we must pause to think about orientations. The surface S is given to be oriented upward, and this $\vec{N}$ is also an upward orientation because $\vec{N}$ has a positive z-coordinate. So this parametrization gave us the proper orientation.

Plugging in to Equation 5.179, we get

$$\begin{aligned} \iint_S \vec{F}(\vec{x}) \cdot d\vec{S} &= \iint_D \vec{F}(\vec{x}(u,v)) \cdot \vec{N} \, du \, dv \\ &= \iint_D \vec{F}(u, v, 4 - u - v) \cdot (1, 1, 1) \, du \, dv \\ &= \iint_D (2v, 4 - u - v, -u) \cdot (1, 1, 1) \, du \, dv \\ &= \int_0^1 \int_0^2 4 - 2u + v \, du \, dv \\ &= 5 \end{aligned}$$

Example 5.12.4. Suppose the surface S is the sphere centered at the origin with radius 3, oriented outward, and let $\vec{F}(x, y, z) = (yz, -3xz, 2xy)$.

Again we will compute the flux of the vector field $\vec{F}$ through the surface S. This time however, a clever observation will allow us to bypass the parametrization.

Because of the symmetries of the sphere, we know that the unit normal vector to S at a point $\vec{x}$ is parallel to the position vector for that point. Given also that it points outward and has unit length, we can then immediately write it down as $\vec{n} = \vec{x}/3 = (x/3, y/3, z/3)$. See Figure 5.110.

The clever observation then is that this unit normal vector is perpendicular to the vector field, which we see by computing $\vec{F} \cdot \vec{n} = xyz/3 - xyz + 2xyz/3 = 0$.

The flux integral then becomes

$$\iint_S \vec{F} \cdot \vec{n} \, dS = \iint_S 0 \, dS = 0$$

It will certainly not always be the case that these dot products will be zero, as the previous examples show. But in cases when one can conveniently write down a normal vector without parametrizing, it is fairly easy then to look at the dot product with the vector field to see if it will be zero. Students would be well advised to keep an eye out for problems where this trick might apply.

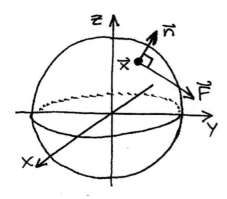

Figure 5.110:

5.12.5 Comparison to line integrals

In Subsection 5.11.5 we noted that we had three different notations representing vector line integrals. Each had its own advantages and uses, although of course all three represent exactly the same quantity. It turns out that we have the same situation here with vector surface integrals.

Even further though, there is a very strong analogy between the different notations for vector line integrals and for vector surface integrals, and also between the related objects. In this section we will draw out that analogy.

Of course both of these types of integrals involve vector fields. As we discussed in Section 5.11, one of the interpretation that can be made for a vector line integral is that it represents the amount of a flow that is moving along (parallel to) the domain curve. In a similar way, our primary interpretation of a vector surface integral is that it represents the amount of a flow that is moving through (perpendicular to) the domain surface. These are different in that one is parallel and one is perpendicular, but in each case these integrals represent the component of the flow in the direction that defines the orientation of the domain – the orientation of a curve is defined by a tangent vector, while the orientation of a surface is defined by a normal vector.

Also, both are computed over domains whose dimensions are not the same as those of the spaces they live in.

Below we have summaries of the quantities and notations that are relevant in discussions of these two types of integrals, and the relationships between them. These are represented in Figure 5.111

The way we have these two summaries laid out on the page, you will note that certain of the quantities relating to vector line integrals play the same roles in those equations as certain other quantities relating to vector surface integrals.

For example, the vectors $d\vec{x}$ and $d\vec{S}$ play corresponding roles. In each of these vectors, the length represents the size of the given piece of the domain, and the direction indicates the orientation of that piece of the domain. In these senses, these two vectors are the most complete algebraic representatives for their corresponding pieces of their domains.

The vectors $\vec{x}'$ and $\vec{N}$ play corresponding roles as well. Each has length that gives the "stretching factor" from the corresponding pieces of the parametrization domain.

(This comparison seems surprising at first because $\vec{x}'$ is a derivative vector and $\vec{N}$ does not appear to be. Remember though that $\vec{N} = \vec{x}_u \times \vec{x}_v$, and $\vec{x}_u$ and $\vec{x}_v$ are derivative vectors. The point then is that $\vec{N}$ encapsulates the relevant characteristics of the derivative vectors $\vec{x}_u$ and $\vec{x}_v$, so in fact it is reasonable that it would relate to $\vec{x}'$.)

The vectors $\vec{T}$ and $\vec{n}$ play corresponding roles, each being unit vectors defining the orientation of the given domain.

The scalars ds and dS play corresponding roles, each representing the scalar size of the given piece of the domain.

We see also that both of these types of integrals have analogous notational options. The first integrals listed for each represent the quantity (work or flux) simply as a sum over pieces.

5.12. VECTOR SURFACE INTEGRALS

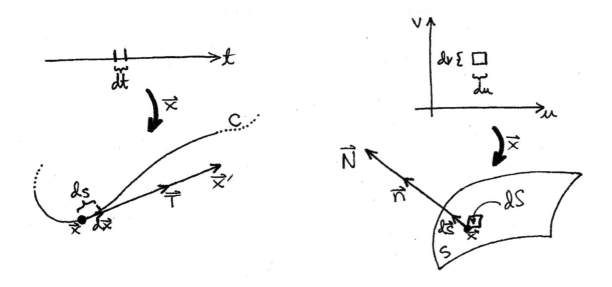

Figure 5.111:

The second integrals represent these in a coordinate free way that communicates very simply how those quantities are computed. In each case, the computation is ultimately just a dot product with the appropriate vector differential. These notations are not useful for direct computations, but they are easy to write down and communicative of what is being represented.

The third integrals rephrase each in terms of a parametrization of the given domain, resulting in forms that can be rewritten as nested integrals. These notations are very useful for doing direct computations.

The fourth integrals rewrite these expressions as scalar integrals. Of course the integrands for these scalar integrals are functions that depend on both the position and orientation of each piece of the domain. But still, these notations are useful in that they give a way to connect these vector line and surface integrals to previously discussed concepts.

The fifth and sixth integrals make an interpretation of the dot products as components. These line integrals then can be interpreted from these expressions as representing the total over the domain of the appropriate component of the vector field.

Vector Line Integrals:

$$d\vec{x} = \vec{x}' \, dt = \vec{T} \, ds$$
$$\|d\vec{x}\| = \|\vec{x}'\| \, dt = ds$$

$$\text{Work} = \int_C dw$$
$$= \int_C \vec{F} \cdot d\vec{x}$$
$$= \int_C \vec{F} \cdot \vec{x}' \, dt$$
$$= \int_C \vec{F} \cdot \vec{T} \, ds$$
$$= \int_C \text{comp}_{\vec{T}} \vec{F} \, ds$$
$$= \int_C \|\vec{F}_\parallel\| \, ds$$

Vector Surface Integrals:

$$d\vec{S} = \vec{N} \, du \, dv = \vec{n} \, dS$$
$$\|d\vec{S}\| = \|\vec{N}\| \, du \, dv = dS$$

$$\text{Flux} = \iint_S d\Phi$$
$$= \iint_S \vec{F} \cdot d\vec{S}$$
$$= \iint_S \vec{F} \cdot \vec{N} \, du \, dv$$
$$= \iint_S \vec{F} \cdot \vec{n} \, dS$$
$$= \iint_S \text{comp}_{\vec{n}} \vec{F} \, dS$$
$$= \iint_S \|\vec{F}_\perp\| \, dS$$

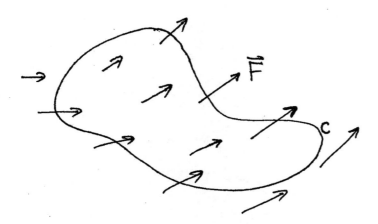

Figure 5.112:

Students might already have noticed that the derivations given in this section for the different forms of vector line integrals and vector surface integrals went in sort of opposite directions. For vector line integrals, the first form we wrote down was the coordinate free form, and from that we motivated the parametric form and then the scalar form. But for vector surface integrals, we first wrote down the scalar form.

The reason for this is merely circumstantial. For vector line integrals, it was easy to talk about the vector $d\vec{x}$, as it can be motivated simply as representing a displacement along the curve. Furthermore we had a pre-existing formula for work in terms of force and displacement, that displacement being this vector $d\vec{x}$. On the other hand, for vector surface integrals, the vector $d\vec{S}$ was not as easily recognizable, so it is easier to start from the scalar form and work in the other direction.

Certainly, this is merely an author's choice, and is not representative of any important difference between these types of integrals.

5.12.6 Flux through a curve

Up to now we have discussed flux only as a notion in $\mathbb{R}^3$. The idea is that if we interpret a vector field as representing some sort of flow, then the "flux through a surface" is intended to represent the rate of that flow passing through the surface.

But note that we can consider an analogous notion in $\mathbb{R}^2$. That is, if we interpret a vector field in $\mathbb{R}^2$ as representing some sort of flow through the plane, then the "flux through a curve" represents the rate of that flow passing through the curve.

It is not difficult to see applications of this analogous concept. For example, movements of a given species around a geographic area could be viewed as a flow and represented by a vector field, in which case the flux of that vector field through a given curve would represent the rate at which their numbers were crossing that curve. If the curve were a boundary as in Figure 5.112, such as a state border for example, then the flux would represent the total rate at which their numbers were leaving the state across the border.

We make this precise in a way that is analogous to what we did in $\mathbb{R}^3$. Very similarly to what we did in Subsection 5.12.1, we write the flux of a constant vector field $\vec{F}$ through a staight line segment of length L and normal vector $\vec{n}$ as

$$\Phi = \left(\vec{F} \cdot \vec{n}\right) L \tag{5.182}$$

Again as in the $\mathbb{R}^3$ case, we extend this to non-constant vector fields and non-straight curves simply by breaking the curve into pieces, which are approximately straight and over which the vector field is approximately constant. This results in the following formula for flux through a curve in $\mathbb{R}^2$,

$$\Phi = \int_C \vec{F} \cdot \vec{n}\, ds \tag{5.183}$$

5.12. VECTOR SURFACE INTEGRALS

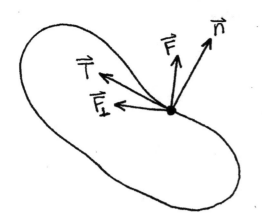

Figure 5.113:

(At this point in our development of flux in $\mathbb{R}^3$, we introduced the vector surface differential $d\vec{S} = \vec{n}\, dS$ as a notational and conceptual convenience. We will not use an analogous $d\vec{s} = \vec{n}\, ds$ notation for flux in $\mathbb{R}^2$ because of the potential for confusion with the vector differential $d\vec{x} = \vec{T}\, ds$ used in line integrals.)

Students will note immediately that this expression looks very much like the expression for a vector line integral. In fact that only difference is that the unit vector used in the dot product is the unit normal, not the unit tangent. This is consistent with the interpretations we already have for these two quantities – the line integral measures the amount the flow moves along the curve, and the flux measures the amount the flow moves perpendicular to the curve.

The above integral can be computed directly by the following observation. First, we note that the unit tangent vector $\vec{T}$ is always perpendicular to the unit normal vector $\vec{n}$. Specifically, if the curve is a positively oriented boundary and the normal is chosen pointing outward, then $\vec{T}$ is always exactly a quarter turn counterclockwise of $\vec{n}$. If we then define a new vector field $\vec{F}_\perp = (-Q, P)$ as a quarter turn counterclockwise of $\vec{F} = (P, Q)$, then geometrically we can see that $\vec{F} \cdot \vec{n} = \vec{F}_\perp \cdot \vec{T}$. See Figure 5.113.

We can then simplify Equation 5.183 as

$$\Phi = \int_C \vec{F} \cdot \vec{n}\, ds = \int_C \vec{F}_\perp \cdot \vec{T}\, ds = \int_C \vec{F}_\perp \cdot d\vec{x} = \int_C (-Q, P) \cdot \vec{x}'\, dt \qquad (5.184)$$

The final expression on the right can then be computed directly from the parametrization of the boundary curve C.

Exercises

Exercise 5.12.1 (exer-Ints-VSI-1). Compute the flux of the vector field $\vec{F}(x, y, z) = (z, x, y)$ through the portion of the graph of the function $f(x, y) = 4 - x^2 - y^2$ with $z \geq 0$, with the upward orientation.

Exercise 5.12.2 (exer-Ints-VSI-2). Compute the flux from Exercise 5.12.1, but this time with the surface oriented downward. How does this change the result?

Exercise 5.12.3 (exer-Ints-VSI-3). Compute the surface integral of the vector field $\vec{F}(x, y, z) = (yz, 2xz, -3xy)$ through the unit sphere, with the inward orientation, without parametrizing the surface. (*Hint: Use geometry to write an explicit formula for the unit normal vector, and then compute $\vec{F} \cdot \vec{n}$ explicitly.*)

Exercise 5.12.4 (exer-Ints-VSI-4). Compute the surface integral of the vector field $\vec{F}(x, y, z) = (x, y, -z)$ through the unit sphere, with the inward orientation.

Exercise 5.12.5 (exer-Ints-VSI-5). Compute the surface integral of the vector field $\vec{F}(x,y,z) = (xy, -2xz, x-z)$ through the portion of the surface with equation $xz - x^2z + y = 0$ that satisfies the inequality $x^2 + z^2 \leq 1$, oriented in the positive y direction. *(Hint: Parametrize this surface by solving for y.)*

Exercise 5.12.6 (exer-Ints-VSI-6). The surface S is the portion of the plane $3x - y + 2z = 0$ sitting above $[1,3] \times [2,4]$ in the xy-plane. The vector field $\vec{F}$ is not known exactly, but it is known that at any point on S we have $\text{comp}_{\vec{n}} \vec{F} = x^2 + y^2$. Compute the total flux of $\vec{F}$ through S.

Exercise 5.12.7 (exer-Ints-VSI-7). Suppose that S is a portion of the graph of the function f. Show that the graph parametrization $g(u,v) = (u, v, f(u,v)) = (x, y, z)$ will always give an "upward" unit normal vector.

Exercise 5.12.8 (exer-Ints-VSI-8). Compute the flux of the vector field $\vec{F}(x,y) = (xy^2, x^2 - y)$ through the portion of the graph of the function $f(x) = x^2 - x$ between $x = 0$ and $x = 2$, with the downward orientation.

Exercise 5.12.9 (exer-Ints-VSI-9). Compute the flux of the vector field $\vec{F}(x,y) = (x, -y)$ through the unit circle, oriented inward.

Exercise 5.12.10 (exer-Ints-VSI-10). We will say that a vector field $\vec{F}(x,y,z)$ has "mirror antisymmetry through the yz-plane" if the x coordinate has even symmetry through the yz-plane and the y and z coordinates have odd symmetry through the yz-plane. Written precisely,

$$\vec{F}(x,y,z) = \begin{bmatrix} F_1(x,y,z) \\ F_2(x,y,z) \\ F_3(x,y,z) \end{bmatrix} \quad \text{with} \quad \begin{aligned} F_1(-x,y,z) &= F_1(x,y,z) \\ F_2(-x,y,z) &= -F_2(x,y,z) \\ F_3(-x,y,z) &= -F_3(x,y,z) \end{aligned}$$

(Unlike with similar vector fields in $\mathbb{R}^2$ discussed in Exercises 5.11.8, 5.11.9, and 5.11.10, note then that the vectors in a mirror antisymmetric vector field appear to be the negatives of mirror images of each other through the yz-plane.)

We consider here the surface S with equation $z = x^2 + y^3$. Note that S is symmetric through the yz-plane, and the vector field $\vec{F}(x,y,z) = (x^2, -xy, xy^3z^2)$ has mirror antisymmetry through the yz-plane.

1. Compute the field vector $\vec{F}$ at the point $(1, 1, 2)$ and its reflection.
2. Compute the unit normal vector $\vec{n}$ to S at each of these points.
3. Compute $\vec{F} \cdot \vec{n}$ at each of these points.
4. Confirm that the value of $\vec{F} \cdot \vec{n}$ at $(1,1,2)$ is exactly the negative of the value of $\vec{F} \cdot \vec{n}$ at its reflection point.

Exercise 5.12.11 (exer-Ints-VSI-11). As is suggested by the result of Exercise 5.12.10, if S is symmetric through the yz-plane and $\vec{F}$ has mirror antisymmetry through the yz-plane, then the value of the surface integral $\iint_S \vec{F} \cdot d\vec{S} = \iint_S \vec{F} \cdot \vec{n}\, dS$ must be zero. Confirm by direct computation that this claim is true when the surface S is the portion of $z = x^2 + y^3$ above $[-1,1] \times [0, 2]$ in the xy-plane, and $\vec{F}(x,y,z) = (x^2, -xy^2z, x^3z)$.

Exercise 5.12.12 (exer-Ints-VSI-12). Use a symmetry argument to compute the flux of the vector field $\vec{F}(x,y,z) = (\cos^3 x, x^3 y e^z, x^3 yz - \sin^3 x)$ through the surface with equation $e^{x^2} + y^4 + z^6 = 5$.

Exercise 5.12.13 (exer-Ints-VSI-13). Using Exercises 5.12.10 and 5.12.11 as a guide, define what it should mean for a vector field to have mirror antisymmetry through the xy-plane, and state a symmetry result for such vector fields.

Exercise 5.12.14 (exer-Ints-VSI-14). Compute the flux of the vector field $\vec{F}(x,y,z) = (yz, xz, xy)$ through the surface $e^x - x + e^y - y + z^4 = 7$.

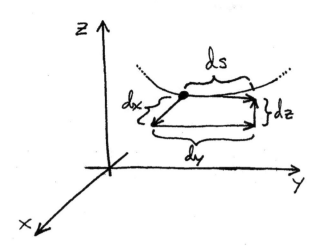

Figure 5.114:

5.13 Coordinate Line Integrals and Surface Integrals

In this section we introduce yet another notation for representing vector line and surface integrals.

There is a more sophisticated view of these types of integrals, using something called differential forms. Differential forms are very powerful and provide a notational option that is very useful and convenient in many instances. However, due to the amount of setup that must be done to give a reasonable discussion of differential forms, they are beyond the scope of this text.

So while we cannot discuss differential forms here, the notation we introduce in this section is somewhat suggestive of that from differential forms. It is also in common use, and students will very likely see the notation presented here in applications of these types of integrals.

5.13.1 Line integrals

Reinterpretation

Consider the vector line integral given by

$$\int_C \vec{F} \cdot d\vec{x}$$

where $\vec{F} = (P, Q, R)$. Recall of course from Section 5.11 that we can write the vector differential as $d\vec{x} = \vec{x}' \, dt$.

The vector $\vec{x}'$ of course can be written in coordinates as $\vec{x}' = (x', y', z') = (\frac{dx}{dt}, \frac{dy}{dt}, \frac{dz}{dt})$. We can then rewrite the vector differential as

$$d\vec{x} = (x', y', z') \, dt = \left(\frac{dx}{dt} dt, \frac{dy}{dt} dt, \frac{dz}{dt} dt \right) \tag{5.185}$$
$$= (dx, dy, dz) \tag{5.186}$$

This result is not surprising. That is, while $d\vec{x}$ represents the total change in the position vector $\vec{x}$ along the piece of the curve C with length ds, its components should represent the change in the individual coordinate variables. Fortunately, this is exactly the pre-existing interpretation we have for the coordinate differentials dx, dy and dz. See Figure 5.114.

This is a fine interpretation, but there is another interpretation that can be made for these coordinate differentials that will end up being useful when we compare with the results we will get in Subsection 5.13.2.

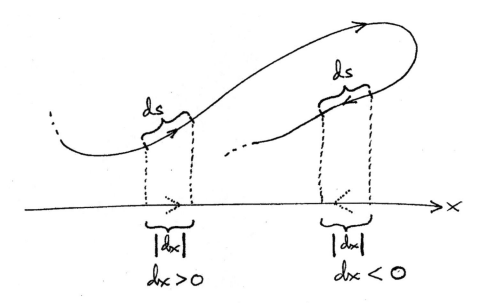

Figure 5.115:

Arbitrarily, we will consider the single coordinate differential dx. Note that the value of dx has two parts – a sign (positive or negative), and a magnitude ($|dx|$). Rather than thinking of these parts as they pertain to the change in the value of the coordinate x as $\vec{x}$ moves along the corresponding piece of the curve C, we will make interpretations of these parts as they pertain to the entire piece of curve itself.

The magnitude $|dx|$ is the size of the projection of the piece of curve (of length ds) to the x-axis.

The sign of dx can be interpreted based on the given orientation of the curve C. The orientation of the curve gives a notion of direction on the curve, and in projecting the piece of the curve onto the x-axis, the direction can be thought of as projecting as well. This projected direction will be either in the same direction as the positive direction along the x-axis, or the opposite direction (in other words, the negative direction). In either case this will be the same as the sign of dx. This is represented in Figure 5.115.

Notation

Having written $d\vec{x} = (dx, dy, dz)$, where $dx = x'(t)\, dt$, $dy = y'(t)\, dt$, and $dz = z'(t)\, dt$, it would then seem natural to rewrite the original line integral as

$$\int_C \vec{F} \cdot d\vec{x} = \int_C \begin{bmatrix} P \\ Q \\ R \end{bmatrix} \cdot \begin{bmatrix} dx \\ dy \\ dz \end{bmatrix} = \int_C P\, dx + \int_C Q\, dy + \int_C R\, dz \qquad (5.187)$$

This can be thought of as an alternate notation for a vector line integral, in writing it as a sum of these three integrals.

Of course these three integrals on the right side of this equation, while they seem to fall naturally out of the notation, are in forms for which we have not yet given definitions. The following definitions come directly from that notational motivation.

Definition 5.13.1. *The coordinate line integrals of a function f along a curve C (parametrized by $\vec{x}$ from $t = a$ to $t = b$)*

5.13. COORDINATE LINE INTEGRALS AND SURFACE INTEGRALS

with respect to coordinate variables are defined by

$$\int_C f(\vec{x})\,dx = \int_a^b f(\vec{x}(t))\,x'(t)\,dt$$

$$\int_C f(\vec{x})\,dy = \int_a^b f(\vec{x}(t))\,y'(t)\,dt$$

$$\int_C f(\vec{x})\,dz = \int_a^b f(\vec{x}(t))\,z'(t)\,dt$$

These definitions make sense out of Equation 5.187. But one can then naturally ask how these newly defined quantities make sense by themselves, not thought of simply as tools in an alternative notation for a vector line integral. The answer is simply a matter of looking at the interpretations we already have for the coordinate differentials.

For example, let's consider the integral $\int_C f(\vec{x})\,dx$. This is a Riemann sum, taken over the entire curve C. On each piece of the curve C, we can evaluate the value of the function f. We can also look at the size and orientation of the projection of the piece to the x-axis, giving us dx. We multiply these two quantities, and then add up over the entire curve.

Certainly there are instances in which one of these coordinate line integrals by itself represents an object of interest. Very often though, it is used simply as an alternate notation for a vector line integral as in Equation 5.187.

5.13.2 Surface integrals

Reinterpretation

We will take the same approach to rewriting vector surface integrals. Ultimately, we will end up with a very similar set of definitions and interpretations.

Consider the vector surface integral given by

$$\iint_S \vec{F} \cdot d\vec{S}$$

where $\vec{F} = (P, Q, R)$. Recall from Section 5.12 that we can write the vector differential as $d\vec{S} = \vec{N}\,du\,dv$.

The vector $\vec{N}$ of course is defined as $\vec{x}_u \times \vec{x}_v$, which can be written in coordinates as $\vec{N} = (x_u, y_u, z_u) \times (x_v, y_v, z_v)$. Students can work out this cross product directly to confirm that this can be rewritten as $\vec{N} = \left(\frac{\partial(y,z)}{\partial(u,v)}, \frac{\partial(z,x)}{\partial(u,v)}, \frac{\partial(x,y)}{\partial(u,v)}\right)$. (Notice that we are using here the notation

$$\frac{\partial(x,y)}{\partial(u,v)} = \det\begin{pmatrix} \frac{\partial x}{\partial u} & \frac{\partial x}{\partial v} \\ \frac{\partial y}{\partial u} & \frac{\partial y}{\partial v} \end{pmatrix} = \det\begin{pmatrix} x_u & x_v \\ y_u & y_v \end{pmatrix}$$

which we first used in Section 5.6 in our discussion of the stretching factor for change of variables functions.)

We can then rewrite the vector differential as

$$d\vec{S} = \vec{N}\,du\,dv = \left(\frac{\partial(y,z)}{\partial(u,v)}\,du\,dv,\ \frac{\partial(z,x)}{\partial(u,v)}\,du\,dv,\ \frac{\partial(x,y)}{\partial(u,v)}\,du\,dv\right) \quad (5.188)$$

This is very similar to the result of Equation 5.185. In that case we simply cancelled the dt's to arrive at a simplification; in this case, we must be a bit more careful.

Of course in the line integral case we could make the cancellation comfortably because it brought us to familiar quantities and notations ("dx", "dy", and "dz"). We are not so prepared in this case of surface integrals. In fact, the way we will rewrite the above expression will require us to define new quantities and notations.

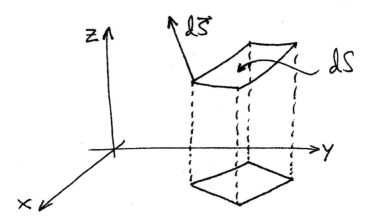

Figure 5.116:

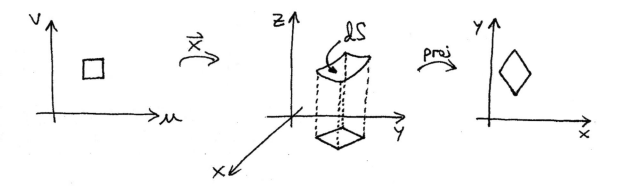

Figure 5.117:

But before we can do that, we need to motivate those definitions, and that motivation will come from understanding a geometric interpretation of those three coordinates of $d\vec{S}$. Similar arguments can be made for all three of these coordinates, but here we will consider only the third coordinate,

$$\frac{\partial(x,y)}{\partial(u,v)}\,du\,dv$$

Recall that in our consideration of line integrals, we came to one interpretation of the components of the $d\vec{x}$ vector by thinking about the projections of the piece of curve that it represents. Similarly, here we will hope to come to an interpretation of the components of $d\vec{S}$ by thinking about projections of the piece of surface that it represents.

Specifically, $d\vec{S}$ represents a piece of surface of area dS, approximately a parallelogram, part of the entire surface S. We will consider the projection of that piece of surface to the xy-plane. That projection will also then be a parallelogram. See Figure 5.116.

But of course the piece of surface that $d\vec{S}$ represents is the image (of a rectangle in the uv-plane) by the parametrization function $\vec{x}$. So the projected parallelogram then is the image the composition of the parametrization function $\vec{x}$ and the projection. The projection serves only to remove the third coordinate to put the point into the xy-plane, so in that composition, the coordinates x and y are the same functions of u and v that are used in defining the parametrization. See Figure 5.117.

At this point we can take a different point of view on this composition. We have two variables x and y that are functions of two other variables u and v, so we can choose to view this as a change of variables function as in Section 5.6. Using the

5.13. COORDINATE LINE INTEGRALS AND SURFACE INTEGRALS

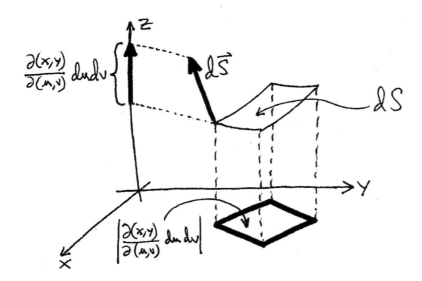

Figure 5.118:

results from that section, we know that the area of the image parallelogram in the xy-plane is in fact exactly $\left|\frac{\partial(x,y)}{\partial(u,v)}\,du\,dv\right|$. That is, it is exactly the absolute value of the third coordinate of $d\vec{S}$. See Figure 5.118.

Before we go on to consider the sign, be sure to note that this is completely analogous to what happened in the line integral case. In that case, the length (the "size") of the projection of the piece of curve to the x-axis was $\left|\frac{dx}{dt}\,dt\right|$ (which we wrote as $|dx|$), which is the absolute value of the first coordinate of the vector differential $d\vec{x}$. In the surface integral case the "size" refers to the area instead of the length, but again it is the absolute value of a coordinate of the vector differential, this time $d\vec{S}$.

Regarding the sign, there are two possibilities, which we can consider separately. If $\frac{\partial(x,y)}{\partial(u,v)}\,du\,dv$ is positive, that means that $d\vec{S}$ is pointing upward, which means that $\vec{n}$ is pointing upward and the surface is oriented in the upward direction. So when the piece of surface projects into the xy-plane, it projects "right side up".

Similarly, if $\frac{\partial(x,y)}{\partial(u,v)}\,du\,dv$ is negative, that means that $d\vec{S}$ is pointing downward, which means that $\vec{n}$ is pointing downward and the surface is oriented in the downward direction. In this case when the piece of surface projects into the xy-plane, it projects "upside down". See Figure 5.119.

Expanding then on the above observation about the analogy between this and the line integral case, we have that both $\frac{dx}{dt}\,dt$ in the line integral case and $\frac{\partial(x,y)}{\partial(u,v)}\,du\,dv$ in the surface integral case have absolute values equal to the size of a corresponding projection of the corresponding vector differential, and sign appropriate to the orientation of that projection.

Motivated by this analogy, and also motivated by the choice of notation is Equation 5.83, we make the following definitions.

Definition 5.13.2. *Given a parametrization $\vec{x}(u,v) = (x(u,v), y(u,v), z(u,v))$ of a surface S, we write*

$$dydz = \frac{\partial(y,z)}{\partial(u,v)}\,du\,dv$$

$$dzdx = \frac{\partial(z,x)}{\partial(u,v)}\,du\,dv$$

$$dxdy = \frac{\partial(x,y)}{\partial(u,v)}\,du\,dv$$

Students should note that the three items defined above are single objects, not products. For example, "$dxdy$" above is

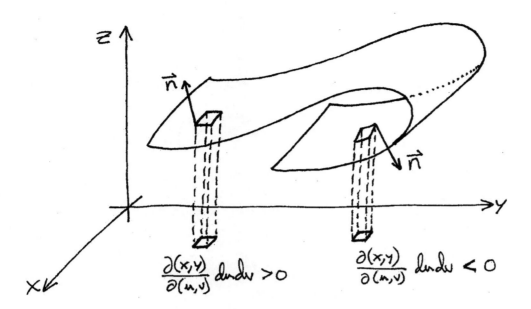

Figure 5.119:

not the product of two separate differentials dx and dy. (This is clear of course from the formula used to define it.) It is a single quantity, even though we write it down with four characters.

Also, a related point, students should note carefully that the order of the variables is critical. Specifically, the order of the variables is the same on both sides of each equation; and on the right side, switching the order of the variables changes the sign of the given determinant. So, "$dxdy$" is NOT to be interpreted as the same thing as "$dydx$" in this usage. In order to avoid such confusion, note that we have not made definitions for $dydx$, $dxdz$ or $dzdy$, and students should not use such things in this context.

Having made the above definitions, we can then finish the simplification of the vector differential $d\vec{S}$ in Equation 5.188 by writing

$$d\vec{S} = \left(\frac{\partial(y,z)}{\partial(u,v)}\,du\,dv,\; \frac{\partial(z,x)}{\partial(u,v)}\,du\,dv,\; \frac{\partial(x,y)}{\partial(u,v)}\,du\,dv\right) = (dydz, dzdx, dxdy) \tag{5.189}$$

Notation

Having written $d\vec{S} = (dydz, dzdx, dxdy)$, it would then seem natural to rewrite the original line integral as

$$\iint_S \vec{F}\cdot d\vec{S} = \iint_S \begin{bmatrix} P \\ Q \\ R \end{bmatrix} \cdot \begin{bmatrix} dydz \\ dzdx \\ dxdy \end{bmatrix} = \iint_S P\,dydz + \iint_S Q\,dzdx + \iint_S R\,dxdy \tag{5.190}$$

This can be thought of as an alternate notation for a vector surface integral, in writing it as a sum of these three integrals.

Of course these three integrals on the right side of this equation, while they seem to fall naturally out of the notation, are again in forms for which we have not yet given definitions. The following definitions come directly from that notational motivation.

Definition 5.13.3. *The coordinate surface integrals of a function f over a surface S (parametrized by $\vec{x}$ with $u \in [a,b]$*

5.13. COORDINATE LINE INTEGRALS AND SURFACE INTEGRALS

and $v \in [c, d]$) with respect to coordinate variables are defined by

$$\iint_S f(\vec{x})\, dy\,dz = \int_c^d \int_a^b f(\vec{x}(u,v)) \frac{\partial(y,z)}{\partial(u,v)}\, du\, dv$$

$$\iint_S f(\vec{x})\, dz\,dx = \int_c^d \int_a^b f(\vec{x}(u,v)) \frac{\partial(z,x)}{\partial(u,v)}\, du\, dv$$

$$\iint_S f(\vec{x})\, dx\,dy = \int_c^d \int_a^b f(\vec{x}(u,v)) \frac{\partial(x,y)}{\partial(u,v)}\, du\, dv$$

These definitions make sense out of Equation 5.190. Again one can then naturally ask how these newly defined quantities make sense by themselves, not thought of simply as tools in an alternative notation for a vector surface integral. The answer is again simply a matter of looking at the interpretations we already have for the coordinate differentials.

For example, let's consider the integral $\iint_S f(\vec{x})\, dx\,dy$. This is a Riemann sum, taken over the entire surface S. On each piece of the surface S, we can evaluate the value of the function f. We can also look at the size and orientation of the projection of the piece to the xy-plane, giving us $dx\,dy$. We multiply these two quantities, and then add up over the entire surface.

Certainly there are instances in which one of these coordinate surface integrals by itself represents an object of interest. Very often though, it is used simply as an alternate notation for a vector surface integral as in Equation 5.190.

Exercises

Exercise 5.13.1 (exer-Ints-Coord-1). The line integral $\int_C x^2 y\, dx - 2xz\, dy + 4xz^2\, dz$ is the line integral of what vector field $\vec{F}$ over the curve C?

Exercise 5.13.2 (exer-Ints-Coord-2). The line integral $\int_C f(x,y,z)\, dy$ is the line integral of what vector field $\vec{F}$ over the curve C?

Exercise 5.13.3 (exer-Ints-Coord-3). The surface integral $\iint_S ze^x\, dy\,dz + xz - y^3\, dz\,dx + 3xy \sin z\, dx\,dy$ is the flux of what vector field $\vec{F}$ through the surface S?

Exercise 5.13.4 (exer-Ints-Coord-4). The surface integral $\iint_S 2e^x y^2 \cos z\, dx\,dy + 3xyz\, dz\,dx - e^{xy}\, dy\,dz$ is the flux of what vector field $\vec{F}$ through the surface S?

Exercise 5.13.5 (exer-Ints-Coord-5). The surface integral $\iint_S f(x,y,z)\, dy\,dz$ is the flux of what vector field $\vec{F}$ through the surface S?

Exercise 5.13.6 (exer-Ints-Coord-6). In this problem we consider the curve C with equation $x^2 + y^4 + y^3 - 3y^2 - 4y - 4 = 25$, oriented counterclockwise. This curve is symmetric over the y-axis, but it is not symmetric over the x-axis. It is a closed curve, roughly circular, with a slight pinch on the sides.

We consider the two points $(5, 2)$ and $(5, -2)$ on this curve, which are the only two points on the curve with $x = 5$.

1. Compute the unit tangent vector $\vec{T}$ to this curve at each of these points. *(Hint: This curve is inconvenient to parametrize, but you can instead use the gradient to find normal vectors and then rotate and scale appropriately.)*

2. Suppose we consider the small pieces of this curve which fall in the thin slice between the lines $x = 5$ and $x = 5.001$ (note that these two small pieces of the curve are at the two points we are considering). On each of these pieces, we must have $|dx| = .001$, but of course the sign of dx depends on the direction of the curve.

   Use the equations $d\vec{x} = (dx, dy)$ and $d\vec{x} = \vec{T}\, ds$ to compute the vector differential $d\vec{x}$ and the length differential ds for each of these two pieces of the curve, at the given points.

3. Is there any relationship between the values of dy or ds on these two pieces of curve?

4. Show that the values of $f(x)\,dx$ on the two pieces of curve we are looking at are exactly negatives of each other. (Note, critically, the function f depends ONLY on the variable x, and not at all on y.)

5. Suppose we are interested in computing the value of the line integral $\int_C f(x)\,dx$. The previous parts of this problem suggest that when we break up the curve into pieces by slicing vertically, the summands in the integral on corresponding pieces of the curve will cancel. In fact this is true in general – if C is any closed curve in the plane, then $\int_C f(x)\,dx = 0$.

Use this fact to compute $\int_C e^x \sin x - 4x^3 e^{x^2}\,dx$, where C is the given closed curve.

Chapter 6

Vector Calculus

6.1 Setup and Approach

6.1.1 The general Stokes's theorem, de Rham cohomology, and compromise

The main theorems discussed in this chapter are some of the most elegant and powerful in analysis.

The tragedy is that a full treatment of these ideas would require the discussion to be in terms of differential forms. In those terms, all of the main theorems of this chapter could be unified, seen as special cases of one single theorem, which we will refer to as the "general Stokes's theorem". And with some additional background some of the other theorems in this chapter could also be unified, seen as special cases of another single theorem, the "de Rham cohomology theorem".

This presents a problem to authors of introductory texts on this subject. That is, discussing differential forms requires more setup and abstraction than would be appropriate in an introductory text, thus making the general Stokes's theorem and the de Rham cohomology theorem completely inaccessible.

Given this problem, many texts simply present the main theorems of this chapter as completely separate theorems. Certainly this can be done, as each of these theorems can be stated and proved independently of the others. This approach has the advantage of not requiring any of the more abstract ideas needed for the full treatment. The disadvantage of course is that students are left without even the vaguest inkling of these more powerful results.

In this text, we take an approach that is a sort of compromise of the above two approaches.

Given that the main theorems of this chapter are in fact all special cases of a single theorem, it is not surprising that there are many strong analogies between those theorems, and a natural structure for relating them to each other. These connections are extremely valuable, as they allow students to use their understanding of one theorem to help in understanding the others. They also give students a rough idea of the forms, though of course not the formal statements, of the larger theorems.

The compromise approach then is to approach the material in such a way as to illuminate these connections as much as possible, while still leaving the main theorems stated as separate theorems. Rather than giving rigorous proofs of these theorems, we will give instead strong motivations, all analogous to each other, that emphasize the conceptual ideas behind the proofs and their relationships to each other. And we will set up some language and framework that will allow students to organize the ideas in their minds in a natural way that is consistent with those connections. (Of course it should be noted that much of the language and framework we present in this chapter, in being merely suggestive of the loftier notions, will not be defined rigorously; that rigor would defeat the purpose of the approach we are taking here.)

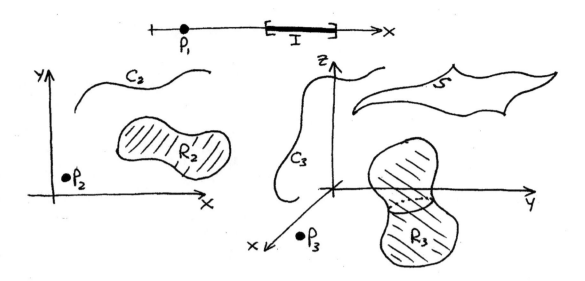

Figure 6.1:

6.1.2 Evaluations, domains and integrands

The main theorems we will discuss in this chapter concern relationships between different types of expressions, most of which are integrals, defined over different types of domains. We will use the general term "evaluations" to refer to all of these expressions.

(Remember, though the choices we will make here might seem arbitrary, they are not. These are in fact not really choices, but the results of considering special cases of a more general construction that we cannot present in this text. Students should accept for the moment that the consideration of these seemingly arbitrary choices of evaluations will lead to useful results, which we will see throughout the remainder of this chapter.)

The fact that most of these evaluations are integrals motivates some of the language we will use in talking about them. Each evaluation we will talk about has two inputs – a "domain" and an "integrand". The domains and integrands for all of these evaluations will be things that we are already familiar with. In fact, in each of the evaluations, the integrand will be either a function or a vector field.

We must break down our discussion into three parts, based on the space serving as the context for the given evaluation. The three spacial contexts we will consider in this text are $\mathbb{R}^1$, $\mathbb{R}^2$ and $\mathbb{R}^3$. In each of these contexts there will be a unique type of domain for each dimension not exceeding that of the context.

In $\mathbb{R}^1$, there are only two types of domains. A 0-dimensional domain in $\mathbb{R}^1$ is a point, such as P_1 in Figure 6.1. A 1-dimensional domain is an interval, such as I in that figure.

In $\mathbb{R}^2$, there are three types of domains. A 0-dimensional domain in $\mathbb{R}^2$ is a point, a 1-dimensional domain in $\mathbb{R}^2$ is a curve, and a 2-dimensional domain in $\mathbb{R}^2$ is a solid region in the plane. These are shown as P_2, C_2 and R_2 (respectively) in Figure 6.1.

In $\mathbb{R}^3$, there are four types of domains. A 0-dimensional domain in $\mathbb{R}^3$ is a point, a 1-dimensional domain in $\mathbb{R}^3$ is a curve, a 2-dimensional domain in $\mathbb{R}^3$ is a surface, and a 3-dimensional domain in $\mathbb{R}^3$ is a solid region in $\mathbb{R}^3$. These are shown as P_3, C_3, S and R_3 (respectively) in Figure 6.1.

We have already encountered all of the evaluations we will discuss in this chapter.

For the evaluations whose domains are points, the corresponding integrands are functions. The evaluation then is computed simply by plugging the point into the function.

6.1. SETUP AND APPROACH

For the evaluations whose domains are of the same dimension as the context (intervals in $\mathbb{R}^1$, solid regions in either $\mathbb{R}^2$ or $\mathbb{R}^3$), the corresponding integrands are again functions. The evaluation then is computed as an integral (of the appropriate dimension) of the function over that domain.

For the evaluations whose domains are curves in either $\mathbb{R}^2$ or $\mathbb{R}^3$, or surfaces in $\mathbb{R}^3$, the corresponding integrands are vector fields. The evaluation of the vector field integrand over a curve domain is computed as the vector line integral of that vector field over the curve; and the evaluation of the vector field integrand over a surface domain is computed as the vector surface integral of that vector field over the surface.

6.1.3 Orientations and boundaries

Orientations

We have already discussed orientations as they pertain to vector line and surface integrals. On a curve, an orientation is a choice of direction along the curve, represented by a consistent choice of unit tangent vector at all points on the curve. On a surface, an orientation is a choice of direction perpendicular to the surface, represented by a consistent choice of normal vector at all points on the surface. These are exactly the notions of orientation that will apply in this setting to curves and surfaces in $\mathbb{R}^2$ and $\mathbb{R}^3$.

We will need to extend the idea of an orientation though to all of the evaluations that we will discuss in this chapter. In order to do this, we make the following definition.

Definition 6.1.1. *An orientation on a domain is a convention regarding sign in evaluations on that domain.*

In the case of vector line and surface integrals, this is exactly what happens with our previously described notions of orientation. That is, it has already been observed in each case that switching the orientation on either a curve or a surface changes the value of the integral only in the sign.

For domains whose corresponding integrands are functions though, we have not yet seen any notion of orientation. For example we have not talked yet about an orientation of a point, or of a solid region in $\mathbb{R}^3$.

To define orientations for these other domains, we simply assign an orientation as either a "+" or a "-".

The notations we will use to associate these orientations to the domains are as simple as we can make them. For example, for a domain D that is a solid region in $\mathbb{R}^2$ or $\mathbb{R}^3$, we write the positive orientation as $+D$; the same set, but with a negative orientation, is written $-D$. If no orientation is written at all, the positive orientation is assumed.

For point domains, the notation we will use is similar, but we put the orientation in quotes to avoid possible confusion with scalar multiplication. For example, "+"(x, y, z) refers to the point (x, y, z) with a positive orientation, and "-"(x, y, z) refers to the point (x, y, z) with a negative orientation (not to be confused with the notation $-(x, y, z)$, referring to the point $(-x, -y, -z)$). Again, if no orientation is written at all, the positive orientation is assumed.

These orientations manifest themselves in evaluations in the expected way in light of Definition 6.1.1. That is,

$$f(\text{``+''}\vec{x}) = f(\vec{x})$$
$$f(\text{``-''}\vec{x}) = -f(\vec{x})$$
$$\int_{\text{``+''}I} f\, dx = \int_I f\, dx$$
$$\int_{\text{``-''}I} f\, dx = -\int_I f\, dx$$
$$\iint_{\text{``+''}R} f\, dA = \iint_R f\, dA$$
$$\iint_{\text{``-''}R} f\, dA = -\iint_R f\, dA$$
$$\iiint_{\text{``+''}R} f\, dV = \iiint_R f\, dV$$
$$\iiint_{\text{``-''}R} f\, dV = -\iiint_R f\, dV$$

We can also use these explicit signs also as a tool regarding orientations on curves and surfaces in $\mathbb{R}^2$ and $\mathbb{R}^3$.

Of course for domains like these there is no intrinsic positive orientation, so the orientation has to be defined along with the curve itself. For example, if we refer simply to the upper unit semicircle in $\mathbb{R}^2$, there is no assumed orientation. So, to be able to view this curve as a domain over which we can compute an evaluation (vector line integral in this case), we have to make an explicit choice for the orientation. For example we might choose to define the curve C as the upper unit semicircle in $\mathbb{R}^2$ with the counterclockwise orientation.

Having made such a choice, explicit signs paired with the oriented domain change the domain in the way as to cause the appropriate change in evaluations, as per Definition 6.1.1. For example, the curve "$-C$" is the upper unit semicircle with the opposite (clockwise) orientation.

In this notation then we can rewrite a previously known result as

$$\int_{-C} \vec{F} \cdot d\vec{x} = -\int_C \vec{F} \cdot d\vec{x} \tag{6.1}$$

Remember, though the choices we will make here might seem arbitrary, they are not. These are in fact not really choices, but the results of considering special cases of a more general construction that we cannot present in this text. Students should accept for the moment that the consideration of these seemingly arbitrary choices of orientations will lead to useful results, which we will see throughout the remainder of this chapter.

Domain arithmetic

With this notation, we can naturally introduce sums and differences of domains.

Very simply, these are new domains that are thought of as combinations of other domains. For example if we have the oriented curves C_1 and C_2, the "sum" of those curves is written as "$C_1 + C_2$", over which line integrals are evaluated by

$$\int_{C_1+C_2} \vec{F} \cdot d\vec{x} = \int_{C_1} \vec{F} \cdot d\vec{x} + \int_{C_2} \vec{F} \cdot d\vec{x} \tag{6.2}$$

Geometrically one can think of this sum domain as being a curve that follows the path C_1, and then subsequently follows the path C_2.

Note that a consequence of this definition is that the curve "$C - C$" will always evaluate to zero, for any integrand. The interpretation of this is that the two curves C and $-C$ "cancel" each other in that arithmetic. Of course this is consistent with the suggestion of the notation.

6.1. SETUP AND APPROACH

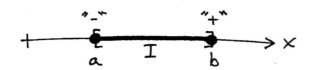

Figure 6.2:

Similar domain arithmetic can be defined for the other different types of domains we have defined.

Boundaries

In any of the three contexts we are considering, there are geometric relationships between the different types of domains. Most importantly we will consider here the idea of a "boundary". Again, we will not make rigorous definitions of these terms, relying instead on the natural geometric intuition that students have for these ideas.

For example, in $\mathbb{R}^1$, the boundary of an interval is a pair of points. In $\mathbb{R}^2$, the boundary of a solid plane region is a curve; and the boundary of a non-closed curve is a pair of points. In $\mathbb{R}^3$, the boundary of a solid region is a surface; the boundary of a non-closed surface is a curve; the boundary of a non-closed curve is a pair of points.

The above examples though have ignored the notion of orientation. Since the idea of a boundary is such a natural geometric notion, it would seem reasonable that if one of these domains has an orientation, then that should naturally give rise to a particular orientation on the boundary domain. In fact there are natural ways to do this.

(Again, we have to note that the conventions we will describe here might seem to be arbitrary choices, but actually they are not. That is, these are simply special cases of the more general notion of boundary which we cannot present in this text. But again, we will see throughout the rest of this chapter many useful results based on these conventions.)

We will use a standard notation for the oriented boundaries we will define below. The partial symbol "∂" is used to represent these boundaries.

For a positively oriented interval in $\mathbb{R}^1$, the boundary is two points. The convention is to give the left endpoint the negative orientation, and the right endpoint the positive orientation, as in Figure 6.2. So for the interval $I = [a, b]$, we write
$$\partial I = \{\text{``-''}a, \text{``+''}b\} \tag{6.3}$$

Of course if the interval has the negative orientation, very naturally the orientations on the boundary are the opposite. That is,
$$\partial(-I) = \{\text{``+''}a, \text{``-''}b\} \tag{6.4}$$

For a connected curve in either $\mathbb{R}^2$ or $\mathbb{R}^3$, the boundary is a pair of points. The orientation on the curve itself defines a direction along the curve, which points "away from" one of the boundary points, and "toward" the other boundary point (these are sometimes referred to as the "starting point" and the "ending point", respectively). The convention then is to give the starting point the negative orientation, and the ending point the positive orientation.

For example, if C is the upper unit semidisk in $\mathbb{R}^2$ oriented counterclockwise, represented in Figure 6.3, then we have
$$\partial C = \{\text{``-''}(1, 0), \text{``+''}(-1, 0)\} \tag{6.5}$$

(Note that if we view a positively oriented interval in $\mathbb{R}^1$ as a curve in $\mathbb{R}^1$ oriented to the right, then these two conventions on the orientation of the boundary are consistent.)

For a positively oriented solid region R in $\mathbb{R}^2$, the boundary will often be a single closed curve, but as is shown in Figure 6.4, it could be several closed curves. Every point on each of these curves has the property that nearby points on one side

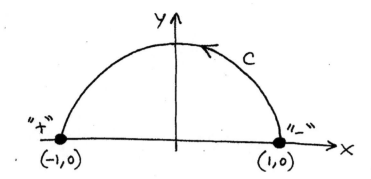

Figure 6.3:

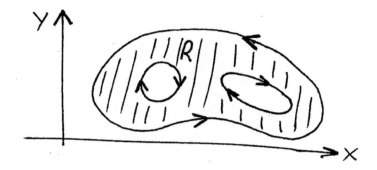

Figure 6.4:

are in the region R, while on the other side nearby points are not in R. The convention then is to orient the boundary curves such that in the direction of the orientation, it is the nearby points on the left that are in the set R, and the nearby points on the right that are not in R.

Of course if the solid region R has no "holes" in it (we call such sets "simply connected"), then the boundary is a single closed curve, and it will always be the case that the orientation on that boundary curve will be counterclockwise. But students should note very carefully from Figure 6.4 that if there are "holes", the corresponding pieces of the boundary are closed curves whose orientations are not counterclockwise. Because of this, it is recommended not to think of this boundary convention in these terms.

For a positively oriented solid region R in $\mathbb{R}^3$, the boundary will often be a single closed surface, but again it could be more than one. As an example of this, Figure 6.5 shows the solid region that is the result of hollowing out a spherical cavity from a larger solid ball. The boundary of this region is two surfaces.

Every point on each of these surfaces has the property that nearby points on one side are in the region R, while on the other side nearby points are not in R. The convention is to orient the boundary surfaces such that the orientation points in the direction in which nearby points are not in the region R.

Again, if the solid region R has no "holes" in it (note, the term "simply connected does not apply in this way for solid domains in $\mathbb{R}^3$), then the boundary is a single closed surface and the orientation on that surface is what we might call "outward". Yet again this will not be the case of boundary surfaces that come from "holes".

For a surface in $\mathbb{R}^3$, the boundary will be a collection of closed curves, as is shown in Figure 6.6.

Of course we quickly note that this is similar to the case of a solid region in $\mathbb{R}^2$. In fact it will turn out that the conventions for the orientations of the boundary will also be similar, but there is a complication in the case of the surface. Specifically, the boundary of a surface in $\mathbb{R}^3$ is still a curve, but there are no longer a mere two sides of that curve as was the case

6.1. SETUP AND APPROACH

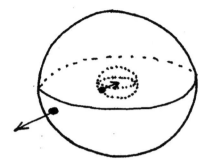

Figure 6.5:

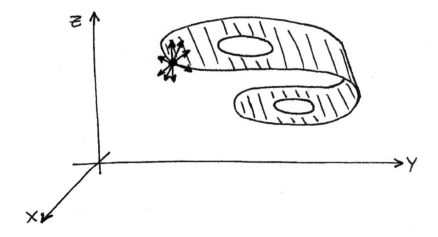

Figure 6.6:

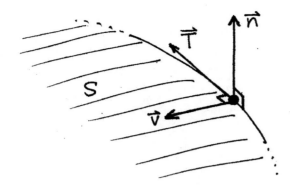

Figure 6.7:

in $\mathbb{R}^2$. Instead there are many different directions in which one can move away from the boundary curve, not just two. This is also shown in Figure 6.6.

To avoid this problem, we take a different point of view in this case. This will involve the notion of "right hand order" that we talked about in Section 1.4.

At any point on the boundary curve, there are three vectors that we will need to consider. Of course at that point there is a normal vector $\vec{n}$ that defines the orientation on the given surface S. There will also be a unit tangent vector to the boundary curve, which we will call $\vec{T}$, indicating the orientation on that boundary curve. Finally, we will consider a vector $\vec{v}$ that points perpendicular to the boundary curve, and parallel to S. Of course there are two such directions; critically, we choose the one that points in the direction in which nearby points are in the surface S, as shown in Figure 6.7.

With this setup, we can define the orientation of the boundary curve as follows. That is, the orientation on the boundary curve is such that the ordered list of vectors $\vec{n}$, $\vec{T}$, $\vec{v}$ is in right hand order. Note that this is the case for the vectors shown in Figure 6.7.

There are different ways that one can think about this. For example, students can convince themselves that, as seen from the head of $\vec{n}$, when proceeding along the boundary curve in the direction of the orientation, the surface S is on the left. In this way we see that it is the orientation on S that provides the connection to the orientation convention for regions in $\mathbb{R}^2$, as we previously suggested.

Another alternative makes an equivalent but seemingly different use of the right hand. Suppose we choose a point $\vec{a}$ in S that is near the boundary curve. Pointing the right thumb in the direction of the normal vector $\vec{n}$ defining the orientation on S, the fingers of the right hand curl around in a specific direction, defining a direction of rotation around $\vec{a}$. The orientation on the boundary curve is that which is consistent with the direction of that rotation. This is shown in Figure 6.8.

Figure 6.9 shows the surface from Figure 6.6, with an orientation indicated by the several given normal vectors. Students can check directly the indicated orientations on the three separate pieces of the boundary.

6.1.4 Diagrams and boundary theorems

There are many objects that have been presented so far in this section. We can organize many of those ideas with the following diagrams. The three diagrams correspond to $\mathbb{R}^1$, $\mathbb{R}^2$ and $\mathbb{R}^3$, respectively. Each collects and organizes the integrands, domains, and evaluations appropriate to its context.

These diagrams will provide a framework for most of the ideas we will discuss in the rest of this chapter. As we develop more ideas, we will represent those ideas in natural places on these diagrams to show how they fit in with other things we have already discussed. Keeping in mind the format and structure of the diagrams will help students to keep track

6.1. SETUP AND APPROACH

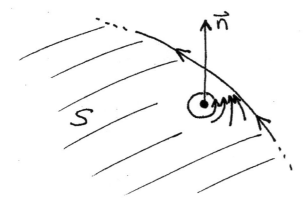

Figure 6.8:

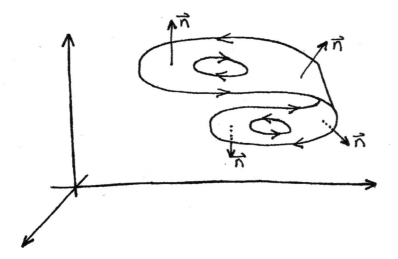

Figure 6.9:

of ideas, to compare and connect with previously discussed concepts, and to diagnose and solve problems later in the chapter.

These diagrams are informal, in these sense that we will use them here only as symbolic reprsentations of things that we do in this chapter. However it should be noted that these diagrams are similar to and strongly suggestive of similar formal structures that would be defined in a robust presentation of vector calculus with differential forms. This relationship is another important advantage to studying these diagrams.

Each column in each diagram corresponds to one of the evaluations we talked about in Subsection 6.1.2. The numbers along the top row of each diagram indicate the dimension of the domains in that column. In the "Evaluations" rows, the integrands listed are just "dummy integrands", used only to allow us to write the form of the corresponding evaluation.

Calculus in $\mathbb{R}^1$

	$\boxed{0}$	$\boxed{1}$
Integrands	(fns)	(fns)
Domains	(points) $\xleftarrow{\partial \text{ bdry}}$	(intervals)
Evaluations	$h(x)$	$\int_a^b h\, dx$

Vector Calculus in $\mathbb{R}^2$

	$\boxed{0}$	$\boxed{1}$	$\boxed{2}$
Integrands	(fns)	(vector fields)	(fns)
Domains	(points) $\xleftarrow{\partial \text{ bdry}}$	(curves) $\xleftarrow{\partial \text{ bdry}}$	(regions)
Evaluations	$h(\vec{x})$	$\int_C \vec{H} \cdot d\vec{x}$	$\iint_R h\, dA$

Vector Calculus in $\mathbb{R}^3$

	$\boxed{0}$	$\boxed{1}$	$\boxed{2}$	$\boxed{3}$
Integrands	(fns)	(vector fields)	(vector fields)	(fns)
Domains	(points) $\xleftarrow{\partial \text{ bdry}}$	(curves) $\xleftarrow{\partial \text{ bdry}}$	(surfaces) $\xleftarrow{\partial \text{ bdry}}$	(regions)
Evaluations	$h(\vec{x})$	$\int_C \vec{H} \cdot d\vec{x}$	$\iint_S \vec{H} \cdot d\vec{S}$	$\iiint_R h\, dV$

As of this moment, the boundary operator is the only connection that we have talked about between the columns of any of these diagrams. We will find that these notions of boundary will give rise to other very strong connections between the columns, including the main theorems of this chapter, all of which are special cases of the general Stokes's theorem. Because the boundary will be our starting point, we will refer to those theorems collectively as "boundary theorems".

6.1.5 The fundamental theorem of calculus

The first boundary theorem we will discuss is one that students should already be highly familiar with – the fundamental theorem of calculus, which we will often abbreviate as "FTC". Students should already have seen in their single variable calculus courses a rigorous proof of this theorem as it applies to continuous functions. As we mentioned in Subsection 6.1.1, we will not present here another proof.

Our purpose for revisiting the fundamental theorem of calculus is based in the fact that it, like the other boundary theorems in this chapter, is a special case of the general Stokes's theorem. With that in mind, the argument that we give here will serve as a template for our discussion of those other boundary theorems.

6.1. SETUP AND APPROACH

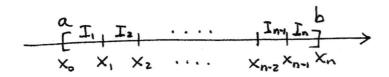

Figure 6.10:

We begin with a continuously differentiable function f, to be considered as an integrand in column zero, and an interval $I = [a, b]$, to be considered as a domain in column one. We represent this in our diagram of vector calculus in $\mathbb{R}^1$ by:

	0	1
Integrands	f	
Domains		I
Evaluations	$h(x)$	$\int_a^b h\, dx$

Of course we can apply the boundary operator to I, giving us $\partial I = \{\text{``+''}b, \text{``-''}a\}$. This then leaves us with both an integrand and a domain in column zero, so we could consider the output of applying the corresponding evaluation, which is $f(b) - f(a)$.

	0	1
Integrands	f	
Domains	$\{\text{``+''}b, \text{``-''}a\}$ $\xleftarrow[\text{bdry}]{\partial}$	I
Evaluations	$h(x)$	$\int_a^b h\, dx$
Outputs	$f(b) - f(a)$	

The existence of the boundary operator allowed us to have both an integrand and a domain with which to apply the evaluation in column zero, giving us a quantity of possible interest. But as of the moment we do not have any integrand in column one, so we cannot apply that evaluation.

What we will do now is try to find a way to view the output of this column zero evaluation as being the result of a column one evaluation. It is not immediately clear that we will be able to do this, but in fact we will. The process of manipulating that expression into such a form will turn the integrand into a new function.

This process involves breaking up the interval I into subintervals, in the usual way. Specifically, we will break it up into n subintervals which we will refer to as I_i, where $i = 1, \ldots, n$. As usual we will use Δx to refer to the size of each subinterval, and we name the endpoints of those subintervals in the usual way, as indicated in Figure 6.10, giving us

$$\partial I_i = \{\text{``-''}x_{i-1}, \text{``+''}x_i\} \tag{6.6}$$

These subintervals have a very important and natural connection to the original interval I. In the language of domain arithmetic that we discussed in Subsection 6.1.3, we can write this connection with the equation

$$\partial I = \sum_i \partial I_i = \partial I_1 + \cdots + \partial I_n \tag{6.7}$$

This equation works because of the great amount of cancellation in the interior of the interval. Specifically, each interior endpoint x_i is the left endpoint of one interval and the right endpoint of another endpoint – thus it appears twice in the summation above, once with a negative sign and once with a positive sign, and therefore cancelling. We can write this out explicitly (using parentheses below only for visual grouping) as

$$\begin{aligned}
\sum_i \partial I_i &= \partial I_1 + \cdots + \partial I_n \\
&= \left\{\left(\text{``-''}x_0, \text{``+''}x_1\right), \left(\text{``-''}x_1, \text{``+''}x_2\right), \cdots, \left(\text{``-''}x_{n-2}, \text{``+''}x_{n-1}\right), \left(\text{``-''}x_{n-1}, \text{``+''}x_n\right)\right\} \\
&= \left\{\text{``-''}x_0, \left(\text{``+''}x_1, \text{``-''}x_1\right), \quad \cdots \quad , \left(\text{``+''}x_{n-1}, \text{``-''}x_{n-1}\right), \text{``+''}x_n\right\} \\
&= \left\{\text{``-''}x_0, \text{``+''}x_n\right\} \\
&= \left\{\text{``-''}a, \text{``+''}b\right\} = \partial I
\end{aligned}$$

We can come to effectively the same result by noting the analogous cancellation in the evaluation over those domains. That is,

$$\Big(f(x_1) - f(x_0)\Big) + \cdots + \Big(f(x_n) - f(x_{n-1})\Big) = f(b) - f(a) \tag{6.8}$$

Having noted this connection, we can begin the computation that will allow us to rewrite the result of the previous column zero evaluation as a column one evaluation.

We start with the summation in Equation 6.8. We can apply the limit as $n \to \infty$ because the first equation shows that the summation does not depend on n. Then, we simply both divide and multiply by Δx, and rewrite $f(x_i) - f(x_{i-1})$, the change in f over the subinterval, as Δf.

$$\begin{aligned}
f(b) - f(a) &= \sum_{i=1}^n \Big(f(x_i) - f(x_{i-1})\Big) & (6.9) \\
&= \lim_{n\to\infty} \sum_{i=1}^n \Big(f(x_i) - f(x_{i-1})\Big) & (6.10) \\
&= \lim_{n\to\infty} \sum_{i=1}^n \left(\frac{f(x_i) - f(x_{i-1})}{\Delta x}\right) \Delta x & (6.11) \\
&= \lim_{n\to\infty} \sum_{i=1}^n \left(\frac{\Delta f}{\Delta x}\right) \Delta x & (6.12)
\end{aligned}$$

We notice that the expression now seems to look very much like a Riemann sum, which is of course an integral – the evaluation that we have associated to column one of our diagram.

To make the next step rigorously, we would need to invoke the mean value theorem. Instead, in order to keep the emphasis on the part of the argument that will serve as a template for future boundary theorems, we will skip the details and claim that we can rewrite the above simply as

$$f(b) - f(a) = \int_a^b \left(\lim \frac{\Delta f}{\Delta x}\right) dx \tag{6.13}$$

At this point we are fortunate to notice that this integrand is familiar, and in fact is exactly the definition of the derivative. So, we can rewrite the above equation as

$$f(b) - f(a) = \int_a^b f' \, dx \tag{6.14}$$

which of course is the fundamental theorem of calculus.

So, as was our goal, we have written the initial output of the column zero evaluation in the form of a column one evaluation. We can take the equality of these two outputs as motivation to consider the derivative operator as a natural operator to turn column zero integrands into column one integrands. Taking this convention, our diagram becomes

6.1. SETUP AND APPROACH

	$\boxed{0}$	$\boxed{1}$
Integrands	$f \xrightarrow{\frac{d}{dx}}$ derivative	f'
Domains	$\{\text{"+"}b, \text{"-"}a\} \xleftarrow{\partial}$ bdry	I
Evaluations	$h(x)$	$\int_a^b h\,dx$
Outputs	$f(b) - f(a)$ ==	$\int_a^b f'\,dx$

This gives a very graceful point of view on the fundamental theorem. That is, given a continuously differentiable function f and an interval I, applying the indicated operators and evaluating in each column always gives equal results.

In the diagram below we rewrite this as an appendix to the original diagram for calculus in $\mathbb{R}^1$.

Calculus in $\mathbb{R}^1$

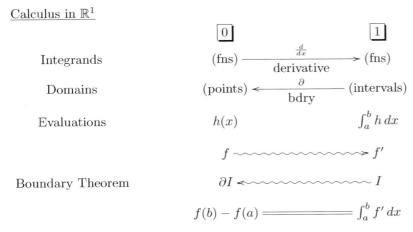

	$\boxed{0}$	$\boxed{1}$
Integrands	(fns) $\xrightarrow{\frac{d}{dx}}$ derivative	(fns)
Domains	(points) $\xleftarrow{\partial}$ bdry	(intervals)
Evaluations	$h(x)$	$\int_a^b h\,dx$
Boundary Theorem	$f \rightsquigarrow f'$ $\partial I \leftsquigarrow I$ $f(b) - f(a) ==\int_a^b f'\,dx$	

As we stated previously, this is not a complete proof of the fundamental theorem. The motivation for showing this argument here is that it will serve as an outline for a general approach that we will apply analogously in the other diagrams. All of the boundary theorems we discuss in this text will be motivated by this approach.

Here is an outline of the approach.

1. Consider a given integrand in column (k) and a given domain in column $(k+1)$ in one of our diagrams.

2. Break up the given domain into small subdomains, and confirm that the sum of the boundaries of the subdomains equals the boundary of the given domain.

3. Apply the column (k) evaluation to the given integrand over the boundary, and rewrite as a sum over the subdomains.

4. Rewrite in the form of the evaluation in column $(k+1)$.

5. Interpret the new integrand as the result of a new operator applied to the given integrand.

6. Interpret the result as the statement that applying the indicated operators and evaluating in each column always gives the same result.

Exercises

Exercise 6.1.1 (exer-VC-SA-1). Suppose that for a given vector field $\vec{F}$ we know that $\int_{C_1} \vec{F} \cdot d\vec{x} = 5$ and $\int_{C_2} \vec{F} \cdot d\vec{x} = 3$. Compute the following.

1. $\int_{-C_1} \vec{F} \cdot d\vec{x}$
2. $\int_{C_2 - C_1} \vec{F} \cdot d\vec{x}$
3. $\int_{3C_1 + 4C_2} \vec{F} \cdot d\vec{x}$

Exercise 6.1.2 (exer-VC-SA-2). For this problem B is the positively oriented unit disk in $\mathbb{R}^2$, and D is the positively oriented disk of radius 4 centered at $(1, 0)$.

1. Give a geometric description (with words and geometric terms) of ∂B, including both the points on the path and the orientation.
2. Give a geometric description of ∂D, including both the points on the path and the orientation.
3. Let A be the set of points in $\mathbb{R}^2$ that are in D but that are not in B. Give a geometric description of A.
4. Give a geometric description of ∂A, including both the points on the path and the orientation.
5. Write an algebraic description for ∂A in terms of ∂B, and ∂D.
6. Thinking then of A as being $D - B$, use the above results to confirm that $\partial(D - B) = \partial D - \partial B$.

Exercise 6.1.3 (exer-VC-SA-3). Let C be the clockwise oriented circle of radius 5 centered at $(2, 4)$, and let P be the part of C that is on or to the left of the line $x = -1$. What is ∂P? (Remember to indicate orientations.)

Exercise 6.1.4 (exer-VC-SA-4). Let S be the portion of the solid ball of radius 5 centered at the origin that is on the plane $y = 3$. Suppose that the orientation of ∂S is clockwise as seen from the positive portion of the y-axis. What is the unit normal vector for S?

Exercise 6.1.5 (exer-VC-SA-5). Let S be the portion of the sphere of radius 10 centered at the origin (oriented outward) that is outside of the solid cylinder $x^2 + y^2 \leq 36$. Give a geometric description of ∂S.

Exercise 6.1.6 (exer-VC-SA-6). Let S be the portion of the sphere of radius 10 centered at the origin (oriented outward) that is inside the solid cylinder $x^2 + y^2 \leq 36$. Give a geometric description of ∂S.

Exercise 6.1.7 (exer-VC-SA-7). Let T be the portion of the ball of radius 10 centered at the origin that is outside of the solid cylinder $x^2 + y^2 \leq 36$. Give a geometric description of ∂T.

Exercise 6.1.8 (exer-VC-SA-8). Let B be the ball of radius 7 centered at the origin, with orientation unknown, and suppose we know that at the point $(2, -3, -6)$ on ∂B the unit normal vector is $(-2/7, 3/7, 6/7)$. Is B oriented positively or negatively? Make sure to explain your answer.

Exercise 6.1.9 (exer-VC-SA-9). Let S be the portion of the cylinder $r = 2$ between $z = 0$ and $z = \pi/2$ for which $\theta = z$, oriented downward. What is ∂S?

Exercise 6.1.10 (exer-VC-SA-10). What is the evaluation of the vector field $\vec{F}(x, y) = (2, 3)$ on the clockwise oriented unit circle in $\mathbb{R}^2$?

Exercise 6.1.11 (exer-VC-SA-11). What is the evaluation of the function $f(x, y) = x^3$ on the region in $\mathbb{R}^2$ bounded by the curves $y = \cos x$ and $y = x^2$?

Exercise 6.1.12 (exer-VC-SA-12). What is the evaluation of the vector field $\vec{F}(x, y, z) = (z, x, y)$ on the straight line path starting at $(3, 2, 1)$ and ending at $(5, 3, 6)$?

Exercise 6.1.13 (exer-VC-SA-13). What is the evaluation of the vector field $\vec{F}(x, y, z) = (z, x, y)$ on the part of the downward oriented plane $x + y + z = 5$ whose projection to the yz-plane is the rectangle $[0, 2] \times [1, 3]$?

Exercise 6.1.14 (exer-VC-SA-14). What is the evaluation of the function $f(x, y, z) = x^2 y^4 (\arctan(1+y) - \arctan(1-y))$ on the solid unit ball?

6.1. SETUP AND APPROACH

Exercise 6.1.15 (exer-VC-SA-15). Let R represent reflection through the yz-plane in $\mathbb{R}^3$. Given an oriented surface S in $\mathbb{R}^3$, we define the orientation on $R(S)$ to be the one that is "mirror symmetric"; in other words, at corresponding points the unit normal vectors are mirror symmetric. Written precisely,

$$\vec{n}(\vec{x}) = \begin{bmatrix} n_1(\vec{x}) \\ n_2(\vec{x}) \\ n_3(\vec{x}) \end{bmatrix} \qquad \text{with} \qquad \begin{aligned} n_1(R(\vec{x})) &= -n_1(\vec{x}) \\ n_2(R(\vec{x})) &= n_2(\vec{x}) \\ n_3(R(\vec{x})) &= n_3(\vec{x}) \end{aligned}$$

Similarly we define mirror symmetric orientation on reflections of curves in $\mathbb{R}^3$ through the yz-plane (using the unit tangent vectors); and analogously for reflections of curves and surfaces through different coordinate planes in $\mathbb{R}^3$, and for curves in $\mathbb{R}^2$ through those coordinate axes.

For domains whose orientations are defined as either positive or negative (points in $\mathbb{R}^2$ or $\mathbb{R}^3$, solid regions in $\mathbb{R}^2$, and solid regions in $\mathbb{R}^3$), we define the orientation on the reflection of the domain to be the same as that for the original domain.

Let C be a symmetric curve in $\mathbb{R}^3$. Do $\partial(R(C))$ and $R(\partial C)$ have the same orientations?

Exercise 6.1.16 (exer-VC-SA-16). Let S be a symmetric surface in $\mathbb{R}^3$. Do $\partial(R(S))$ and $R(\partial S)$ have the same orientations?

Exercise 6.1.17 (exer-VC-SA-17). Let T be a symmetric solid region in $\mathbb{R}^3$. Do $\partial(R(T))$ and $R(\partial T)$ have the same orientations?

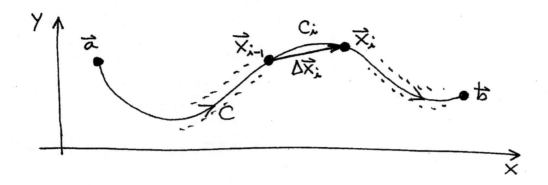

Figure 6.11:

6.2 Fundamental Theorem of Line Integrals

6.2.1 Motivation

Let's apply the approach used in Subsection 6.1.5 in the $\mathbb{R}^2$ context. We begin with an integrand in column zero, and a domain in column one. The column zero integrand is a function, which we will again call f (which we again assume to be continuously differentiable), and the column one domain is an oriented curve, which we will call C. So in our $\mathbb{R}^2$ diagram we have

Vector Calculus in $\mathbb{R}^2$

	0	1	2
Integrands	f		(fns)
Domains		C	(regions)
Evaluations	$h(\vec{x})$	$\int_C \vec{H} \cdot d\vec{x}$	$\iint_R h\, dA$

Again we break up the domain C into subdomains, curves which we call C_i, where $i = 1, \ldots, n$. We will use $\Delta \vec{x}_i$ to refer to the total vector displacement along each of these subdomains. Note of course that unlike Δx used in the fundamental theorem of calculus, $\Delta \vec{x}_i$ is not a constant, but depends on the direction of each piece of curve.

We name the endpoints of those subdomains as indicated in Figure 6.11, giving us

$$\partial C_i = \{\text{``-''}\vec{x}_{i-1}, \text{``+''}\vec{x}_i\} \tag{6.15}$$

Again the boundaries of these subdomains relate to the boundary of the original domain, by

$$\partial C = \sum_i \partial C_i = \partial C_1 + \cdots + \partial C_n \tag{6.16}$$

Just as in the case of an interval in $\mathbb{R}^1$, there is a great amount of cancellation in the interior. Specifically, each interior endpoint $\vec{x}_i$ is the left endpoint of one subdomain and the right endpoint of another subdomain – thus it appears twice in the summation above, once with a negative sign and once with a positive sign, and therefore cancelling. We can write

6.2. FUNDAMENTAL THEOREM OF LINE INTEGRALS

this out explicitly (using parentheses below only for visual grouping) as

$$\begin{aligned}
\sum_i \partial C_i &= \partial C_1 + \cdots + \partial C_n \\
&= \left\{ \left(\text{``-''}\vec{x}_0, \text{``+''}\vec{x}_1\right), \left(\text{``-''}\vec{x}_1, \text{``+''}\vec{x}_2\right), \cdots, \left(\text{``-''}\vec{x}_{n-2}, \text{``+''}\vec{x}_{n-1}\right), \left(\text{``-''}\vec{x}_{n-1}, \text{``+''}\vec{x}_n\right) \right\} \\
&= \left\{ \text{``-''}\vec{x}_0, \left(\text{``+''}\vec{x}_1, \text{``-''}\vec{x}_1\right), \quad \cdots \quad , \left(\text{``+''}\vec{x}_{n-1}, \text{``-''}\vec{x}_{n-1}\right), \text{``+''}\vec{x}_n \right\} \\
&= \left\{ \text{``-''}\vec{x}_0, \text{``+''}\vec{x}_n \right\} \\
&= \left\{ \text{``-''}a, \text{``+''}b \right\} = \partial C
\end{aligned}$$

And again we can come to effectively the same result by noting the analogous cancellation in the evaluation over those domains. That is,

$$\left(f(\vec{x}_1) - f(\vec{x}_0)\right) + \cdots + \left(f(\vec{x}_n) - f(\vec{x}_{n-1})\right) = f(\vec{b}) - f(\vec{a}) \tag{6.17}$$

So just as we did in Subsection 6.1.5, we can apply the evaluation in column zero and use Equation 6.17 to start rewriting that output. Our diagram becomes

	0	1	2
Integrands	f		(fns)
Domains	$\{\text{``+''}\vec{b}, \text{``-''}\vec{a}\} \xleftarrow{\partial \text{ bdry}} C$		(regions)
Evaluations	$h(\vec{x})$	$\int_C \vec{H} \cdot d\vec{x}$	$\iint_R h\, dA$
Outputs	$f(\vec{b}) - f(\vec{a})$		

Using Equation 6.17 we can rewrite this output as

$$f(\vec{b}) - f(\vec{a}) = \sum_{i=1}^n \left(f(\vec{x}_i) - f(\vec{x}_{i-1})\right) \tag{6.18}$$

$$= \lim_{n\to\infty} \sum_{i=1}^n \left(\Delta f_i\right) \tag{6.19}$$

The expression Δf_i represents the amount that the function f changes over C_i, over which the position $\vec{x}$ is displaced by the amount $\Delta \vec{x}_i$. This is approximated by the expression $\nabla f \cdot \Delta \vec{x}_i$. Again we will not present the details here, but it should be plausible that this is sufficiently accurate in the limit, allowing us to rewrite the above expression into the form of the definition of a vector line integral.

$$f(\vec{b}) - f(\vec{a}) = \lim_{n\to\infty} \sum_{i=1}^n \nabla f \cdot \Delta \vec{x}_i \tag{6.20}$$

$$= \int_C \nabla f \cdot d\vec{x} \tag{6.21}$$

We state this as a new theorem, called the fundamental theorem of line integrals (often we will abbreviate this theorem as "FTLI").

Theorem 6.2.1 (Fundamental Theorem of Line Integrals in $\mathbb{R}^2$)**.** *Let* $f : \mathbb{R}^2 \to \mathbb{R}^1$ *be a continuously differentiable function, and C a curve in $\mathbb{R}^2$ from $\vec{a}$ to $\vec{b}$. Then*

$$\int_C \nabla f \cdot d\vec{x} = f(\vec{b}) - f(\vec{a}) \tag{6.22}$$

Looking back at the diagram, we have again been able to write the original column zero evaluation as a column one evaluation. The new integrand for that column one evaluation is ∇f; taking then the gradient operator as a natural choice for an operator that turns column zero integrands into column one integrands, our diagram becomes

	[0]	[1]	[2]
Integrands	$f \xrightarrow{\nabla,\ \text{grad}} \nabla f$		(fns)
Domains	$\{\text{"+"}\vec{b},\text{"-"}\vec{a}\} \xleftarrow{\partial,\ \text{bdry}} C$		(regions)
Evaluations	$h(\vec{x})$	$\int_C \vec{H}\cdot d\vec{x}$	$\iint_R h\, dA$
Outputs	$f(\vec{b}) - f(\vec{a}) =\!=\!= \int_C \nabla f \cdot d\vec{x}$		

Just as with the fundamental theorem of calculus, this addition to the diagram gives us a graceful point of view on the fundamental theorem of line integrals. That is, given an appropriate function f and curve C, applying the indicated operators and evaluating in each column always gives equal results.

Again we can rewrite this as an appendix to the original diagram for calculus in $\mathbb{R}^2$.

Vector Calculus in $\mathbb{R}^2$

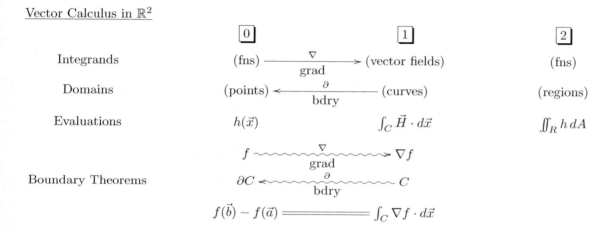

6.2.2 Examples

As with the fundamental theorem of calculus, the fundamental theorem of line integrals is usually used to compute an integral; in this case, a line integal. That is, the domain curve will be given, along with an integrand vector field $\vec{F}$. The function f that corresponds to that vector field, if it exists, very often will not be known initially.

Example 6.2.1. Suppose we want to compute the value of the line integral

$$\int_C \vec{F}\cdot d\vec{x}$$

where C is the upper unit semicircle with the counterclockwise orientation, and $\vec{F} = (2,3)$.

Of course we could parametrize the curve and compute the line integral directly from that parametrization. But the fundamental theorem of line integrals gives us another possible option.

Of course in order to apply that theorem we critically need for the integrand vector field to be the gradient of a function. We will soon see that certainly not all vector fields have this property; but in this case, we are fortunate that our vector field does. Casual inspection reveals that $\vec{F} = \nabla f$ where $f = 2x + 3y$.

6.2. FUNDAMENTAL THEOREM OF LINE INTEGRALS

We can then apply the theorem, giving us

$$\begin{aligned}\int_C \vec{F} \cdot d\vec{x} &= \int_C \nabla f \cdot d\vec{x} \\ &= f(-1, 0) - f(1, 0) \\ &= -4\end{aligned}$$

The previous discussion of the fundamental theorem of line integrals was presented entirely in the context of $\mathbb{R}^2$. But a quick glance through the discussion reveals that nowhere in the argument is there any need for any reference to the dimension. In fact, the exact same argument can be made in $\mathbb{R}^3$, giving us the following similar theorem.

Theorem 6.2.2 (Fundamental Theorem of Line Integrals in $\mathbb{R}^3$). *Let $f : \mathbb{R}^3 \to \mathbb{R}^1$ be a continuously differentiable function, and C a curve in $\mathbb{R}^3$ from $\vec{a}$ to $\vec{b}$. Then*

$$\int_C \nabla f \cdot d\vec{x} = f(\vec{b}) - f(\vec{a}) \tag{6.23}$$

Again this can be appended to the diagram for vector calculus in $\mathbb{R}^3$, giving us

<u>Vector Calculus in $\mathbb{R}^3$</u>

	0	1	2	3
Integrands	(fns) $\xrightarrow{\nabla \text{ grad}}$	(vector fields)	(vector fields)	(fns)
Domains	(points) $\xleftarrow{\partial \text{ bdry}}$	(curves)	(surfaces)	(regions)
Evaluations	$h(\vec{x})$	$\int_C \vec{H} \cdot d\vec{x}$	$\iint_S \vec{H} \cdot d\vec{S}$	$\iiint_R h \, dV$

Boundary Theorems:

$$f \underset{\text{grad}}{\overset{\nabla}{\rightsquigarrow}} \nabla f$$

$$\partial C \underset{\text{bdry}}{\overset{\partial}{\leftsquigarrow}} C$$

$$f(\vec{b}) - f(\vec{a}) = \int_C \nabla f \cdot d\vec{x}$$

Example 6.2.2. Suppose we want to compute the line integral of the vector field $\vec{F} = (4x, 2y, 2z)$ over the curve that is parametrized by $\vec{x}(t) = (e^{t^2 - t}, \cos(\pi t), t^5)$ with $t \in [0, 1]$.

This time we already have the parametrization; and certainly we could plug in and compute the value of this integral directly. But again, the line integral theorem gives us a more convenient alternative.

Again, we need for the vector field $\vec{F}$ to be a gradient, and again we are fortunate to note that we can write $\vec{F} = \nabla f$ with $f = 2x^2 + y^2 + z^2$. The starting and ending points of the curve are $(1, 1, 0)$ and $(1, -1, 1)$, respectively. The fundamental theorem of line integrals then gives us

$$\begin{aligned}\int_C \vec{F} \cdot d\vec{x} &= \int_C \nabla f \cdot d\vec{x} \\ &= f(1, -1, 1) - f(1, 1, 0) \\ &= 4 - 3 = 1\end{aligned}$$

6.2.3 Antigradients

One of the beautiful things about computations done with the fundamental theorem of line integrals is that one can ignore most of the curve. Only the endpoints matter. For example, certainly the curve in Example 6.2.2 is not a recognizable curve, and the parametrization is not particularly convenient; but we were still able to get to the answer by considering only the boundary points. In this way the line integral theorem is very much like the fundamental theorem of calculus.

The catch in the case of the fundamental theorem of line integrals, thought of as a tool for computing line integrals, is that it only applies to certain vector fields – namely, vector fields that are in fact the gradient of some function. As we noted previously this is not the case of all vector fields.

In Examples 6.2.1 and 6.2.2 we were fortunate in that casual inspection allowed us to find functions whose gradients were the given vector fields. We will call these functions "antigradients" of the given vector fields. In the application of the line integral theorem, these antigradients play a role analogous to that played by antiderivatives in the fundamental theorem of calculus.

Despite our good fortune in Examples 6.2.1 and 6.2.2, finding antigradients is usually not easy enough that it can be done by casual inspection. And it should be noted again that many, even most vector fields do not have antigradients at all.

In Section 6.8 we will see that there is a very convenient test for determining if a given vector field has an antigradient or not. For now, we will present a process by which an antigradient can be found if in fact it does exist. We will present the technique in application to a vector field in $\mathbb{R}^3$; the technique is easily adapted to vector fields in $\mathbb{R}^2$.

Suppose we are given a vector field $\vec{F} = (P, Q, R)$, and that it does indeed have an antigradient function f. That is, $\nabla f = \vec{F}$. (Note of course that the antigradient cannot be unique because clearly $f + c$ has the same gradient field as f.) We can rewrite this as

$$\begin{bmatrix} \frac{\partial f}{\partial x} \\ \frac{\partial f}{\partial y} \\ \frac{\partial f}{\partial z} \end{bmatrix} = \begin{bmatrix} P \\ Q \\ R \end{bmatrix} \tag{6.24}$$

which then gives us three component equations

$$\frac{\partial f}{\partial x} = P \qquad \frac{\partial f}{\partial y} = Q \qquad \frac{\partial f}{\partial z} = R \tag{6.25}$$

Each of these equations refers to a partial derivative of f. To each we can apply the corresponding partial antiderivative to solve for f, but critically we must remember in each equation to include the appropriate constants.

$$f = \int P\,dx + c_1(y,z) \qquad f = \int Q\,dy + c_2(x,z) \qquad f = \int R\,dz + c_3(x,y) \tag{6.26}$$

Note that the constants of integration are not actually constants in this case, but rather they are functions. But they are function only of the other variables, so the appropriate partials are still zero. That is,

$$\frac{\partial}{\partial x}c_1(y,z) = 0 \qquad \frac{\partial}{\partial y}c_2(x,z) = 0 \qquad \frac{\partial}{\partial z}c_3(x,y) = 0 \tag{6.27}$$

Given the component function P, Q and R, we can compute the partial antiderivatives, giving us three forms that we know the function f must fit. Very often then the function f can be guessed without too much trouble.

Example 6.2.3. Suppose we want to find the antigradient of the vector field $\vec{F} = (2xy + z^3, x^2 + 2z, 3xz^2 + 2y)$.

Applying the above method, we have

$$f = \int (2xy + z^3)\, dx + c_1(y, z)$$
$$f = \int (x^2 + 2z)\, dy + c_2(x, z)$$
$$f = \int (3xz^2 + 2y)\, dz + c_3(x, y)$$

which becomes

$$f = x^2 y + xz^3 + c_1(y, z)$$
$$f = x^2 y + 2yz + c_2(x, z)$$
$$f = xz^3 + 2yz + c_3(x, y)$$

The desired function f must fit all three of these forms. We notice quickly that the term $x^2 y$ is in both of the first two forms, but it is not explicitly present in the third. We conclude then that it must be a part of the function c_3. Similarly we conclude that xz^3 must be part of c_2 and $2yz$ must be part of c_1.

Given these observations, we note that the function

$$f = x^2 y + xz^3 + 2yz \tag{6.28}$$

satisfies all three forms; taking the gradient of this function reveals that it is indeed an antigradient for $\vec{F}$.

Exercises

Exercise 6.2.1 (exer-VC-FTLI-1). Use the fundamental theorem of line integrals to compute $\int_C \vec{F} \cdot d\vec{x}$, where $\vec{F} = \nabla f$, $f(x, y, z) = x^2 y - e^z$, and C is the first quadrant portion of the unit circle in the yz-plane, oriented counterclockwise as seen from the positive portion of the x-axis. Confirm your result by computing the same line integral directly by parametrizing the curve.

Exercise 6.2.2 (exer-VC-FTLI-2). Compute the line integral $\int_C \vec{F} \cdot d\vec{x}$, where $\vec{F} = (yz, xz, xy + 2z)$ and C is parametrized by $\vec{x}(t) = (t^3 e^t, -t^2, t^2 - t)$ with $t \in [0, 3]$.

Exercise 6.2.3 (exer-VC-FTLI-3). Compute the line integral $\int_C \vec{F} \cdot d\vec{x}$, where $\vec{F} = (y \cos(xy) + z^2, x \cos(xy), 2xz)$ and C is the intersection of $x = y$, $y = z$, and $x^2 + y^2 + z^2 \leq 12$, oriented in the direction with y increasing.

Exercise 6.2.4 (exer-VC-FTLI-4). Compute the line integral $\int_C \vec{F} \cdot d\vec{x}$, where $\vec{F} = (z, x, y)$, and C is the straight line path from $(1, 1, 1)$ to $(2, 4, 3)$.

Exercise 6.2.5 (exer-VC-FTLI-5). In this problem we consider the vector field $\vec{F}(x, y) = (-y, x)$.

1. Compute the line integral of $\vec{F}$ around the curve C that is the unit circle, oriented counterclockwise.

2. Suppose that this vector field were the gradient of some function f, and suppose we consider the curve C as starting at $(1, 0)$ and ending at $(1, 0)$. In terms of values of f, what would the fundamental theorem of line integrals say about the line integral of $\vec{F}$ around the curve C?

3. Is it possible that such a function f might exist?

4. Suppose you need to compute the line integral of this same vector field $\vec{F}$ over some other curve; might you consider trying to use the fundamental theorem of line integrals in your computation? Why or why not?

Exercise 6.2.6 (exer-VC-FTLI-6). Compute the line integral $\int_C \vec{F} \cdot d\vec{x}$, where $\vec{F} = (2xy, x^2 + 2yz, y^2)$, and C is the piecewise straight path that starts at the origin, then goes in succession to the points $(2, 3, 4)$, $(7, 1, 3)$, $(4, 5, 2)$, $(-2, 5, 4)$, and then back to the origin.

Exercise 6.2.7 (exer-VC-FTLI-7). Compute the line integral $\int_C \vec{F} \cdot d\vec{x}$, where $\vec{F} = (2xy, x^2 + 2yz, y^2)$, and C is the boundary of the upward oriented surface that is the intersection of $z = e^{-x^2}$ with the unit ball.

Exercise 6.2.8 (exer-VC-FTLI-8). Compute the line integral $\int_C \vec{F} \cdot d\vec{x}$, where $\vec{F} = (x - y, y - z, x - z)$, and C is the curve in the plane $6x - 2y + 2z = 10$ whose projection to the xy-plane is the unit circle oriented counterclockwise.

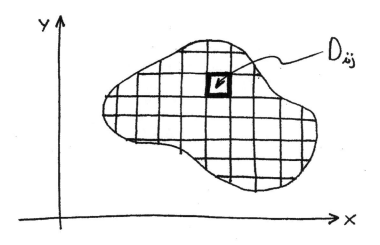

Figure 6.12:

6.3 Green's Theorem

In Section 6.2, we saw a development of a new theorem, the fundamental theorem of line integrals, by applying in a new context ideas taken from a development of the fundamental theorem of calculus. Both of these theorems are boundary theorems, relating an evaluation on a set (a curve or an interval, respectively) to an evaluation on the boundary (a pair of points). In each of these cases the initial set was one-dimensional.

In this section we will do something similar, and will end up with a new boundary theorem. But in this case the dimensions involved will not be the same. That is, rather than beginning with a one-dimensional set, we will begin with a two dimensional set. In this section we consider only two-dimensional domains in $\mathbb{R}^2$.

6.3.1 Motivation

In this case we begin with an integrand in column one, and a domain in column two. The column one integrand is a vector field, which we will call $\vec{F} = (P, Q)$ (which we again assume to be continuously differentiable), and the column two domain is an oriented region in $\mathbb{R}^2$, which we will call D. So in our $\mathbb{R}^2$ diagram we have

Vector Calculus in $\mathbb{R}^2$

	$\boxed{0}$	$\boxed{1}$	$\boxed{2}$
Integrands	(fns)	$\vec{F}$	
Domains	(points)		D
Evaluations	$h(\vec{x})$	$\int_C \vec{H} \cdot d\vec{x}$	$\iint_R h \, dA$

Again we break up the domain D into subdomains, in the same way that we did when setting up the double integral in Section 5.2. (Of course the domain D need not be a rectangle, so not all of the subdomains are rectangles; but this turns out not to be a problem.)

We will refer to the individual subdomains as D_{ij}, where $i, j = 1, \ldots, n$. We will use ΔA_{ij} to refer to the area of each of these subdomains. See Figure 6.12.

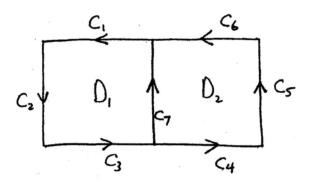

Figure 6.13:

We will soon see that, again, the boundaries of these subdomains relate to the boundary of the original domain, by

$$\partial D = \sum_{i,j} \partial D_{ij} \qquad (6.29)$$

While the picture is very different than in the previous cases we have considered, again this equation works because of cancellation of the pieces in the interior.

To see this, let's first consider the sum of the boundaries of only two subdomains, which are contiguous. In order to be able to write down the algebra, we will label and assign arbitrary orientations to the individual boundary curve segments, in Figure 6.13. (Note that these chosen orientations are arbitrary designations.)

We can write the boundaries of these two subdomains as

$$\partial D_1 = \left(C_1 + C_2 + C_3 + C_7\right)$$
$$\partial D_2 = \left(C_4 + C_5 + C_6 - C_7\right)$$

We must pay very careful attention to the signs in the above summation. Remember that by convention the boundary of a rectangular region in $\mathbb{R}^2$ goes counterclockwise around the rectangle. Conveniently the segments making up ∂D_1 are all oriented in the way that is counterclockwise around D_1, so all of those signs are positive. But this is not so for the segments making up ∂D_2 – the segment C_7 is oriented in a way that points *clockwise* around D_2. Of course then $-C_7$ points counterclockwise around D_2, so we must make sure to include this negative sign when writing down that boundary.

Adding these together, we get

$$\begin{aligned} \partial D_1 + \partial D_2 &= \left(C_1 + C_2 + C_3 + C_7\right) + \left(C_4 + C_5 + C_6 - C_7\right) \\ &= C_1 + C_2 + C_3 + C_4 + C_5 + C_6 \\ &= \partial\left(D_1 \cup D_2\right) \end{aligned}$$

(Note, this calculation would have arrived at the same conclusion if the orientations of the boundary segments had been assigned differently, even though the signs in the intervening algebra would have been different.)

Alternatively we could represent the above cancellation as in Figure 6.14. Here we do not label any of the boundary segments, and we do not write down any of the associated algebra as above. Instead we use arrows to represent the orientations of each segment, thought of as part of the boundary of a respective subdomain. With this representation we see that the middle piece (C_7 from Figure 6.13) appears twice, but with opposite orientations, therefore cancelling in the summation – just like they did in the algebra above.

Representing the boundaries in this way, we can now consider what happens when we consider all of the subdomains of D taken together. There are many boundary segments for all of these subdomains, some of which are part of the boundary of D, and most of which are not. See Figure 6.15

6.3. GREEN'S THEOREM

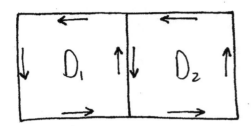

Figure 6.14:

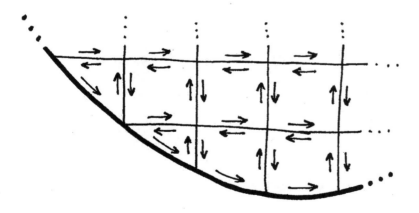

Figure 6.15:

In a wonderful piece of good fortune, any of these boundary segments which is not part of the boundary of D must be part of the boundary of exactly two subdomains – and with opposite orientations as part of each. Therefore, in the summation, those interior segments all cancel, leaving us with the result claimed in Equation 6.29.

(Note that even though the picture is very different from those for the fundamental theorem of calculus and the fundamental theorem of line integrals, there is still a very strong analogy in what happened in this part of the process. Again, the boundary of the entire domain is the sum of the boundaries of the subdomains. Again this happens because the interior pieces cancel. And again, the cancellation happens because of the conventions on the way we assign orientations on boundaries.)

Having motivated Equation 6.29, we can continue on with our template by considering that boundary of D, applying the appropriate evaluation in column one, and then using Equation 6.29 to begin rewriting that output. Our diagram becomes

	$\boxed{0}$	$\boxed{1}$	$\boxed{2}$
Integrands	(fns)	$\vec{F}$	
Domains	(points)	$\partial D \underset{\text{bdry}}{\overset{\partial}{\longleftarrow}} D$	
Evaluations	$h(\vec{x})$	$\int_C \vec{H} \cdot d\vec{x}$	$\iint_R h\, dA$
Outputs		$\int_{\partial D} \vec{F} \cdot d\vec{x}$	

Using Equation 6.29 as suggested, we rewrite the output in column one as

$$\int_{\partial D} \vec{F} \cdot d\vec{x} = \sum_{i,j} \left(\int_{\partial D_{ij}} \vec{F} \cdot d\vec{x} \right) \tag{6.30}$$

$$= \lim_{n \to \infty} \sum_{i,j} \left(\int_{\partial D_{ij}} \vec{F} \cdot d\vec{x} \right) \tag{6.31}$$

Analogously to what we did at this point in our discussion of the fundamental theorem of calculus, we divide and multiply by ΔA_{ij}, giving us

$$\int_{\partial D} \vec{F} \cdot d\vec{x} = \lim_{n \to \infty} \sum_{i,j} \left(\frac{\int_{\partial D_{ij}} \vec{F} \cdot d\vec{x}}{\Delta A_{ij}} \right) \Delta A_{ij} \tag{6.32}$$

The good news here is that the right side of Equation 6.32 appears to be exactly a double integral over the domain D. So, continuing the analogy with the development of the previous theorems, we have turned a column one evaluation into a column two evaluation. Again skipping the details, we claim plausibly that this can be rewritten as

$$\int_{\partial D} \vec{F} \cdot d\vec{x} = \iint_D \left(\lim_{n \to \infty} \frac{\int_{\partial D_{ij}} \vec{F} \cdot d\vec{x}}{\Delta A_{ij}} \right) dA \tag{6.33}$$

Unlike in the development of the previous two theorems, the expression in the integrand this time is not something that we recognize from previous experience. Still, the expression can be simplified, though we will not show those details here. Instead, we simply state the result, that

$$\lim_{n \to \infty} \frac{\int_{\partial D_{ij}} \vec{F} \cdot d\vec{x}}{\Delta A_{ij}} = \frac{\partial Q}{\partial x} - \frac{\partial P}{\partial y} \tag{6.34}$$

We can then rewrite Equation 6.33 as

$$\int_{\partial D} \vec{F} \cdot d\vec{x} = \iint_D \left(\frac{\partial Q}{\partial x} - \frac{\partial P}{\partial y} \right) dA \tag{6.35}$$

6.3. GREEN'S THEOREM

Having arrived now at a workable double integral, which is the evaluation we have associated to column two, we can use this to motivate a natural choice for an operator that turns column one integrands into column two integrands. We name this operator the "Green operator", which we abbreviate as "grn", defined by

$$\mathrm{grn}\vec{F} = \frac{\partial Q}{\partial x} - \frac{\partial P}{\partial y} \tag{6.36}$$

With this choice, our diagram becomes

	⓪	①	②
Integrands	(fns)	$\vec{F} \xrightarrow{\mathrm{grn}} \mathrm{grn}\vec{F}$	
Domains	(points)	$\partial D \xleftarrow{\partial}{\mathrm{bdry}} D$	
Evaluations	$h(\vec{x})$	$\int_C \vec{H} \cdot d\vec{x}$	$\iint_R h\, dA$
Outputs		$\int_{\partial D} \vec{F} \cdot d\vec{x} =\!=\!= \iint_D \mathrm{grn}\vec{F}\, dA$	

We now state our result as a theorem, called Green's theorem.

Theorem 6.3.1 (Green's Theorem). *Let $\vec{F}$ be a continuously differentiable vector field, and D a domain in $\mathbb{R}^2$. Then*

$$\begin{aligned}\int_{\partial D} \vec{F} \cdot d\vec{x} &= \iint_D \left(\frac{\partial Q}{\partial x} - \frac{\partial P}{\partial y}\right) dA \\ &= \iint_D \mathrm{grn}\vec{F}\, dA\end{aligned}$$

Just as with the previous boundary theorems, the diagram gives us a graceful point of view on Green's theorem. That is, given an appropriate vector field $\vec{F}$ and domain D in $\mathbb{R}^2$, applying the indicated operators and evaluating in each column always gives the same results.

Again we can rewrite this as an appendix to the original diagram for calculus in $\mathbb{R}^2$.

<u>Vector Calculus in $\mathbb{R}^2$</u>

	⓪	①	②
Integrands	(fns)	(vector fields) $\xrightarrow{\mathrm{grn}}$	(fns)
Domains	(points)	(curves) $\xleftarrow{\partial}{\mathrm{bdry}}$	(regions)
Evaluations	$h(\vec{x})$	$\int_C \vec{H} \cdot d\vec{x}$	$\iint_R h\, dA$
Boundary Theorems		$\vec{F} \xrightarrow{\mathrm{grn}} \mathrm{grn}\vec{F}$ $\partial D \xleftarrow{\partial}{\mathrm{bdry}} D$ $\int_{\partial D} \vec{F} \cdot d\vec{x} =\!=\!= \iint_D \mathrm{grn}\vec{F}\, dA$	

6.3.2 Examples

As was previously noted, the fundamental theorem of calculus and the fundamental theorem of line integrals are usually used to evaluate integrals. That is, they are used to reduce dimension, from a column one evaluation to a column zero evaluation.

We will soon see that there are cases when Green's theorem is used to reduce dimension as well. But the most immediate applications of Green's theorem involve going the other way, increasing the dimension. That is, we will use the theorem to compute a line integral around a boundary curve by turning it into a double integral over the interior.

Of course in order to be able to do this, it is critical that the curve domain for the line integral actually be a boundary of some region – and certainly not all curves have this property. This is something that must be checked before attempting to apply the theorem in this way.

Example 6.3.1. Suppose we wish to compute the line integral of the vector field $\vec{F} = (e^x + y, \arctan y + 3x)$ over the curve C that traces out the unit circle going counterclockwise.

Rather than parametrizing the curve, we note that this curve is in fact exactly the boundary of a solid region in $\mathbb{R}^2$, namely, the unit disk D. Thus we can apply Green's theorem, which tells us

$$\begin{aligned}
\int_C \vec{F} \cdot d\vec{x} &= \int_{\partial D} \vec{F} \cdot d\vec{x} \\
&= \iint_D \operatorname{grn} \vec{F}\, dA \\
&= \iint_D \left(\frac{\partial}{\partial x} (\arctan y + 3x) - \frac{\partial}{\partial y} (e^x + y) \right) dA \\
&= \iint_D (3 - 1)\, dA \\
&= 2 \iint_D dA \\
&= 2\pi
\end{aligned}$$

Example 6.3.2. Suppose we wish to compute the line integral of the vector field $\vec{F} = (3x + 4y, 5x - 7y)$ over the curve C that goes in straight line segments from the origin, to $(2, 2)$, to $(2, 0)$, to $(0, 2)$, and then back to the origin.

The curve starts and ends at the same location, but it crosses over itself, so it is not immediately clear that it is a boundary. But as can be seen in Figure 6.16, in fact this curve is exactly the boundary of the difference $D_1 - D_2$ of the regions indicated in that figure – because it goes counterclockwise around D_1, and clockwise around D_2.

So we can apply Green's theorem. First noting that

$$\operatorname{grn} \vec{F} = \frac{\partial}{\partial x} (5x - 7y) - \frac{\partial}{\partial y} (3x + 4y) = 1 \tag{6.37}$$

the computation becomes

$$\begin{aligned}
\int_C \vec{F} \cdot d\vec{x} &= \int_{\partial(D_1 - D_2)} \vec{F} \cdot d\vec{x} \\
&= \iint_{D_1 - D_2} \operatorname{grn} \vec{F}\, dA \\
&= \iint_{D_1} 1\, dA - \iint_{D_2} 1\, dA \\
&= 0
\end{aligned}$$

The last step is a consequence of the fact that the two regions D_1 and D_2 have the same area.

6.3. GREEN'S THEOREM

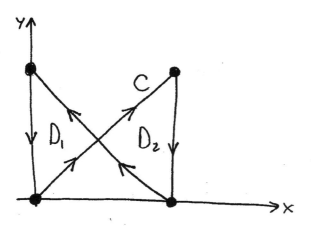

Figure 6.16:

Example 6.3.3. In Example 6.3.1, we were very fortunate that some of the more complicated terms in the vector field disappeared completely when we applied Green's theorem. We can see why this happened by looking explicitly at the formula for the Green operator – specifically, the components P and Q of $\vec{F} = (P, Q)$ appear in the expression for $\mathrm{grn}\vec{F}$ only as part of specific partial derivatives. So, any part of one of those component functions that does not involve the corresponding coordinate variable will disappear completely.

The following are generalizations of this observation.

$$\int_{\partial D} \begin{bmatrix} P(x) \\ 0 \end{bmatrix} \cdot d\vec{x} = \int_{\partial D} P(x)\, dx = 0$$

$$\int_{\partial D} \begin{bmatrix} 0 \\ Q(y) \end{bmatrix} \cdot d\vec{x} = \int_{\partial D} Q(y)\, dy = 0$$

Again, note that these results apply only when the curve in question is a boundary.

6.3.3 Computing area

All of the examples in Subsection 6.3.2 show how to use Green's theorem to turn a line integral over a boundary curve into a double integral over the interior region. Sometimes however Green's theorem can be thought of in the other direction.

For example, suppose we want to compute the area of some region D. That area can be interpreted as a double integral, with integrand 1. Of course Green's theorem involves an integral, with integrand $\mathrm{grn}\vec{F}$. If we can find a vector field $\vec{F}$ with $\mathrm{grn}\vec{F} = 1$, then these integrands will be the same.

Fortunately, there are many such vector fields. The most convenient one is $\vec{F} = (0, x)$.

This gives us the following theorem as an immediate consequence of Green's theorem.

Theorem 6.3.2. *For a domain D in $\mathbb{R}^2$ with boundary curve ∂D, the area of D can be computed as*

$$area = \int_{\partial D} \begin{bmatrix} 0 \\ x \end{bmatrix} \cdot d\vec{x} = \int_{\partial D} x\, dy \tag{6.38}$$

As previously suggested, instead of using a double integral to compute a line integral, here Green's theorem allows us to use a line integral to compute a double integral (specifically, one representing area).

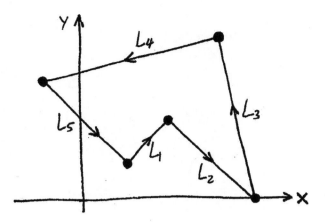

Figure 6.17:

In some cases, the line integrals that come out of Theorem 6.3.2 can be computed by this next result.

Theorem 6.3.3. *If the curve L is a straight line segment from the point (x_1, y_1) to the point (x_2, y_2), then*

$$\int_L \begin{bmatrix} 0 \\ x \end{bmatrix} \cdot d\vec{x} = \int_L x\, dy = \left(\frac{x_1 + x_2}{2}\right)(y_2 - y_1) = \bar{x}\Delta y \tag{6.39}$$

In Example 6.3.4 we show how to use these two results together to compute an area.

Example 6.3.4. The polygon P has vertices (going counterclockwise around the boundary) $(1,1)$, $(2,2)$, $(4,0)$, $(3,4)$, $(-1,3,)$. What is the area of this polygon?

As can be seen from Figure 6.17, it would be tedious to compute this area directly.

Instead, we can use Theorem 6.3.2 to rewrite the area as

$$\begin{aligned}
\text{area} &= \int_{\partial P} x\, dy \\
&= \int_{L_1} x\, dy + \int_{L_2} x\, dy + \int_{L_3} x\, dy + \int_{L_4} x\, dy + \int_{L_5} x\, dy \\
&= \bar{x}_1 \Delta y_1 + \bar{x}_2 \Delta y_2 + \bar{x}_3 \Delta y_3 + \bar{x}_4 \Delta y_4 + \bar{x}_5 \Delta y_5 \\
&= \left(\frac{1+2}{2}\right)(2-1) + \left(\frac{2+4}{2}\right)(0-2) + \left(\frac{4+3}{2}\right)(4-0) + \left(\frac{3+(-1)}{2}\right)(3-4) + \left(\frac{(-1)+1}{2}\right)(1-3) \\
&= \frac{17}{2}
\end{aligned}$$

Exercises

Exercise 6.3.1 (exer-VC-Green-1). Let D_1, D_2, D_3, and D_4 be the parts of $[-1, 1] \times [-1, 1]$ in the first, second, third, and fourth quadrants, respectively. We label the edges of these four squares $E_1, \ldots, E_{12}$ in a left to right, top to bottom fashion (so that E_1 and E_2 are the two horizontal edges on $y = 1$). All horizontal edges are oriented to the right, and all vertical edges are oriented upward.

6.3. GREEN'S THEOREM

1. Write out explicit formulas for the boundaries of D_1, D_2, D_3, and D_4 in terms of $E_1, \ldots, E_{12}$.

2. Confirm directly that $\partial(D_1 \cup D_2) = \partial D_1 + \partial D_2$.

3. Confirm directly that $\partial(D_1 \cup D_4) = \partial D_1 + \partial D_4$.

4. Confirm directly that $\partial\left(\bigcup_{i=1}^{4} D_i\right) = \sum_{i=1}^{4} \partial D_i$.

Exercise 6.3.2 (<u>exer-VC-Green-2</u>). Slicing horizontally and vertically at the integers divides the disk of radius 2 centered at the origin into 16 pieces (only some of which are squares). Label these as $D_1, \ldots, D_{16}$ in a left to right, top to bottom fashion. There are 24 interior edges, to be labeled $E_1, \ldots, E_{24}$ in a left to right, top to bottom fashion (all horizontal edges oriented to the right and all vertical edges oriented upward); and there are 12 boundary edges, to be labeled $E_{25}, \ldots, E_{36}$ starting with the edge just above $(2,0)$ and going counterclockwise (all oriented counterclockwise).

Confirm directly that $\partial\left(\bigcup_{i=1}^{16} D_i\right) = \sum_{i=1}^{16} \partial D_i$

Exercise 6.3.3 (<u>exer-VC-Green-3</u>). Compute the line integral $\int_C \vec{F} \cdot d\vec{x}$, where $\vec{F}(x,y) = (2xy - y^2, y^2 - x^2)$ and C is the boundary of $[2,4] \times [1,3]$.

Exercise 6.3.4 (<u>exer-VC-Green-4</u>). Compute the line integral $\int_C \vec{F} \cdot d\vec{x}$, where $\vec{F}(x,y) = (x\cos(x^2) - 3y, 4y - 2x - e^{y^2})$ and C is the curve parametrized by $\vec{x}(t) = (2\cos t, 4\sin t)$, with $t \in [0, 2\pi]$.

Exercise 6.3.5 (<u>exer-VC-Green-5</u>). Compute the line integral $\int_C \vec{F} \cdot d\vec{x}$, where $\vec{F}(x,y) = (2x - 3y, 4x - 7y)$ and C is the part of the unit circle with $y \leq 0$, oriented counterclockwise.

Exercise 6.3.6 (<u>exer-VC-Green-6</u>). Use Green's theorem, with a clever choice of vector field $\vec{F}$, to show that for any continuous function f and any path C that is the boundary of some region, we have $\int_C f(x)\, dx = 0$. (Recall that this result was suggested by Exercise 5.13.6.)

Exercise 6.3.7 (<u>exer-VC-Green-7</u>). Use Green's theorem to show that for any continuous function f and any path C that is the boundary of some region, we have $\int_C f(y)\, dy = 0$.

Exercise 6.3.8 (<u>exer-VC-Green-8</u>). Use Green's theorem and a parametrization of the unit circle to confirm that the area of the unit disk is equal to π.

Exercise 6.3.9 (<u>exer-VC-Green-9</u>). Use Green's theorem and a parametrization of the ellipse with equation $\frac{x^2}{a^2} + \frac{y^2}{b^2} = 1$ to find a formula for the area inside this ellipse. *(Hint: An easy parametrization comes from viewing this ellipse as the result of stretching the unit circle, for which there is an easy parametrization.)*

Exercise 6.3.10 (<u>exer-VC-Green-10</u>). Compute the area of the polygon whose boundary has vertices (in counterclockwise order) at $(0,0)$, $(-1,1)$, $(1,2)$, $(-1,-3)$, $(-5,-1)$, $(-6,-2)$, $(1,-7)$, $(5,1)$, $(0,4)$, $(-3,-2)$, and then back to the origin.

Exercise 6.3.11 (<u>exer-VC-Green-11</u>). Prove Theorem 6.3.3 by direct computation of the line integral from a parametrization of the line segment.

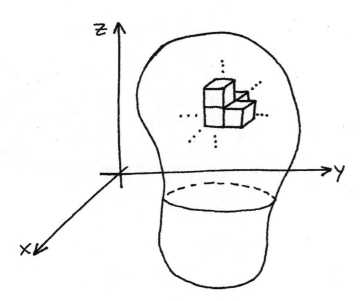

Figure 6.18:

6.4 Gauss's Divergence Theorem

Again, here we will apply in a different context the same ideas that have led up to the other boundary theorems. The context is $\mathbb{R}^3$, and we consider 3-dimensional solid domains.

Students will note that the development we give here leading up to Gauss's divergence theorem is particularly similar to that for Green's theorem.

6.4.1 Motivation

We begin with an integrand in column two, and a domain in column three. The column two integrand is a vector field, which we will call $\vec{F} = (P, Q, R)$ (which we again assume to be continuously differentiable), and the column three domain is an oriented region in $\mathbb{R}^3$, which we will call D. So in our $\mathbb{R}^3$ diagram we have

<u>Vector Calculus in $\mathbb{R}^3$</u>

	$\boxed{0}$	$\boxed{1}$	$\boxed{2}$	$\boxed{3}$
Integrands	(fns)	(vector fields)	$\vec{F}$	
Domains	(points)	(curves)		D
Evaluations	$h(\vec{x})$	$\int_C \vec{H} \cdot d\vec{x}$	$\iint_S \vec{H} \cdot d\vec{S}$	$\iiint_R h\, dV$

Again we break up the domain D into subdomains, this time in the same way that we did when setting up the triple integral in Section 5.5. (Of course the domain D need not be rectangular, so not all of the subdomains are rectanglar; but this turns out not to be a problem.)

We will refer to the individual subdomains as D_{ijk}, where $i, j, k = 1, \ldots, n$. We will use ΔV_{ijk} to refer to the volume of each of these subdomains. See Figure 6.18.

6.4. GAUSS'S DIVERGENCE THEOREM

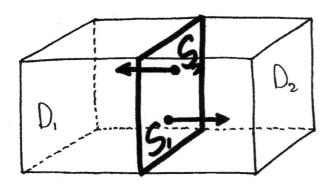

Figure 6.19:

We will soon see that, again, the boundaries of these subdomains relate to the boundary of the original domain, by

$$\partial D = \sum_{i,j,k} \partial D_{ijk} \qquad (6.40)$$

Again this equation works because of cancellation of the pieces in the interior. The picture justifying this is different, but actually still very similar to the picture used in the corresponding discussion of Green's theorem.

Again, the surface pieces in the summation in Equation 6.40 fall into two categories. Some are part of the boundary of D, contributing to the total of ∂D.

All of the remaining pieces of surface, in being in the interior of D, must be part of the boundary of exactly two of the subdomains D_{ijk}, and thus must appear exactly twice in the summation. Furthermore, the orientations of those two instances of that piece of surface must be opposite, and so they cancel.

For example, in Figure 6.19, the rectangular surface between the subdomains labeled D_1 and D_2 appears in two instances in the summation – as part of the boundaries of each of those subdomains. As part of the boundary of D_1 we label it S_1, and as part of the boundary of D_2 we label it S_2. By convention, the orientation on S_1 points outward from D_1, which is to the right in the figure. But by the same convention, the orientation on S_2 points outward from D_2, which is to the left in the figure. So while S_1 and S_2 cover the same set of points in space, they have opposite orientations, and so we have $S_1 + S_2 = 0$.

Since all of the interior pieces of surface in the summation in Equation 6.40 cancel, the boundary pieces are all that remain. This confirms that equation. (Students should note a very strong analogy between this calculation and the corresponding calculation in the discussion of Green's theorem.)

Having motivated Equation 6.40, we can continue on with our template by considering that boundary of D, applying the appropriate evaluation in column two, and then using Equation 6.40 to begin rewriting that output. Our diagram becomes

	0	1	2	3
Integrands	(fns)	(vector fields)	$\vec{F}$	
Domains	(points)	(curves)	$\partial D \xleftarrow{\partial}_{\text{bdry}} D$	
Evaluations	$h(\vec{x})$	$\int_C \vec{H} \cdot d\vec{x}$	$\iint_S \vec{H} \cdot d\vec{S}$	$\iiint_R h\, dV$
Outputs			$\iint_{\partial D} \vec{F} \cdot d\vec{S}$	

Using Equation 6.40 as suggested, we rewrite the output in column two as

$$\iint_{\partial D} \vec{F} \cdot d\vec{S} = \sum_{i,j,k} \left(\iint_{\partial D_{ijk}} \vec{F} \cdot d\vec{S} \right) \tag{6.41}$$

$$= \lim_{n \to \infty} \sum_{i,j,k} \left(\iint_{\partial D_{ijk}} \vec{F} \cdot d\vec{S} \right) \tag{6.42}$$

Analogously to what we did at this point in our discussion of Green's theorem, we divide and multiply by ΔV_{ijk}, giving us

$$\iint_{\partial D} \vec{F} \cdot d\vec{S} = \lim_{n \to \infty} \sum_{i,j,k} \left(\frac{\iint_{\partial D_{ijk}} \vec{F} \cdot d\vec{S}}{\Delta V_{ijk}} \right) \Delta V_{ijk} \tag{6.43}$$

The good news here is that the right side of Equation 6.43 appears to be exactly a triple integral over the domain D. So, continuing the analogy with the development of the previous theorems, we note that we have turned a column two evaluation into a column three evaluation. Again skipping the details, we claim plausibly that this can be rewritten as

$$\iint_{\partial D} \vec{F} \cdot d\vec{S} = \iiint_D \left(\lim_{n \to \infty} \frac{\iint_{\partial D_{ijk}} \vec{F} \cdot d\vec{S}}{\Delta V_{ijk}} \right) dV \tag{6.44}$$

As in the development of Green's theorem, the expression in the integrand is not something that we recognize from previous experience. Still, the expression can be simplified, though we will not show those details here. Instead, we simply state the result, that

$$\lim_{n \to \infty} \frac{\iint_{\partial D_{ijk}} \vec{F} \cdot d\vec{S}}{\Delta V_{ijk}} = \frac{\partial P}{\partial x} + \frac{\partial Q}{\partial y} + \frac{\partial R}{\partial z} \tag{6.45}$$

We can then rewrite Equation 6.44 as

$$\iint_{\partial D} \vec{F} \cdot d\vec{S} = \iiint_D \left(\frac{\partial P}{\partial x} + \frac{\partial Q}{\partial y} + \frac{\partial R}{\partial z} \right) dV \tag{6.46}$$

Having arrived now at a workable triple integral, which is the evaluation we have associated to column three, we can use this to motivate a natural choice for an operator that turns column two integrands into column three integrands. We name this operator the "divergence" operator, which we abbreviate as "div", defined by

$$\text{div}\vec{F} = \frac{\partial P}{\partial x} + \frac{\partial Q}{\partial y} + \frac{\partial R}{\partial z} \tag{6.47}$$

(An alternative and very common notation for this operator involves the "∇" symbol used with gradients. Since one might write

$$\nabla f = \begin{bmatrix} \frac{\partial f}{\partial x} \\ \frac{\partial f}{\partial y} \\ \frac{\partial f}{\partial z} \end{bmatrix} = \begin{bmatrix} \frac{\partial}{\partial x} \\ \frac{\partial}{\partial y} \\ \frac{\partial}{\partial z} \end{bmatrix} f$$

it is common to write

$$\nabla = \begin{bmatrix} \frac{\partial}{\partial x} \\ \frac{\partial}{\partial y} \\ \frac{\partial}{\partial z} \end{bmatrix}$$

6.4. GAUSS'S DIVERGENCE THEOREM

which then suggests that we might rewrite divergence as

$$\begin{aligned}
\text{div}\vec{F} &= \frac{\partial P}{\partial x} + \frac{\partial Q}{\partial y} + \frac{\partial R}{\partial z} \\
&= \frac{\partial}{\partial x}P + \frac{\partial}{\partial y}Q + \frac{\partial}{\partial z}R \\
&= \begin{bmatrix} \frac{\partial}{\partial x} \\ \frac{\partial}{\partial y} \\ \frac{\partial}{\partial z} \end{bmatrix} \cdot \begin{bmatrix} P \\ Q \\ R \end{bmatrix} \\
&= \nabla \cdot \vec{F}
\end{aligned}$$

So very often "div$\vec{F}$" is instead written as "$\nabla \cdot \vec{F}$".)

With this choice, our diagram becomes

	[0]	[1]	[2]	[3]
Integrands	(fns)	(vector fields)	$\vec{F} \xrightarrow[\text{div}]{\nabla \cdot} \nabla \cdot \vec{F}$	
Domains	(points)	(curves)	$\partial D \xleftarrow[\text{bdry}]{\partial} D$	
Evaluations	$h(\vec{x})$	$\int_C \vec{H} \cdot d\vec{x}$	$\iint_S \vec{H} \cdot d\vec{S}$	$\iiint_R h\, dV$
Outputs			$\iint_{\partial D} \vec{F} \cdot d\vec{S} =\!\!= \iiint_D \nabla \cdot \vec{F}\, dV$	

We now state our result as a theorem, called Gauss's divergence theorem (or sometimes simply "the divergence theorem" or "Gauss's theorem").

Theorem 6.4.1 (Gauss's Divergence Theorem). *Let $\vec{F}$ be a continuously differentiable vector field, and D a domain in $\mathbb{R}^3$. Then*

$$\iint_{\partial D} \vec{F} \cdot d\vec{S} = \iiint_D \nabla \cdot \vec{F}\, dV$$

Just as with the previous boundary theorems, the diagram gives us a graceful point of view on Gauss's divergence theorem. That is, given an appropriate vector field $\vec{F}$ in column two and domain D in $\mathbb{R}^3$, applying the indicated operators and evaluating in each column always gives the same results.

Again we can rewrite this as an appendix to the original diagram for calculus in $\mathbb{R}^3$.

<u>Vector Calculus in $\mathbb{R}^3$</u>

	[0]	[1]	[2]	[3]
Integrands	(fns)	(vector fields)	(vector fields) $\xrightarrow[\text{div}]{\nabla \cdot}$	(fns)
Domains	(points)	(curves)	(surfaces) $\xleftarrow[\text{bdry}]{\partial}$	(regions)
Evaluations	$h(\vec{x})$	$\int_C \vec{H} \cdot d\vec{x}$	$\iint_S \vec{H} \cdot d\vec{S}$	$\iiint_R h\, dV$
Boundary Theorems			$\vec{F} \xrightarrow[\text{div}]{\nabla \cdot} \nabla \cdot \vec{F}$ $\partial D \xleftarrow[\text{bdry}]{\partial} D$ $\iint_{\partial D} \vec{F} \cdot d\vec{S} =\!\!= \iiint_D \nabla \cdot \vec{F}\, dV$	

As was the case for Green's theorem, the most immediate applications of Gauss's divergence theorem involve increasing the dimension. That is, we will use it to convert a flux integral into a triple integral.

Again, in order for this to be possible it is critical that the surface domain for the flux integral actually be a boundary of a solid region – and again, certainly not every surface has this property. This must be confirmed before attempting to apply the theorem in this way.

Example 6.4.1. **Suppose we wish to compute the flux of the vector field** $\vec{F} = (3x + y^2, e^x, y - z)$ **through the unit sphere** S **in** $\mathbb{R}^3$ **oriented outward.**

Of course this could be computed directly by parametrizing the sphere. On the other hand, since the sphere is exactly the boundary of the solid unit ball B**, we can instead apply Gauss's divergence theorem.**

$$\begin{aligned}
\iint_S \vec{F} \cdot d\vec{S} &= \iiint_B \nabla \cdot \vec{F} \, dV \\
&= \iiint_B \frac{\partial}{\partial x}(3x + y^2) + \frac{\partial}{\partial y}(e^x) + \frac{\partial}{\partial z}(y - z) \, dV \\
&= \iiint_B 2 \, dV \\
&= 2\left(\frac{4\pi}{3}\right) = \frac{8\pi}{3}
\end{aligned}$$

6.4.2 Flux density

We will soon see applications of Gauss's divergence theorem to physics, in a way that is strongly suggestive of another interpretation of divergence, and of flux itself. This interpretation can already be taken based on the motivation we gave for this theorem.

In Section 5.12, the definition of a flux integral indicates that it is computed as a sum that is taken over the surface domain for that integral. So, very reasonably, we have been thinking of flux as something that happens on the surface itself. Of course this is the most important point of view on flux.

However, in the case of a surface that is the boundary of a solid region, we can take a different point of view in light of Equation 6.41. That equation suggests that we can think of flux (the left side of the equation) as a sum of fluxes computed in the interior region, and not only on the boundary.

Said differently, "flux through the boundary" can be thought of as something that accumulates over the interiors of regions, in the same way that mass accumulates. Specifically, the mass in a given region is certainly the sum of the masses in its subregions; similarly, we have just noted that Equation 6.41 states the "flux through the boundary" of a region is the sum of the "fluxes through the boundary" in its subregions.

This suggests we can think of flux itself in a way that is similar to mass, as a type of "stuff" that occupies space, and not just as something computed on a surface.

We can apply this idea to Gauss's divergence theorem. Writing the flux through the boundary as Φ, the theorem states

$$\Phi = \iint_{\partial R} \vec{F} \cdot d\vec{S} = \iiint_R \nabla \cdot \vec{F} \, dV \tag{6.48}$$

Now thinking of flux as an accumulation over the region R, we could write

$$\Phi = \iiint_R d\Phi \tag{6.49}$$

which leaves us with

$$d\Phi = \nabla \cdot \vec{F} \, dV \tag{6.50}$$

6.4. GAUSS'S DIVERGENCE THEOREM

Of course a similar consideration of mass gives us

$$\iiint_R dm = m = \iiint_R \delta\, dV \tag{6.51}$$

and thus

$$dm = \delta\, dV \tag{6.52}$$

where δ represents "mass density" – the amount of mass per unit volume in a given region of space.

We see a strong similarity between Equations 6.50 and 6.52. In the former, the divergence $\nabla \cdot \vec{F}$ plays a role analogous to that of what we call "mass density" in the latter. Given this analogy, we sometimes refer to divergence as "flux density".

So, in the same way that a triple integral of mass density over a region gives the mass inside of that region, Gauss's divergence theorem can now be interpreted as stating that a triple integral of flux density over a region gives the total flux that accumulates in that region.

Density interpretations in other boundary theorems

Not surprisingly, the strong similarities between Gauss's divergence theorem, Green's theorem and the fundamental theorem of calculus allow for similar reinterpretations of ideas used in those other theorems.

In the same way that Gauss's divergence theorem uses a triple integral to compute flux, the fundamental theorem of calculus uses a single variable integral to compute what we will casually refer to here as "change"; that is, the amount the function in question changes over the given domain. Writing this change as "Δ", we have

$$\Delta = f(b) - f(a) = \int_a^b f'\, dx \tag{6.53}$$

Now thinking of "change" as an accumulation over the interior interval, we could write

$$\Delta = \int_a^b d\Delta \tag{6.54}$$

which leaves us with

$$d\Delta = f'\, dx \tag{6.55}$$

Of course a similar consideration of the mass of a rod over that interval gives us

$$\int_a^b dm = m = \int_a^b \delta\, dx \tag{6.56}$$

and thus

$$dm = \delta\, dx \tag{6.57}$$

where δ represents a one-dimensional "mass density" – the amount of mass per unit length near a given point on the number line.

Noting the similarities then between Equations 6.55 and 6.57, we can interpret the derivative f' as representing what one might call the "change density" of the function f. With this in mind, we can reinterpret Equation 6.53 as saying that an integral of "change density" yields the total change of the function.

To make a similar interpretation of Green's theorem, we first need to come up with a label on what Green's theorem actually computes. Of course the theorem uses a double integral to compute the line integral of a vector field around a boundary curve; and a line integral, roughly speaking, computes how much a vector field points along the curve over the length of the curve. So we could view Green's theorem as computing how much a vector field points along the boundary curve as it makes a complete turn counterclockwise. That is, how much the vector field itself is turning around counterclockwise. Viewing the vector field as representing a flow, this conjures the image of the fluid "circulating", such as when water has been stirred in a pot.

So we refer here to the line integral computed in Green's theorem as "circulation". Writing this circulation as "σ", Green's theorem then says

$$\sigma = \int_{\partial D} \vec{F} \cdot d\vec{x} = \iint_D \mathrm{grn}\vec{F}\, dA \tag{6.58}$$

Now thinking of circulation as an accumulation over the interior area, we could write

$$\sigma = \iint_D d\sigma \tag{6.59}$$

which leaves us with

$$d\sigma = \mathrm{grn}\vec{F}\, dA \tag{6.60}$$

Of course a similar consideration of the mass of a sheet over the domain D gives us

$$\iint_D dm = m = \iint_D \delta\, dA \tag{6.61}$$

and thus

$$dm = \delta\, dA \tag{6.62}$$

where δ represents this time two-dimensional "mass density" – the amount of mass per unit area near a given point in the plane.

Noting the similarities then between Equations 6.60 and 6.62, we can interpret $\mathrm{grn}\vec{F}$ as representing what one might call the "circulation density" of the vector field $\vec{F}$. (This "circulation density" is also called "vorticity".)

With this in mind, we can reinterpret Equation 6.58 as saying that a double integral of circulation density yields the total circulation of the flow around the boundary curve.

6.4.3 Applications to physics

Fluid flow

Recall that one of our primary interpretations of vector fields is as representing the flow of a fluid. With that interpretation in mind, flux is then interpreted as the rate (quantity per unit time) of flow of the fluid through a given surface.

The flux through a boundary of a region then represents the net rate of fluid flowing out of the region; that is, the rate of flow of fluid out of the region minus the rate of flow of fluid into the region.

There are two ways that we can interpret this rate of flow into or out of a region.

First, for the moment we assume that the fluid represented by the vector field cannot simply disappear, nor spontaneously generate. In this case then the flux must represent the rate of change of the total amount of the fluid in the given region. That is, a positive flux through a boundary must indicate that the total amount of fluid in the region is decreasing (remember that boundaries are oriented outward, not inward), and likewise a negative flux through a boundary must indicate that the total amount of fluid in the region is increasing.

Since the volume of the region is a constant, this means that the average density of fluid in the region must be changing correspondingly. We could say for example that with negative flux through a boundary, the fluid is being "compressed", and likewise for positive flux through a boundary we could say the fluid is being "decompressed".

Thinking now to our recent interpretation of divergence as flux density, we can restate that as corresponding to the rate at which the corresponding fluid is being compressed. We might even in this case think of divergence as "local decompression rate".

On the other hand let's now assume the opposite. That is, we retract our assumption that the fluid cannot simply disappear or spontaneously generate, and instead assume that the fluid in question cannot be compressed or decompressed.

6.4. GAUSS'S DIVERGENCE THEOREM

Then if there is a positive flux through a boundary of a region, the fluid flowing out of that region must be being generated in that region; we refer to this as being a "fluid source". Similarly, if there is a negative flux through the boundary, then the fluid flowing into that region must be disappearing somehow; we refer to this as being a "fluid sink".

Again then thinking of divergence as flux density, we can think of positive divergence as representing "source density", and negative divergence as representing "sink density".

Gravity

Students who have had a course in Newtonian physics will recall that the magnitude of the gravitational force caused by an object of mass m_1 on an object of mass m_2 at a distance d is given by

$$\|\text{force}\| = \frac{Gm_1m_2}{d^2} \tag{6.63}$$

where G is the universal gravitational constant.

We can rewrite this as a vector field created by the first object by defining its location as the origin, and noting that the gravitational force always points in the direction directly toward that object. That unit vector direction at a point $\vec{x}$ is given by $-\vec{x}/\|\vec{x}\|$. And of course the distance d from that point to the first object is just $\|\vec{x}\|$. Treating the second mass m_2 as 1 so that the resulting vector field can be viewed as being intrinsic to the mass m_1, we can write the gravitational field as

$$\vec{F}_g = \frac{-Gm_1}{\|\vec{x}\|^3}\vec{x} \tag{6.64}$$

An amazing fact about this vector field is that, except at the origin where the vector field is not defined, the divergence of this vector field is exactly zero! This can be checked by direct computation.

$$\nabla \cdot \vec{F}_g(\vec{x}) = 0 \quad \text{when } \vec{x} \neq \vec{0} \tag{6.65}$$

The consequences of this fact are enormous. Suppose for example that we wish to compute the flux Φ_g through the boundary surface ∂R (of a region R in $\mathbb{R}^3$) of the gravitational field created by an object M. We must consider two cases.

Example 6.4.2. First, we consider the case in which the object M is outside the region R. In that case of course the divergence of $\vec{F}_g$ is zero in the entire region R, so Gauss's theorem gives us that

$$\Phi_g = \iint_{\partial R} \vec{F}_g \cdot d\vec{S} = \iiint_R \nabla \cdot \vec{F}_g \, dV = 0 \tag{6.66}$$

We can conclude immediately from this that the gravitational flux through a boundary surface is completely independent of mass that is outside of that region! See Figure 6.20.

Example 6.4.3. As a second case, we consider what happens if the object M is inside the region R. In that case the vector field is not continuous and so we cannot apply Gauss's theorem directly as was done in Equation 6.66.

However, we can make a different use of Gauss's theorem. Suppose we let S be a small sphere around the mass M, and let R_0 be the solid region that is obtained by removing from R the region inside of S. See Figure 6.21.

Note that $\partial R_0 = \partial R - S$, and of course also $\nabla \cdot \vec{F}_g = 0$ in the entire region R_0. Applying Gauss's theorem to the region R_0 then gives us

$$\iint_{\partial R} \vec{F}_g \cdot d\vec{S} - \iint_S \vec{F}_g \cdot d\vec{S} = \iiint_{R_0} \nabla \cdot \vec{F}_g \, dV = 0 \tag{6.67}$$

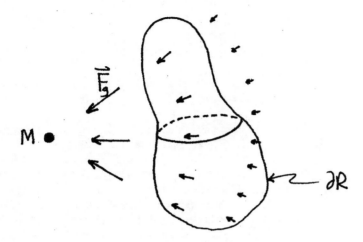

Figure 6.20:

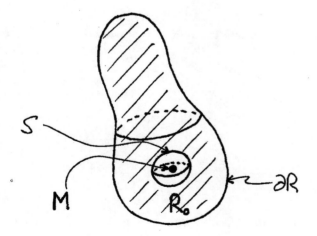

Figure 6.21:

6.4. GAUSS'S DIVERGENCE THEOREM

from which we can conclude

$$\Phi_g = \iint_{\partial R} \vec{F}_g \cdot d\vec{S} = \iint_S \vec{F}_g \cdot d\vec{S} \tag{6.68}$$

Of course this can be done no matter where inside of R the mass M is, and the resulting flux on the right side of Equation 6.68 will be the same, depending only on the mass of M.

The conclusion then is that the gravitational flux through a boundary surface is completely independent of how that mass is distributed! It depends only on the total mass that is enclosed in the region.

These are remarkable conclusions themselves, but we can do even more. Specifically, suppose we have mass distributed in a continuous manner over a region of space, and that we would like to compute the divergence of the gravitational field at a given point P. We will arrive at a conclusion by computing a particular flux, and applying Gauss's divergence theorem.

Because the mass is distributed continuously, we can treat the mass density as being roughly constant near P; as usual we write that density as δ. Let S be a small sphere of radius r around the point, and let us consider the flux Φ_g of the gravitational field through S.

As we observed in Example 6.4.2, for the purposes of computing flux we need only consider the mass that is inside S; we will write that mass as M. And as we observed in Example 6.4.3, for the purposes of computing flux we can treat all of that mass as if it were located exactly at the center of the sphere. In that case, the gravitational field would be

$$\vec{F}_g = \frac{-GM}{\|\vec{x}\|^3} \vec{x} \tag{6.69}$$

Of course being a sphere, the unit normal vector $\vec{n}$ can be written as $\vec{x}/\|\vec{x}\|$, and also we have $\|\vec{x}\| = r$ everywhere on the sphere. So we can compute this flux directly as

$$\begin{aligned}\Phi_g &= \iint_S \vec{F}_g \cdot d\vec{S} = \iint_S \vec{F}_g \cdot \vec{n}\, dS = \iint_S \frac{-GM}{\|\vec{x}\|^3} \vec{x} \cdot \frac{\vec{x}}{\|\vec{x}\|}\, dS & (6.70)\\ &= \frac{-GM}{r^2} \iint_S dS = \frac{-GM}{r^2}(4\pi r^2) & (6.71)\\ &= -4\pi GM & (6.72)\end{aligned}$$

Now we compute the same flux again, but this time with Gauss's divergence theorem and using that the divergence is constant near the given point. We get

$$\Phi_g = \iiint_R \nabla \cdot \vec{F}_g\, dV = (\nabla \cdot \vec{F}_g)(V) \tag{6.73}$$

Equating the results of these two computations, we get

$$\nabla \cdot \vec{F}_g = -4\pi G \delta \tag{6.74}$$

That is, the divergence of the gravitational field at a given point is directly proportional to the mass density at that point.

This result provides another wonderful motivation for thinking of divergence as being "flux density", as discussed in Subsection 6.4.2. With that interpretation, Equation 6.74 states simply that flux density is proportional to mass density.

We can apply this result about gravitational divergence to any boundary surface. Suppose for example that we wish to compute the gravitational flux through a boundary surface ∂R. The divergence theorem gives us

$$\Phi_g = \iint_{\partial R} \vec{F}_g \cdot d\vec{S} \tag{6.75}$$

$$= \iiint_R \nabla \cdot \vec{F}_g \, dV \tag{6.76}$$

$$= \iiint_R -4\pi G \delta \, dV \tag{6.77}$$

$$= -4\pi G \iiint_R \delta \, dV \tag{6.78}$$

$$= -4\pi G M \tag{6.79}$$

Example 6.4.4. Let the surface S be an outward oriented sphere with center at the center of the sun, and with a radius of 100 million miles. What is the total gravitational flux through S?

Equation 6.79 tells us that the answer is simply $\Phi_g = -4\pi G M$, where M is the total mass inside of the given surface. The only objects of significant size inside the surface are the sun, Mercury, Venus, Earth, and the Earth's moon, so the sum of those masses gives us a very accurate estimate of the value in that equation.

The other planets and other objects in the solar system do not count in this sum, because they are outside of the boundary surface and therefore to not contribute to the flux.

Example 6.4.5. Suppose we have a large, hollow spherical shell with uniform mass density. Can we determine the gravitational field on the inside of the shell?

To answer this question we will have to begin with an intuitive claim. That is, because the shell is spherically symmetric, we argue that the gravitational field $\vec{F}_g$ must also be spherically symmetric.

This gives us two results. First, the gravitational field at a point P in the interior of the shell must point along the radial line from the center of the sphere through that point, as indicated in Figure 6.22. Second, the magnitude of the vector field must depend only on the radial distance ρ to the center.

Algebraically we can write these observations as

$$\vec{F}_g = f(\rho)\,\vec{x} \tag{6.80}$$

In order to compute the vector field strength at the given point P, we will consider the sphere ∂B of radius ρ passing through that point. Of course this sphere is also the boundary of the ball B that it encloses. We will compute the gravitational flux through this sphere in two different ways.

First, we compute it directly from Equation 6.80, taking advantage of the facts that $\vec{x}$ and $\vec{n}$ are parallel, and ρ is constant on the boundary sphere.

$$\Phi_g = \iint_{\partial B} \vec{F}_g \cdot d\vec{S} = \iint_{\partial B} f(\rho)\,\vec{x}\cdot\vec{n}\,dS = f(\rho) \iint_{\partial B} \rho\,dS = 4\pi\rho^3 f(\rho) \tag{6.81}$$

Second, we compute the flux immediately from Equation 6.79 – since there is no mass in the region B, we must have

$$\Phi_g = 0 \tag{6.82}$$

Combining these results gives us that $f(\rho) = 0$, and thus that $\vec{F}_g = \vec{0}$. Amazingly, the gravitational field is identically zero everywhere inside the shell!

6.4. GAUSS'S DIVERGENCE THEOREM

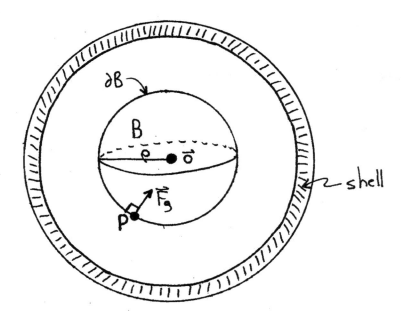

Figure 6.22:

It should be noted carefully though that this remarkable result hinges critically on our initial observation that the shell is in fact spherically symmetric. For shells that are not spherically symmetric, the above reasoning does not apply.

Electric and magnetic fields

Students who have had a course in electricity and magnetism will recall that the magnitude of the electrostatic force caused by an object of charge q_1 on an object of charge q_2 at a distance d is given by

$$F_e = \frac{q_1 q_2}{4\pi \epsilon_0 d^2} \tag{6.83}$$

where ϵ_0 is a universal constant known as the "permittivity of free space".

This formula of course is very similar to that in Equation 6.63, most importantly in that the dependence on distance is as the inverse square. Not surprisingly then, there are many extremely similar results.

1. The divergence of the electric field created by a point charge is zero, away from that charge.

2. The flux of the electric field of a charge through a boundary surface is zero if the charge is located entirely outside of the region bounded by that surface.

3. The flux of the electric field of a charge through a boundary surface is completely independent of the distribution of charges inside of the region, and depends only on the total charge in the region.

4. The divergence of the electric field is directly proportional to the charge density.

Students can work through arguments very similar to those done for gravitational fields in the last section to arrive at the following very similar results for electric fields.

The electric field created by a charge Q is given by

$$\vec{E} = \frac{Q}{4\pi\epsilon_0 \|\vec{x}\|^3}\vec{x} \tag{6.84}$$

The divergence of the electric field at a given point in space is given by

$$\nabla \cdot \vec{E} = \frac{\rho}{\epsilon_0} \tag{6.85}$$

where ρ represents the "charge density" at that point (the total amount of electric charge per unit volume near that point).

The total electric field flux through a boundary surface ∂R is given by

$$\Phi_e = \frac{Q}{\epsilon_0} \tag{6.86}$$

where Q is the total charge in the region R.

Students are strongly encouraged to work through and arrive at the above results by the methods previously applied to the gravitational field.

Example 6.4.6. Equation 6.85 is one version of one of four of Maxwell's equations, describing properties and relationships of electric and magnetic fields. Specifically, this is called the "differential version" of that equation.

The other version of this equation, called the "integral version", is equivalent to Equation 6.86. It is often written out more explicitly as

$$\iint_{\partial R} \vec{E} \cdot d\vec{S} = \frac{Q}{\epsilon_0} \tag{6.87}$$

Upon first inspection it would appear that these two forms, Equations 6.85 and 6.87, are completely different statements. Of course we already know that these two equations are directly related by way of Gauss's divergence theorem.

Example 6.4.7. Another of the four of Maxwell's equations can be viewed in two different forms, this one involving the magnetic field $\vec{B}$. In its differential form, it is written

$$\nabla \cdot \vec{B} = 0 \tag{6.88}$$

and in its integral form it is written

$$\iint_{\partial R} \vec{B} \cdot d\vec{S} = 0 \tag{6.89}$$

Again, these two forms are immediately related by Gauss's divergence theorem.

6.4.4 Gauss's theorem in $\mathbb{R}^2$

In Section 6.3 we motivated and discussed Green's theorem. It came about by considering a line integral of a vector field around a boundary curve, and applying the same sorts of ideas that have given us several other boundary theorems.

6.4. GAUSS'S DIVERGENCE THEOREM

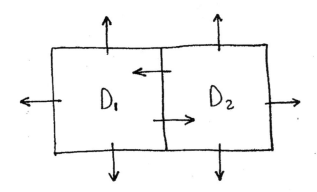

Figure 6.23:

We began with a line integral in that argument for reasons that were never explained. Along with all of the other evaluations that we settled on in Section 6.1, we claimed simply that the line integral is a special case of a more general construction involving things called "differential forms" which we cannot discuss in this text, and left it at that.

The majority of this chapter remains, still, the consideration of theorems that come from the setup that we settled on in Section 6.1. However, there is an interesting result that comes about if we change things around just a bit. Here in this subsection we will discuss this result.

The change we make is in the starting point of the argument. Rather than beginning with a line integral of a vector field around a boundary, as we did in Section 6.3, giving us Green's theorem, we instead consider the flux of a vector field through a boundary, as was discussed briefly in Subsection 5.12.6. Applying our usual boundary theorem arguments, we find that everything goes through just fine.

First, we observe that the flux through the boundary ∂R is indeed the sum of the fluxes through the boundaries of the subdomains.

$$\int_{\partial D} \vec{F} \cdot \vec{n} \, ds = \sum_{i,j} \int_{\partial D_{ij}} \vec{F} \cdot \vec{n} \, ds \tag{6.90}$$

This is analogous to Equation 6.30, and is true for similar though geometrically different reasons.

Equation 6.30 works because each boundary segment from the right side of the equation that is in the interior is part of the boundary of exactly two subdomains, and with opposite orientations as part of each of those two boundaries. This opposition happens because, as part of each boundary, the notions of "counterclockwise" are opposite. This is represented in Figure 6.14.

Similarly, Equation 6.90 works because each boundary segment from the right side of the equation that is in the interior is, again, part of the boundary of exactly two subdomains – and again, with opposite orientations as part of each of those two boundaries. In this case the opposition happens because, as part of each boundary, the notions of "outward" are opposite. This is represented in Figure 6.23.

Again we can take the limit as the size of the subdomains approaches zero, and the right hand side again becomes an integral.

$$\int_{\partial D} \vec{F} \cdot \vec{n} \, ds = \iint_{D} \left(\lim_{n \to \infty} \frac{\int_{\partial D_{ij}} \vec{F} \cdot \vec{n} \, ds}{\Delta A_{ij}} \right) dA \tag{6.91}$$

The limit in the integrand can again be computed (again we skip those details here), resulting in

$$\lim_{n \to \infty} \frac{\int_{\partial D_{ij}} \vec{F} \cdot \vec{n} \, ds}{\Delta A_{ij}} = \frac{\partial P}{\partial x} + \frac{\partial Q}{\partial y} \tag{6.92}$$

and thus Equation 6.91 becomes

$$\int_{\partial D} \vec{F} \cdot \vec{n}\, ds = \iint_D \left(\frac{\partial P}{\partial x} + \frac{\partial Q}{\partial y} \right) dA \qquad (6.93)$$

Interestingly, even though this discussion takes place in $\mathbb{R}^2$, this new integrand looks very similar to the formula for divergence that came up in our recent derivation of Gauss's divergence theorem. Given that similarity and the fact that both theorems concern flux through boundaries, we refer to this new expression also as divergence, for a 2-dimensional vector field,

$$\text{div}\vec{F} = \frac{\partial P}{\partial x} + \frac{\partial Q}{\partial y} = \nabla \cdot \vec{F} \qquad (6.94)$$

and refer to the following restatement of this result as Gauss's divergence theorem in $\mathbb{R}^2$.

Theorem 6.4.2 (Gauss's Diverence Theorem in $\mathbb{R}^2$). *Let $\vec{F}$ be a continuously differentiable vector field, and D a domain in $\mathbb{R}^2$. Then*

$$\int_{\partial D} \vec{F} \cdot \vec{n}\, ds = \iint_D \nabla \cdot \vec{F}\, dA$$

(All of the previous boundary theorems we have discussed motivated a new operator to be included into the diagrams that we have been developing throughout this chapter. We will not do so for this theorem because, as we observed at the beginning of this subsection, the starting point for this theorem is not consistent with all of the other choices that are represented in the diagrams; as such, it simply does not fit in very well with those other theorems on those diagrams. Students should consider this result simply as a separate result.)

Example 6.4.8. Suppose we wish to compute the flux of the vector field $\vec{F} = (3x^2 + 2y^3, e^x - y)$ through the unit circle C oriented outward.

Of course this could be computed directly by parametrizing the circle, but since the circle is the boundary of the unit disk D, we can instead use Gauss's divergence theorem in $\mathbb{R}^2$, which gives us

$$\begin{aligned}
\int_C \vec{F} \cdot \vec{n}\, ds &= \iint_D \nabla \cdot \vec{F}\, dA \\
&= \iint_D \frac{\partial}{\partial x}(3x^2 + 2y^3) + \frac{\partial}{\partial y}(e^x - y)\, dA \\
&= \iint_D 6x - 1\, dA \\
&= -\pi
\end{aligned}$$

Exercises

Exercise 6.4.1 (**exer-VC-Gauss-1**). Suppose we consider the small box D in $\mathbb{R}^3$ defined by $D = [x_0, x_0 + \Delta x] \times [y_0, y_0 + \Delta y] \times [z_0, z_0 + \Delta z]$, and the vector field $\vec{F}(x, y, z) = (P, Q, R)$. The boundary of this box has six faces; we label the two perpendicular to the x-direction S_1 (in the plane $x = x_0$) and S_2; similarly for S_3 and S_4 perpendicular to the y-direction and S_5 and S_6 perpendicular to the z-direction.

1. What are the normal vectors for these six boundary faces?
2. What are the areas of these six boundary faces?

6.4. GAUSS'S DIVERGENCE THEOREM

3. If we assume that the vector field is approximately constant on S_1 with value (P_0, Q_0, R_0), what is the resulting approximation of the flux of $\vec{F}$ through S_1?

4. The points on S_2 are a distance Δx away from the points on S_1, in the x-direction; if we assume that the vector field is approximately constant on S_2 also, a reasonable approximation of the value of the vector field would be $(P_0 + (\frac{\partial P}{\partial x})(\Delta x), Q_0 + (\frac{\partial Q}{\partial x})(\Delta x), R_0 + (\frac{\partial R}{\partial x})(\Delta x))$. What is the resulting approximation of the flux of $\vec{F}$ through S_2?

5. What is the sum of these approximations of the fluxes of $\vec{F}$ through S_1 and S_2?

6. Using an approach similar to that in the last three parts of this problem, find a reasonable approximation of the sum of the fluxes of $\vec{F}$ through S_3 and S_4. Also, find a reasonable approximation of the sum of the fluxes of $\vec{F}$ through S_5 and S_6.

7. What is the resulting approximation of $\frac{\iint_{\partial D} \vec{F} \cdot d\vec{S}}{\Delta V}$ for this box? Compare the result of this calculation to Equation 6.45.

Exercise 6.4.2 (**exer-VC-Gauss-2**). Compute $\iint_S \vec{F} \cdot d\vec{S}$, where $\vec{F}(x,y,z) = (y^2 z, e^x - 3y, 3x^2 y^2)$, and S is the unit sphere with outward orientation.

Exercise 6.4.3 (**exer-VC-Gauss-3**). Compute $\iint_S \vec{F} \cdot d\vec{S}$, where $\vec{F}(x,y,z) = (y^2 z, e^x - 3y, 3x^2 y^2)$, and S is the unit sphere with inward orientation.

Exercise 6.4.4 (**exer-VC-Gauss-4**). Compute $\iint_S \vec{F} \cdot d\vec{S}$, where $\vec{F}(x,y,z) = (x,y,z)$, and S is the boundary of the region T bounded by $z = x^2 + y^2$ and $z = 8 - x^2 - y^2$.

Exercise 6.4.5 (**exer-VC-Gauss-5**). Show that for any vector field of the form $\vec{F}(x,y,z) = (P(y,z), Q(x,z), R(x,y))$ and any surface S that is a boundary of a solid region, we must have $\iint_S \vec{F} \cdot d\vec{S} = 0$.

Exercise 6.4.6 (**exer-VC-Gauss-6**). Compute $\iint_S \vec{F} \cdot d\vec{S}$, where $\vec{F}(x,y,z) = (x^2 + 2xz - 2xy, y^2 - 2xy, -z^2)$, and S has spherical equation $\rho = 4 + \sin\phi \sin\theta - \cos(4\phi)$.

Exercise 6.4.7 (**exer-VC-Gauss-7**). Use the Gauss divergence theorem and a clever choice of vector field to show that for a solid region R, we can compute the volume of R as $\iint_{\partial R} x \, dydz$.

Exercise 6.4.8 (**exer-VC-Gauss-8**). Suppose we consider the vector field $\vec{F}(x,y,z) = (x^2 y - yz, xy - y^2, xz)$. Is the point $(1,2,1)$ a vector field source or a vector field sink? Explain your answer.

Exercise 6.4.9 (**exer-VC-Gauss-9**). Suppose there are point masses at $(3,4,1)$ and $(2,7,3)$, each with mass equal to 5 kg. Suppose also that there are no other masses nearby. What is the flux of the total gravitational field through the outward oriented unit sphere?

Exercise 6.4.10 (**exer-VC-Gauss-10**). With the masses distributed the same way as in Exercise 6.4.9, what is the flux of the total gravitational field through the outward oriented sphere of radius 6 centered at the origin?

Exercise 6.4.11 (**exer-VC-Gauss-11**). A gas cloud is in the shape of a ball of radius R centered at the origin. The density of the cloud is a constant δ. What is the formula for the gravitational field $\vec{F}_g(x,y,z)$? (Hint: You may assume, as was done in Example 6.4.5, that this vector field is spherically symmetric. You can then compute the flux through a sphere of radius ρ by considering only the gas inside that sphere.)

Exercise 6.4.12 (**exer-VC-Gauss-12**). A region of space has a charged particle (with constant positive charge Q) at every point with integer coordinates. What is the flux of the electric field through the sphere of radius 2.5 centered at the origin, oriented outward?

Exercise 6.4.13 (**exer-VC-Gauss-13**). Beginning with Equation 6.85, take a triple integral of both sides and then apply the Gauss divergence theorem to derive the result of Example 6.4.6.

Exercise 6.4.14 (**exer-VC-Gauss-14**). As is suggested in Example 6.4.7, use Equation 6.88 to derive Equation 6.89.

Exercise 6.4.15 (**exer-VC-Gauss-15**). Compute the flux of the vector field $\vec{F}(x,y) = (2x - 3y, x - 2y)$ through the rectangle $R = [2,6] \times [3,8]$.

Exercise 6.4.16 (exer-VC-Gauss-16). The flow of locusts is given by $\vec{F}(x,y) = (5x - y, x - 2y)$ (in millions of locusts per mile per day). Square county is located over the region $S = [0,2] \times [2,4]$. If we measure the rate of change of the number of locusts in a region as the flux through the boundary curve of the vector field $\vec{F}$, how fast is the number of locusts in Square county changing?

Exercise 6.4.17 (exer-VC-Gauss-17). Suppose we have a region $D \in \mathbb{R}^2$ with boundary curve C. Apply the 2-d Gauss divergence theorem to the vector field $\vec{G} = (Q, -P)$ and use the result to derive Green's theorem for the vector field $\vec{F} = (P, Q)$. (Hint: If the outward normal vector $\vec{n}$ has components $\vec{n} = (a, b)$, then the counterclockwise unit tangent vector $\vec{T}$ has components $\vec{T} = (-b, a)$; use this to show that $\vec{G} \cdot \vec{n} = \vec{F} \cdot \vec{T}$.)

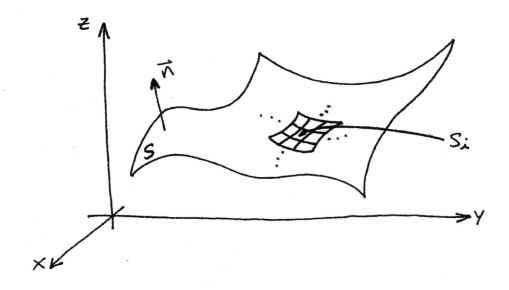

Figure 6.24:

6.5 Stokes's Curl Theorem

Again, here we will apply in a different context the same ideas that have led up to the other boundary theorems. The context is again $\mathbb{R}^3$, but now we consider oriented surfaces as our domains.

6.5.1 Motivation

We begin with an integrand in column one, and a domain in column two. The column one integrand is a vector field, which we will call $\vec{F} = (P, Q, R)$ (which we again assume to be continuously differentiable), and the column two domain is an oriented surface in $\mathbb{R}^3$, which we will call S. So in our $\mathbb{R}^3$ diagram we have

Vector Calculus in $\mathbb{R}^3$

	$\boxed{0}$	$\boxed{1}$	$\boxed{2}$	$\boxed{3}$
Integrands	(fns)	$\vec{F}$		(fns)
Domains	(points)		S	(regions)
Evaluations	$h(\vec{x})$	$\int_C \vec{H} \cdot d\vec{x}$	$\iint_S \vec{H} \cdot d\vec{S}$	$\iiint_R h\, dV$

Again we break up the domain S into subdomains, this time in the same way that we did when setting up the triple integral in Section 5.9. Each of these subdomains is oriented in the same way as the original surface on that part of the surface.

We will refer to the individual subdomains as S_i, where $i = 1, \ldots, n$. We will use ΔS_i to refer to the area of each of these subdomains. See Figure 6.24.

We will soon see that, again, the boundaries of these subdomains relate to the boundary of the original domain, by

$$\partial S = \sum_i \partial S_i \tag{6.95}$$

Again this equation works because of cancellation of the pieces in the interior. The picture justifying this is different,

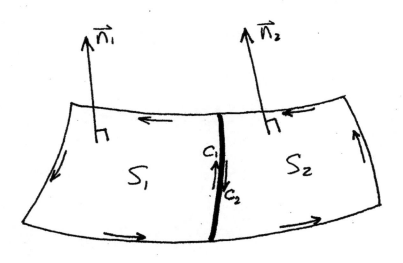

Figure 6.25:

but actually still very similar to the picture used in the corresponding discussion of Green's theorem.

Again, the surface pieces in the summation in Equation 6.95 fall into two categories. Some are part of the boundary of S, contributing to the total of ∂S.

All of the remaining pieces of surface, in being in the interior of S, must be part of the boundary of exactly two of the subdomains S_i, and thus must appear exactly twice in the summation. Furthermore, the orientations of those two instances of that piece of surface must be opposite, and so they cancel.

For example, in Figure 6.25, the curve between the subdomains labeled S_1 and S_2 appears in two instances in the summation – as part of the boundaries of each of those subdomains. As part of the boundary of S_1 we label it C_1, and as part of the boundary of S_2 we label it C_2. By convention, given the orientation vector $\vec{n}_1$, the orientation on C_1 points along the curve in the direction which is upward in the figure. But by the same convention, given the orientation vector $\vec{n}_2$, the orientation on C_2 points along the curve in the direction which is downward in the figure. So while C_1 and C_2 cover the same set of points in space, they have opposite orientations, and so we have $C_1 + C_2 = 0$.

Since all of the interior pieces of surface in the summation in Equation 6.95 cancel, the boundary pieces are all that remain. This confirms that equation. (Students should note a very strong analogy between this calculation and the corresponding calculation in the discussions of the other boundary theorems, particularly Green's theorem and Gauss's theorem.)

Having motivated Equation 6.95, we can continue on with our template by considering that boundary of S, applying the appropriate evaluation in column one, and then using Equation 6.95 to begin rewriting that output. Our diagram becomes

	[0]	[1]	[2]	[3]
Integrands	(fns)	$\vec{F}$		(fns)
Domains	(points)	$\partial S \xleftarrow{\partial\text{ bdry}} S$		(regions)
Evaluations	$h(\vec{x})$	$\int_C \vec{H} \cdot d\vec{x}$	$\iint_S \vec{H} \cdot d\vec{S}$	$\iiint_R h\, dV$
Outputs		$\int_{\partial S} \vec{F} \cdot d\vec{x}$		

6.5. STOKES'S CURL THEOREM

Using Equation 6.95 as suggested, we rewrite the output in column one as

$$\int_{\partial S} \vec{F} \cdot d\vec{x} = \sum_i \left(\int_{\partial S_i} \vec{F} \cdot d\vec{x} \right) \tag{6.96}$$

$$= \lim_{n \to \infty} \sum_i \left(\int_{\partial S_i} \vec{F} \cdot d\vec{x} \right) \tag{6.97}$$

Analogously to what we did at this point in our discussion of the other boundary theorems, we divide and multiply by ΔS_i, giving us

$$\int_{\partial S} \vec{F} \cdot d\vec{x} = \lim_{n \to \infty} \sum_i \left(\frac{\int_{\partial S_i} \vec{F} \cdot d\vec{x}}{\Delta S_i} \right) \Delta S_i \tag{6.98}$$

The good news here is that the right side of Equation 6.98 appears to be exactly a surface integral over the domain S. So, continuing the analogy with the development of the previous theorems, we note that we have turned a column one evaluation into a column two evaluation. Again skipping the details, we claim plausibly that this can be rewritten as

$$\int_{\partial S} \vec{F} \cdot d\vec{x} = \iint_S \left(\lim_{n \to \infty} \frac{\int_{\partial S_i} \vec{F} \cdot d\vec{x}}{\Delta S_i} \right) dS \tag{6.99}$$

As in the developments of Green's theorem and Gauss's theorem, the expression in the integrand is not something that we recognize from previous experience. Still, the expression can be simplified, though we will not show those details here. Instead, we simply state the result, that

$$\lim_{n \to \infty} \frac{\int_{\partial S_i} \vec{F} \cdot d\vec{x}}{\Delta S_i} = \begin{bmatrix} \frac{\partial R}{\partial y} - \frac{\partial Q}{\partial z} \\ \frac{\partial P}{\partial z} - \frac{\partial R}{\partial x} \\ \frac{\partial Q}{\partial x} - \frac{\partial P}{\partial y} \end{bmatrix} \cdot \vec{n} \tag{6.100}$$

We can then rewrite Equation 6.99 as

$$\int_{\partial S} \vec{F} \cdot d\vec{x} = \iint_S \left(\begin{bmatrix} \frac{\partial R}{\partial y} - \frac{\partial Q}{\partial z} \\ \frac{\partial P}{\partial z} - \frac{\partial R}{\partial x} \\ \frac{\partial Q}{\partial x} - \frac{\partial P}{\partial y} \end{bmatrix} \cdot \vec{n} \right) dS \tag{6.101}$$

$$= \iint_S \begin{bmatrix} \frac{\partial R}{\partial y} - \frac{\partial Q}{\partial z} \\ \frac{\partial P}{\partial z} - \frac{\partial R}{\partial x} \\ \frac{\partial Q}{\partial x} - \frac{\partial P}{\partial y} \end{bmatrix} \cdot d\vec{S} \tag{6.102}$$

$$\tag{6.103}$$

Having arrived now at a workable vector surface integral, which is the evaluation we have associated to column two, we can use this to motivate a natural choice for an operator that turns column one integrands into column two integrands. We name this operator the "curl" operator, defined by

$$\text{curl}\vec{F} = \begin{bmatrix} \frac{\partial R}{\partial y} - \frac{\partial Q}{\partial z} \\ \frac{\partial P}{\partial z} - \frac{\partial R}{\partial x} \\ \frac{\partial Q}{\partial x} - \frac{\partial P}{\partial y} \end{bmatrix} \tag{6.104}$$

(As was the case with divergence, there is an alternative and very common notation for the curl operator involving the "∇" symbol used with gradients. Again using the notation

$$\nabla = \begin{bmatrix} \frac{\partial}{\partial x} \\ \frac{\partial}{\partial y} \\ \frac{\partial}{\partial z} \end{bmatrix}$$

note that very conveniently it turns out that

$$\begin{aligned}\nabla \times \vec{F} &= \begin{vmatrix} \vec{e}_1 & \vec{e}_2 & \vec{e}_3 \\ \frac{\partial}{\partial x} & \frac{\partial}{\partial y} & \frac{\partial}{\partial z} \\ P & Q & R \end{vmatrix} \\ &= \begin{bmatrix} \frac{\partial}{\partial y}R - \frac{\partial}{\partial z}Q \\ \frac{\partial}{\partial z}P - \frac{\partial}{\partial x}R \\ \frac{\partial}{\partial x}Q - \frac{\partial}{\partial y}P \end{bmatrix} \\ &= \text{curl}\vec{F} \end{aligned}$$

So very often "curl$\vec{F}$" is instead written as "$\nabla \times \vec{F}$".)

With this choice, our diagram becomes

	[0]	[1]	[2]	[3]
Integrands	(fns)	$\vec{F} \xrightarrow[\text{curl}]{\nabla \times} \nabla \times \vec{F}$		(fns)
Domains	(points)	$\partial S \xleftarrow[\text{bdry}]{\partial} S$		(regions)
Evaluations	$h(\vec{x})$	$\int_C \vec{H} \cdot d\vec{x}$	$\iint_S \vec{H} \cdot d\vec{S}$	$\iiint_R h\, dV$
Outputs		$\int_{\partial S} \vec{F} \cdot d\vec{x} =\!=\!= \iint_S \left(\nabla \times \vec{F} \right) \cdot d\vec{S}$		

We now state our result as a theorem, called Stokes's curl theorem, or sometimes simply "the curl theorem". (Note, students should avoid referring to this theorem as "Stokes's theorem". The problem is that this label might be interpreted as referring to the "general Stokes's theorem" involving differential forms. Of course the Stokes's curl theorem is a special case of the general Stokes's theorem, as are all of the boundary theorems in this chapter, but still the ambiguity is undesirable.)

Theorem 6.5.1 (Stokes's curl theorem). *Let $\vec{F}$ be a continuously differentiable vector field, and S an oriented surface in $\mathbb{R}^3$. Then*

$$\int_{\partial S} \vec{F} \cdot d\vec{x} = \iint_S \left(\nabla \times \vec{F} \right) \cdot d\vec{S} \tag{6.105}$$

Just as with the previous boundary theorems, the diagram gives us a graceful point of view on Stokes's curl theorem. That is, given an appropriate vector field $\vec{F}$ in column one and surface S in column two, applying the indicated operators and evaluating in each column always gives the same results.

Again we can rewrite this as an appendix to the original diagram for calculus in $\mathbb{R}^3$.

<u>Vector Calculus in $\mathbb{R}^3$</u>

	[0]	[1]	[2]	[3]
Integrands	(fns)	(vector fields) $\xrightarrow[\text{curl}]{\nabla \times}$ (vector fields)		(fns)
Domains	(points)	(curves) $\xleftarrow[\text{bdry}]{\partial}$ (surfaces)		(regions)
Evaluations	$h(\vec{x})$	$\int_C \vec{H} \cdot d\vec{x}$	$\iint_S \vec{H} \cdot d\vec{S}$	$\iiint_R h\, dV$
Boundary Theorems		$\vec{F} \xrightarrow[\text{curl}]{\nabla \times} \nabla \times \vec{F}$ $\partial S \xleftarrow[\text{bdry}]{\partial} S$ $\int_{\partial S} \vec{F} \cdot d\vec{x} =\!=\!= \iint_S \left(\nabla \times \vec{F} \right) \cdot d\vec{S}$		

6.5.2 Examples

We usually use Stokes's curl theorem to compute line integrals around boundaries, by turning them into flux integrals.

Example 6.5.1. Suppose we wish to compute the line integral of the vector field $\vec{F} = (x^2 - y, e^y + x, \ln(z^2 + 1))$ along the curve C that is parametrized by $\vec{x}(t) = (\cos t, \sin t, 0)$, where $t \in [0, 2\pi]$.

First, we note that the curve is exactly the unit circle in the xy-plane sitting inside of xyz-space. Therefore it is in fact a boundary. Of course there are many surfaces for which this curve is a boundary, and we are within our rights to choose any of those surfaces to be our surface S. In this case, it is convenient to use the simplest of these surfaces, namely, the unit disk in the xy-plane oriented upward.

We can then apply Stokes's curl theorem, giving us

$$\begin{aligned} \int_C \vec{F} \cdot d\vec{x} &= \int_{\partial S} \vec{F} \cdot d\vec{x} \\ &= \iint_S \left(\nabla \times \vec{F}\right) \cdot d\vec{S} \\ &= \iint_S (0, 0, 2) \cdot d\vec{S} \end{aligned}$$

Here we have a flux integral to compute, which we can certainly do by parametrizing the surface and following the methods of Section 5.12. But, even better in this case, because we chose the surface S as a very simple surface we can immediately write down the unit normal vector $\vec{n}$. Because the surface is entirely within the xy-plane in xyz-space and is oriented upward, the unit normal vector is $(0, 0, 1)$. So we get

$$\begin{aligned} \int_C \vec{F} \cdot d\vec{x} &= \iint_S (0, 0, 2) \cdot \vec{n} \, dS \\ &= \iint_S (0, 0, 2) \cdot (0, 0, 1) \, dS \\ &= 2 \, (\text{area of } S) \\ &= 2\pi \end{aligned}$$

Example 6.5.2. Suppose we want to compute the line integral of the vector field $\vec{F} = (y, z, x)$ along the curve C that is the intersection of the plane with equation $x - z = 0$ and the cylinder with equation $x^2 + y^2 = 1$, oriented counterclockwise as seen from above.

As shown in Figure 6.26, the curve C is the boundary of a surface, specifically the one labelled S in the figure. Then Stokes's curl theorem gives us

$$\begin{aligned} \int_C \vec{F} \cdot d\vec{x} &= \int_{\partial S} \vec{F} \cdot d\vec{x} \\ &= \iint_S \left(\nabla \times \vec{F}\right) \cdot d\vec{S} \\ &= \iint_S (-1, -1, -1) \cdot d\vec{S} \end{aligned}$$

Again we are fortunate that the unit normal vector can be determined directly. In this case we know that the surface S is entirely within the plane with equation $x - z = 0$; and given that C is oriented counterclockwise as seen

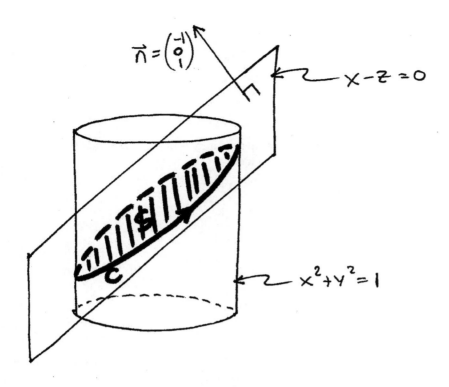

Figure 6.26:

from above, the normal vector $\vec{n}$ must point upward to give the same orientation on $\partial S = C$. So we must have $\vec{n} = \frac{1}{\sqrt{2}}(-1, 0, 1)$, and again we can use this to compute the above flux.

$$\begin{aligned}\int_C \vec{F} \cdot d\vec{x} &= \iint_S (-1, -1, -1) \cdot d\vec{S} \\ &= \iint_S (-1, -1, -1) \cdot \vec{n}\, dS \\ &= \frac{1}{\sqrt{2}} \iint_S (-1, -1, -1) \cdot (-1, 0, 1)\, dS \\ &= 0\end{aligned}$$

It turns out that without too much trouble we can use Stokes's curl theorem to prove Green's theorem. This is a useful exercise also because by relating the two theorems, we will see a natural relationship in their interpretations.

Example 6.5.3. Suppose we have a two-dimensional vector field $\vec{F}$ and a closed curve C in the xy-plane, that is the boundary ∂D of a solid region in the xy-plane. We begin by considering the line integral $\int_{\partial D} \vec{F} \cdot d\vec{x}$ that is one side of the equation in Green's theorem.

Our first step is to take a different point of view on our setting. That is, instead of thinking of this as a problem in $\mathbb{R}^2$, let us instead think of the xy-plane as being part of $\mathbb{R}^3$ – that is, the plane $z = 0$ in xyz-space. Along those same lines we write the vector field as $\vec{F} = (P, Q, 0)$. And instead of thinking of the curve C merely as the boundary of a solid region in $\mathbb{R}^2$, let us think of that region in $\mathbb{R}^2$ as then being a surface in the xy-plane in $\mathbb{R}^3$. Note that if we give the surface the upward orientation, then the corresponding orientation on its boundary is exactly what it should be. Of course since the surface D is entirely contained in the xy-plane, that upward unit normal vector must be $\vec{n} = (0, 0, 1)$. See **Figure 6.27**.

6.5. STOKES'S CURL THEOREM

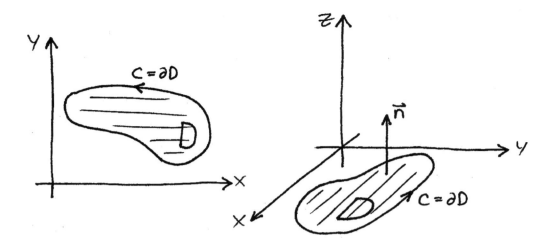

Figure 6.27:

At this point the line integral of interest, $\int_{\partial D} \vec{F} \cdot d\vec{x}$, can now be considered as a line integral in $\mathbb{R}^3$, and we can apply Stokes's curl theorem, giving us

$$\begin{aligned}
\int_{\partial D} \vec{F} \cdot d\vec{x} &= \iint_D \left(\nabla \times \vec{F}\right) \cdot d\vec{S} \\
&= \iint_D \left(\begin{bmatrix} \frac{\partial R}{\partial y} - \frac{\partial Q}{\partial z} \\ \frac{\partial P}{\partial z} - \frac{\partial R}{\partial x} \\ \frac{\partial Q}{\partial x} - \frac{\partial P}{\partial y} \end{bmatrix}\right) \cdot \vec{n}\, dS \\
&= \iint_D \left(\begin{bmatrix} \frac{\partial R}{\partial y} - \frac{\partial Q}{\partial z} \\ \frac{\partial P}{\partial z} - \frac{\partial R}{\partial x} \\ \frac{\partial Q}{\partial x} - \frac{\partial P}{\partial y} \end{bmatrix}\right) \cdot \begin{bmatrix} 0 \\ 0 \\ 1 \end{bmatrix} dS \\
&= \iint_D \frac{\partial Q}{\partial x} - \frac{\partial P}{\partial y}\, dS
\end{aligned}$$

Of course the surface area element dS is exactly $dA = dx\, dy$ because the surface is entirely in the xy-plane, and so this is exactly the right side of Green's theorem.

6.5.3 Interpretation of curl

An "oriented density"

Like the other boundary theorems, Stokes's curl theorem can be interpreted as an observation that there is something that accumulates over the interior to give a total on the boundary. This is seen by interpreting Equation 6.99.

The difference here is that in this case, the quantity that accumulates depends not only on the original integrand (the vector field $\vec{F}$), but also on the orientation of the domain surface. That is, it depends also on the vector $\vec{n}$, as we see in Equation 6.100. So we must keep this in mind when making an interpretation of the eventual new integrand, $\text{curl}\vec{F} = \nabla \times \vec{F}$.

Just like in Green's theorem, Stokes's theorem relates the value of a line integral. As we have already argued in Subsection 6.4.2, this can be interpreted as measuring how strongly the vector field bends around with the boundary curve; we used

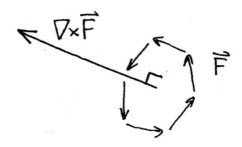

Figure 6.28:

the term "circulation" to describe this. The line integral in Stokes's curl theorem can be interpreted in this same way. Using the same notation and terminology from that subsection then, we can write Stokes's curl theorem as

$$\sigma = \int_{\partial S} \vec{F} \cdot d\vec{x} = \iint_S \left(\nabla \times \vec{F} \right) \cdot \vec{n} \, dS \tag{6.106}$$

Again interpreting this as something that accumulates over the surface, we write

$$\sigma = \iint_S d\sigma \tag{6.107}$$

and thus

$$d\sigma = \left(\nabla \times \vec{F} \right) \cdot \vec{n} \, dS \tag{6.108}$$

Again comparing this to previously discussed notions of density, the interpretation that we take from Equation 6.108 is that the curl, $\nabla \times \vec{F}$, of the vector field is a sort of an "oriented density" – by itself it is not a density exactly, but upon a dot product with $\vec{n}$ the result is the density that accumulates over the surface.

This is actually a necessary complication. While in $\mathbb{R}^2$ we can discuss simply the *amount* that the vector field circulates, in $\mathbb{R}^3$ we must also keep track of the *direction* around which the vector field is circulating, and how this relates to the surface. The curl vector, $\nabla \times \vec{F}$, has magnitude representing the amount that the vector field is circulating, and the direction of the curl is the axis around which that vector field is circulating. See Figure 6.28.

Combining these two previous observations, we could think of the curl, $\nabla \times \vec{F}$, as "oriented circulation density". (It is also called "vorticity".)

In Figure 6.29 we show how this relates to the surface integral. Consider first a case in which the curl and the normal vectors are pointing in similar directions. This means that the vector field is circulating approximately around the normal vector, and thus it is also circulating approximately parallel to the corresponding piece of surface. The dot product $\left(\nabla \times \vec{F} \right) \cdot \vec{n}$ is large, as is the amount of circulation that accumulates over that piece of surface.

On the other hand, now consider the case in which the curl and the normal vectors are pointing in nearly perpendicular directions. The vector field is circulating, but in a direction that is not along the corresponding piece of surface. The dot product $\left(\nabla \times \vec{F} \right) \cdot \vec{n}$ is small, as is the amount of circulation that accumulates over that piece of surface.

A previous "oriented density"

As it had not been discussed until now, this notion of an oriented density might seem to be a interpretation peculiar to Stokes's theorem. But it is not, and in fact it applies very nicely to another previously discussed boundary theorem – the fundamental theorem of line integrals.

The fundamental theorem of line integrals can be thought of as computing "change", as in the discussion of the fundamental theorem of calculus in Subsection 6.4.2. Using that same notation and terminology, we can write the line integral

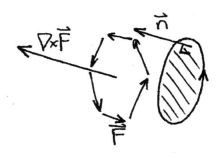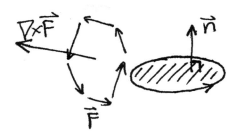

Figure 6.29:

theorem as
$$\Delta = f(\vec{b}) - f(\vec{a}) = \int_C \nabla f \cdot d\vec{x} = \int_C \nabla f \cdot \vec{T}\, ds$$

Again interpreting as an accumulation, this time over the curve C, we write
$$\Delta = \int_C d\Delta$$

and thus conclude
$$d\Delta = \nabla f \cdot \vec{T}\, ds$$

Just as in our recent analogous discussion of curl, the gradient, ∇f, can be seen here as an oriented density – by itself not a density, but upon a dot product with the unit tangent vector $\vec{T}$ the result is the density.

Along the lines of the curl discussion then, we could choose to think of the gradient vector, ∇f, as "oriented change density".

While this point of view is new, it is perfectly consistent with what we already know about the gradient. The gradient of a potential function f has both a magnitude (how fast the the potential function changes) and an associated direction (the direction in which the function is changing the fastest).

Taking this point of view, we see a strong analogy between curl and gradient. As oriented densities, both have a magnitude (curl's measuring how much the vector field circulates, gradient's measuring how much the potential function changes), and both have a direction that is used to measure how much these measured quantities are realized on the domain of the integrals in question.

Students might find it helpful to think about this analogy and their previous intuition for the gradient to develop some intuition for curl.

6.5.4 Applications to physics

Just as Gauss's divergence theorem related the two forms of two of Maxwell's equations, so does Stokes's curl theorem relate two forms of the other two of Maxwell's equations.

Magnetic fields and current

Students who have taken an introductory course in electricity and magnetism will recall that electric currents create magnetic fields. This is usually presented in the context of current flowing through a wire; a current in the wire creates a magnetic field flowing around the wire.

Here we will consider a generalization, in which we consider current to be a flow through space. That is, instead of thinking of charged particles as being confined to a wire, we imagine them flowing through space, much like any other substance flowing through space, such as air for example, or any other fluid.

In the same way that we would use a vector field to represent the flow of a fluid, we will also use a vector field to represent this flow of charged particles. We refer to this as the "current field", denoted by "$\vec{J}$".

The differential form of one of Maxwell's equations states simply that, in the case of a steady state flow,

$$\nabla \times \vec{B} = \mu_0 \vec{J} \tag{6.109}$$

where μ_0 is a universal constant known as the "permeability of free space".

(This can be related to the presentation that students might have seen in their introductory physics courses. Specifically, again it is often stated that a current creates a magnetic field that "flows" (or "circulates") around the current. Given our interpretation of curl as something that measures the amount a vector field is circulating, we can view Equation 6.109 as being a generalization of that previous idea.)

Now, suppose we have a surface S with boundary curve ∂S, and we consider the flux of each side of Equation 6.109 through S. Applying Stokes's curl theorem to the left side, we get

$$\iint_S \left(\nabla \times \vec{B}\right) \cdot d\vec{S} = \iint_S \left(\mu_0 \vec{J}\right) \cdot d\vec{S}$$

$$\int_{\partial S} \vec{B} \cdot d\vec{x} = \mu_0 \iint_S \vec{J} \cdot d\vec{S}$$

The integral on the right side, being the flux of a flow field, represents the total flow rate through the surface. In this case then, it is simply the total current through the surface. So we have

$$\int_{\partial S} \vec{B} \cdot d\vec{x} = \mu_0 I \tag{6.110}$$

where I is the current through the surface S. This is the integral version of Equation 6.109. The interpretation is that the line integral of the magnetic field around a boundary depends only on the total current through the surface, and does not depend on how that current flows through the surface.

Electric fields and magnetic flux

Another fact about electric and magnetic fields, also one of Maxwell's equations, is that a changing magnetic field creates an electric field. They too are related by curl, as expressed in the differential form by

$$\nabla \times \vec{E} = -\frac{\partial \vec{B}}{\partial t} \tag{6.111}$$

Again we can take the flux of both sides through a surface S and apply Stokes's curl theorem to the left side. We get

$$\iint_S \left(\nabla \times \vec{E}\right) \cdot d\vec{S} = \iint_S \left(-\frac{\partial \vec{B}}{\partial t}\right) \cdot d\vec{S}$$

$$\int_{\partial S} \vec{E} \cdot d\vec{x} = -\frac{d}{dt} \iint_S \vec{B} \cdot d\vec{S}$$

We refer to the flux integral on the right side of the equation as "magnetic flux", which we will write as Φ_B. So we have

$$\int_{\partial S} \vec{E} \cdot d\vec{x} = -\frac{d\Phi_B}{dt} \tag{6.112}$$

This is the integral version of Equation 6.111.

6.5. STOKES'S CURL THEOREM

We can make an interpretation of this if we consider a surface S and suppose that there is a wire along the boundary ∂S. If the magnetic flux through the surface is changing, then Equation 6.112 tells us that the line integral of the electric field around the wire will be nonzero. Of course the electric field represents force applied to charged particles, so this means that in total around the curve, there will be a nonzero force pushing the charged particles through the wire. This creates a current in the wire.

To summarize then, a changing magnetic flux can be used to create a current in a wire. This is the fundamental idea behind the design of an electric generator. A similar idea allows for the design of an electric motor.

Exercises

Exercise 6.5.1 (exer-VC-Stokes-1). The unit sphere, oriented outward, is divided into eight pieces by the eight octants of $\mathbb{R}^3$. We label the four above the four quadrants of the xy-plane as S_1, S_2, S_3, S_4, respectively; and we label the four below the four quadrants of the xy-plane as S_5, S_6, S_7, S_8, respectively. The four edges in the xy-plane are labeled E_1, E_2, E_3, E_4, again according to quadrant, and oriented counterclockwise; and similarly E_5, E_6, E_7, E_8 for the four edges in the zx-plane, and $E_9, E_{10}, E_{11}, E_{12}$ for the four edges in the yz-plane.

1. Write out the boundary of each of the eight pieces $S_1, \ldots, S_8$ as combinations of the edges.

2. Write out $\partial \left(\bigcup_{i=1}^{4} S_i \right)$ (the boundary of the upper half of the sphere), and confirm that this is equal to the sum $\sum_{i=1}^{4} \partial(S_i)$ of the boundaries of those four pieces.

3. What is the boundary of the entire sphere? Is this equal to the sum of the boundaries of the eight pieces?

Exercise 6.5.2 (exer-VC-Stokes-2). Compute the line integral of the vector field $\vec{F}(x, y, z) = (3x - 2y + 4z, 2x + 9z, x + y + 8z)$ over the curve that is the boundary of the union of the rectangles $[1, 9] \times [3, 5]$ and $[2, 4] \times [2, 7]$ in the xy-plane, oriented clockwise as seen from the positive part of the z-axis.

Exercise 6.5.3 (exer-VC-Stokes-3). Compute the line integral of the vector field $\vec{F}(x, y, z) = (xe^x + xy, y^2 e^y + xz, z \cos z)$ over the curve parametrized by $\vec{x}(t) = (\cos t e^{\sin t}, 5, \sin t e^{\cos t})$, with $t \in [0, 2\pi]$.

Exercise 6.5.4 (exer-VC-Stokes-4). Show that if $\vec{F}(x, y, z) = (P(x), Q(y), R(z))$ and if S is an oriented surface, then the line integral of F around the boundary of S must equal zero.

Exercise 6.5.5 (exer-VC-Stokes-5). Show that if $\vec{F}(x, y, z) = (P(x, z), Q(y, z), R(x, y, z))$ and if S is an oriented surface inside of a plane $z = c$, then the line integral of F around the boundary of S must equal zero.

Exercise 6.5.6 (exer-VC-Stokes-6). Compute the line integral of the vector field $\vec{F}(x, y, z) = (x^2 - z^3, y^3 - e^y, z^2 - xe^z)$ around the curve that is the boundary of the union of the rectangle $[0, 5] \times [-2, 1]$ in the xy-plane oriented upward and the rectangle $[-2, 1] \times [0, 3]$ in the yz-plane oriented in the direction of the positive part of the x-axis.

Exercise 6.5.7 (exer-VC-Stokes-7). Compute the flux of the vector field $\vec{F} = \nabla \times (y^2 z - z^2 e^y, xe^z, xy + xz^2)$ through the part of the surface $x = 4 - y^2 - z^2$ with $x \geq 0$, oriented in the direction of the positive part of the x-axis.

Exercise 6.5.8 (exer-VC-Stokes-8). Suppose that the steady state current through a wire on the x-axis is $I_0 = 6$, in the positive direction on the x-axis. What is the line integral of the magnetic field that current creates around the unit circle in the yz-plane oriented counterclockwise as seen from the positive part of the x-axis? What is the line integral of the magnetic field around the curve parametrized by $\vec{x}(t) = (3, 4 - \cos t, 3 + 2 \sin t)$?

Exercise 6.5.9 (exer-VC-Stokes-9). A wire is in the shape of the unit circle in the xy-plane. A magnet is brought close to the wire, moving at a moment in time $t = 0$ in such a way that the magnetic field is approximately uniform near the wire, given by $\vec{B}(t) = (3 - t, 4 + 2t, 2 + 5t)$. What is the line integral of the electric field in a counterclockwise direction around the wire at the time $t = 0$?

Exercise 6.5.10 (exer-VC-Stokes-10). Suppose that there is a steady state current through a wire on the z-axis. At any nearby point $\vec{x}$, the magnetic field created by this current points in a direction that is tangent to the circle centered on the z-axis that is parallel to the xy-plane and that passes through $\vec{x}$.

Show that the magnitude of the magnetic field at a point (not on the z-axis) is inversely proportional to the distance from that point to the wire. *(Hint: Consider the line integral of the magnetic field around one of the circles described above, of radius r.)*

6.6 Path Independence

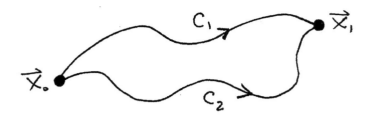

Figure 6.30:

6.6.1 Gradient fields

Suppose that we have a path C in $\mathbb{R}^n$ (here we consider primarily $\mathbb{R}^2$ and $\mathbb{R}^3$) starting at the point $\vec{x}_0$ and ending at the point $\vec{x}_1$, and a continuously differentiable function f defined on $\mathbb{R}^n$. As we have already discussed, the fundamental theorem of line integrals then tells us that

$$\int_C \nabla f \cdot d\vec{x} = f(\vec{x}_1) - f(\vec{x}_0) \tag{6.113}$$

One of the remarkable things about this equation, and all of the boundary theorems for that matter, is that each side of this equation involves computations that are performed on different domains – the left side is evaluated on the curve itself, while the right side is evaluated only on the boundary points.

A critical conclusion can be taken from this observation. The right side of the equation, since it does not involve the interior part of the curve, clearly does not depend on anything that the curve does between the endpoints. Therefore, the left side must also have this property. That is, even though the integral is performed over the entire curve, the value of the integral does not depend on what the curve does between the endpoints.

A different way to say this is that if two curves C_1 and C_2 both start at the same point $\vec{x}_0$, and both end at the same point $\vec{x}_1$, as in Figure 6.30, then the line integral of a gradient over each curve will always give the same value. That is,

$$\int_{C_1} \nabla f \cdot d\vec{x} = \int_{C_2} \nabla f \cdot d\vec{x} \tag{6.114}$$

We view this equation as being a property of gradient fields, ∇f. That is, these gradient fields have the property that their line integrals depend only on the endpoints of the paths, and not on the paths themselves. We call this property "path independence", and make the following definition.

Definition 6.6.1. *A continuous vector field $\vec{F}$, on either $\mathbb{R}^2$ or $\mathbb{R}^3$, is said to be "path independent" if, for any two points $\vec{x}_0$ and $\vec{x}_1$ and for any two curves C_1 and C_2 that both start at $\vec{x}_0$ and both end at $\vec{x}_1$, we then have*

$$\int_{C_1} \vec{F} \cdot d\vec{x} = \int_{C_2} \vec{F} \cdot d\vec{x} \tag{6.115}$$

The result expressed in Equation 6.114 can then be restated as a theorem.

Theorem 6.6.1. *If a continuous vector field $\vec{F}$ is the gradient of a function f defined on $\mathbb{R}^n$, then $\vec{F}$ is path independent.*

We can represent this theorem in a shorthand with the diagram

$$\left(\vec{F} = \nabla f\right) \longrightarrow \left(\vec{F} \text{ is path independent}\right)$$

One might reasonably ask then if there are any other vector fields that have this property of path independence. As it turns out, there are not, and we can write this also as a theorem.

Theorem 6.6.2. *If $\vec{F}$ is a path independent vector field, then there exists some function f such that $\vec{F}$ is the gradient of f.*

The proof of this theorem comes very nicely from the fundamental theorem of line integrals.

We begin with a path independent vector field $\vec{F}$. Given this vector field, we define a function, f, by writing

$$f(\vec{x}_0) = \int_{\vec{0}}^{\vec{x}_0} \vec{F} \cdot d\vec{x} \tag{6.116}$$

(Of course we have not yet used this notation when writing line integrals; it means simply that we compute the line integral over a curve that begins at $\vec{0}$ and ends at $\vec{x}_0$. Upon first thought this would appear not to be well-defined since there are many curves that begin at $\vec{0}$ and end at $\vec{x}_0$ – but of course by assumption $\vec{F}$ is path independent, so the choice of curve does not matter.)

The fundamental theorem of line integrals of course tells us that

$$f(\vec{x}_0) - f(\vec{0}) = \int_{\vec{0}}^{\vec{x}_0} \nabla f \cdot d\vec{x} \tag{6.117}$$

and we can use Equation 6.116 to see easily that $f(\vec{0}) = 0$. Then equating the integrals in Equations 6.116 and 6.117, we have

$$\int_{\vec{0}}^{\vec{x}_0} \vec{F} \cdot d\vec{x} = \int_{\vec{0}}^{\vec{x}_0} \nabla f \cdot d\vec{x} \tag{6.118}$$

Since Equation 6.118 must be true for all points $\vec{x}_0$ and all curves from $\vec{0}$ to $\vec{x}_0$, we conclude that we must have $\vec{F} = \nabla f$, as desired.

We can then update the shorthand diagram to include both Theorems 6.6.1 and 6.6.2,

$$\left(\vec{F} = \nabla f\right) \longleftrightarrow \left(\vec{F} \text{ is path independent}\right)$$

So we have the very elegant result that (among continuous vector fields on $\mathbb{R}^n$) path independent vector fields are exactly gradient fields.

Example 6.6.1. The vector field $\vec{F} = (P, Q, R) = (z\,p(\vec{x}), Q(\vec{x}), R(\vec{x}))$ is known to be the gradient of the function f, but none of the functions p, Q, R or f is given. Suppose we wish to compute the line integral $\int_C \vec{F} \cdot d\vec{x}$, where C is parametrized by $\vec{x}(t) = (t^3 e^t, t^2 - t^3, t^6 - t^2)$ with $t \in [0, 1]$.

Of course we cannot compute this line integral directly from the parametrization because the component functions of the vector field are not given.

However, since we are given that the vector field is a gradient, we immediately conclude that the vector field is also path independent. So we can change the path without affecting the value of the integral, as long as the starting and ending points do not change.

The starting and ending points are $(0, 0, 0)$ and $(e, 0, 0)$, respectively. Instead of using the given curve, we will instead invoke the path independence and consider instead the curve C_2 parametrized by $\vec{x}(t) = (et, 0, 0)$ with $t \in [0, 1]$.

6.6. PATH INDEPENDENCE

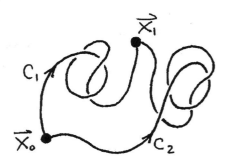

Figure 6.31:

We then compute directly

$$\int_C \vec{F} \cdot d\vec{x} = \int_{C_2} \vec{F} \cdot d\vec{x}$$
$$= \int_0^1 \vec{F} \cdot \vec{x}' \, dt$$
$$= \int_0^1 \begin{bmatrix} z\,p(\vec{x}) \\ Q(\vec{x}) \\ R(\vec{x}) \end{bmatrix} \cdot \begin{bmatrix} e \\ 0 \\ 0 \end{bmatrix} \, dt$$
$$= \int_0^1 e z\, p(\vec{x}) \, dt$$

and because $z = 0$ on the entire curve C_2, this must be zero.

Gradient fields are sometimes called "conservative" vector fields because, thought of as force fields, these are exactly the vector fields for which there can be a notion of potential energy, which then allows for the notion of the conservation of energy. The notion of potential energy turns out to come directly from the antigradient function f for which $\nabla f = \vec{F}$, so that function is sometimes called the "potential function".

6.6.2 Curl

There are also important connections between path independence and some of the other operators that were derived in our discussions of boundary theorems. In order to discuss these connections we must separate the discussions of $\mathbb{R}^2$ and $\mathbb{R}^3$. Here we will discuss what happens in $\mathbb{R}^3$; the arguments in $\mathbb{R}^2$ are entirely analogous and will be left as exercises to the reader.

Suppose we have two points $\vec{x}_0$ and $\vec{x}_1$ and two curves C_1 and C_2 that both start at $\vec{x}_0$ and end at $\vec{x}_1$.

We must begin by acknowledging an awkwardness. That is, it is tempting to say that there must be an oriented surface S for which $\partial S = C_2 - C_1$. This would certainly be the case in the first situation drawn in Figure 6.31.

(Note that the minus sign is a result of how we define orientations on boundaries. Here we have drawn the curves C_1 and C_2 with orientations such that their boundaries have the same orientation. And in doing so, the consequence is that they themselves have the opposite orientations when thought of as part of the boundary of S.)

But in the second case in that figure it is harder to see how this would be the case at all. The curves are tied in knots, and they do not appear to be the boundary of anything.

It turns out that this complication does not effect the results that we will conclude in this section. That is, there are more sophisticated techniques that allow us to get around this problem and arrive at the same results. We will not discuss those techniques in this text; instead, we will simply argue as if these problems do not exist. So, we will state simply that, "effectively", there is a surface S for which $\partial S = C_2 - C_1$, and claim that we can safely ignore the complication.

Given a surface S with $\partial S = C_2 - C_1$, and a continuously differentiable vector field $\vec{F}$ on $\mathbb{R}^3$, we can apply Stokes's curl theorem and rewrite the result as

$$\iint_S \left(\nabla \times \vec{F}\right) \cdot d\vec{S} = \int_{\partial S} \vec{F} \cdot d\vec{x} = \int_{C_2} \vec{F} \cdot d\vec{x} - \int_{C_1} \vec{F} \cdot d\vec{x} \tag{6.119}$$

We can now consider the consequences of Equation 6.119.

First suppose that $\nabla \times \vec{F} = \vec{0}$. That is, that the curl vector field itself is identically zero, everywhere. Then certainly the left side of Equation 6.119 would always be zero, which immediately implies that the right side must always be zero. That is, for any points $\vec{x}_0$ and $\vec{x}_1$ and curves C_1 and C_2 that both start at $\vec{x}_0$ and end at $\vec{x}_1$, those two integrals must be equal. This is exactly what it means for the vector field to be path independent.

Now we consider the argument in reverse. Suppose that a vector field is known to be path independent. Then for any surface S with $\partial S = C_2 - C_1$, the right side of Equation 6.119 would always be zero, which immediately implies that the left side must always be zero. And of course if an integral of $\nabla \times \vec{F}$ is zero over every possible surface, then we must have that the integrand $\nabla \times \vec{F}$ is zero itself.

These arguments give us two more results relating to path independence. We can state them with the following theorem.

Theorem 6.6.3. *A continuously differentiable vector field $\vec{F}$ on $\mathbb{R}^3$ is path independent if and only if the curl vector field $\nabla \times \vec{F}$ is the zero vector field.*

We can summarize this theorem by appending to the summary diagram from Subsection 6.6.1,

$$\left(\vec{F} = \nabla f\right) \longleftrightarrow \left(\vec{F} \text{ is path independent}\right) \longleftrightarrow \left(\nabla \times \vec{F} = \vec{0}\right)$$

Recall of course that in Section 6.5 we interpreted the curl as representing how much the vector field suggests a circulation, or rotating around an axis. In light of this interpretation and Theorem 6.6.3, we sometimes refer to a path independent vector field as "irrotational".

Exercises

Exercise 6.6.1 (exer-VC-PI-1). Show that a vector field $\vec{F}$ on $\mathbb{R}^2$ is path independent iff $\operatorname{grn}\vec{F} = 0$ everywhere. *(Hint: You can use an argument similar to that used in this section for vector fields in $\mathbb{R}^3$.)*

Exercise 6.6.2 (exer-VC-PI-2). The vector field $\vec{F} = (y, x)$ is path independent, so all paths from $(0, 1)$ to $(0, -1)$ should give the same line integral of that vector field. Confirm that this is the case for the straight line path, the path going counterclockwise along the unit circle, and the path going clockwise along the unit circle.

Exercise 6.6.3 (exer-VC-PI-3). Use path independence, but not the fundamental theorem of line integrals, to compute the line integral $\int_C \nabla f \cdot d\vec{x}$, with $f(x, y) = x^2 + y^2$ and C parametrized by $\vec{x}(t) = ((\pi t - t^2 + 1) \cos t, (\pi t - t^2 + 1) \sin t)$ for $t \in [0, \pi]$.

Exercise 6.6.4 (exer-VC-PI-4). Confirm that $\vec{F} = (ye^{x+y} + 2xy \sin(x^2 y), (1+y)e^{x+y} + x^2 \sin(x^2 y))$ is path independent. Then use that to compute the line integral of this vector field over the path parametrized by $\vec{x}(t) = ((\pi t - t^2 + 1) \cos t, (\pi t - t^2 + 1) \sin t)$ for $t \in [0, \pi]$.

Exercise 6.6.5 (exer-VC-PI-5). For each of the following, determine if the vector field is path independent. If it is, then use path independence (without using the fundamental theorem of line integrals) to compute the line integral of the vector field over the curve that is parametrized by $\vec{x}(t) = (te^{t^2-t}, \cos \pi t, t^3)$, with $t \in [0, 1]$.

6.6. PATH INDEPENDENCE

1. $\vec{F}(x,y,z) = (yz, xz, xy)$
2. $\vec{F}(x,y,z) = (y, 0, xyz)$
3. $\vec{F}(x,y,z) = (2xyz, x^2z, x^2y + x)$

Exercise 6.6.6 (**exer-VC-PI-6**). Use Equation 6.116 directly to find a function f that is an antigradient for the path independent vector field $\vec{F}(x,y,z) = (3x^2, z^2, 2yz)$. Confirm by taking the gradient of this f that this gave a valid result.

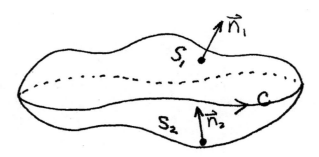

Figure 6.32:

6.7 Surface Independence

6.7.1 Curl fields

Most of the ideas in Section 6.6 can conveniently translate up one dimension. That is, instead of considering two curves that have the same boundary points, we can instead consider two surfaces that have the same boundary curve. Most of the arguments from that section can then be adapted to give analogous results.

Suppose that we have a surface S in $\mathbb{R}^3$ with boundary curve C, and a continuously differentiable vector field $\vec{F}$ defined on $\mathbb{R}^3$. Stokes's curl theorem then tells us that

$$\iint_S \left(\nabla \times \vec{F}\right) \cdot d\vec{S} = \int_C \vec{F} \cdot d\vec{x} \tag{6.120}$$

Again, we have that each side of this equation involves computations that are performed on different domains – the left side is performed on the surface itself, while the right side is evaluated only on the boundary curve.

Since the right side involves only the boundary, any two surfaces with the same boundary will certainly result in the same value on the right side, and so therefore also on the left side. Said precisely then, if S_1 and S_2 have the same boundary curve $C = \partial S_1 = \partial S_2$ as in Figure 6.32, then for any continuously differentiable vector field $\vec{F}$ we will have

$$\iint_{S_1} \left(\nabla \times \vec{F}\right) \cdot d\vec{S} = \iint_{S_2} \left(\nabla \times \vec{F}\right) \cdot d\vec{S} \tag{6.121}$$

We view this equation as being a property of curl fields, $\nabla \times \vec{F}$. That is, these curl fields have the property that their flux integrals depend only on the boundary of the surface, and not on the surface itself. We call this property "surface independence", and make the following definition.

Definition 6.7.1. *A continuous vector field $\vec{G}$ on $\mathbb{R}^3$ is said to be "surface independent" if, for any two surfaces S_1 and S_2 with $\partial S_1 = \partial S_2$, we then have*

$$\iint_{S_1} \vec{G} \cdot d\vec{S} = \iint_{S_2} \vec{G} \cdot d\vec{S} \tag{6.122}$$

The result expressed in Equation 6.121 can then be restated as a theorem.

Theorem 6.7.1. *If a continuous vector field $\vec{G}$ is the curl of a vector field $\vec{F}$ on $\mathbb{R}^3$, then $\vec{G}$ is surface independent.*

We can represent this theorem in a shorthand with the diagram

$$\left(\vec{G} = \nabla \times \vec{F}\right) \longrightarrow \left(\vec{G} \text{ is surface independent}\right)$$

6.7. SURFACE INDEPENDENCE

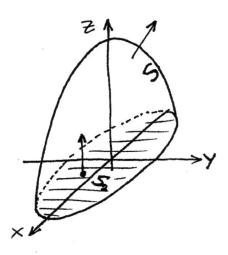

Figure 6.33:

One might reasonably ask then if there are any other vector fields that have this property of surface independence. As it turns out, there are not, and we can write this also as a theorem.

Theorem 6.7.2. *If $\vec{G}$ is a surface independent vector field, then there exists some vector field $\vec{F}$ such that $\vec{G}$ is the curl of $\vec{F}$.*

The proof of this theorem unfortunately is not conveniently analogous to what we did at this point in Section 6.6. We will not show here the proof of this result.

We can update the shorthand diagram to include both Theorems 6.7.1 and 6.7.2,

$$\left(\vec{G} = \nabla \times \vec{F}\right) \longleftrightarrow \left(\vec{G} \text{ is surface independent}\right)$$

So we have the very elegant result that (among continuous vector fields on $\mathbb{R}^3$) surface independent vector fields are exactly curl fields.

Example 6.7.1. Suppose we are given that the vector field $\vec{G} = (y, x + z^2, xy)$ is the curl of an unknown vector field $\vec{F}$, and we want to compute the flux of $\vec{G}$ through the upward oriented surface S that is the part of the elliptical paraboloid with equation $z = 36 - 4x^2 - 9y^2$ that is above the xy-plane. See **Figure 6.33**.

Given that $\vec{G}$ is the curl of another vector field, we can conclude that the vector field is surface independent. This means that we can change the domain of the flux integral to any surface that has the same boundary as S. The boundary of course is an ellipse entirely within the xy-plane, so very reasonably we choose the surface S_2 that is the solid ellipse in the xy-plane with the same boundary. We must again choose the upward orientation so that the boundary is oriented correctly.

We can then rewrite the desired flux as

$$\iint_S \vec{G} \cdot d\vec{S} = \iint_{S_2} \vec{G} \cdot d\vec{S}$$
$$= \iint_{S_2} \vec{G} \cdot \vec{n}\, dS$$

Very conveniently though, because our new surface S_2 is entirely within the xy-plane and oriented upward, we can

write down the unit normal vector very easily as $\vec{n} = (0, 0, 1)$. So our integral becomes

$$\iint_S \vec{G} \cdot d\vec{S} = \iint_{S_2} \begin{bmatrix} y \\ x + z^2 \\ xy \end{bmatrix} \cdot \begin{bmatrix} 0 \\ 0 \\ 1 \end{bmatrix} dS$$

$$= \iint_{S_2} xy \, dS$$

Of course this integrand has odd symmetry over the xz-plane and the domain is symmetric through that same plane, so the integral is equal to zero by symmetry.

6.7.2 Divergence

There are also important connections between surface independence and divergence.

Suppose we have two surfaces S_1 and S_2 with the same boundary curve C, as in Figure 6.32. Again, "effectively" (ignoring awkward cases that can be shown not to be relevant to the conclusions we will draw here), there must be an oriented region R for which $\partial R = S_1 - S_2$.

(Note that the minus sign is again a result of how we define orientations on boundaries. Here we have drawn the surfaces S_1 and S_2 with orientations such that their boundaries have the same orientation. And in doing so, the consequence is that they themselves have the opposite orientations when thought of as part of the boundary of R.)

Given a region R with $\partial R = S_1 - S_2$, and a continuously differentiable vector field $\vec{G}$ on $\mathbb{R}^3$, we can apply Gauss's divergence theorem and rewrite the result as

$$\iiint_R \nabla \cdot \vec{G} \, dV = \iint_{\partial R} \vec{G} \cdot d\vec{S} = \iint_{S_1} \vec{G} \cdot d\vec{S} - \iint_{S_2} \vec{G} \cdot d\vec{S} \qquad (6.123)$$

We can now consider the consequences of Equation 6.123.

First suppose that $\nabla \cdot \vec{G} = 0$. That is, that the divergence function itself is identically zero, everywhere. Then certainly the left side of Equation 6.123 would always be zero, which immediately implies that the right side must always be zero. That is, for any two surfaces S_1 and S_2 with the same boundary curve C, those two flux integrals must be equal. This is exactly what it means for the vector field to be surface independent.

Now we consider the argument in reverse. Suppose that a vector field $\vec{G}$ is known to be surface independent. Then for any region R with $\partial R = S_1 - S_2$, the right side of Equation 6.123 would always be zero, which immediately implies that the left side must always be zero. And of course if an integral of $\nabla \cdot \vec{G}$ is zero over every possible region, then we must have that the integrand $\nabla \cdot \vec{G}$ is identically zero itself.

These arguments give us two more results relating to surface independence. We can state them with the following theorem.

Theorem 6.7.3. *A continuously differentiable vector field $\vec{G}$ on $\mathbb{R}^3$ is surface independent if and only if the divergence function $\nabla \cdot \vec{G}$ is identically zero.*

We can summarize this theorem by appending to the summary diagram from Subsection 6.7.1,

$$\left(\vec{G} = \nabla \times \vec{F} \right) \longleftrightarrow \left(\vec{G} \text{ is surface independent} \right) \longleftrightarrow \left(\nabla \cdot \vec{G} = 0 \right)$$

Example 6.7.2. Suppose we are given a vector field $\vec{G} = (y^3, x^3 + z, xy)$, and we want to compute the flux of $\vec{G}$ through the same surface from Example 6.7.1, that is, the upward oriented surface S that is the part of the elliptical paraboloid with equation $z = 36 - 4x^2 - 9y^2$ that is above the xy-plane. See Figure 6.33.

6.7. SURFACE INDEPENDENCE

This example is very similar to that in Example 6.7.1, but, critically, we are not given that the vector field $\vec{F}$ is a curl field. So we cannot immediately conclude that this vector field is surface independent.

We can take advantage of Theorem 6.7.3 though, by computing the divergence of $\vec{G}$. It is easy to compute that $\nabla \cdot \vec{G} = 0$, and so the theorem then tells us that the vector field is indeed surface independent.

This means that we can again change the domain of the flux integral to any surface that has the same boundary as S. Proceeding from here just as in Example 6.7.1, we compute that the flux is

$$\iint_S \vec{G} \cdot d\vec{S} = \iint_{S_2} \begin{bmatrix} y^3 \\ x^3 + z \\ xy \end{bmatrix} \cdot \begin{bmatrix} 0 \\ 0 \\ 1 \end{bmatrix} dS$$
$$= \iint_{S_2} xy \, dS$$
$$= 0$$

As we observed at the beginning of this section, this idea of surface independence is highly analogous to that of path independence discussed in Section 6.6. But in fact, from a certain point of view surface independence can be argued to be even more useful as a computational tool than path independence. The reason for this is simply a matter of available options.

For path independent vector fields, we know that the vector field is a gradient. Then the line integral in question can often be computed with the fundamental theorem of line integrals by taking the antigradient of the given vector field. This is discussed in detail in Subsection 6.2.3.

In theory the analogous thing could be said about surface independent vector fields, and given flux integrals could be computed with Stokes's curl theorem by taking an "anticurl". Unfortunately, it just turns out that computing anticurl is not as convenient as computing antigradients. So, since the alternative options are not as attractive, the use of surface independence is correspondingly more valuable.

We end this section with some motivation for another way to think of a surface independent vector field.

Suppose that we think of a given vector field as representing a fluid flow. Recalling our discussion of fluid flow in Subsection 6.4.3, one interpretation of divergence is that it represents the way in which the fluid is being compressed or decompressed (assuming that it cannot spontaneously generate or disappear).

We have just seen though that a surface independent vector field has divergence equal to zero. Thinking of the above interpretation then, a surface independent vector field could be thought of as representing the flow of an incompressible fluid that has no sources or sinks, such as the flow of water in a closed fish tank.

Exercises

Exercise 6.7.1 (exer-VC-SI-1). Determine which of the following vector fields is surface independent.

1. $\vec{F}(x, y, z) = (e^{xy} + 3xy^2, x^2 - y^3, -yze^{xy})$
2. $\vec{F}(x, y, z) = (x - 3y + 4z, 2x - y - z, 6x + 4y + z)$
3. $\vec{F}(x, y, z) = (x^2y - 3y^3, y^3 - 4xy^2z, 7xyz)$

4. $\vec{F}(x,y,z) = \nabla \times (x^2 e^y, e^{y^3 - x^3}, \sin(xyz))$

Exercise 6.7.2 (exer-VC-SI-2). In each of the following, the two given surfaces S_1 and S_2 have the same boundary, so that $S_1 - S_2$ is the boundary of some solid region T. In each case, describe the region T, making sure to note orientations as needed.

1. S_1 is the upper half of the unit sphere, oriented upward ; S_2 is the lower half of the unit sphere, oriented upward.

2. S_1 is the upper half of the unit sphere, oriented downward ; S_2 is the unit disk in the xy-plane, oriented downward.

3. S_1 is the part of the graph of the function $f(x,y) = x(1 - x^2 - y^2)$ with $x^2 + y^2 \leq 1$, oriented upward; S_2 is the unit disk in the xy-plane, oriented upward.

Exercise 6.7.3 (exer-VC-SI-3). Compute the flux of the vector field $\vec{F} = (z - x^2 + 3xy^2, 2xy - y^3, 2xye^x)$ through the surface S that is the part of the surface $x^2 + z^4 + y = 5$ with $y \geq 0$, oriented in the positive y direction.

Exercise 6.7.4 (exer-VC-SI-4). Compute the flux of the vector field $\vec{F} = (x^2 + 5x - yz, 6 - 2xy, e^y - 5z)$ through the surface S that is the part of the surface $x = y^2 + z^2 - 4$ with $x \leq 0$, oriented in the negative x direction.

Exercise 6.7.5 (exer-VC-SI-5). Compute the flux of the vector field $\vec{F} = (e^z - 2xy, y^2 - e^z, 2xy - y^2)$ through the surface S that is the part of the surface $z = x^4 + e^{y^2}$ with $x + y + z \leq 1$, oriented upward.

Exercise 6.7.6 (exer-VC-SI-6). Let C be the unit circle in the xy-plane, oriented counterclockwise as seen from above. The vector field $\vec{F} = (z, x, y)$ is surface independent, so the flux through every surface with boundary C should be the same. Confirm that this is the case with the upper half of the unit sphere, the lower half of the unit sphere, and the unit disk in the xy-plane.

6.8 Lifetime Theorems

Each of sections 6.2 through 6.5 is involved with the discussion of a single boundary theorem. All of the motivations given for these theorems are analogous, showing strong connections between these theorems. We summarize these theorems, and also underscore their similarities, by representing them all together on the diagrams below.

Vector Calculus in $\mathbb{R}^2$

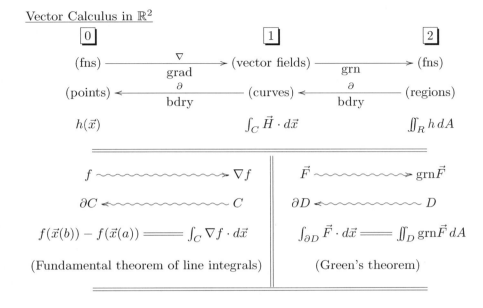

Vector Calculus in $\mathbb{R}^3$

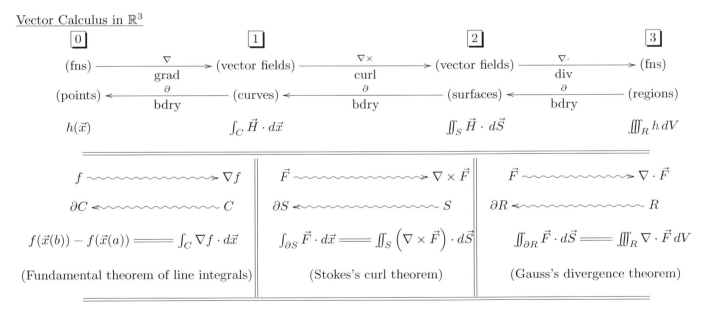

But in fact there are even more connections between these theorems. We will find in this section that the operators created in the developments of these theorems interact with each other in natural ways; and the boundary theorems themselves can be used together to demonstrate these relationships.

Even more, these relationships that we will discuss in this section are themselves all related. That is, in the same way that the boundary theorems are all special cases of the "general Stokes's theorem" that we cannot discuss in this text, also the theorems we will discuss in this section are all special cases of a much larger theorem that we cannot discuss in this text. Specifically, they can all be viewed as coming from the "de Rham cohomology theorem".

In the same way that we use the term "boundary theorems" to refer to all of the special cases of the general Stokes's theorem, likewise we will use a term to refer to all of these corollaries of the de Rham cohomology theorem. The term we will use here is "lifetime theorems", as in the title of this section. We will see the motivation for this term in Subsection 6.8.1.

6.8.1 "Life cycle" of an integrand

In this subsection we set up metaphors that will allow for a convenient discussion of the theorems later in this section.

We consider here the compositions of the various operators acting on the integrands in the above diagrams. That is, we consider an integrand at some position on one of the diagrams, apply the appropriate operator, then apply the next operator to the result, and continue. With the application of each operator, we will interpret the passing of one unit of "time".

If an integrand is nonzero, we think of that integrand as being "alive". Applying the appropriate operator, if the result is still nonzero, then we think of the original integrand as having "survived", to the next column.

An "alive" integrand might perhaps have survived from a previous column, but it might not necessarily have done so. If it is "alive" without having "survived" from a previous column, we say that the operator is "born" at that column.

Continuing on, at some point the result of applying an operator might be zero. Certainly then the result of all subsequent operators will continue to be zero. We think of the integrand then as being "dead".

Example 6.8.1. Consider the function $f(x,y,z) = x^2 + y^2 + z^2$, thought of as a column zero integrand in $\mathbb{R}^3$. Obviously this function has not "survived" from a previous column because there are no previous columns, so we say that this integrand is "born" here in the first column

The operator to apply to this integrand is the gradient operator; applying this operator, we get

$$\nabla f = \begin{bmatrix} 2x \\ 2y \\ 2z \end{bmatrix} = \vec{F} \tag{6.124}$$

This resulting integrand, the vector field $\vec{F}$ thought of in column one, is not zero, so we say that the original integrand f from column zero has "survived" to column one.

Applying now the next operator, the curl operator, we get

$$\nabla \times \vec{F} = \begin{bmatrix} 0 \\ 0 \\ 0 \end{bmatrix} = \vec{G} \tag{6.125}$$

This time the resulting integrand, the vector field $\vec{G}$ thought of in column two, is in fact the zero vector field. So we say that the original integrand from column zero "died" here in column two.

We can represent this process with the diagram

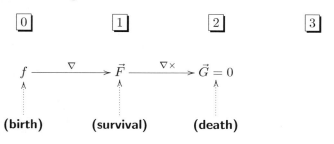

6.8. LIFETIME THEOREMS

Example 6.8.2. Consider the vector field $\vec{F} = (-y, x, 0)$, thought of as a column one integrand in $\mathbb{R}^3$.

The operator to apply to this integrand is the curl operator; applying this operator, we get

$$\nabla \times \vec{F} = \begin{bmatrix} 0 \\ 0 \\ 2 \end{bmatrix} = \vec{G} \qquad (6.126)$$

This resulting integrand, the vector field $\vec{G}$ in column two, is not zero. This tells us two things. Most obviously of course, the original integrand $\vec{F}$ survived to column two.

Secondly though, the fact that the curl is not zero tells us that the vector field is not path independent, and therefore also that the vector field cannot be the gradient of a function. This means that our original vector field $\vec{F}$ could not have been a survivor from column zero. So, this means that the integrand $\vec{F}$ was indeed born in column one.

Applying the next operator, the divergence operator, we get

$$\nabla \cdot \vec{G} = 0 = h \qquad (6.127)$$

This time the resulting integrand, the function h thought of in column three, is in fact the zero function. So we say that the original integrand "died" here in column three.

We can represent this process with the diagram

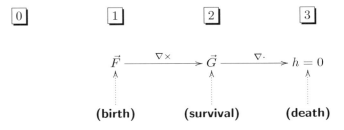

6.8.2 Lifetime theorems

Having set up the language as in the previous subsection, we can conveniently summarize many results in one single sentence. That is, for integrands that are defined on either $\mathbb{R}^2$ or $\mathbb{R}^3$: **"Every integrand has a lifetime of 2."**. (There is a single exception to this statement; we will mention this later.)

Examples 6.8.1 and 6.8.2 illustrate this statement – in each of those examples, the integrand in question dies exactly two columns after the one in which it is born.

We now enumerate and discuss the specific results that are summarized above. Each of these can be thought of as coming from the de Rham cohomology theorem, and will be referred to in this text as a "lifetime theorem". (Note, for convenience, all functions and vector fields referred to in the theorems below are assumed to be continuously differentiable, and defined on all of either $\mathbb{R}^2$ or $\mathbb{R}^3$.)

Theorem 6.8.1. *If a vector field $\vec{F}$ on $\mathbb{R}^2$ is the gradient of a function f, then $\operatorname{grn}\vec{F} = 0$.*

Theorem 6.8.2. *If a vector field $\vec{F}$ on $\mathbb{R}^3$ is the gradient of a function f, then $\nabla \times \vec{F} = \vec{0}$.*

Theorem 6.8.3. *If a vector field $\vec{G}$ on $\mathbb{R}^3$ is the curl of a vector field $\vec{F}$, then $\nabla \cdot \vec{G} = 0$.*

Theorem 6.8.4. *If a vector field $\vec{F}$ on $\mathbb{R}^2$ has $\operatorname{grn}\vec{F} = 0$, then it is the gradient of some function f.*

Theorem 6.8.5. *If a vector field $\vec{F}$ on $\mathbb{R}^3$ has $\nabla \times \vec{F} = \vec{0}$, then it is the gradient of some function f.*

Theorem 6.8.6. *If a vector field $\vec{G}$ on $\mathbb{R}^3$ has $\nabla \cdot \vec{G} = 0$, then it is the curl of some vector field $\vec{F}$.*

Theorem 6.8.7. *Every function f on $\mathbb{R}^3$ is the divergence of some vector field $\vec{F}$.*

Theorem 6.8.8. *Every function f on $\mathbb{R}^2$ is the grn of some vector field $\vec{F}$.*

Theorems 6.8.2 and 6.8.5 follow immediately from our discussions of path independence in Section 6.6. Theorems 6.8.1 and 6.8.4 follow from the corresponding arguments in $\mathbb{R}^2$. And Theorem 6.8.3 follows immediately from our discussions of surface independence in Section 6.7. (Of course many of our conclusions about path independence and surface independence come from the boundary theorems, so many of these lifetime theorems can be thought of in terms of the boundary theorems.)

Theorems 6.8.6, 6.8.7 and 6.8.8 do not follow conveniently from any discussions we have given here, but they are still special cases of the de Rham cohomology theorem and fit in nicely with the other theorems; we will not prove them here.

Theorems 6.8.1, 6.8.2 and 6.8.3 all refer to integrands "dying"; together, they say that no integrand can have a lifetime greater than two. (Note, in addition to proving these theorems with arguments from Sections 6.6 and 6.7, these three theorems can also be checked by direct computation from the formulas for these operators. Students should make sure to do this.)

Similarly, Theorems 6.8.4, 6.8.5, 6.8.6, 6.8.7 and 6.8.8 all refer to integrands having "survived". These theorems together say that every integrand, with one exception, must have a lifetime of at least two.

The exception is simply the only circumstance that is not addressed by those five "survival theorems". That is, the case of functions in column zero for either $\mathbb{R}^2$ or $\mathbb{R}^3$. Of course if the gradient of such a function is not zero then indeed it does still achieve a lifetime of two. But if the gradient of such a function is zero, there is simply no previous column for it possibly to have survived from. So its lifetime would only be one – not two.

But these cases are all easily remembered. Because, of course if a function has a gradient of zero, then that function must be a constant. So constant functions are the only exceptions to our rule.

We can append Theorems 6.8.1 through 6.8.6 to our diagrams, illustrating their strong similarities. Each theorem is represented on its own line. Each theorem is an implication ("If <*hypothesis*>, then <*conclusion*>") and each hypothesis and conclusion involves the application of an operator. For the hypotheses, the operator is represented with a solid arrow; for the conclusions, the operator is represented by a dotted arrow. We leave off Theorems 6.8.7 and 6.8.8 because they will not be used significantly outside of this section.

With these additions, we now have the final form of our diagrams, in Figure 6.34. Students should make sure to note this page number (page 453) for convenient future reference.

As a final comment, note that in the statements of the above theorems, we require all of the functions and vector fields to be defined on the entire space ($\mathbb{R}^2$ or $\mathbb{R}^3$). We make these requirements in order to be able to draw the results from the de Rham cohomology theorem, as desired. In fact, without making this assumption, some of these results would not be true. For example, if a vector field $\vec{F}$ is defined on the "punctured plane" (this includes all of the points in $\mathbb{R}^2$ except for $\{\vec{0}\}$), then it is possible to have $\text{grn}\vec{F} = 0$ even when $\vec{F}$ is not the gradient of any function. Such examples are why we must require in Theorem 6.8.4 that the vector field in question be defined on all of $\mathbb{R}^2$. (An explicit example of such a vector field is constructed in Exercise 6.8.11.)

6.8.3 Examples

Example 6.8.3. Suppose we are interested in computing a line integral of the vector field $\vec{F} = (xy, x^2 z, yz)$. Ideally we would like to use the fundamental theorem of line integrals, but note that we are not given any information as to whether this vector field is a gradient, or not.

Of course we could simply attempt directly to compute the antigradient, but it is easier to use the lifetime theorems

6.8. LIFETIME THEOREMS

Figure 6.34:

Vector Calculus in $\mathbb{R}^2$

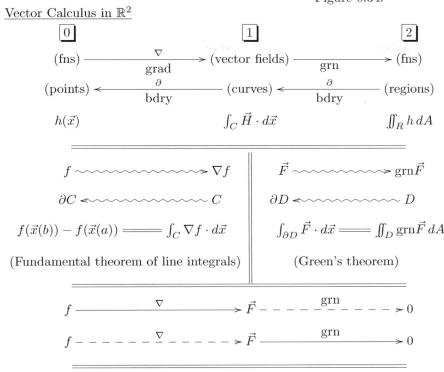

Vector Calculus in $\mathbb{R}^3$

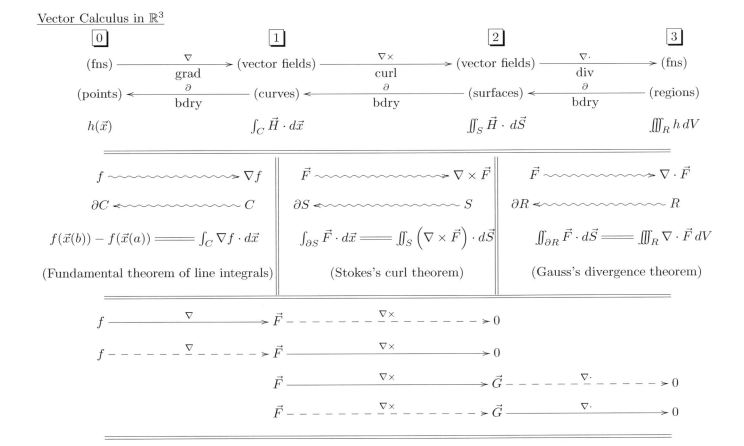

to answer this question. Computing the curl of this vector field, we get

$$\nabla \times \vec{F} = (y - x^2, 0, 2xz - x) \neq \vec{0} \qquad (6.128)$$

This is not the zero vector field, so Theorem 6.8.2 then tells us that $\vec{F}$ cannot be a gradient of any function and therefore we cannot use the fundamental theorem of line integrals. (In fact, notice that we could make the determination that the curl is not zero after computing only the first component of the curl... the computation of the other two components was not even necessary.)

We can also think of this simply in terms of lifetimes. Since $\nabla \times \vec{F} \neq \vec{0}$ in column two, it could not have been alive in column zero, else its lifetime would be at least three. So $\vec{F}$ cannot be a gradient.

Example 6.8.4. Suppose we want to compute a flux integral of the vector field $\vec{F} = (yz^3, e^{xz}, x - \sin y)$ over a given surface.

We can quickly notice that

$$\nabla \cdot \vec{F} = 0 + 0 + 0 = 0 \qquad (6.129)$$

(This is evident even by casual inspection upon noting that there are no x's in the first component, no y's in the second component, and no z's in the third component.)

They by Theorem 6.8.6, we note that this vector field must be the curl of some other vector field, the determination of which would allow for the use of Stokes's curl theorem. Also of course we know that this means the vector field is surface independent.

Again we can think of this in terms of lifetimes. Since the vector field dies in column three, it must have been born in column one to have the necessary lifetime of two. That is, the given vector field must be the curl of a vector field in column one.

Example 6.8.5. Suppose we wish to compute the line integral of the vector field $\vec{F} = (2xy^3 + 9x^2y, 3x^3 + 3x^2y^2 + 4y^3)$ over the curve C parametrized by $\vec{x}(t) = (t^3 e^{t-1}, \cos(\pi t))$, with $t \in [0, 1]$.

Again we would like to use the fundamental theorem of line integrals, but do not know if the given vector field is a gradient, or not. As we are in $\mathbb{R}^2$, we compute $\operatorname{grn}\vec{F}$, giving us:

$$\begin{aligned}
\operatorname{grn}\vec{F} &= \frac{\partial}{\partial x}(3x^3 + 3x^2y^2 + 4y^3) - \frac{\partial}{\partial y}(2xy^3 + 9x^2y) \\
&= (9x^2 + 6xy^2) - (6xy^2 + 9x^2) \\
&= 0
\end{aligned}$$

We conclude from this that indeed this vector field must be the gradient of some function f, and then can proceed to attempt to find this function.

$$\begin{aligned}
f &= \int 2xy^3 + 9x^2y \, dx & f &= \int 3x^3 + 3x^2y^2 + 4y^3 \, dy \\
&= x^2y^3 + 3x^3y + k_1(y) & &= 3x^3y + x^2y^3 + y^4 + k_2(x)
\end{aligned}$$

So we must have that $f(x, y) = 3x^3y + x^2y^3 + y^4 + c$. The curve begins at the point $(0, 1)$ and ends at the point

6.8. LIFETIME THEOREMS

$(1, -1)$, so by the fundamental theorem of line integrals we have

$$\begin{aligned}\int_C \vec{F} \cdot d\vec{x} &= f(\vec{x}(1)) - f(\vec{x}(0)) \\ &= f(1, -1) - f(0, 1) \\ &= (-3) - (1) = -4\end{aligned}$$

Example 6.8.6. In a steady state (when the electric and magnetic fields are constant), one of Maxwell's laws states that the curl of the electric field must be zero (see Subsection 6.5.4).

According to the corresponding lifetime theorem then, this means that in a steady state the electric field must itself be the gradient of a function. We call this function the "electric potential function", written $(-V)$ (the negative is an acknowledgement that the electric field points away from higher potentials, not toward).

As per the fundamental theorem of line integrals then, the values of this electric potential function give us information about line integrals of the electric field.

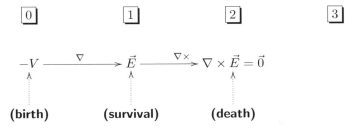

(And since the electric field is a force field, these line integrals relate to the amount of work (energy) performed by the vector field on a charged particle. Thus we also have that the values of this electric potential function can be thought of as relating to the potential energy that comes from the electric field.)

Example 6.8.7. Another of Maxwell's laws says that the divergence of the magnetic field $\vec{B}$ is always zero (see Example 6.4.7).

According to the corresponding lifetime theorem then, this means that the magnetic field $\vec{B}$ must itself be the curl of some other vector field. We call this vector field the "magnetic potential field", written $\vec{A}$.

As per Stokes's curl theorem then, line integrals of this magnetic potential field give us information about flux integrals of the magnetic field.

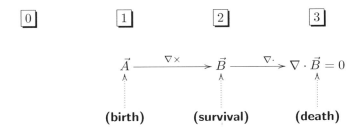

6.8.4 Domains

Students might find it interesting to note that the idea of "lifetimes" can also be applied to domains and their boundaries. Unfortunately it is not quite true then to say that every domain has a lifetime of exactly two... but it is "usually" true, with only familiar exceptions.

Certainly every domain lifetime must be at most two. That is, the boundary of the boundary of any domain is always zero.

For example, suppose you have a surface with a boundary curve; then that curve must have boundary equal to zero. That is, that curve cannot have "endpoints", but rather it must close up on itself in order to be a boundary.

Similarly, suppose you have a solid region in $\mathbb{R}^3$, with a boundary surface; then that boundary surface cannot itself have a boundary curve, but rather it must also close up on itself in order to be a boundary.

Going in the other direction, it is tempting (and almost true) to say that if the boundary of a domain is zero, then that domain must itself be the boundary of another domain. Certainly this is true in many cases.

Unfortunately there are counterexamples involving knots, similar to those discussed in Section 6.6. That is, a closed loop tied in a knot has no boundary, but at the same time, because it is tied in a knot, it is not the boundary of a surface in $\mathbb{R}^3$.

The good news here is that students will not need to state or apply any theorems regarding these ideas. We mention them partially because of the near symmetry to the lifetime theorems for integrands, and partly also so that students can contemplate the cases when this is true, and then hopefully be more confident in future attempts to consider boundaries of domains.

We conclude this brief discussion with a terse phrasing of this idea that is usually true for domains. That is, with certain exceptions, "Every domain either *has* a boundary, or *is* a boundary."

Exercises

Exercise 6.8.1 (exer-VC-LT-1). For each of the following functions $f : \mathbb{R}^2 \to \mathbb{R}^1$, confirm that $\mathrm{grn}(\nabla f) = 0$.

1. $f(x, y) = x^2 y^5 - xy^3 \sin x$
2. $f(x, y) = e^{xy}$
3. $f(x, y) = \cos^3(x^2 - y^2)$

Exercise 6.8.2 (exer-VC-LT-2). Show that for every function $f : \mathbb{R}^2 \to \mathbb{R}^1$, it is always true that $\mathrm{grn}(\nabla f) = 0$.

Exercise 6.8.3 (exer-VC-LT-3). For each of the following functions $f : \mathbb{R}^3 \to \mathbb{R}^1$, confirm that $\nabla \times (\nabla f) = \vec{0}$.

1. $f(x, y, z) = x^3 y^4 z^2 - x^2 y^3 z^3$
2. $f(x, y, z) = xyz^2 e^{xy^2 z}$
3. $f(x, y, z) = \sin^3(xy^2 - yz)$

Exercise 6.8.4 (exer-VC-LT-4). Show that for every function $f : \mathbb{R}^3 \to \mathbb{R}^1$, it is always true that $\nabla \times (\nabla f) = \vec{0}$.

Exercise 6.8.5 (exer-VC-LT-5). For each of the following vector fields $\vec{F}$ in $\mathbb{R}^3$, confirm that $\nabla \cdot (\nabla \times \vec{F}) = 0$.

1. $\vec{F} = (xy^3, xyz^2, x^2 y^3 z)$

6.8. LIFETIME THEOREMS

2. $\vec{F} = (e^{xyz}, e^{x+y-z}, 2xe^y)$

3. $\vec{F} = (\cos(xy - z), 3x^2y, \cos z)$

Exercise 6.8.6 (exer-VC-LT-6). Show that for every vector field $\vec{F}$ in $\mathbb{R}^3$, it is always true that $\nabla \cdot (\nabla \times \vec{F}) = 0$.

Exercise 6.8.7 (exer-VC-LT-7). For each of the following vector fields, determine if it is the gradient of some function.

1. $\vec{F} = (xy^3 e^{xy}, x^2 e^{xy})$

2. $\vec{F} = (2xy^4, 4x^2y^3)$

3. $\vec{F} = (yz^2 e^{xyz^2}, xz^2 e^{xyz^2}, 2xyze^{xyz^2})$

4. $\vec{F} = (2x\cos(x^2y - z^3), xy\cos(x^2y - z^3), xz\cos(x^2y - z^3))$

Exercise 6.8.8 (exer-VC-LT-8). For each of the following vector fields, determine if it is the curl of some other vector field.

1. $\vec{F} = (yz^4 e^{yz}, x^2z - x\cos(x^2 - z^3), xy)$

2. $\vec{F} = (4x^2y^3 - 2xz, 4 - 2xy^4, z^2)$

3. $\vec{F} = \nabla \times (e^{xyz}, xe^{xyz}, ye^{xyz})$

Exercise 6.8.9 (exer-VC-LT-9). Suppose that the static charge density in a region of space is given by a continuous function $\rho(x, y, z)$. What is the divergence of the electric field that this charge density generates? (Recall Equation 6.85.) Use this result to give an explanation for why Theorem 6.8.7 should be true.

Exercise 6.8.10 (exer-VC-LT-10). Suppose that a steady state current field in a region of space is given by the continuous vector field $\vec{J}$. (Note, in order for this to be a steady state, the divergence of this vector field must be zero.) What is the curl of the magnetic field that this current field generates? (Recall Equation 6.109.) Use this result to give an explanation for why Theorem 6.8.6 should be true.

Exercise 6.8.11 (exer-VC-LT-11). In this exercise we construct and consider a vector field that is defined on the punctured plane.

1. In the "right half plane" (this is the part of the xy-plane with $-\pi/2 < \theta < \pi/2$), we have $\theta = \arctan(y/x)$. We can view this as defining a function θ_1 on this right half plane. Show that on this half plane we have $\nabla \theta_1 = \left(\frac{-y}{x^2+y^2}, \frac{x}{x^2+y^2}\right)$.

2. In the "upper half plane" (this is the part of the xy-plane with $0 < \theta < \pi$), we have $\theta = \text{arccot}(x/y)$. We can view this as defining a function θ_2 on this upper half plane. Show that on this half plane we have $\nabla \theta_2 = \left(\frac{-y}{x^2+y^2}, \frac{x}{x^2+y^2}\right)$.

3. In the "left half plane" (this is the part of the xy-plane with $\pi/2 < \theta < 3\pi/2$), we have $\theta = \pi + \arctan(y/x)$. We can view this as defining a function θ_3 on this left half plane. Show that on this half plane we have $\nabla \theta_3 = \left(\frac{-y}{x^2+y^2}, \frac{x}{x^2+y^2}\right)$.

4. In the "lower half plane" (this is the part of the xy-plane with $\pi < \theta < 2\pi$), we have $\theta = \pi + \text{arccot}(x/y)$. We can view this as defining a function θ_4 on this lower half plane. Show that on this half plane we have $\nabla \theta_4 = \left(\frac{-y}{x^2+y^2}, \frac{x}{x^2+y^2}\right)$.

5. Note that the above four vector fields defined on half planes all have the same formula, and so they all agree where they overlap. We can then use that formula to define a new vector field $\vec{F} = \left(\frac{-y}{x^2+y^2}, \frac{x}{x^2+y^2}\right)$ defined on the entire punctured plane (that is, everywhere in the xy-plane except for the origin).

   Show by direct computation that $\text{grn}\vec{F} = 0$ at every point in the punctured plane.

6. Use the fundamental theorem of line integrals and the function θ_1 to show that if $\vec{x}_1$ and $\vec{x}_2$ are both points in the right half plane and if C is any curve in the right half plane from $\vec{x}_1$ to $\vec{x}_2$, then $\int_C \vec{F} \cdot d\vec{x}$ gives the difference between the values of θ at these two points.

7. Suppose that $\vec{x}_1$ and $\vec{x}_3$ are in the first and third quadrants, respectively, and that C is any curve from $\vec{x}_1$ to $\vec{x}_3$ that goes from the first quadrant to the second quadrant and then into the third quadrant (and does not go through either the origin or the fourth quadrant). Show that $\int_C \vec{F} \cdot d\vec{x}$ gives the difference between the values of θ at these two points.

    (Hint: The curve must go through the second quadrant; choose some point $\vec{x}_2$ on the curve in the second quadrant, dividing the curve C into two parts, C_1 and C_2. Using the function θ_2 and the idea from the previous part of this problem you can compute the amount that θ changes between $\vec{x}_1$ and $\vec{x}_2$, and with a similar approach θ_3 can be used to compute the amount that θ changes between $\vec{x}_2$ and $\vec{x}_3$. You can then put these together to answer the question.)

8. Suppose that a curve C starts and ends at a point in the first quadrant, making a counterclockwise loop around the origin by moving first into the second quadrant, then into the third, then into the fourth, and then returning into the first (never touching the origin and never moving back into a previous quadrant). What is the value of $\int_C \vec{F} \cdot d\vec{x}$? (Hint: Use points in each of the other quadrants to break this curve into pieces, over each of which the fundamental theorem of line integrals can be used.)

9. Use the result of the previous part to explain why this vector field $\vec{F}$ cannot be the gradient of any function.

6.9 Computing Line Integrals and Surface Integrals

In this section we summarize and provide examples of the use of all of the theorems that we have studied in this chapter. Specifically, we will discuss how one can easily determine which of the theorems can be used when attempting to compute a line integral or a surface integral.

6.9.1 Diagnosis

When faced with the computation of a line integral or a surface integral, it is not immediately obvious which of the many theorems we have studied, if any, can or should be used in the computation. Of course we have seen many examples at this point of computing these types of expressions with boundary theorems, but in most of those cases the boundary theorem to be used was suggested by the topic being discussed at that point in the text. Without such strong hints, one can find oneself not knowing what theorem to try to apply.

Making this determination is the first step in any such problem. (And of course we must acknowledge that it is entirely possible that none of the boundary theorems will apply, in which case we must compute the given integral directly from a parametrization. Still, we hope that a boundary theorem will apply, since very often that can lead to a simpler computation.)

Certain things are fairly apparent upon casual inspection. Still, students must make a conscious effort to notice these things to avoid wasting time with nonsense. For example, what is the context of the problem ($\mathbb{R}^2$ or $\mathbb{R}^3$)? What type of evaluation are we trying to compute? Upon noting the answers to these questions, students can identify where on which diagram (from page 453) the problem in question fits in.

The position in the diagram then leads to possible boundary theorems that might be able to apply.

Example 6.9.1. If one is trying to compute a line integral in $\mathbb{R}^3$, this puts us in column one of the $\mathbb{R}^3$ diagram. There are two boundary theorems contiguous with this column – the fundamental theorem of line integrals and Stokes's curl theorem. Each of these theorems then is a candidate that might be useful in the computation.

Noting this, students should not waste any time thinking about other boundary theorems irrelevant to the problem, such as Green's theorem or Gauss's divergence theorem in this case.

There are then at most two boundary theorems that might be useful – one on the right side of the relevant column, and one on the left side. Each boundary theorem relates the given evaluation to another evaluation, whose integrand and domain relate to the given integrand and domain by two arrows on the diagram – one representing the boundary operator and one representing the appropriate integrand operator. Of course these arrows always point in opposite directions on the diagram, so in this process we will always need to proceed "backwards" through exactly one of those two arrows.

Example 6.9.2. Suppose we are interested in a line integral in $\mathbb{R}^3$. We can try to relate this to a column zero evaluation with the fundamental theorem of line integrals, as in Figure 6.35. To do so, note that we are moving one column to the left in the diagram. The boundary operator naturally moves to the left, so that part of the problem is simple. But the integrand operator there (the gradient operator) moves to the right, meaning that in order to proceed we would need to find an antigradient for the given integrand. This is then the hard part of the problem, and as we have seen it may or may not even be possible.

On the other hand, suppose we are again interested in a line integral in $\mathbb{R}^3$, and we wish to relate this to a column two evaluation with Stokes's curl theorem, as in Figure 6.36. Now we are moving one column to the right in the diagram. This time we are moving the same direction that the relevant integrand operator (the curl operator) moves, so that part of the problem is simple; we just compute the curl of the given vector field. But now we are moving in the direction opposite to that of the boundary operator, so in fact we need to "un-boundary" the given

Figure 6.35:

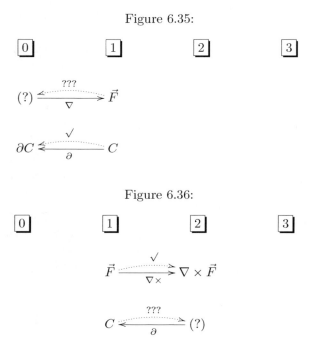

Figure 6.36:

domain... that is, we need to find a new domain whose boundary is the given domain. With geometric intuition this is usually not too hard if it is possible, but we must note that it is certainly not always possible.

We must then determine, for the given problem, which of the "backwards" operations is possible.

For the integrands, this is easily done with the lifetime theorems. That is, if pushing forward with the integrand operation gives zero, then the lifetime theorems tell us that it will be possible to push backward through the previous operator.

Example 6.9.3. Considering again a line integral in $\mathbb{R}^3$, to determine if we can compute the antigradient, we compute first the curl. If the curl is zero, then there will be an antigradient; that is, we can push backwards through the gradient arrow. In this case we can then apply the fundamental theorem of line integrals. See Figure 6.37.

If the curl is not zero, then there is no antigradient, and we cannot push backwards through the gradient arrow. In this case we cannot apply the fundamental theorem of line integrals. See Figure 6.38.

For boundaries we can try to use a similar strategy. Of course the problem here is that the lifetime theorems for

Figure 6.37:

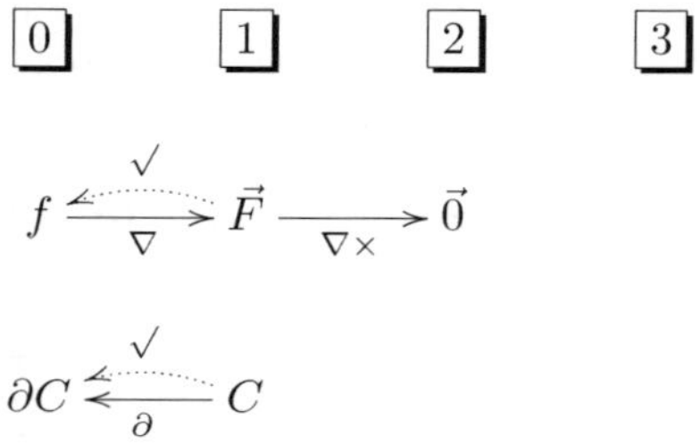

Figure 6.38:

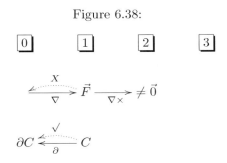

boundaries are not as simple as for integrands as we discussed previously. The good news is that our geometric intuition is usually good enough to determine without too much effort whether the given domain is a boundary or not.

Summarizing then, when faced with computing an integral for which a boundary theorem might be useful:

1. Identify the relevant location on the appropriate diagram.

2. Apply the appropriate operator to the integrand.

3. If the result is zero, then you can use the boundary theorem on the left.

4. If the domain is a boundary, then you can use the boundary theorem on the right.

5. If neither of the potential boundary theorems applies, then the integral must be computed directly, using techniques from Chapter 5.

6.9.2 Examples

Example 6.9.4. Suppose we wish to compute the line integral of the vector field $\vec{F} = (y^2 + 2xz, 2xy + z, y + x^2)$ over the curve C parametrized by $\vec{x}(t) = (t\cos(\pi t), e^{\sin(\pi t)}, e^{t^2+t})$, with $t \in [0,1]$.

First we identify that this expression fits in column one of the $\mathbb{R}^3$ diagram. So immediately we know that the two possible boundary theorems we might be able to use are the fundamental theorem of line integrals and Stokes's curl theorem.

As suggested by the diagrams, the next thing we do is to compute $\nabla \times \vec{F}$. The result is

$$\nabla \times \vec{F} = \vec{0} \tag{6.130}$$

This means that, as in Figure 6.37, this vector field must be the gradient of some function, allowing us to use the fundamental theorem of line integrals.

Noting that the endpoints of the given curve, $\vec{x}(0)$ and $\vec{x}(1)$, are not the same, we see immediately that this curve is not closed, and therefore it cannot possibly be a boundary of any surface. So we know we cannot use Stokes's curl theorem.

So the fundamental theorem of line integrals is the only boundary theorem that we can apply. The solution then requires computing the antigradient of the given vector field.

Example 6.9.5. Suppose we wish to compute the line integral of the vector field $\vec{F} = (x^2 e^x, y \sin y, e^{z^2})$ over the curve C parametrized by $\vec{x}(t) = (t \sin(\pi t), e^{\sin(\pi t)}, e^{t^2 - t})$, with $t \in [0, 1]$.

Again we identify that this expression fits in column one of the $\mathbb{R}^3$ diagram. So again we immediately know that the two possible boundary theorems we might be able to use are the fundamental theorem of line integrals and Stokes's curl theorem.

Again as suggested by the diagrams, the next thing we do is to compute $\nabla \times \vec{F}$. The result is

$$\nabla \times \vec{F} = \vec{0} \tag{6.131}$$

This means that, again as in Figure 6.37, this vector field must be the gradient of some function, allowing us to use the fundamental theorem of line integrals.

This time however we note that the endpoints of this new curve, $\vec{x}(0)$ and $\vec{x}(1)$, are in fact the same point. So this curve is a closed curve, and so it is the boundary of some surface. This means we could also use Stokes's curl theorem to compute this line integral.

In fact it turns out that each of these computations is very easy. Using the fundamental theorem of line integrals gives us

$$\int_C \vec{F} \cdot d\vec{x} = f(\vec{x}(1)) - f(\vec{x}(0)) = 0 \tag{6.132}$$

because, as we just observed, the endpoints $\vec{x}(0)$ and $\vec{x}(1)$ are the same point.

Instead using Stokes's curl theorem, we can also compute

$$\int_C \vec{F} \cdot d\vec{x} = \iint_S \nabla \times \vec{F} \cdot d\vec{S} = 0 \tag{6.133}$$

because, as we computed at the start of this problem, $\nabla \times \vec{F} = \vec{0}$. (Note that because the curl is zero, we do not even need to be able to write down the surface S that is the domain for this new integral.)

Example 6.9.6. Suppose we wish to compute the line integral of the vector field $\vec{F} = (y, z, x)$ over the curve C that is the intersection of the surfaces with equations $x + y + z = 0$ and $x^2 + y^2 = 1$, oriented counterclockwise as seen from above.

Again we identify that this expression fits in column one of the $\mathbb{R}^3$ diagram. So again we immediately know that the two possible boundary theorems we might be able to use are the fundamental theorem of line integrals and Stokes's curl theorem.

We compute the curl, giving us

$$\nabla \times \vec{F} = (-1, -1, -1) \tag{6.134}$$

This is not zero, so we know that this vector field is not the gradient of any function. So we will not be able to use the fundamental theorem of line integrals for this computation.

The curve, as the intersection of a cylinder and a tilted plane, is an ellipse, which specifically can be seen as the boundary of a flat elliptical surface, which we will call S. See Figure 6.39.

The surface S is entirely contained in the given plane, so its normal vector is $\vec{n} = \pm(1, 1, 1)/\sqrt{3}$. Because the curve is oriented counterclockwise as seen from above, the proper boundary orientation comes about from choosing $\vec{n} = +(1, 1, 1)/\sqrt{3}$.

Applying Stokes theorem then to rewrite the line integral as a flux integral, we note that $\nabla \times \vec{F}$ and $\vec{n}$ are both constant, so their dot product is also constant. The computation of the flux integral then comes down to computing the area of the elliptical surface.

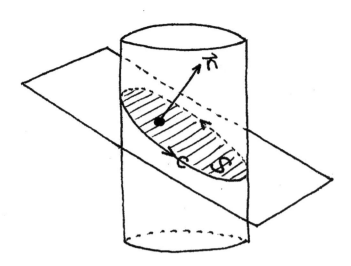

Figure 6.39:

Example 6.9.7. Suppose we wish to compute the line integral of the vector field $\vec{F} = (y, z, x)$ over the curve C that is the intersection of the surfaces with equations $x - z = 0$ and $x^2 + y^2 = 1$, oriented counterclockwise as seen from above.

We note that this is extremely similar to the problem in Example 6.9.6. Again, Stokes's curl theorem is the only boundary theorem that can apply.

This time however we end up with $\vec{n} = (-1, 0, 1)/\sqrt{2}$, and so $\nabla \times \vec{F} \cdot \vec{n} = 0$. So we can immediately conclude that the flux integral arrived at by Stokes's curl theorem is zero.

Example 6.9.8. Suppose we wish to compute the line integral of the vector field $\vec{F} = (x^2 + y, 3x + ye^y)$ over the boundary of the unit square D.

The vector field is two dimensional, so this expression fits in column one of the $\mathbb{R}^2$ diagram. So our only candidate boundary theorems are the fundamental theorem of line integrals and Green's theorem.

Computing the Green operator, we get
$$\operatorname{grn}\vec{F} = 2 \tag{6.135}$$

This is not zero so we know that this vector field is not a gradient, which means we cannot use the fundamental theorem of line integrals.

The curve is explicitly given however as a boundary, that of the unit square in $\mathbb{R}^2$. Applying Green's theorem then gives us
$$\int_{\partial D} \vec{F} \cdot d\vec{x} = \iint_D \operatorname{grn}\vec{F}\, dA = \iint_D 2\, dA = 2 \tag{6.136}$$

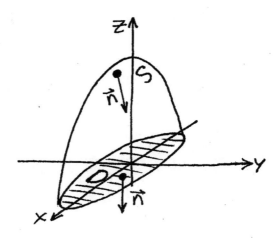

Figure 6.40:

Example 6.9.9. Suppose we wish to compute the flux of the vector field $\vec{F} = (xz, -yz, 0) = \nabla \times (0, 0, xyz)$ through the surface S that is the part of the elliptical paraboloid with equation $z = 4 - x^2 - 4y^2$ that is above the xy-plane, oriented downward.

As a flux integral, we note that this fits in column two of the $\mathbb{R}^3$ diagram. So the two possible boundary theorems are Stokes's curl theorem and Gauss's divergence theorem.

We would normally compute the divergence of the given vector field, but the vector field is explicitly given as a curl so we know that we can indeed apply Stokes's curl theorem. The surface S is seen from Figure 6.40 not to be a boundary, so we cannot use Gauss's divergence theorem.

Applying Stokes's curl theorem, we get

$$\iint_S \vec{F} \cdot d\vec{S} = \iint_S (\nabla \times (0, 0, xyz)) \cdot d\vec{S} = \int_{\partial S} (0, 0, xyz) \cdot d\vec{x} \tag{6.137}$$

Of course as we see from Figure 6.40 the boundary curve is entirely within the xy-plane, so we can write $d\vec{x} = (dx, dy, dz) = (dx, dy, 0)$. The dot product in the integral is thus zero, making the integral zero.

An alternative approach to this problem makes use of surface independence. As a curl of another vector field we know that $\vec{F}$ is surface independent, so we can choose to replace the given surface S with the surface D shown in Figure 6.40, having the same boundary. Because it is entirely contained in the xy-plane and oriented downward, we know that we can write $\vec{n} = (0, 0, -1)$ for the surface D.

The integral then becomes

$$\iint_S \vec{F} \cdot d\vec{S} = \iint_D \vec{F} \cdot d\vec{S} = \iint_D (xz, -yz, 0) \cdot (0, 0, -1)\, dS = 0 \tag{6.138}$$

Example 6.9.10. Suppose we wish to compute the flux of the vector field $\vec{F} = (x + y^2 z, x^2 z + y, xy - z)$ through the unit sphere S, oriented outward.

Again, as a flux integral, we know this fits in column two of the $\mathbb{R}^3$ diagram, making Stokes's curl theorem and Gauss's divergence theorems the only candidate boundary theorems.

Computing the divergence, we get

$$\nabla \cdot \vec{F} = 1 + 1 - 1 = 1 \tag{6.139}$$

This is not zero, so we know this vector field is not a curl, and therefore we cannot make use of Stokes's curl theorem. The surface S however is a boundary, that of the solid unit ball B, so we can use Gauss's divergence theorem.

Applying Gauss's divergence theorem we get

$$\iint_S \vec{F} \cdot d\vec{S} = \iiint_B \nabla \cdot \vec{F}\, dV = \iiint_B 1\, dV = \frac{4\pi}{3} \tag{6.140}$$

Example 6.9.11. Suppose we wish to compute the integral

$$\iiint_B \left(\nabla \cdot \begin{bmatrix} -yze^{x^2y^3\sin(z)} \\ 3xze^{x^2y^3\sin(z)} \\ -2xye^{x^2y^3\sin(z)} \end{bmatrix} \right) dV \tag{6.141}$$

where B is the solid ball of radius 5, centered at the origin.

This triple integral fits into column three of the $\mathbb{R}^3$ diagram, so Gauss's divergence theorem is the only boundary theorem that we can consider using.

The integrand function is explicitly given as a divergence, so we can apply Gauss's divergence theorem to get

$$\iiint_B \left(\nabla \cdot \begin{bmatrix} -yze^{x^2y^3\sin(z)} \\ 3xze^{x^2y^3\sin(z)} \\ -2xye^{x^2y^3\sin(z)} \end{bmatrix} \right) dV = \iint_{\partial B} \begin{bmatrix} -yze^{x^2y^3\sin(z)} \\ 3xze^{x^2y^3\sin(z)} \\ -2xye^{x^2y^3\sin(z)} \end{bmatrix} \cdot d\vec{S} = \iint_{\partial B} \left(\begin{bmatrix} -yz \\ 3xz \\ -2xy \end{bmatrix} (e^{x^2y^3\sin(z)}) \right) \cdot \vec{n}\, dS \tag{6.142}$$

The surface ∂B of course is a sphere of radius 5, so we can write $\vec{n} = (x,y,z)/5$, **giving us**

$$= \iint_{\partial B} \left(\begin{bmatrix} -yz \\ 3xz \\ -2xy \end{bmatrix} (e^{x^2y^3\sin(z)}) \right) \cdot \frac{1}{5}\begin{bmatrix} x \\ y \\ z \end{bmatrix} dS = \iint_{\partial B} \left(\frac{e^{x^2y^3\sin(z)}}{5} \right) \begin{bmatrix} -yz \\ 3xz \\ -2xy \end{bmatrix} \cdot \begin{bmatrix} x \\ y \\ z \end{bmatrix} dS = 0 \tag{6.143}$$

Exercises

Exercise 6.9.1 (exer-VC-CLISI-1). For each of the items below, indicate which boundary theorems might be useful in computing the given integral.

1. $\int_C \vec{F}(x,y) \cdot d\vec{x}$
2. $\int_C (P,Q,R) \cdot d\vec{x}$
3. $\iint_S \vec{F} \cdot d\vec{S}$
4. $\iint_S f(x,y,z)\, dz\, dx$

Exercise 6.9.2 (exer-VC-CLISI-2). Suppose we are interested in computing a line integral of the vector field $\vec{F} = (x^2y, xyz^2, xy^2)$. Without finding an antigradient, determine if the fundamental theorem of line integrals can give us the answer to such a question. (Make sure to explain your reasoning.)

Exercise 6.9.3 (**exer-VC-CLISI-3**). Suppose we are interested in computing a line integral of the vector field $\vec{F} = (3x^2y, x^3 - y^3)$. Without finding an antigradient, determine if the fundamental theorem of line integrals can give us the answer to such a question. (Make sure to explain your reasoning.)

Exercise 6.9.4 (**exer-VC-CLISI-4**). Suppose we are interested in computing a flux integral of the vector field $\vec{F} = (x^2 - y^2, -2xyz, xz^2 - 2xz)$. Does there exist a vector field for which Stokes's curl theorem could be used to compute this? (Make sure to explain your reasoning, but note that you are not being asked to find this vector field if it does exist.)

Exercise 6.9.5 (**exer-VC-CLISI-5**). Suppose we are interested in computing a line integral over the curve C which is the intersection of the surface $x - y - 3z = 0$ with the surface $2x - 7y + z = 6$. Is it possible that Stokes's curl theorem could allow us to compute this integral? (Make sure to explain your reasoning.)

Exercise 6.9.6 (**exer-VC-CLISI-6**). Suppose we are interested in computing a line integral over the curve C which is the intersection of $11x - 3y = 0$ with the unit disk in the xy-plane. Is it possible that Green's theorem could allow us to compute this integral? (Make sure to explain your reasoning.)

Exercise 6.9.7 (**exer-VC-CLISI-7**). Suppose that $\vec{F}$ is the gradient of a function f, and that C is the boundary of a surface S in $\mathbb{R}^3$. Show that $\int_C \vec{F} \cdot d\vec{x} = 0$.

Exercise 6.9.8 (**exer-VC-CLISI-8**). Suppose that $\vec{F}$ is the gradient of a function f, and that C is the boundary of a solid region D in $\mathbb{R}^2$. Show that $\int_C \vec{F} \cdot d\vec{x} = 0$.

Exercise 6.9.9 (**exer-VC-CLISI-9**). Suppose that $\vec{F}$ is the curl of a vector field $\vec{G}$, and that S is the boundary of a solid region T in $\mathbb{R}^3$. Show that $\iint_S \vec{F} \cdot d\vec{S} = 0$.

Exercise 6.9.10 (**exer-VC-CLISI-10**). Compute the line integral $\int_C \vec{F} \cdot d\vec{x}$, where $\vec{F} = (2xe^{yz}, x^2ze^{yz}, x^2ye^{yz})$, and C is a curve that starts at the origin and ends at the point $(4, 2, 0)$.

Exercise 6.9.11 (**exer-VC-CLISI-11**). Compute the flux of the vector field $\vec{F} = (6y - z^2, x^2z - z^3, e^{xy})$ through the surface S that is the boundary of the solid formed as the union of the three solid balls of radius 2 that are centered at $(0,0,0)$, $(0,3,0)$, and $(0,0,3)$.

Exercise 6.9.12 (**exer-VC-CLISI-12**). Compute the flux of the vector field $\vec{F} = \frac{k}{\|\vec{x}\|^3}\vec{x}$ through the surface S that is the portion of $x^2 + y^2 + (z-4)^2 = 25$ that is above the xy-plane, with an "outward" orientation.

Exercise 6.9.13 (**exer-VC-CLISI-13**). Compute the line integral of the vector field $\vec{F} = (4x^3e^x - y^2, x^3 + y^3)$ around the clockwise oriented unit circle in the xy-plane.

Exercise 6.9.14 (**exer-VC-CLISI-14**). Compute the line integral $\int_C \vec{F} \cdot d\vec{x}$, where $\vec{F} = (3y + 2x, 2z, 2x - 4y)$, and C is the boundary of the surface S (oriented upward) that is the portion of the paraboloid $z = 9 - x^2 - 4y^2$ that is above the plane $2x - 3y - 2z = 0$.

Exercise 6.9.15 (**exer-VC-CLISI-15**). Compute the flux of the vector field $\vec{F} = (x, y, z)$ through the downward oriented surface which is the portion of $z = 8x^2 + 5y^2$ that is over $[1, 2] \times [0, 2]$ in the xy-plane.

Index

acceleration, 139, 140
antigradient, 398, 441
anticurl, 447
arrow, 4

boundary, 198, 383
boundary theorems, 388, 449
bounded, 199

Cauchy-Schwarz inequality, 16, 17
centripetal acceleration, 141
centroid, 263, 283
chain rule, 182, 183
change of variables, 290, 296
charge density, 422, 457
clockwise order, 22
closed set, 199
component, 18
conservative, 441
constrained local maximum, 209
constraint function, 209
continuous, 108, 110
continuously differentiable, 164, 165
coordinates, 43
coordinate line integrals, 372
coordinate surface integrals, 376
counterclockwise order, 22
critical point, 201, 211, 215
cross product, 28
cross section, 54
curl, 159, 429, 433
current field, 436, 457
curvature, 140, 140
cylindrical coordinates, 45
cylindrical coordinates function, 303

de Rham cohomology theorem, 379, 449
degeneracy condition, 211, 216
derivative, single variable, 133
derivative of a parametric curve, 136
derivative transformation, 162
determinant, 25
differentiable, 162, 165
differential vector, 347, 362, 371, 371
direction of fastest increase, 176
direction vector, 37
directional derivative, 142, 175, 163
directionally differentiable, 165

directional linearity, 161, 165
divergence, 159, 412
domain, 49, 380, 382
dot product, 15
double integral, 234

electric field, 422
electric potential function, 455
equation, 53
evaluations, 380

flow field, 339
flux, 359, 360, 362, 368
flux density, 414, 419
force field, 338
function, 49, 53
fundamental theorem of calculus, 388
fundamental theorem of line integrals, 395, 397, 453

Gauss-Weierstrass heat kernel, 159
Gauss's divergence theorem, 413, 424, 453
global maximum point, 220
gradient, 175
gradient field, 339
graph, 51, 71, 72
graph parametrization, 91
gravitational field, 417
Green operator (grn), 405
Green's theorem, 405, 432, 453

heat equation, 159
helix, 88
Hessian matrix, 204
hyperboloid, 67, 68

image, 49, 83
implicit function theorem, 195, 195
incompressible fluid, 447
integral conditions, 229
integrand, 380
interior, 198
intermediate variables, 184
irrotational, 442
iterated integral, 244

Jacobian matrix, 169

Lagrange multiplier theorem, 211, 216
Laplacian, 159
law of cosines, 17

left-hand order, 23
level set, 76
lifetime theorems, 450, 451
limit, 103, 104, 109
line, 35, 37
line integral, 321, 325, 347, 350
linear combination, 114
linear transformation, 114
local maximum point, 199

magnetic potential field, 455
magnitude, 9
matrix, 122
matrix addition, 126
matrix multiplication, 127
matrix-vector multiplication, 124
Maxwell's equations, 422, 436
moment of inertia, 284

nested integral, 244, 268
normal component of acceleration, 140
normal vector, 35, 36

objective function, 209
octant, 1
odd symmetry, 297, 298
open ball, 198
open set, 198
optimization, 198
orientation, 360, 381, 383
orthogonal, 18, 28

Pappus, theorem of, 264
paraboloid, 56, 56, 58, 66
parallel, 7
parallelepiped, 26
parallelogram, 26, 30
parametric curve, 83, 84
parametric surface, 83, 89
parametrize, 39, 83
partial derivative, 155
partial differentiability, 165
path independence, 440, 442
perpendicular, 18
plane, 36
polar coordinates, 44
polar coordinates function, 302
position vector, 6
potential function, 441
pre-image, 76
principal unit normal vector, 140
punctured plane, 452, 457

range, 49
reflection, 22, 23, 61
Riemann sum, 227, 234, 250, 267
right-hand order, 23
rotation, 66, 119

rotational symmetry, 64

saddle point, 203
scaling transformation, 61
second derivative test, 205
shearing transformation, 120
simply connected, 384
sink, 417
solution set, 53
source, 417
speed, 137
spherical coordinates, 46
spherical coordiantes function, 305
standard basis vector, 12, 116
standard position, 6
Stokes's theorem, curl, 430, 453
Stokes's theorem, general, 379
stretching factor, 287, 290, 296
subdomain, 220
substitution rule, 285
surface independent, 444, 445, 446
surface integral, 328, 333, 362, 363
symmetric equations, 40
symmetry theorems, 297, 298, 299, 327, 336, 357, 370, 393

tangent vector, 199
tangential component of acceleration, 140
target, 49
Taylor polynomial, 204
translation, 61
triple integral, 267

unconstrained local maximum, 200
unit directional derivative, 149
unit vector, 12
unit normal vector, 360
unit tangent vector, 140, 353

vector, 4
vector calculus diagrams, 453, 459
vector integral analogies, 367
vector field, 341
velocity vector, 137

wave equation, 159
work, 19, 346

xy-plane, 1
xyz-space, 1

Made in the USA
San Bernardino, CA
11 December 2015